THE REAL WORLD

THE REAL WORLD

UNDERSTANDING THE MODERN WORLD THROUGH THE NEW GEOGRAPHY

GENERAL EDITOR: BRUCE MARSHALL

PRINCIPAL WRITER: PHILIP BOYS

HOUGHTON MIFFLIN COMPANY
BOSTON LONDON MELBOURNE

A MARSHALL EDITION
in association with Marc Jaffe
and Houghton Mifflin Company

Conceived, edited, and designed by
Marshall Editions
170 Piccadilly
London W1V 9DD

Consultant Editor: Dr. Risa Palm

Special Advisers: Len Brown
Dr. David Green
Dr. Linda Newson

Typeset by Servis Filmsetting Limited, Manchester, UK
Originated by Reprocolor Llovet SA, Barcelona, Spain
Printed and bound by Printer Industria Gráfica SA, Barcelona, Spain

Library of Congress Cataloging-in-Publication Data

The Real world / Bruce Marshall, general editor.
p. cm.
Includes bibliographical references and index.
ISBN 0-395-52450-4
1. Geography. I. Marshall, Bruce, date.
G128.R42 1991
910 — dc20 90-28891
CIP

Page 1 *Camel caravan, Sahara Desert.*
Page 2 *Ndebele house painting, South Africa* (left).
Gondolas, Venice (right).
Angmagssalik, Greenland (bottom).
Page 3 *Night skyline, Dallas, Texas.*
Page 4 *Ceremonial headdress, Philippines* (top).
Intersection, Tokyo (bottom).
Page 5 *Stock exchange, Hong Kong.*
Page 6 *Farm, Oshkosh, Wisconsin.*

CONTENTS

One World, a Multitude of Places

Geography began with the realization that there was another side to the mountain and other people across the sea. The early Mediterranean explorers, the Greeks particularly, made maps of the lands they found and brought home stories about the people they met.

The same curiosity still enthralls. New maps are drawn, informed by satellite photographs which fill in minute and ever-changing details. To physical geographers these are vital data which may help to forecast weather, feed the world, and preserve the environment.

And the story-telling continues. Human geographers analyze the groupings of populations and migration patterns, the distribution of languages, religions, and customs, the growth and decline of cities and states, and trends in politics, justice, even warfare.

Modern geography therefore addresses issues at the heart of today's problems and opportunities. How do we choose where to live and work? How does the environment affect the quality of our lives? What is the significance of place in political instability?

Geographers should be more clearly heard in the great debates. By relating what people do to where they do it, they have compelling insights into the social and political problems of our time. The day's headlines result as much from geographical factors as political ones – the scarcity or abundance of natural resources, the arbitrariness of national frontiers, the possession of strategic features such as port cities and oil wells.

Geography can help us understand not only what is where, but why it is there – why cities are circles, squares, or patchwork designs; why neighborhoods evolve in a certain sequence; why food may be plentiful or scarce and cost more or less.

The Real World demonstrates the riches of information and insight that can be located through geography. It charts a clear, well-lit path across a hugely complex landscape.

Risa Palm

Risa Palm
Professor of Geography, University of Colorado
Past President of the Association of American Geographers

MICROCOSM OF THE REAL WORLD

Switzerland's Rhone Valley

At first glance, it appears as if Switzerland should not exist as a sovereign state, let alone enjoy standards of affluence that make it preeminent in the world. Its geography seems so dead set against it. It is tiny, mountainous, landlocked, and has few natural resources. Large areas of the country are covered in snow for much of the year.

Neither does its social mix seem auspicious. With a total population smaller than that of many big cities – 5.5 million nationals, speaking four languages, and not quite a million resident foreigners, speaking many more – the Swiss nation is fragmented not only by language but also by religion and culture. Yet differences that in the rest of Europe have led to protracted and bitter conflict – for example, between Catholics and Protestants, or German-speakers and French-speakers – have no such significance here.

Does this mean that physical and social geography have no importance in the modern world? Not at all. Switzerland has not become a great country by defying location. On the contrary, it owes everything to it. The Swiss have entered into a dynamic partnership with their geography. They have been challenged

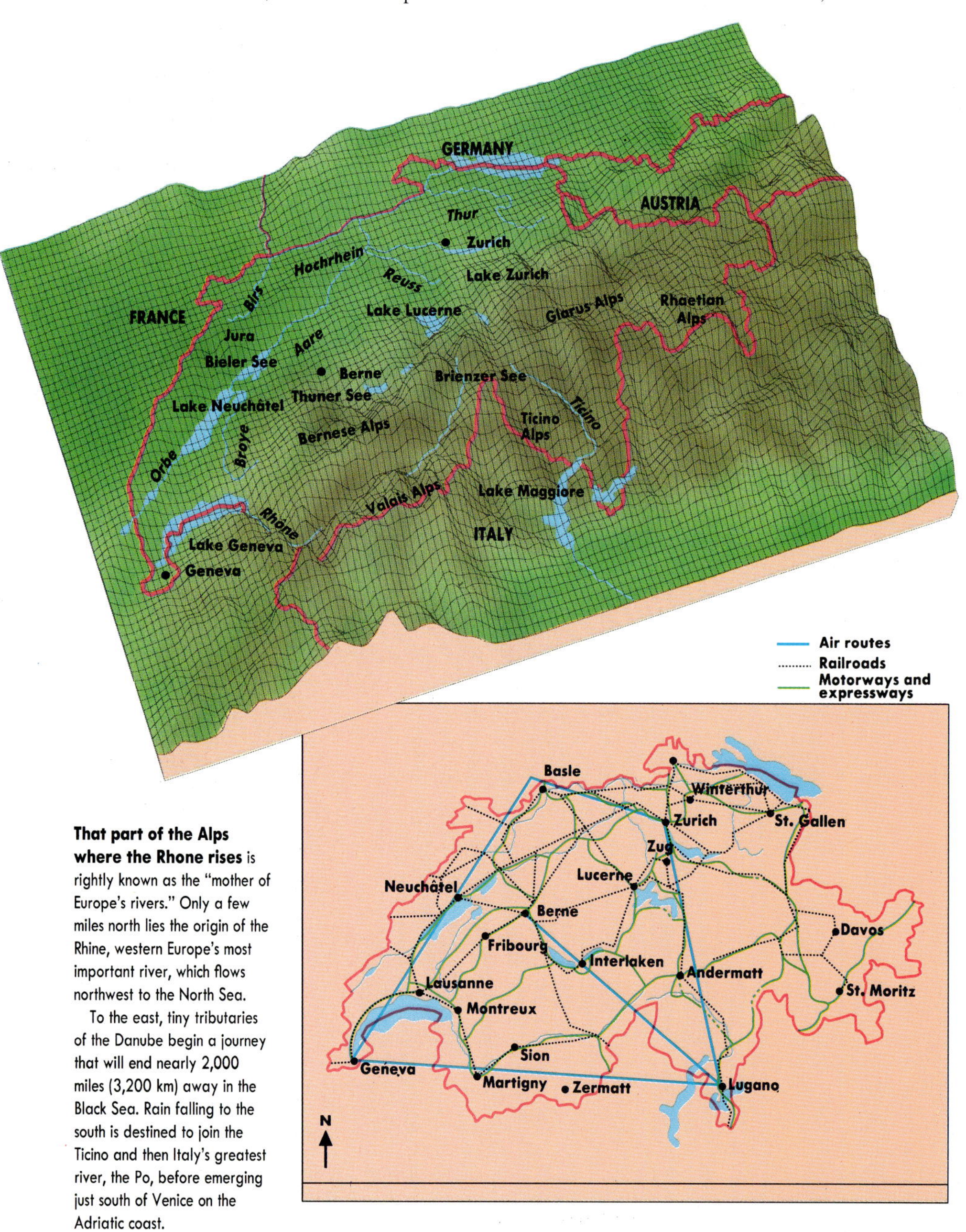

That part of the Alps where the Rhone rises is rightly known as the "mother of Europe's rivers." Only a few miles north lies the origin of the Rhine, western Europe's most important river, which flows northwest to the North Sea.

To the east, tiny tributaries of the Danube begin a journey that will end nearly 2,000 miles (3,200 km) away in the Black Sea. Rain falling to the south is destined to join the Ticino and then Italy's greatest river, the Po, before emerging just south of Venice on the Adriatic coast.

Even politics are dictated by geography in Switzerland: cantons are usually defined by drawing lines from mountain peak to mountain peak.

In such difficult terrain, transport is of prime importance. There are 34 road passes at elevations of over 3,300 ft (1,000 m), of which the Simplon, built by Napoleon, is the best known. These passes enable motorists to cross the mountains for most of the year. The Swiss are justifiably proud of their ability to maintain transport links whatever the weather.

The Swiss railroads are a masterpiece of engineering, particularly the tunnels of the Gotthard railroad, which wind through the mountains and thus overcome the problem of differences in altitude from one valley to another.

There are three major international airports – at Zurich, Geneva, and Basle-Mulhouse (the last-named shared with the French) – and Swissair is one of the world's largest commercial carriers.

and changed by it, but they have also modified it – and in the process once again have subtly changed themselves. The interaction continues to this day.

In particular, Switzerland owes its existence to the Alps, the great arc of mountains that separates Mediterranean Italy from cooler, wetter northern Europe. Although the main alpine superstructure was thrown up millions of years ago by great movements of the Earth's crust, in geological terms these are very young mountains, which have hardly been eroded and are therefore still sharp and jagged. And they are still growing, by around one twenty-fifth of an inch (1 mm) a year.

The landforms we see today are the product of countless natural and human agencies working on their fabric – among them sun,

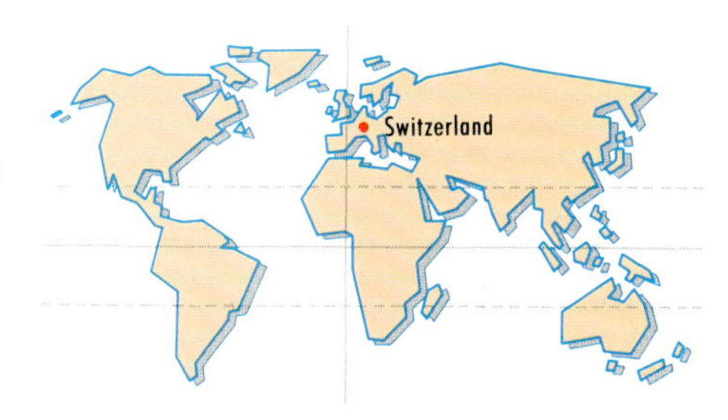

The Rhone valley was carved out by a glacier during the last ice age and exposed when the glacier began to retreat around 10,000 years ago. Today the glacier stops at around 5,000 ft (1,500 m), and a trickle of milky-looking water issues from an ice cave at its base. This is the source of the Rhone.

The Swiss have a lot to thank their many glaciers for. The deep, broad, wide-bottomed valleys they carved out lace the mountains with natural routes. They are also well watered and filled with fertile silt.

Until the Swiss canalized it, the upper Rhone flowed erratically across a marshy flood plain. But control of the river has transformed the valley floor into a neat farming mosaic. In the absence of other energy resources, the Swiss turned to the river for hydroelectric power.

But now the Rhone valley is beginning to pay the price of its intensive exploitation and the good life its citizens have earned. Traffic exhaust fumes kill the trees that hold the winter snow in place. Avalanches and landslides threaten the old villages. This most humanized of landscapes could die of self-inflicted wounds.

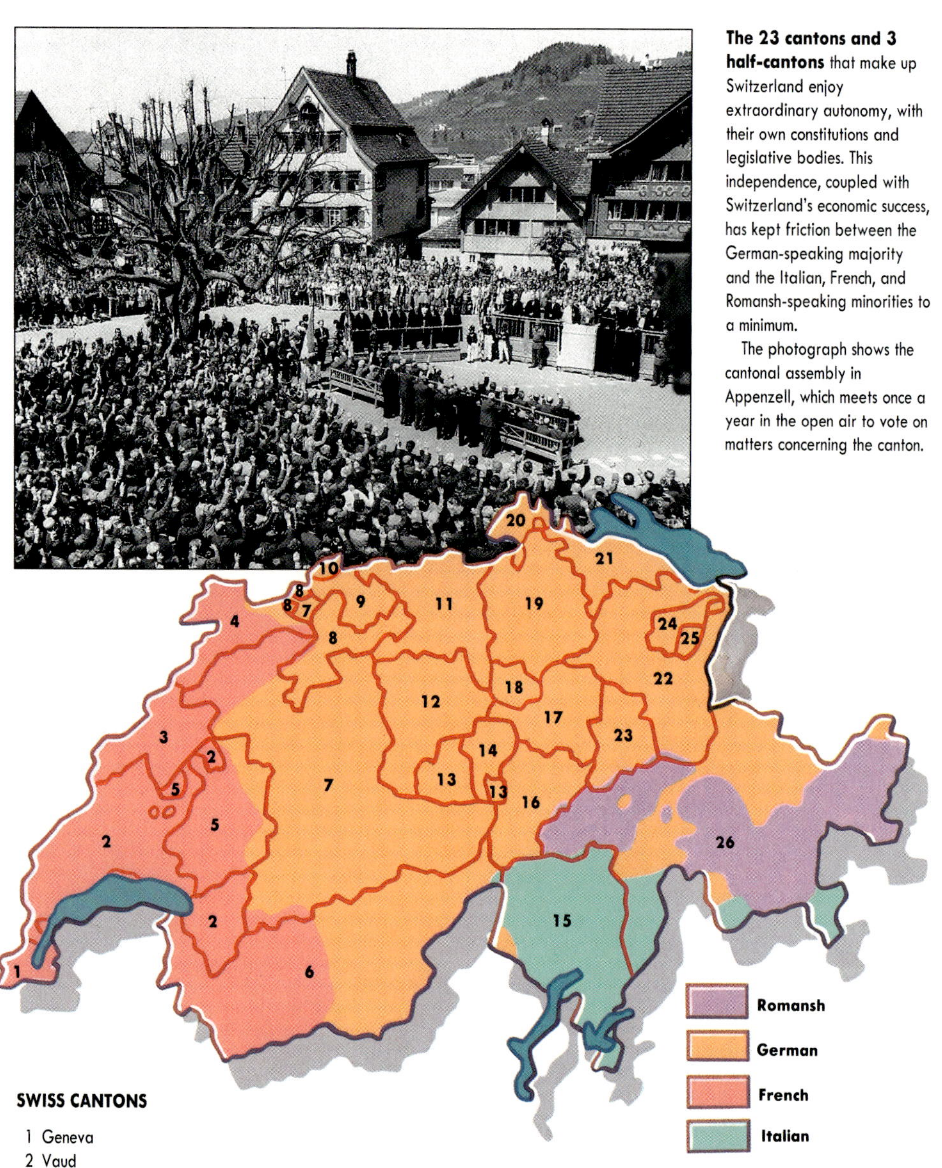

The 23 cantons and 3 half-cantons that make up Switzerland enjoy extraordinary autonomy, with their own constitutions and legislative bodies. This independence, coupled with Switzerland's economic success, has kept friction between the German-speaking majority and the Italian, French, and Romansh-speaking minorities to a minimum.

The photograph shows the cantonal assembly in Appenzell, which meets once a year in the open air to vote on matters concerning the canton.

SWISS CANTONS

1 Geneva
2 Vaud
3 Neuchâtel Unterwalden
4 Jura
5 Fribourg
6 Valais
7 Berne
8 Solothurn
9 Basle-Land
10 Basle-Stadt
11 Aargau
12 Lucerne
13 Obwalden
14 Nidwalden
15 Ticino
16 Uri
17 Schwyz
18 Zug
19 Zurich
20 Schaffhausen
21 Thurgau
22 St. Gallen
23 Glarus
24 Appenzell-Ausserrhoden
25 Appenzell-Innerrhoden
26 Grisons

Switzerland is a composite nation based on the equality of its different groups. Its official name is *Confederation helvetica* – Swiss confederation – the origin of the initials CH that appear on car number plates and postal codes.

Regional and national referenda are widely used to decide major issues: should Switzerland join the United Nations (no), should the church and state be separated (no), should there be tolls on highways (yes), should there be legal equality between men and women (yes). Women were only given the vote in 1971, making Switzerland one of the last countries in Europe to grant this right.

wind, atmosphere, plants, animals, and running water. But this is also a thoroughly humanized environment, a landscape shaped by a partnership between people and ice.

Many highland nations elsewhere in the world – like Tibet, Albania, Ethiopia, and Bolivia – are notoriously deprived economically. The climate and terrain in these places make farming a problem, while poor communications prevent the free flow of raw materials, goods, and new ideas, encouraging social isolation and division.

But mountains in Switzerland have conferred important advantages. For millennia, small groups of people colonized the Alps from every direction, typically infiltrating up river valleys. Using ax and fire to clear the forest, they formed numerous isolated communities. (The word Swiss probably originates from the Old German *suedan*, to burn.)

In 1291, three of the most remote of these forest districts, or cantons, took an oath of perpetual alliance to defend one another from all imposed tyranny. It is from this pact that modern Swiss political unity stems. Over the next few hundred years neighboring areas escaped from the dominance of the great powers and joined the confederation.

For outsiders, though, these rugged, cold highlands have been impossibly difficult to control. The Alps make the perfect terrain for guerrilla warfare. For a thousand years Switzerland has been eyed covetously from the west by France, from the south by Italy, from the north by German states, and from the east by Austria. But as the British discovered in Afghanistan in the last century (and the Soviet Union forgot to its cost in this century), areas as mountainous as this are too dangerous to occupy.

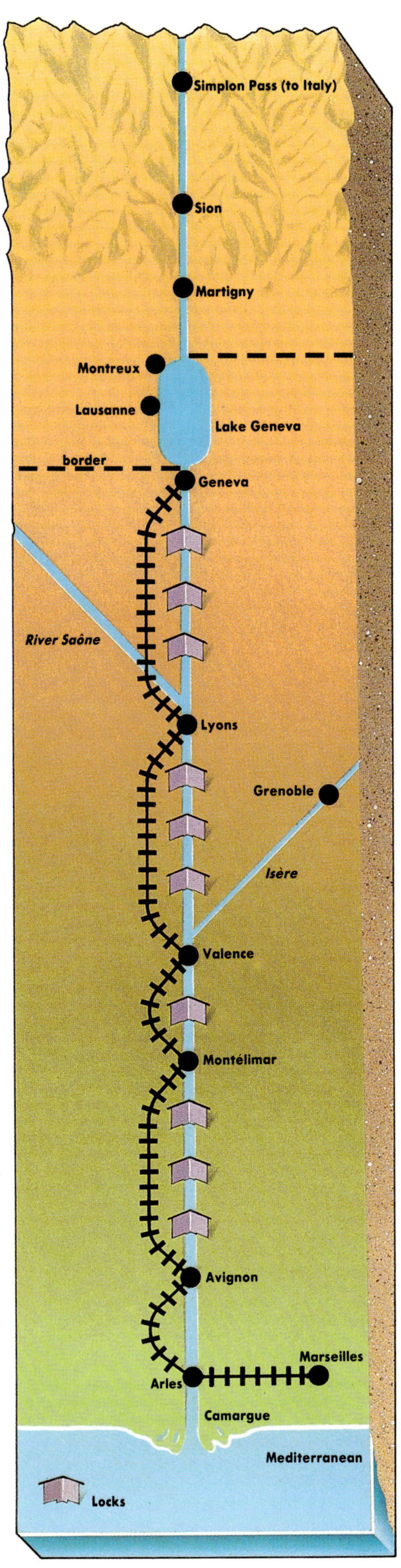

With the exception of a brief period after 1798, when Napoleon annexed part of Switzerland and imposed a centralized French-style government, Switzerland has remained an independent federation. It was neutral during both world wars, taking in the sick and wounded from both sides and giving assistance to all who needed it.

This neutrality has made it a natural home for international agencies like the Red Cross, the World Council of Churches, and the World Health Organization. Indeed, Switzerland is so determined to maintain its independence that it will not join the European Community, whose members practically surround it. Neither does it belong to the United Nations, although the U.N.'s European headquarters are in Geneva.

The Alps, then, form a remote and elevated fastness protecting the nation's political integrity, and promoting the distinct identity of

The Romans brought the vine with them from Italy. Some say that it was wine more than legionaries that opened up the roads north. Rome used wine in much the same way as the English and French used spirits in North America, to demoralize indigenous peoples.

The Swiss government subsidizes agriculture heavily – they have an ideology of self-reliance here. In the photograph above, vines give way to neat fields and greenhouses, typical of the agricultural mix found in the Rhone valley.

The Rhone (as shown in the illustration right) has always been an important trade and migration route, as place names along the river testify. Reading the names of villages, towns, and cities up the river shows that it was the conduit through which French culture penetrated the Alps. Above Lyons, on the northern shores of Lake Geneva lie such famous cities as Geneva, Lausanne, and Montreux, all of which bear French names. Higher still, there are small towns like Sion and passes like the Simplon, which leads into Italy. Then farther up the names change, becoming German in origin – Munster, St. Gothard, Andermatt.

At Vevey on the edge of Lake Geneva, through which the Rhone passes as it leaves Switzerland for France, lies the headquarters of a major transnational corporation, Nestlé.

Switzerland has been identified with fine chocolate for nearly two centuries. The use of chocolate spread very slowly through Europe from Spain, which for a long time kept the secret of chocolate-making to itself, having learned it from the Aztecs around 1500.

Swiss manufacturers are forced to use Swiss raw materials such as milk, but the chocolate itself comes from many countries. This has made the industry vulnerable to foreign competition, notably from the U.S.-based Mars company, which captured a sizable slice of the children's candy market.

The Swiss have recently put a lot of effort into the export of high-class chocolates to countries only now developing as markets, such as Japan and Saudi Arabia.

Nestlé also produces milk products, such as evaporated and condensed milk, milk powder, and baby food. The founder of Nestlé, Henri Nestlé, made baby food from condensed milk. By combining this with chocolate, Daniel Peter made the first milk chocolate in 1875.

But perhaps Nestlé's greatest contribution has been the invention of soluble coffee – instant coffee – now almost universally known as Nescafé.

Machinery 21.6
Chemicals 15
Precision instruments (watches, etc) 11.5
Metals 5.7
Textiles 4.4
Others 9.8
Tourism 7
Figures are in billions of Swiss francs
US$1 = 1.25 francs

The lack of raw materials caused the Swiss economy to concentrate on value-added goods – those in which the profit comes from the investment of workmanship and expertise. The Swiss are also renowned for the efficiency – and discretion – of their banking services.

The diagram shows revenue earned by leading Swiss exports in a year.

its separate cultural groupings. But how has geography shaped Switzerland's economy?

Switzerland has very little arable land – certainly nothing on the same scale as the great wheat-growing fields of North America or the Ukraine, or even Britain's fenlands. Yet it produces much food, and is famed throughout the world for its agricultural produce, particularly cheese, and the infant formula, chocolate, and coffee that are the mainstays of the major transnational corporation Nestlé.

Once again, Switzerland has the Alps to thank. Swiss agriculture originally developed to take advantage of their gentle upland slopes. Cleared of woodland and kept free of new tree-growth by grazing animals, these became phenomenally productive meadows. Swiss farmers developed a seasonally nomadic lifestyle, coming up the mountainsides in the spring, after the snows had melted, bringing their cattle, sheep, and goats, and staying until the following fall. There, they made cheese – a traditional method of preserving for the winter the produce of the spring-summer pastures.

They overwintered in the warmer, sheltered valley floors that had been formed during past ice ages by glaciers that gouged and rasped away the rock face. About 10,000 years ago, the glaciers retreated, leaving behind them U-shaped, wide-bottomed valleys that rapidly filled with fertile silt. The valley floors were highly productive, although flooding was an ever-present threat. When the rivers that

Remarkably for a country so determined to maintain its neutrality that it does not even belong to the United Nations, Switzerland has compulsory national service with regular refresher courses. All adult males are automatically on the reserve list until the age of 50, or 55 if they have attained officer status. A unique aspect of the system is that every man keeps his arms and ammunition at home.

The neutral stance does not mean that Switzerland is unaware of the threat posed by the outside world. Official policy insists that every new home must have an underground nuclear shelter, and work is proceeding to provide shelters for all citizens by the year 2000.

meandered across the valleys were canalized during the last century, the floors were successfully turned into some of the world's most productive truck farms.

The steep valley walls, too, have been adapted for agriculture. The south-facing slopes have been stepped or terraced; they are warm and dry, and well-protected from winds. This explains the apparently extraordinary fact that Switzerland has productive vineyards growing only a few miles away from permanent snow.

Using the virtues of their mountainous terrain, then, the Swiss have made more than three-fourths of their land agriculturally productive – 46 percent is meadowland and pasture, 25 percent is forest, 6 percent is arable. Even the land that is unproductive from the point of view of agriculture draws nature-lovers, mountaineers, and skiers from the whole world.

It may seem a paradox, but the Alps, by presenting a formidable obstacle to trade between southern and northern Europe, have in fact enhanced Swiss manufacturing, trading, and financial interests. It was doubtless control of the trade that passed across these Alps that formed the basis for the Swiss economy from the late Middle Ages right up to the present day.

Compelled to cross via a limited number of passes and lowland routes, merchants had to meet Swiss terms. They were forced to pay high taxes and tributes. A Swiss service industry flourished, providing guides, guards, carriers, and accommodation. The establishment of banking and insurance facilities set Switzerland on the road to becoming one of the world's foremost financial centers.

The Swiss did not restrict themselves to mercantile services. Their lack of exploitable mineral deposits and inability to grow industrial crops did not hold them back. Lack of coal, for example, was no handicap. Using "white coal," water cascading down their mountain slopes, the Swiss pioneered hydroelectric power.

Rather like Japan (another mountainous country with few natural resources), Switzerland developed an industrial base that specialized in lightweight, high-value, miniaturized products. Above all, it is a high-skill economy. For centuries the expression "Swiss-made," whether applied to watches or scientific instruments, has been synonymous with precision and quality. More recently this reputation for product sophistication has been extended to a thriving pharmaceuticals industry.

Today Switzerland sits right at the business heart of the new Europe, the center of the richest area, which extends roughly from Stuttgart and Munich through Zurich, to Milan in the south and Lyons in the west. It enjoys industrial and commercial wealth and standards of living far above those of most of the rest of the world, including much of Europe. It has used its physical and social geography to become one of the most durable, economically sophisticated, and socially harmonious societies. According to practically any measure of well-being – standard of living, longevity, low infant mortality – Switzerland is among the world's leaders.

The story of Switzerland reflects in microcosm the central precept of this book – that the history and social and economic development of a people or an area demonstrate in every aspect the intimate, dynamic interaction between humans and their broader physical environment. Switzerland's geography, which seems to conspire against it, explains its success and its position in today's real world.

Planet and People

Planet Earth

It is wholly remarkable that our planet is habitable at all. A few miles beneath our feet, a nuclear-powered inferno rages. Another impact with a mile-wide meteor, and a new ice age might ensue. A little closer to the Sun, or a slower spin, or a few percentage points more of oxygen in the atmosphere, and all living things would burn away to ash.

Yet we can easily ignore how balanced, finely tuned, and self-sustaining this system is – until we are faced with the consequences of global warming, or "holes" appearing in the ozone layer that protects us from the Sun's lethal ultraviolet light.

Humans Emergent

Although environmental hazards – earthquakes, volcanic eruptions, storms, droughts, or floods – remain an ever-present, scarcely avoidable threat, the Earth has also held out a promise, a glorious opportunity for human creative development.

Humanlike creatures first evolved some ten million years ago in tropical forests. Between then and 100,000 years ago, a variety of related beings acquired many of the physical, intellectual, and social attributes that make us the only fully global species.

Since then we have multiplied our powers, our numbers, and our mutual interdependence – particularly in the last 500 years, since Christopher Columbus and his fellow navigators tied hitherto separate hemispheres together to create the single world system that now connects us all.

Planet Earth

If we could hold the Earth in the palm of our hands like a fruit, its highest mountains and deepest valleys would feel no rougher to the touch than the skin of a peach. Its atmosphere and immense oceans, some parts of which could drown Mount Everest under a mile of water, would scarcely register to the senses even as a film of moisture. Yet it is on this fine tissue of earth, water, and air no more than 10 miles (15 km) thick, from the height of an intercontinental flight to the depths of a diamond mine, that all life depends and owes its origins.

Peeling away that skin of solid ground, water, and air would reveal that the planet seethes in the heat and violence of a nuclear-powered furnace, where even diamonds melt and flow like molasses. New York City rides on a raft of rock that is a mere 22 miles (35 km) above this furnace. In the deep ocean trenches, the crust is thinner still, a bare 3 miles (5 km). On occasion, the planet is so restless that earthquakes kill hundreds of thousands of people.

But taken as a whole, conditions in the "biosphere" are astonishingly benign, certainly in comparison with our neighbors in the solar system, frigid Mars, or superheated Venus and Mercury.

A series of remarkable coincidences has made this planet a fit place to live on. The Earth's distance from the Sun, its slight tilt and 24-hour rotation, its very small variation in relief, all mean that the Sun's energy is fairly evenly spread in area and duration. The Earth's delicate atmosphere is transparent to the Sun's most useful light, yet it blocks out most of the meteorites, ultraviolet rays, and cosmic particles whose fearsome energy would destroy life. As a result, a high proportion of the Earth's surface is habitable.

It is an extraordinary balancing act. So narrow is the range of conditions in which life can exist that quite small changes, if uncorrected, would have catastrophic results. An increase of a couple of percent in the amount of oxygen in the atmosphere and all plants would burst into flame. A very few degrees' rise in global temperature and much of the world would be drowned. The whole is so interdependent, finely tuned, and self-sustaining that many now believe that the Earth is itself alive. They are calling it Gaia, after the Greek goddess who, according to legend, gave birth to the Earth.

The planet that humans have inherited has always been a difficult one to cope with. But we have done very much better than simply cope. Earth has also held out a promise, a glorious opportunity for creative human development. We have not just accommodated ourselves to our planet, we have actually transformed it. And in the process we have transformed ourselves.

Surtsey, 1963: a glimpse into prehistory. An island is born in the North Atlantic, near Iceland, when an immense upwelling of molten rock from deep within our planet erupts through a split in the Earth's crust. Within a few months, 10.5 billion cubic ft (300 million cubic m) of molten lava, twice that much ash, and huge amounts of carbon dioxide, and clouds of steam had poured out of the sea.

Events like these throw some light on the fundamental history of the Earth. The Sun and its family of planets formed around five billion years ago when a primeval cloud of gas and dust began to condense.

Within a billion years, heat generated from within the Earth by radioactive decay was transforming the planet. A liquid nickel-iron core was surrounded by a semimolten rocky mantle and a brittle surface crust. This fragile shell, in places only a few miles thick, is all that separates us from the internal inferno.

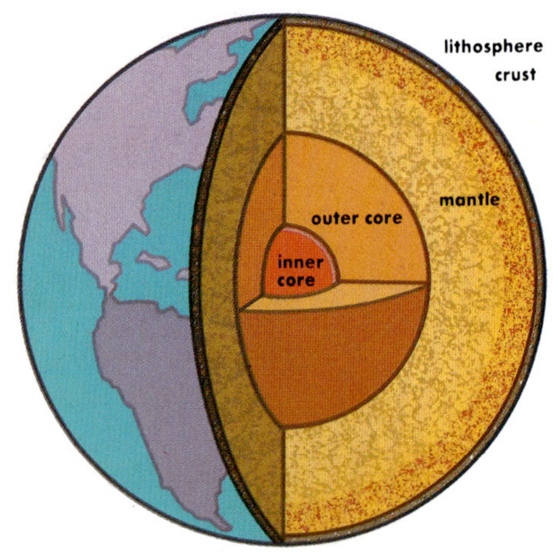

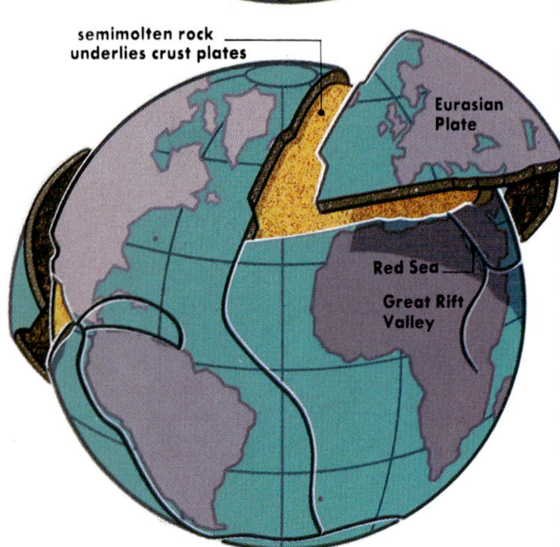

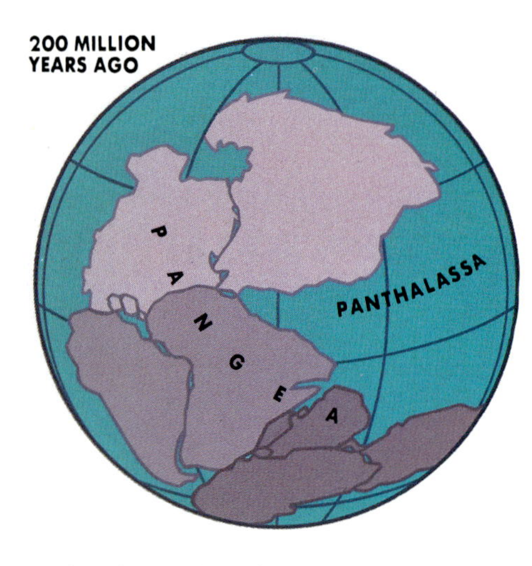

*A thin, shattered shell of rock is all that insulates the Earth's surface from a nuclear-powered inferno (*top*). The heat beneath is so intense that rocks and metals flow and swirl like boiling molasses, endlessly rearranging the jigsaw of plates on the surface of the globe (*middle*). Embedded in some of these plates, today's landmasses once formed a single supercontinent, Pangea, its shores lapped by a universal sea, Panthalassa (*bottom*). About 200 million years ago, Pangea started to split apart.*

EARTH FIRE

EARLY IN ITS HISTORY, PERHAPS FOUR BILLION YEARS ago, the Earth's substance began to sort itself into layers. Heated from within to around 5,400°F (3,000°C) by the natural decay of radioactive elements such as uranium, the heavier materials tended to sink while the lighter ones rose. Eventually a very dense, formidably hot "core" of iron and nickel, about 4,500 miles (7,000 km) across, was formed. As the Earth spins through space, fluid parts of this core tend to lag behind solid parts. These stirring movements are equivalent to an electric dynamo, and generate the Earth's magnetic field.

The core came to be surrounded by a mantle of rather lighter, semiplastic rock 1,800 miles (2,900 km) thick. As the Earth slowly cooled, a crust of the lightest matter rose to the surface, hardened, and floated like a brittle scum. The thickness of this crust appears to vary from 3 to 30 miles (5 to 50 km). It is thinnest under the deepest trenches in the ocean floor and thickest under the highest mountains. Like the Earth's solid crust, the envelope of water and air surrounding it is also ultimately the product of the magma. Outpourings of a great variety of gases, including water vapor and carbon dioxide, have been lavish throughout the life of the planet.

The Earth's gravity was sufficient to hold these gases 'around itself like a cloak, creating an atmosphere that could sustain life. Had the Earth been much smaller, or much less dense, gases and water vapor would simply have evaporated away into space. Presumably the Moon enjoyed much the same early geological history as Earth, yet with only one-sixth of the gravity, it has been unable to accumulate any atmosphere. When the Earth cooled and water vapor condensed, more than 70 percent of the planet's crust was drowned to an average depth of more than 2 miles (3.5 km). The oceans were formed.

THE MOVING EARTH

Although the crust covers the entire planet, it is neither seamless nor stationary. In fact it is fragmented into mobile, semirigid plates composed partly of ocean floor and partly of continental crust. According to the theory of plate tectonics, the Earth's internal heat causes such powerful convection currents in the molten magma at its core that the raftlike plates are carried along passively on its swirls and eddies.

Notions about the movement of the continental landmasses emerged in the early 17th century. The English philosopher Francis Bacon observed on the new detailed maps of his day that South America fitted so neatly around the bulge of the West African coastline that the two continents might once have been joined.

An inspection of any world map will show many other examples of astonishingly intimate "fits" between now-separate landmasses. Around Africa, these include Madagascar and the land on either side of the Red Sea – the Arabian peninsula and the Horn of Africa. In North America, the Baja California peninsula just south of Los Angeles seems to have come adrift from the mainland of Mexico. Most curiously of all, India seems to fill a gap between Antarctica and Africa.

Visual comparison of geographical shape is not the only way to bring pieces together. Comparison of rock samples, fossil types, the directions of mountain ranges, and even the evidence of former ice caps all seem to confirm the links. As a result, scholars eventually pieced together all the world's landmasses to reconstruct whole primitive "supercontinents" to which they gave names like Gondwana (the southern continent) and Laurasia (the northern continent).

But all attempts at explanation of the astonishingly precise matching of coastlines were for one reason or another unsatisfactory. Some writers in the Judeo-Christian tradition thought that God might have hurled the continents apart at the time of the fall of Adam and Eve. Others with a more mechanistic turn of mind speculated that the Earth must have expanded as it cooled down, causing the original landmasses to crack and separate.

One problem with this latter argument was that although the parts do fit, they do not fit well enough to suit the theory. In addition, the fragments of the former continent are all awry; if the process had been one of simple expansion, their relative positions should have been very precisely maintained. In short, each continent-fragment seems to have wandered across the face of the planet almost at random.

In 1912 the German geologist and meteorologist Alfred Wegener proposed a unifying theory of extraordinary comprehensiveness. This not only claimed to be able to account for the origin and movements of the continental masses, it also shed light on the great landscape features of the world (the mountains, plains, deep submarine trenches, and rift valleys), the distribution of many of the world's mineral resources, and the areas where earthquakes and volcanoes occurred. Yet until the 1960s, Wegener's hypotheses seemed too fantastic to take seriously. It was only then, 30 years after his death, that the modern theory of plate tectonics supported some of his ideas, explaining the basic shapes and relative positions of all the major landmasses, oceans, and mountains, as well as the causes of earthquakes and volcanic activity.

According to the theory, it is not continents that "drift" around the globe, but the plates of which they are a part. Some of these plates are relatively small, but others are very large indeed. The North American Plate, for example, stretches

6,000 miles (9,500 km) from the Pacific coast to the middle of the Atlantic. Many of the plates include areas of dense, but thin, ocean floor as well as areas of lighter, thicker continental material.

At present there are 12 major plates and many minor fragments, but their number, size, and configuration change all the time. The plates continually jostle for position, and at their margins, volcanoes and earthquakes make life potentially dangerous for humans. The plate centers are, by contrast, geologically quiescent, uneventful places in which to live.

Plate movements affect human life in many other ways. In establishing the distribution, position, and size of the continents, they have influenced the climate, the sea level, and the direction of ocean currents. By making and breaking land bridges between continents and islands, plate movements have powerfully influenced the pathways of evolution and even human distribution around the globe.

Basing their deductions chiefly on the pattern of rocks and fossils, geologists believe that they now have a very good idea of how the crust fragments may once have been joined together. They suggest that about 220 million years ago there may have been a single supercontinent, which they call Pangea ("All-earth"), isolated in a great ocean Panthalassa ("All-sea").

Pangea then divided into two landmasses. The northern supercontinent, Laurasia, embraced most of modern Eurasia and North America. The southern supercontinent, Gondwana, contained all the other landmasses, including Africa, Antarctica, India, and Australia. Later movements of the plates eventually split these landmasses into the present-day continents.

Moving Margins

Nearly all of the most dynamic geological activity takes place at the margins of the plates. According to the geological events they engender, margins are described as "constructive" or "destructive."

At constructive margins, plates are moving apart from one another, creating a rift between them. Molten magma rushes up to plug the gap, solidifies, and attaches itself to each side of the spreading plates, so forming a volcanic, mountainous ridge. As the plates continue to move apart, they are constantly added to from below – the gap between the margins acts rather like an elongated, continuous volcano spewing out toffee-like magma. In time, the widening rift is flooded to become sea and eventually ocean. Today, most constructive margins are located on ocean floors.

The Atlantic Ocean began to form in this way around 180 million years ago, and is still growing. A typical midocean ridge runs down its entire length. Here, the plates are being driven apart at a rate of about 1 in (2.5 cm) a year – slower than the growth of fingernails, but still sufficient over a long period to have created an ocean some 3,000 miles (5,000 km) wide. The youngest rifts are still dry, as in the East African Great Rift Valley, or as yet quite narrow, like the Red Sea. At its current rate of expansion, the Red Sea will become as wide as the Atlantic Ocean in 200 million years.

On occasion, the magma erupted from the midocean ridge builds up to such an extent that new islands emerge from out of the ocean. Such an event occurred as recently as 1963, when the island of Surtsey emerged off the coast of Iceland during a prolonged volcanic eruption. The Mid-Atlantic Ridge has spawned many such islands in the past, including the Azores and Ascension Island. Continued upwelling has pushed some of these islands far from their point of origin. Bermuda, now more than 1,000 miles (1,600 km) from the ridge, was formed in this way about 36 million years ago. Tristan da Cunha, which still sits astride the ridge, is a mere one million years old.

Detailed mapping of the ocean floor using sonar, an echo-sounding technique, has revealed that the Atlantic ridge does not finish in the South Atlantic. In fact, it continues around the Cape of Good Hope, up the eastern coast of Africa, and through the Red Sea. A branch in the middle of the Indian Ocean sweeps around under Australia and the South Pacific before passing in a gigantic circle toward California and eventually finishing in Alaska. In all, the ridge is around 40,000 miles (64,000 km) long.

In a form of geological compensation, if moving plates are being added to at some of their margins, elsewhere they must be coming together and being destroyed or powerfully modified. These are the so-called destructive margins.

When moving plates collide, the impact can have awesome consequences. In some cases, the edges of the continents are massively deformed, upheaved, and welded together. The same forces that once separated the continental areas before their slow and inevitable collision distort volcanic lava and sediments from the ocean bed.

When the Indian Plate became detached from Gondwana around 180 million years ago, it migrated northward toward Eurasia. When the two finally collided, about four million years ago, they fused and threw up the greatest mountain range in the world today, the Himalayas. A similar process, which is not unlike scuffing up a carpet, formed the Alps when the African Plate collided with Europe.

Mountains such as the Rockies and Andes on the west coast of the Americas were also formed at the margins of two plates, but in a rather different way. Here the much denser Pacific Ocean floor dives beneath the lighter land and descends into the mantle – a process known as "subduction." Offshore, a deep ocean trough or trench is created. Onshore, mountains are thrown up. The scrape of rock against rock builds up so much heat that the

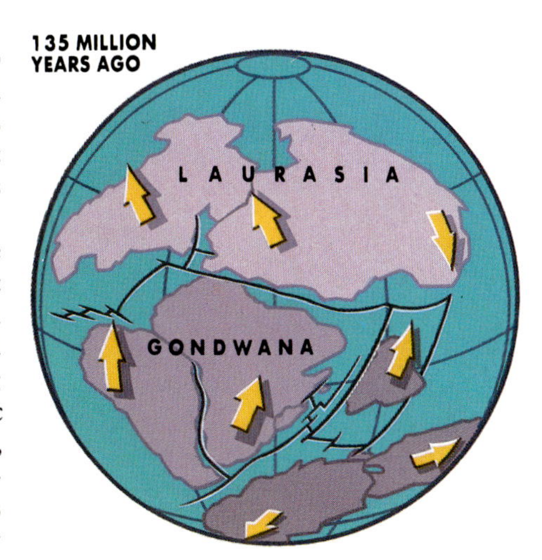

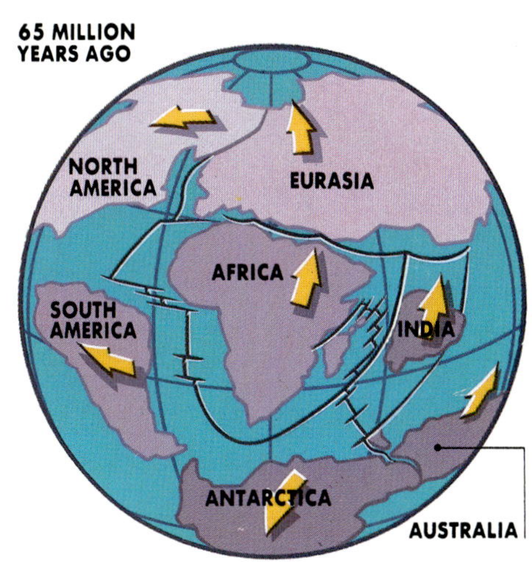

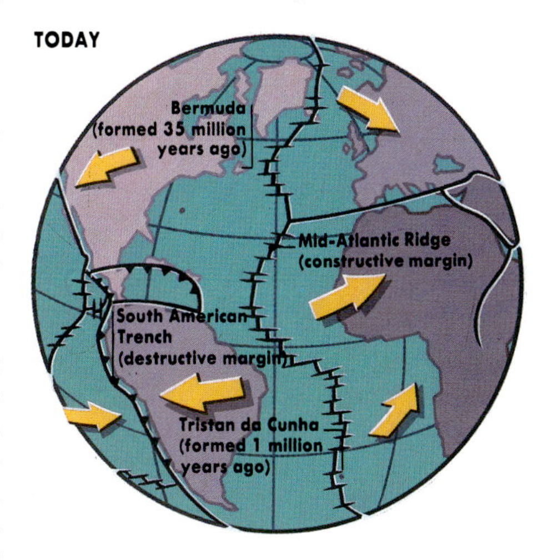

*During the 70 years of an average human lifetime, Europe and Africa will have grown apart from the Americas by a little less than 6 ft (1.8 m). The cause of this relative motion helps to explain the long recognized but mysterious "fit" between the Old and New worlds (*top*). The plates in which the continents are embedded are moving apart from a rift in the Earth's crust, through which molten magma is pouring. The flooding of this widening gap formed the Atlantic Ocean (*middle, bottom*).*

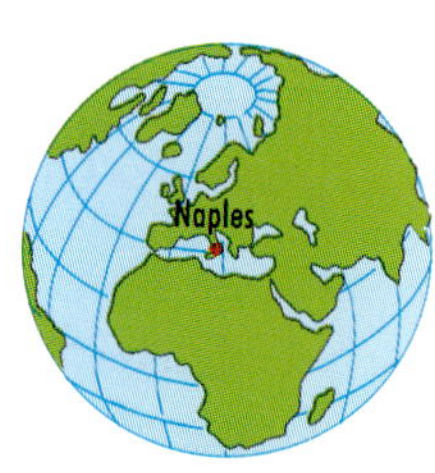

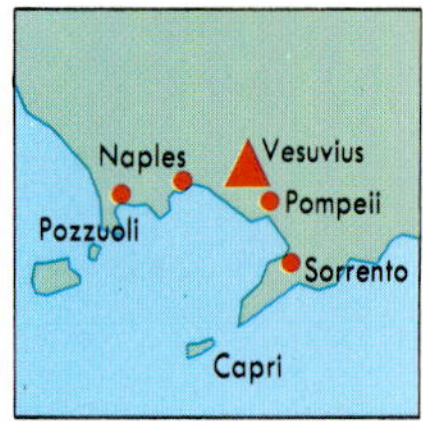

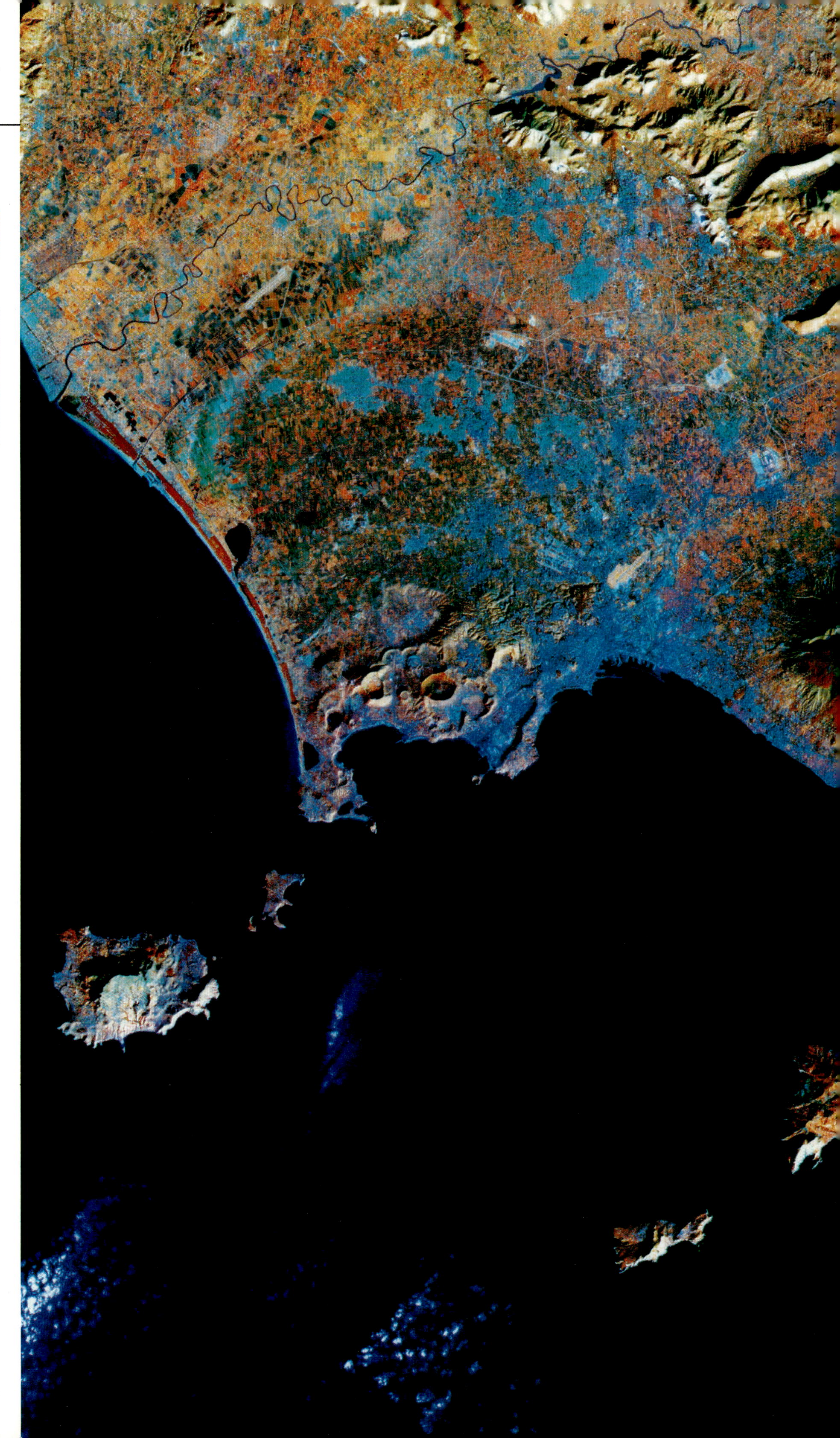

The conical volcano Mount Vesuvius dominates the pockmarked face of the Naples region of southern Italy, seen here in a Landsat image. Populated areas are colored pale blue.

Like many dormant volcanoes, Vesuvius was always a potential killer. In August of AD 79, Vesuvius erupted, burying the region beneath hot cinders, pumice stone, and burning rock. The town of Pompeii was covered to a height of 10 ft (3 m), killing thousands but preserving buildings, paintings, and other features of Roman life.

Since then Vesuvius has erupted many times, notably in 1906 and 1944. Superheated groundwater constantly threatens to explode like a pressure cooker.

The Naples region is also regularly shaken by earthquakes. One in 1980 killed more than 2,700 people and left 100,000 homeless. Houses everywhere are undermined by land movement. The port of Pozzuoli mysteriously rises or falls by up to 1 in (2.5 cm) a week.

rocks melt. Material then attaches itself to the underside of the landmass above, causing it to rise.

Although most parts of the Rockies are now geologically stable, the Andes, as well as volcanic island chains around the Pacific "Rim of Fire," are still being added to in this way. When the build-up of molten matter is particularly great, it forces its way upward and erupts as lava from a volcano. The most powerful volcanic eruptions occur when water seeps down and is trapped in a chamber within the molten rock. The superheated steam can then exert so much pressure that the volcanic mountain cone is blown off, as happened at Krakatoa in 1883.

Where the contact between plates is only glancing, as in the infamous San Andreas Fault in southern California, the plates grate jerkily past each other. For most of the time they may seem stationary, but enormous sheering pressures are constantly building up. Stresses of such power develop within the rocks that only a cataclysmic rupture of the crust – an earthquake – can relieve the tension. The breaking and sudden shifting of rocks deep within the crust causes the whole planet to vibrate like a tuning fork, which is why the quake can be registered on instruments all around the world.

The Earth's moving crust does more than mold the pattern of the planet's surface. Just a few miles underground, the temperature reaches that of boiling water. Indeed, in Iceland water heated by rocks provides the central heating in most homes. But the overall amount of heat arriving at the surface is surprisingly small, and contributes a mere 1 percent to the heat present above ground. The crust acts as splendid insulation, with much of the intense heat being dissipated before it reaches the surface.

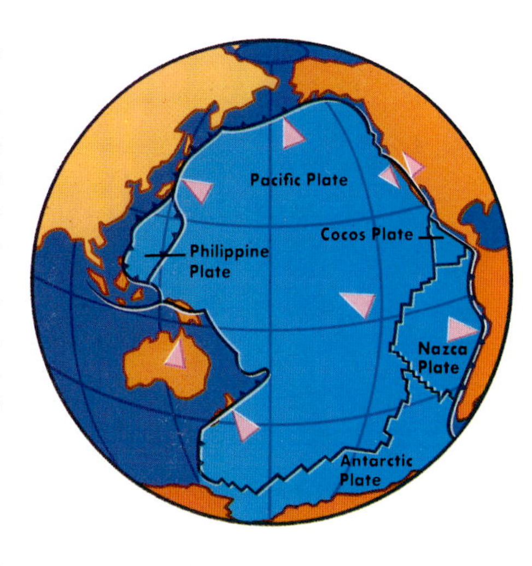

Most of the world's 600 active volcanoes and many of its earthquake-prone zones girdle the Pacific or lie on a branch across Indonesia, precisely where human population is growing fastest. This forms the Pacific "Rim of Fire."

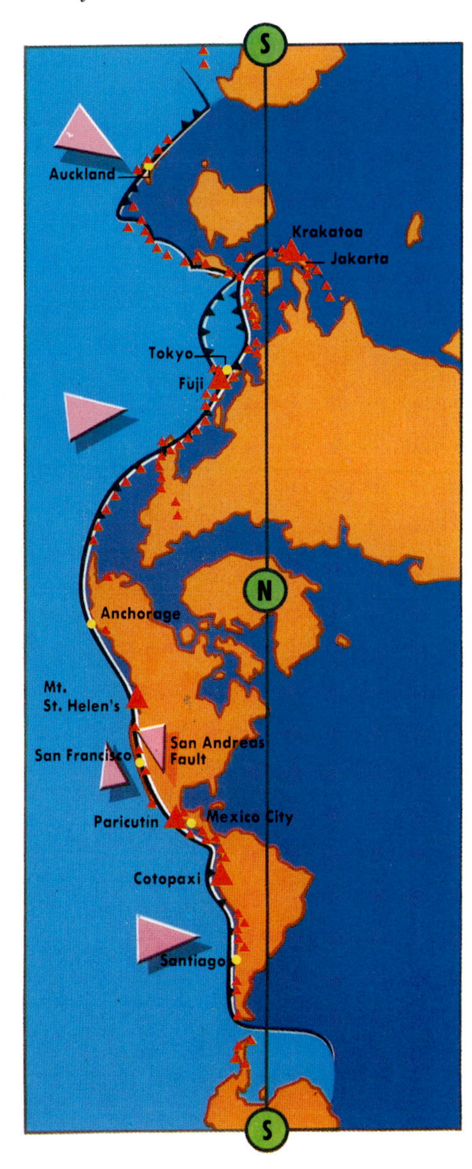

SUN POWER

IN THE DISTANT PAST, MOST OF THE ENERGY THAT reached the surface of the Earth was generated from within. But today, with a gradually cooling interior and the insulation provided by the Earth's crust, almost all the surface energy is derived from the Sun's incoming radiation. The Sun's surface temperature is around 11,000°F (6,000°C), but the Earth's small size and great distance from the Sun – 93 million miles (150 million km) – mean that only a tiny proportion (0.002 percent) of the Sun's total radiated energy arrives at the Earth. Even so, this is still equivalent to an average 1,160 watts on every square yard (1,390 W per square m).

The Sun's rays are essential to power living processes, but within limits. For Earth to sustain life – to be green and pleasant – equability is the key. It must be neither too hot nor too cold, nor must the temperature fluctuate too wildly. The

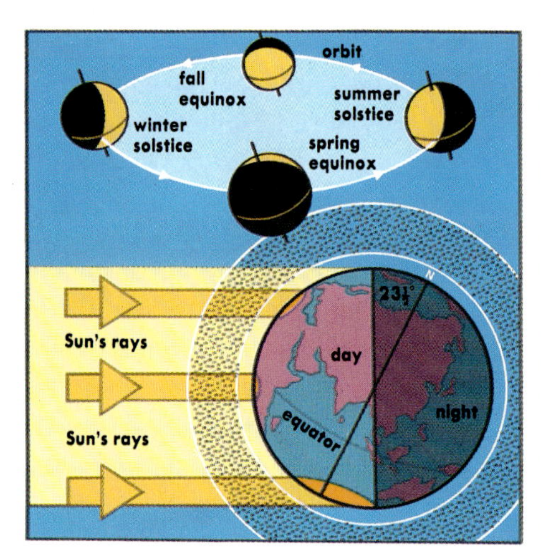

*By creating day, night, and the seasons, the Earth's shape and movements all help to moderate temperatures. As the Earth is a tilted sphere, the equator receives most incoming solar radiation (insolation), although this varies throughout the year. Seasonal effects are increasingly pronounced nearer the Arctic and Antarctic circles (*top*), with "permanent daylight" in summer and "permanent night" in winter.*

*Even so, conditions would fluctuate between lethal extremes without additional protection in the form of complex processes involving the atmosphere, land surfaces, and the oceans. Oxygen, for example, helps to moderate the effect of ultraviolet light by forming a layer of ozone about 10–20 miles (15–30 km) above the Earth's surface (*bottom*).*

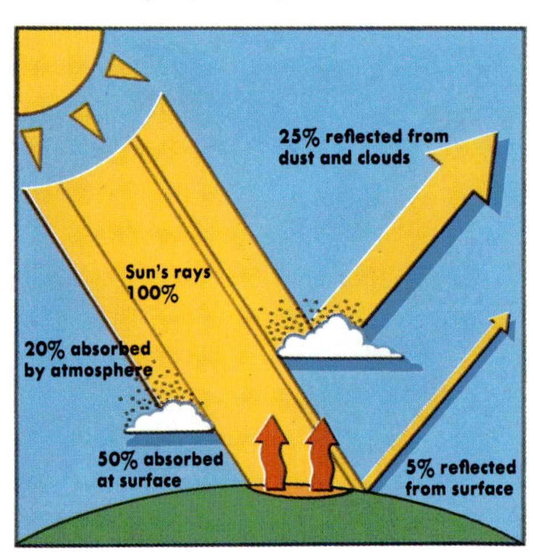

delicate balancing needed to meet these demands depends on the planet's shape and its complex rotations in space, the atmosphere, and the oceans. Through these and their interactions, the Sun's potentially overwhelming incident energy is avoided, reflected, filtered, absorbed, radiated, and moved about the planet so effectively that life is made possible virtually everywhere.

The Earth alone in our solar system enjoys these conditions. It has a relatively comfortable average temperature of around 63°F (17°C), compared to 900°F (480°C) on Venus, or – 310°F (– 190°C) on Uranus. Oscillations in temperature are also much smaller than elsewhere. Even during the ice ages, the average temperature at the equator was only a couple of degrees lower than it is now.

Globe Dynamics

The mechanisms that serve to keep the Earth's temperature "just right" are many. One set is provided by the spherical shape of the Earth and its movements in space.

The Earth's movements comprise two elements. The first is the Earth's year-long revolution around the Sun, along an elliptical path. The other is the Earth's daily rotation around an axis that is more or less constant but slightly tilted (by nearly 24 degrees) with respect to the Sun. Taken together, these give day and night, and the seasons.

From the point of view of its vulnerability to the Sun's assault, the Earth can be seen as an outclassed but wily boxer, doing all it can to deflect any blows. As it is very nearly a perfect sphere, the Earth does not present itself flat on. This means that insolation (*in*coming *sol*ar radi*ation*) is not uniformly powerful over its whole surface.

Thus there is a fundamental link between latitude and climate. Insolation is at its most intense at the tropics, where the Sun is overhead or high in the sky for most of the year. It is at its lowest at the poles, where the "slope" of the Earth's surface is greatest and the Sun's rays have to penetrate a greater thickness of atmosphere.

The Earth's daily spin prevents it from being scorched on its facing side and deep frozen on the other, for no area has to face the full intensity of the Sun's heat for more than a few hours at a stretch. Except in equatorial areas, the Earth's tilted axis, which gives the planet its seasons, also limits the periods of greatest insolation to a definite time of the year – summer. At other times, the surface (including its plant and animal life and its freshwater reserves) may have a chance to recuperate before the Sun's next onslaught.

Atmosphere

Even with the defense against the Sun provided by planetary movement, if the Earth were completely transparent to all incoming and outgoing wavelengths of energy, conditions would be extremely harsh. The ground temperature would be much higher during the day, but would always plummet well below freezing at night. However, the Earth does not experience these extremes because of the moderating effect of the atmosphere and the oceans.

Early in the life of the planet, its surface was protected from a lot of the Sun's energy by dense fumes, ash, and clouds of water vapor belched out from volcanoes. Much of the Sun's energy was reflected back into space, or scattered.

At night, the primitive atmosphere had the opposite effect, keeping the Earth warmer than it would otherwise have been. The energy that had penetrated through the blanket was kept from radiating straight back into space by the very high concentrations of water vapor, carbon dioxide, and methane – the so-called greenhouse effect. Like the glass panes in a greenhouse, the gases let the light in, but do not allow heat to escape.

Water vapor and clouds absorbed outgoing heat at night and radiated some of the warmth they had acquired back to the ground. The condensation of water vapor into rain also released heat. The gases carbon dioxide and methane added to the effects of water vapor, but in a more selective way.

Carbon dioxide is a clear gas that allows visible light through. But it strongly absorbs the invisible infrared bands of the spectrum that are associated with heating. (Infrared radiation is the heat felt from a radiator.) Since much of the energy being radiated back into space from Earth is in the infrared bands, the carbon dioxide heats up and itself begins to radiate energy – both out into space and back toward the Earth.

The oceans came to have a similar effect. They eventually covered more than two-thirds of the planet, and were capable of absorbing massive amounts of heat from sunshine. At night, they released some of this back to their surroundings.

The oceans continue to play an important role in heat storage. Although the composition of the atmospheric gases is somewhat different – there is much less methane, and plants have absorbed much of the carbon dioxide, building up oxygen in return – the same processes of reflection, filtration, absorption, evaporation, condensation, and radiation are still at work.

It is not just the total amount of incoming energy that is reduced by the processes of reflection and absorption. On the whole, the wavelengths that penetrate tend to be useful rather than harmful to current life-forms. Ultraviolet light

Seen from space, the Earth's vulnerability to the Sun's rays is all too clear. The Sahara and Arabian deserts in northern Africa and the Namib in the south are almost permanently cloud free. Long and intense exposure to the Sun's rays means that temperatures in these areas can rise to 100°F (38°C) during the day, yet fall to freezing point at night.

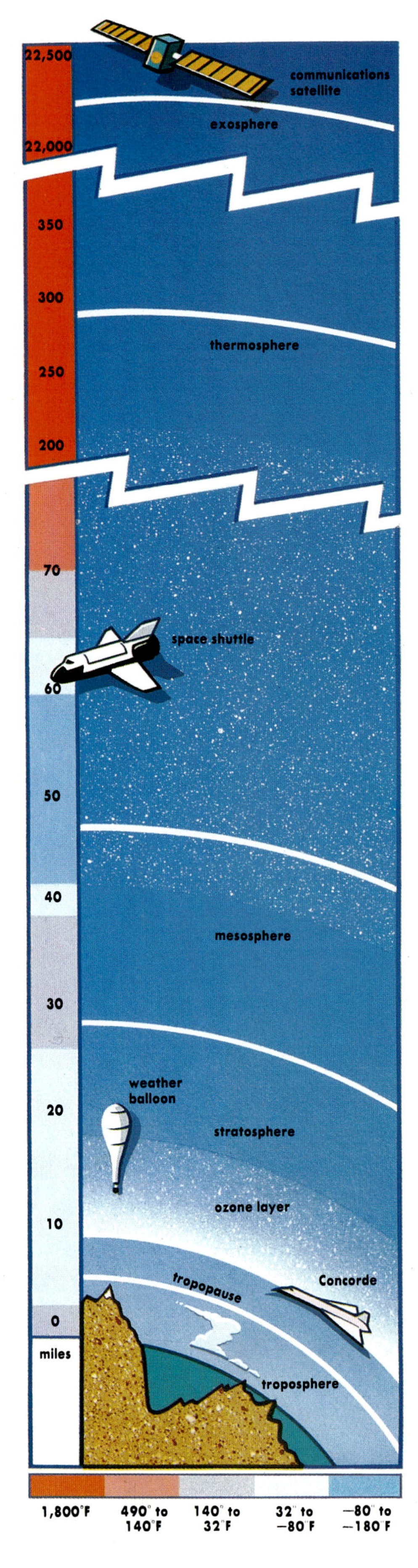

(UV), for example, is largely filtered out by a layer of oxygen-generated ozone in the upper atmosphere. UV is lethal in large amounts (it is widely used to sterilize surgical equipment), but even at quite low levels it can cause sunburn, skin cancers, and eye cataracts.

When this ozone barrier developed, around 420 million years ago, life was able to emerge from the sea. Previously, it had been confined there by the amounts of UV light that sterilized the land surface. (Sea water absorbs UV and renders it harmless, at least below the surface.) Rapid depletion of the ozone layer caused by atmospheric pollution has now become a serious environmental issue, with major repercussions for health and the stability of plant and animal populations.

Of the Sun's incident energy that reaches the ground, a proportion of it is immediately reflected back. A measure of the reflectivity of a surface is called its albedo: the higher the reflectivity, the higher the albedo. Snow, ice, and desert all have a high albedo: they absorb very little of the incoming energy. Freshly fallen snow, for example, reflects back around 95 percent of the energy on it. Dark bare soil, however, absorbs virtually all of the light that shines on it.

Even plants reflect significant amounts of energy. This prevents them from overheating. Green fields reflect from 10–15 percent of visible light, but dark green coniferous forest may reflect only 5–10 percent.

Winds and Currents

If latitude alone decided climate, our planet would scarcely be habitable at all. The tropics would broil, while everywhere north of about 38 degrees – the latitude of Washington, D.C. and Athens – would be very much colder. London, New York, and Tokyo would be frozen all year.

Living conditions are tolerable only because planetwide processes transfer immense amounts of heat. The net transfer is from the tropics – where, in spite of barriers in the atmosphere, heat arrives from the Sun in dangerously large amounts – to the temperate and polar regions, most of which are thereby rendered more benign.

Taking the globe's earth, water, and atmosphere as a whole, the amount of heat received

Thin as the Earth's atmosphere is, it exerts pressure of around 1 ton per square ft. Its gases are vital for respiration and photosynthesis, and as a filter from the Sun's most lethal ultraviolet rays. Much of the Sun's energy is absorbed before it reaches the surface, which accounts for the rise and fall in temperature at different heights.

The name Earth for our planet, when seen from space, is a misnomer. More than 70 percent of its surface is covered in ocean to an average depth of 2 miles (3 km). Slow to heat, but also slow to cool, and with an immense capacity to absorb the Sun's energy, oceans are vital in global heat transactions and the creation of the world's diverse climates.

Since warm winds and currents generally travel toward the poles, to be replaced by cooler ones, the process acts like an energy conveyor belt. The equatorial regions are effectively cooled while the higher latitudes are warmed.

In addition, the process involves the evaporation, condensation, and precipitation – as rain or snow – of incredible volumes of water. Nearly one-fourth of the world's ocean is recycled every year, with around 104 million billion gal (394 million billion l) of fresh water being distilled from the oceans – enough to cover the parched Sahara to a depth of about 13 ft (4 m).

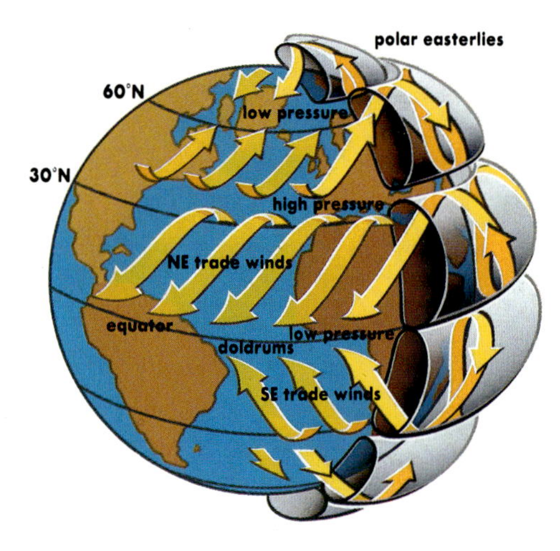

Without an energy conveyor belt of warm winds and currents moving heat from the equator poleward, tropical lands would be little more than scorched earth while Vancouver, London, and Moscow might be frozen all year round.

The overall process is simple. Heated air rises above the equator to a great height, diverges north and south, gives up its heat, and then descends to the ground, after which it returns to the equator at ground level as cool air that is ready to be heated again (top).

In fact the transfer, though continuous, cannot take place in a single equator-pole-equator circuit because of the Earth's spinning. Thermodynamics dictates a series of self-contained stages, or "cells." The basic pattern involves three such cells. Air rises above the equator and descends at around 30 degrees north and south, the high pressure zones where most of the world's deserts are found. In the temperate cell, air circulates in the area between 30 and 60 degrees north and south. Air also rises and diverges at around 60 degrees north and south, with compensating cool airflows coming in at ground level from the poles.

from the Sun in a year is matched by the amount of heat escaping back into space.

Yet if heat income and heat expenditure for different parts of the world are compared, there is a striking discrepancy. The temperate and polar latitudes seem to be radiating very much more than they receive, while the tropics receive more than they radiate.

The solution to this seeming paradox lies in the conveyor belts of warm winds, water vapor, and warm ocean currents that move heat poleward. Since nature cannot tolerate a vacuum, there is a compensating flow that fills the gap these conveyors leave behind. The returning flow of cold air or water can pick up a new consignment of heat and start the cycle again. Differences in temperature persist, of course, but as a result of the continuous global heat transfer, the equatorial areas are a mere 80°F (45°C) warmer than the polar regions – not the 180°F (100°C) or so that would otherwise be the case.

The patterns of winds, currents, and rainfall that global circulation sets up are astonishingly regular, and it is these that give every region its characteristic climate or "average weather."

WINDS

In principle, the overall process that cools the equatorial regions and warms the poles is a single convection current. Hot air rises at the equator, diverges north and south at altitude, cools, and then descends to the ground at higher latitudes. Where air rises, atmospheric pressure is low. Where it falls, pressure is high. Winds are driven by these pressure differences: the greater the difference, the stronger the wind. The circuit is complete because cold winds from the polar high-pressure areas transfer back along the ground toward the tropical low-pressure areas.

In fact, winds cannot travel directly from the equator to the pole and back again in one cycle. Instead, they move in three huge loops, or "cells." Neither can the winds travel in straight lines north and south from areas of high pressure to areas of low pressure. Because of the Earth's spin, winds veer to the right in the northern hemisphere and to the left in the southern hemisphere.

In the tropical cell, hot air rises over the equator, forming a low-pressure belt. It cannot rise forever, so it is pushed sideways to the north and south. As it does so, it cools and starts to sink, falling back to Earth between 25 and 30 degrees north and south, creating high-pressure zones below which are the major deserts of the world.

On the ground at the equator, air converges from south and north to replace any that has migrated away. At sea this equatorial convergence zone is so stable that it became infamous with sailors as the "calms" or "doldrums." Ships could be stranded there for weeks at a time with insufficient wind to move.

Outside of the doldrums, the air flows within the tropics are so strong and constant in direction that sailors call them "trade winds." The term has nothing to do with commerce; it is nautical slang for "regular track." Owing to the Earth's spin, the trades in the northern hemisphere come not from the north but from the northeast – hence they are called north easterlies. Trade winds come from the southeast in the southern hemisphere.

The second, or temperate, cell circulates between the high-pressure calms at 30 degrees (the "horse latitudes") and the low-pressure "midlatitudes" of around 60 degrees north and south. Winds rushing toward the midlatitudes are strongly veered to the west, and can be very powerful, particularly in the southern hemisphere where there are fewer major landmasses to interfere with them. Mariners in the southern seas referred to them as the "Roaring Forties."

At around 60 degrees north and south, cold, dense, polar air slides in beneath the poleward-bound warmer air and pushes it upward, creating another low-pressure band. This "polar front" is much more erratic than other zones, however, which is why places like the British Isles have such changeable weather. Localized, mobile, low-pressure areas tend to develop. In these "depressions," the cold polar air swirls around and inward. Where moist air rises over the cold air, rain is a frequent result.

The third, or polar, cell involves the warmer air that rises in the low-pressure system at 60 degrees. Some of this cycles back toward the horse latitudes, as part of the temperate cell. The rest moves over the top of the advancing polar front, before plunging down into the chill high-pressure area over the pole. It then returns in the form of cold air to the midlatitudes.

OCEAN CURRENTS

The movement of ocean currents is another very efficient mechanism of heat transfer. Ocean currents are created by the friction on the water surface from warm winds persistently traveling in the same direction. Like winds, currents away from the equator are deflected to the right in the northern hemisphere and to the left in the southern hemisphere. This accounts for their chiefly circular routes oceanwide, and the thin compensating currents, the equatorial drifts, that travel in a straight line along the equator from east to west.

Major warm ocean currents, such as the Gulf Stream, involve enormous volumes of water and profoundly influence the climates of countries even thousands of miles away. Cold currents, too, can have important climatic effects. For example, cold currents that descend from around Greenland bring icebergs with them that have "calved" from the ends of glaciers. Where such currents meet with warm currents, as off the coast of Newfoundland, persistent fogs may occur.

The Gulf Stream originates in the warm, shallow waters of the Caribbean. From there it flows up the coast of North America, leaves the coast at Cape Cod, and moves northeast across the Atlantic. It then divides into two. Part flows via the Canary Islands and the west coast of Africa, and then back along the equator. The other branch, now called the North Atlantic Drift, continues northward past the British Isles and Scandinavia, into the Arctic Circle.

The Gulf Stream is often likened to "a river in the ocean," but it is much larger than any river on land. At its start it is 50 miles (80 km) wide and a quarter of a mile (0.4 km) deep. It moves at speeds of up to 4 miles (6.5 km) per hour. It carries a thousand times as much water each second as the Mississippi – indeed, 25 times more water than all the rivers in the world put together.

Such a mighty, warm "river" cannot fail to have profound effects on the lands it passes. The current brings such great amounts of heat along with it that the northwestern coasts of Europe are warmed. An isotherm (a line connecting points that have the same temperature) plotted for 32°F (0°C) in January produces what has been called the "Gulf of Winter Warmth." Without the Gulf Stream and North Atlantic Drift, northwestern Europe would probably be icebound in the winter. The spectacular effects of the current are obvious in the subtropical gardens growing on the west coast of Scotland, at the same latitude as southern Siberia or Hudson Bay, Canada.

Clouds and Rain

A third critical mechanism for global heat transfer involves the immense amounts of heat taken up and given off by the evaporation and condensation of water. Wet wind carries very much more heat with it than dry wind. Water evaporates readily into the air from land, ocean, or vegetation, particularly when wind blows over the surface. Since a considerable amount of energy must be absorbed in order for water to evaporate, its surroundings are cooled. High-energy water vapor gives up that heat again wherever it condenses to form clouds. Consequently water takes heat from where it evaporates and warms those areas where it condenses.

Climate

Climate plays a fundamental role in life on Earth. The characteristic patterns of temperature, rainfall, winds, and humidity do not merely affect the way people feel and the clothes they wear. They also influence vegetation, the crops that can be grown (and hence what is eaten or exported), the forms of the landscape, the distribution of fresh water, and the buildings people live in. Extreme weather events such as storms, tornadoes, and hurricanes also constitute a serious threat to lives and property.

Of course, climates in the real world are very much more complicated than simplified models of latitude and global circulation suggest – as a glance at global maps of temperatures or rainfall, or the directions of winds and ocean currents, readily shows. Nevertheless, much of the variation from the expected climate patterns can be explained with the addition of only a few basic principles.

During April, May, and June, the position of the Sun moves north of the equator, and climate zones alter accordingly. At its northernmost point, the Sun lies over the Tropic of Cancer, and the climate of that region becomes more equatorial during this period. The North Pole, however, remains intensely cold, and the climate zones of the northern hemisphere are squeezed into the much reduced space between. Climate zones in the southern hemisphere are correspondingly stretched. In October, November, and December, the Sun moves south toward the Tropic of Capricorn, and the squeezing occurs in the southern hemisphere.

Altitude above sea level also influences climate. In general, temperature falls quite rapidly with increasing height – about 2°F (1°C) for every 100–150 yards. This helps to explain how a mountain in an equatorial region, like Kilimanjaro in Tanzania, can be covered in snow throughout the year, while nearby Mombasa on the coast can be 99°F (37°C). Similarly, the climates of highland cities such as Quito in Ecuador are much cooler than the lowland plains that surround them.

A third climate principle relates to the different rates at which land and ocean take up and give off heat. The ocean has an enormous thermal capacity: it takes up heat slowly, but loses it slowly as well. Land absorbs and releases heat much faster. The differences can have profound global as well as local effects.

In winter, "maritime" countries that are bathed in a warm current will experience much warmer, wetter conditions than landlocked, "continental" regions. In summer, however, maritime climates are much cooler (but still much wetter) than continental climates. In the heart of Canada, average winter temperatures of −50°F (−45°C) are not uncommon, with temperatures rising to in excess of 104°F (40°C) in summer. By contrast, Vancouver, situated on the western seaboard, has average monthly temperatures ranging merely from 37°–63°F (3–17°C).

The surface of land is also much rougher than that of the ocean. This slows winds down and causes a slight change of direction, especially close to the ground. Land also puts obstructions in the path of winds, in the form of mountains. Since air cools with height, when warm wet air rises over a mountain it tends to shed its load of water vapor as

The world's great ocean currents move heat from the equator toward the poles. Conditioned by the Earth's spin and the distribution and shape of the continents, and varying somewhat in intensity and location according to season, they travel clockwise in the northern hemisphere and counterclockwise in the southern hemisphere. The Gulf Stream, for instance, cools the Caribbean and southern states of the U.S. but creates a "Gulf of Winter Warmth" that serves to keep northwestern Europe from descending into a virtual ice age.

Ocean currents have had a profound influence on human history beyond their effects on global climates. The world's finest fishing grounds are associated with oxygen- and nutrient-rich cold currents, such as those off the coast of Peru, and around Greenland and Labrador. Exploration and transport, too, have been conditioned by currents. The Canary and North Equatorial currents greatly assisted Spain and Portugal in their first voyages of discovery around Africa and to the New World of the Americas.

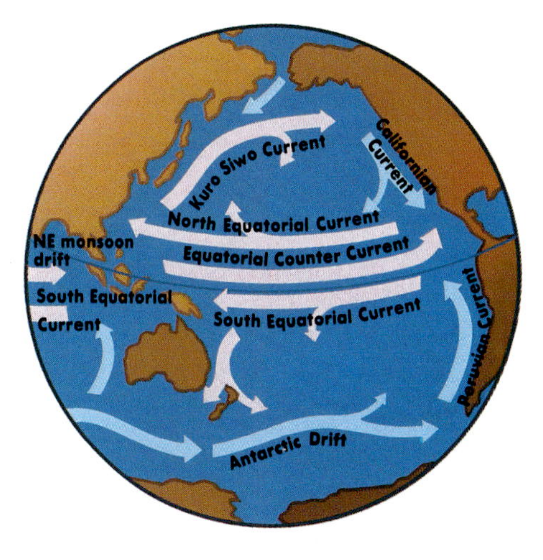

rain. If the mountains are high enough, most of the water contained in the prevailing winds falls as rain on the windward side. Descending on the leeward slope, the wind is warm but dry; the drier area thus created is known as a "rain shadow."

This effect is shown clearly in the U.S. in the dry central states that lie inside the rain shadow of the Rockies. Similarly, in spite of its proximity to areas that experience heavy monsoon rains, Tibet is extremely dry because it lies in the rain shadow of the immense Himalayas. In such cases, only a few miles may separate areas that receive 200 in (5,000 mm) of rainfall a year from places that receive no more than 20 in (500 mm).

A Changing Climate

The earth's climates have not always been as they are today. As continents have moved, so climates have changed. When landmasses crossed the equator they experienced different bands of climate. Deserts have overwhelmed tropical forests, yet they, in turn, have been transformed into temperate zones. Such changes, associated with plate movements, usually occur over many millions of years and are relatively long lasting.

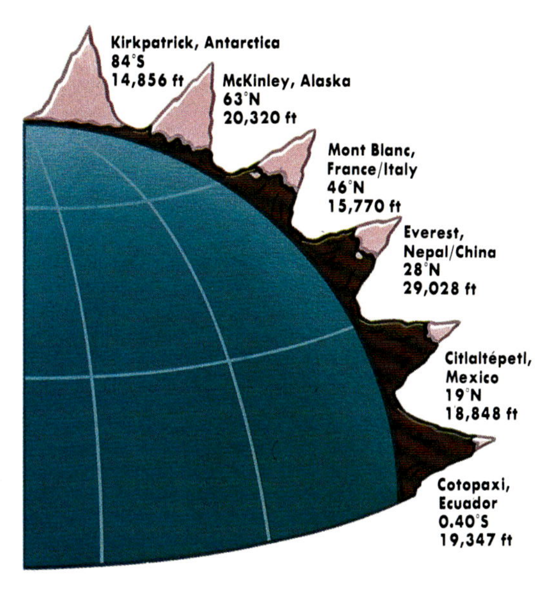

The height of the snow line – the altitude at which snow and ice never completely melt in summer – varies from sea level at the poles to a height of around 18,000 ft (5,500 m) at the equator. Latitude is the principal influence, although the exact height also depends upon local and regional factors. The snow line is on average appreciably higher in the French than the Austrian Alps, thanks to the prevailing warm westerly winds. In addition, snow falls can be later and more uncertain, with serious consequences for the ski industry.

Less than 250 miles (400 km) south of the equator, the jagged 14,980-ft (4,565-m) peak of Mount Meru in Arusha National Park, Tanzania, towers over the ruptured crater of a great dormant volcano.

Like the summit of its famous neighbor, Kilimanjaro, the peak of Mount Meru is covered in snow all year. The vegetation on the upper slopes of the great mountains of East Africa is alpine, with moorland and then forest zones lower down.

Temperatures on the summits vary so much that they are said to have summer every day and winter every night. On a fine day the temperature may reach 104°F (40°C), yet it may remain below freezing for up to ten hours at night.

Other climate changes are cyclical: every 100,000 years or so there is a regular cooling known as an ice age. Although at the tropics temperatures are only slightly diminished – perhaps by 2–4°F (1–2°C) – global average temperatures fall by about 9–18°F (5–10°C). This sounds modest, but it has dramatic effects.

Over the last two to three million years, each of the cooler phases has resulted in the extension of massive polar ice sheets down into the middle latitudes of Eurasia and North America. Ice, sometimes several miles thick, has reached as far south as London and Warsaw in Europe, and well beyond the Great Lakes in North America. Major mountainous regions south of the ice sheets, such as the Alps, also generated their own local ice domes and glaciers. Immediately in front of ice sheets, frozen tundra conditions prevailed.

With so much fresh water locked up in ice, sea levels were probably more than 400 feet (120 m) lower than now. Continent margins were exposed for up to 65 miles (100 km), and land bridges formed that allowed animals such as woolly mammoths and saber-toothed cats – as well as hunter-gatherer peoples – to travel freely between lands that have since become widely separated by sea. Asia and America were joined across the Bering Strait; Indonesia and the Philippines were

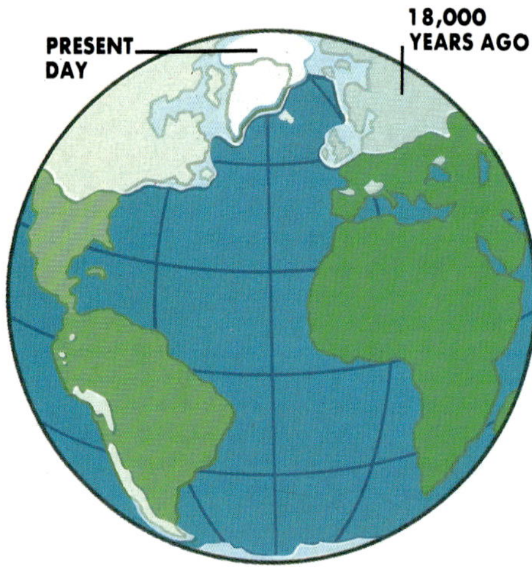

Although ice cover has been considerably reduced since the end of the last ice age around 10,000 years ago, it still amounts to 6 million cubic miles (25 million cubic km) of water. What would happen if this were to melt? The loss of ice from the North Pole would make no difference to sea levels because, in the absence of land, the ice cap floats on the sea. As the ice melted it would simply restore the volume it currently displaces.

Greenland and Antarctica, however, are very different. Because the ice caps are on land, all the meltwater would be added to existing sea levels. Oceans would therefore probably rise by around 200 ft (60 m), more than enough to drown virtually every major city around the coasts of North America, Australia, or Japan.

*The illustration (*above*) shows the extent of the polar ice cap today and its extent 18,000 years ago.*

continuous with Southeast Asia; and Australia, Tasmania, and New Guinea formed a single enormous island continent.

Since ice reflects back more than 90 percent of the Sun's energy, temperatures remain low during ice ages. Rainfall, too, decreases by as much as one-third, and all the circulation patterns that give rise to specific climate and vegetation zones are squeezed down toward the equator. There is so little rainfall that deserts and savanna regions expand to occupy up to one-third of the ice-free landmass. Even the tropical rainforests shrink.

At the peak of the last cold phase, about 18,000 years ago, the Amazonian rainforest may have existed only in small isolated patches. It is believed that our human ancestors first evolved from jungle-dwelling apelike animals between ten and one million years ago because of these periodic contractions in forests and the consequent growth of savanna regions. Many specifically human adaptations, such as relative hairlessness, erect posture, and walking on two feet, can be related to life on lightly wooded grasslands.

After the Ice

Ice ages are followed by warmer – but generally rather shorter – interludes. The last such interlude began about 10,000 years ago. As the ice melted, the water level rose once more by about 300 feet (100 m). Much dry land was inundated. The land bridge at the Bering Strait between North America and Asia was drowned, as were the North Sea and the English Channel. People who had followed the receding ice sheet into Britain were cut off from the mainland of Europe.

In the cool, wet period that followed the thaw, many of today's deserts were verdant. As recently as 5,000 years ago, as cave paintings testify, cattle herders roamed the Sahara desert. Savanna animals such as giraffes and elands were abundant. Yet today the Sahara is the world's largest, hottest, and most barren desert.

It seems that there are various lengths to natural climate cycles, some lasting as long as 200–500 million years, and some as short as 11–27 years. Each seems to have specific causes. The very long phase changes (which occur every 200–500 million years and every 30–60 million years) seem to be associated with the changing positions of landmasses as they are pushed around the globe by convection currents in the mantle. The greatest ice ages only occur when there is a concentration of landmasses close to the poles.

Medium-term cycles (every 100,000 years, 40,000 years, and 20,000 years) are caused by slight but regular and predictable changes in the orbit of the Earth around the Sun, the angle of tilt of the Earth's axis, and the direction in which the axis points. All of these factors affect both the length and the intensity of the seasons. The Earth may then amplify these changes. For example, when ice starts to grow it increases the albedo of the land surface. The climate then cools even more as a greater proportion of the incoming sunlight is reflected back.

Why should this cooling process ever stop? One theory put forward suggests that ice grows until rain-bearing winds can no longer penetrate the high-pressure area that has developed over the ice. Precipitation ceases, preventing any more ice from accumulating.

In addition to long and medium cycles, there are shorter periods of cooling that still have considerable effects. The shortest term changes (every 2,500 years, 100–400 years, and 11–27 years) are most probably related to changing energy outputs from the Sun, and include sunspot cycles. Between 100 and 500 years ago (the "Little Ice Age"), temperatures were rather cooler than

they are now. The Greenland and north polar ice caps descended farther south and glaciers expanded down their valleys. Settlements disappeared from Greenland and Iceland, which had been colonized by the Vikings between the 10th and the 14th centuries during a warmer spell.

In contrast to cyclical events, some climate changes occur at random, or in accordance with no obvious pattern. Some volcanic eruptions, for example, eject vast quantities of dust so high that they swiftly enter the jet streams over the subtropics and polar front, and are dispersed around the planet. As a result, the dust scatters back more of the incoming radiation and its particles act as nuclei around which water droplets condense before they fall as rain. Volcanic activity is often associated with haze and cooler, wetter conditions that may last for months or even years.

Global Warming

Natural climatic change is still taking place, but there is a danger that it will be rapidly eclipsed by human induced changes in the Earth's climates. Global warming induced by atmospheric pollution has become one of the world's gravest environmental threats. The burning of fossil fuels, such as coal and oil, is adding greatly to the total volume of "greenhouse gases" in the atmosphere. In addition, the destruction of large areas of tropical forest is not only itself contributing to atmospheric carbon dioxide, it is reducing the total capacity of plants to consume the excess carbon dioxide and replace it with oxygen. We are adding so much insulation at such a fast rate that the planet may overheat over the coming decades, with drastic consequences for global weather, agriculture, sea levels, and natural ecosystems.

Cotopaxi Glacier, Ecuador, a mere 65 miles (100 km) from the equator. Worldwide, there are around 200,000 glaciers. Ice covers more than 10 percent of our planet's land surface, while permafrost – ground perennially frozen, sometimes to 1,000 ft (300 m) or more – accounts for another 20 percent. Such zones present formidable problems for settlement, mineral extraction, and communications.

Were it not for the peculiarities of coral, people could never inhabit the millions of islands it has created throughout the world.

Coral is porous, so it soaks up rain that falls on it. Since fresh water is slightly less dense than sea water the two do not mix and fresh water is readily available.

Today, on coral islands like the Maldives, tourists are making such great demands on water supply that desalination plants may have to be installed, but at a cost that will far outrun any income the tourists bring with them.

The Fresh Water Story

Fresh water is vital to all life. Our own bodies are almost entirely water – 95 percent – and we need to replenish our losses from excretion, sweating, and evaporation every day if the rest of our bodily functions are not to be impaired.

Yet fresh water makes up only 3 percent of the world's total water, which is overwhelmingly salty ocean. Moreover, of the 3 percent that is fresh water, 2.1 percent is semipermanently locked up in ice sheets and glaciers. Any water that falls as snow in these regions is likely to be trapped there for tens of thousands of years.

This leaves less than 0.9 percent of global water potentially available for life processes, and the majority of that is stored below the surface in groundwater or in natural reservoirs (aquifers). At any one time, then, all the world's streams, rivers, and lakes contain a mere 0.29 percent of total global water. Only another 0.01 percent is in the atmosphere, the equivalent of just 1 in (2.5 cm) of rainfall for the whole surface of the planet, or ten days' water supply.

Fortunately, water is recycled quickly. Continual evaporation of sea and land water, followed by condensation into clouds and precipitation as rain and snow returns water to the land, maintaining the global stock of fresh water, on which all living things rely.

Drought and Inundation

But the supply of rainfall is by no means constant for any one region. Indeed, there are such great variations in the amount, intensity, and timing of rainfall in some parts of the world that lives and livelihoods are often placed at risk. From May to September, intense and repeated monsoon deluges often lead to flooding and soil erosion in Southeast Asia. For much of the rest of the year, after the ground has dried out, the soil is baked so hard that it is incapable of growing anything without careful irrigation.

The areas receiving most rainfall lie in the tropics, although some parts of the midlatitude belt (between the tropics and the polar circles) also receive heavy rain. In the equatorial regions rainfall is often regular and frequent, clouds building up rapidly during the day before being discharged in the afternoon. Forest trees protect the soils from the deluge because little water penetrates directly to the ground, and much of the rain evaporates off the canopy leaves.

By contrast, the high-pressure subtropics tend to experience negligible or unreliable rains. These are the savanna and desert regions. As a result, sustained agricultural exploitation is made much more difficult.

Even temperate countries, normally well supplied with fresh water, may suffer serious rainfall problems. For example, the grasslands that have formed in the midwestern states of the United States, in the rain shadow of the Rocky Mountains, are prone to severe drought.

Water that falls as snow presents another set of problems. Snow and ice reflect much of the Sun's energy, which keeps temperatures low. Snow can therefore persist for months, making land unusable for a large part of the year. Later, if a thaw is fast, flooding and avalanches may result.

Groundwater and Aquifers

Humans rely for their water supplies not only on rainwater that remains on the surface, collected in rivers and lakes or in reservoirs, but also on groundwater supplies, sometimes piped from far away. Groundwater is formed when water is soaked up by the soil and percolates down into the porous rocks beneath. Not all the water remains here, though, some simply flows downhill until it emerges into streams and rivers and even directly into the ocean.

Rainwater percolating downward creates a "water table." This is the level below which the rock is completely saturated with groundwater. Wells dug to below this level will therefore fill with water. The water table may descend during hot spells but will generally rise up again during rainy seasons. A succession of dry years and excessive use of the groundwater, however, can cause the water table to drop below the level at which it can naturally replenish itself.

Aquifers – underground reservoirs often covering many hundreds of square miles – accumulate their stores of groundwater over thousands of years or longer. Some of the water they contain may even date back to earlier, wetter climate phases. Aquifers occur in the most surprising places. Many desert regions, such as Libya, Oman, and central Australia, are able to tap deep reserves of water that accumulated over 40,000 years ago before the deserts formed. In such dry regions, though, these bodies of "fossil" water are unlikely to be replenished.

Sculpting the Planet

Water is not only a necessity for most plant and animal life; it also plays a key role in shaping the Earth's surface.

The great landscape features making up the Earth's shape – such as the mountains and submarine trenches – are created directly by movements of the Earth's plates. Once created, these initial shapes are like rough sculpting blocks ready to be chiseled, smoothed, worn away, and broken up by the actions of ice, rain, wind, sunlight, and the sea. Slowly and inevitably, over millions of years, mountains are reduced to little more than hills, and hills, in their turn, become plains. Much of the material eroded from them ends up as sediment on the ocean bed. Eventually, these deposits themselves turn to rock and may be pushed up onto the land surface once again in a new cycle of creation and destruction of the landscape.

The gradual processes of whittling, removal, and redumping of land materials create a range of typical landscapes: wind-sculpted dunes in deserts, ice-carved hollows in the high mountains, coastal cliffs and beaches, river valleys and gorges, and many more. Sometimes the shaping occurs suddenly, as when a massive river flood cuts a new channel, or when part of a hillside collapses, but more often it is a slow and imperceptible process that takes place over thousands or millions of years. Because it takes so long we can often see in the landscape the product of gradually changing climates, changing sea levels and, increasingly, the influence of humans.

Weathering and Soil Formation

Without the processes of change, much of the landscape would be composed of bare, hard rock surfaces offering little in the way of sustenance or support for plants. Instead, after a few years of exposure to the elements, rock surfaces develop a rudimentary soil, containing essential nutrients in a form that can be assimilated by plants.

Ice, wind, and sunlight help in the disintegration of the rock surface, mostly by physically breaking up the rocks. Water, too, can force fragments of rock apart by freezing in cracks and fissures and exerting great pressure on the rocks. It also acts chemically. Water, particularly rainwater, which is slightly acidic, can dissolve limestone. More importantly, it can bring about the chemical breakdown of the very minerals that make up the rock.

Chemical breakdown is extremely useful as it leaves behind a residue of resistant minerals while creating clays from the disintegration of less resistant minerals. In the process, many of the

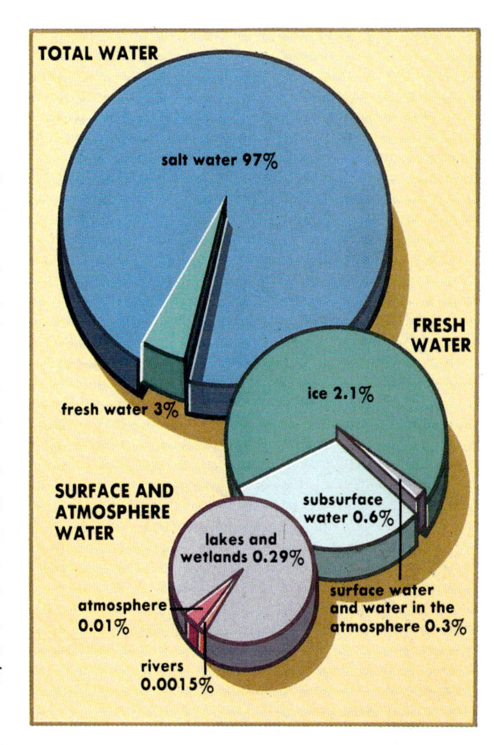

At any one time, there is only enough moisture in the atmosphere to keep the world alive for ten days. Without constant replenishment, most surface water in rivers, lakes, and reservoirs would soon evaporate. Fortunately, water is recycled very quickly.

The hydrological cycle is almost entirely powered by the Sun. Most moisture is evaporated from the oceans, to which the greater part of it soon returns. However, a small amount is carried by winds over the land, where, together with moisture evaporated there or transpired by plants, it falls as rain or snow.

Once returned to the ground, water makes its way back toward the ocean under the pull of gravity, passing through soil, or along streams and rivers. On this journey, much of the water is evaporated or transpired back into the atmosphere. Ultimately, though, runoff from the land to the ocean precisely matches the volume of water that is brought from the oceans as vapor.

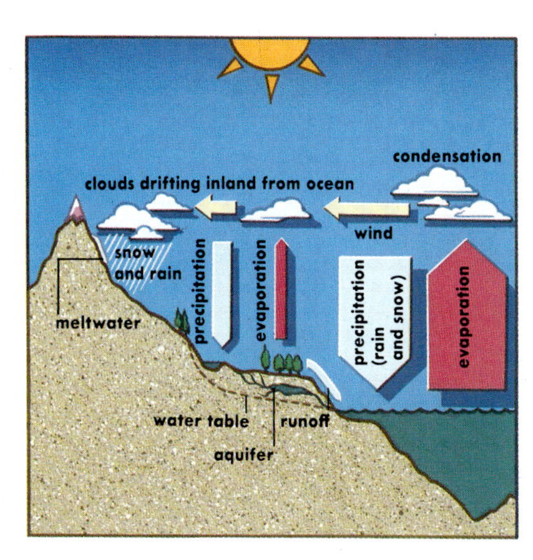

nutrients locked up in the solid rock are released. Some nutrients remain in the water, while others attach themselves to the outer surfaces of the clay particles. These can be taken up when they are needed by the plant.

In a natural recycling system, nutrients pass from rocks, to soil, to plants and animals, and then eventually back into the soil or even into the longer-term rock store again. Recycling from the plants and animals back into the soil relies heavily upon bacteria that decompose the organic matter, thereby releasing the nutrients locked within it.

Soil Variation

Soils vary enormously, depending upon their maturity, the rocks from which they have formed, and the climate. Very wet climates often have poor soils, since most of the valuable nutrients are "leached out." Much of the tropical rainforest owes its great lushness more to the speed with which nutrients are recycled than to the fertility of the soil. Clearance of the forest all too often results in an accelerated loss of nutrients in rainwater that now beats unimpeded onto the ground. Typically these areas cannot sustain a productive crop for more than two or three years.

Areas that have dry climates may possess salty soils formed when water containing dissolved mineral salts is drawn to the surface and evaporated in the hot, dry conditions. There is little rain to remove the excess salts and little water with which to sustain the plants. Both the lack of water and the high salinity severely limit the agricultural potential of such areas.

Bryce Canyon, Utah, a spectacular array of limestone and sandstone pillars or "hoodoos" sculpted by millions of years of weathering and erosion.

This fantastic scene demonstrates how the continuous operation of basic natural processes transforms, transports, and recycles the Earth's materials. Sixty million years ago, the entire region was a shallow sea. Sediment in this sea – "skeletons" of incalculable generations of minute sea creatures, together with eroded materials brought down from surrounding hills – eventually became compressed into solid rock. Uplifted thousands of feet by subterranean forces, the rocks became vulnerable to the action of sun, flowing water, ice, and wind.

Differences in the degree of hardness of the rocks account for the weird shapes they have adopted. Oxidation of metals such as iron and manganese in the rocks has tinted the scene red, yellow, brown, and lavender blue.

Ecosystems

Less than one-thousandth of the Sun's radiant energy that reaches the Earth's atmosphere is captured by plants, yet it is on this minute fraction that every living thing relies.

In its raw form, the Sun's energy is of little use to most organisms; it certainly cannot help in cell-building processes, movement, growth, or reproduction. For these functions, all organisms need fuel in the form of sugars – chemicals that act as biological "storage batteries."

Green plants are the only source of such sugars. They create them out of carbon dioxide and water using solar energy, in a chemical process known as photosynthesis. Most of this biologically stored energy is released within a short time by oxidation (respiration) to provide energy in a form useful to organisms' life processes.

Since animals cannot produce their own energy sources, they rely on eating plants or on eating animals that have themselves eaten plants. In this way they, too, can release the energy that is stored in the plants to perform their various activities,

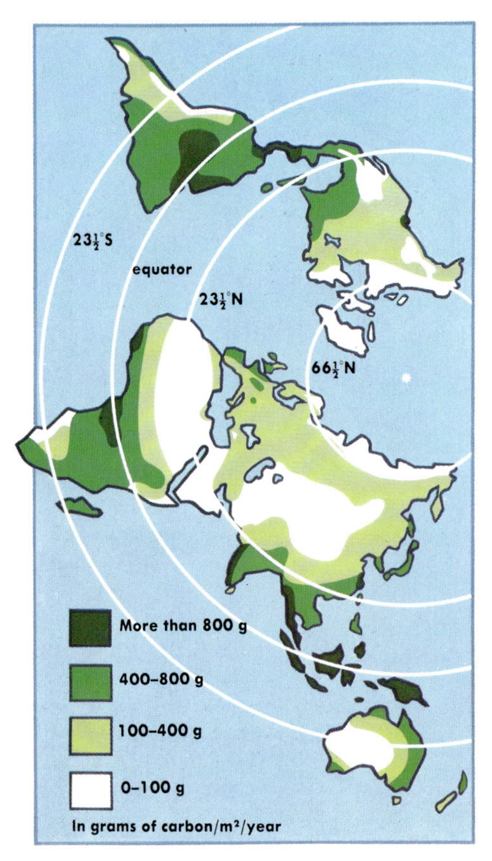

All animals, carnivores as well as herbivores, ultimately rely on plants for their nutritional needs. Plant productivity – especially the rate at which vegetation converts carbon dioxide and water into sugars using the Sun's energy – varies greatly around the world, depending largely on climate. Three factors in particular closely control the total weight of new material added to organisms in an ecosystem (biomass): temperature, water supply, and hours of sunlight.

Tropical rainforests, mangrove swamps, coral reefs, and estuaries are the most productive environments, both of biomass and of diverse species, animal as well as plant. As a result, energy transfers and nutrient cycles are at their most complex and rapid here.

Deserts of ice, rock, and sand are the least productive, and generally the most simple, ecosystems. Open oceans, too, create relatively little biomass for a given area, partly because key nutrients such as iron may be lacking. But their immense size means their total production is enormous nevertheless.

The diagram (above) *indicates the productivity of ecosystems globally, measuring this productivity in grams of carbon produced per square meter in one year.*

including muscular work and the generation of nerve impulses.

It is this energy cascade through the entire chain of life that maintains every plant and animal community. The characteristics of any one community depend on the amount of sunlight reaching the ground, the temperature near the ground, and the supply of water. When there is much sunlight and high temperatures but little water, deserts are the result. Good supplies of water but short growing seasons and low temperatures result in tundra. But when there is sunlight, water, and warmth in great abundance, the outcome is tropical rainforest, the land's most productive ecosystem.

Food Chains and Webs

Each plant community has its own rate of fixing energy as sugars and using them up again in its life processes. The balance of the energy store remains available for animals to consume and is passed along through the ecosystem in a series of steps of eating and being eaten – a "food chain." Although some organisms are fussy eaters and only ever eat one particular foodstuff, others will accept more or less anything that is on the menu. Most are somewhere in between: they have their favorite foods but on occasion will take something else. Chains, therefore, are often interlinked to create complex "food webs."

Plant eaters such as sap-sucking insects and grazing mammals are efficient organisms, but even they are able to convert only a part of their plant food into animal tissue. A good deal is lost as gas, urine, and feces, and much of the energy that is assimilated is burned up in movement or other activities and dissipated into the environment as heat. Only a small part remains to be converted into bodily tissue – the sole source of energy for the next level of organisms, the meat eaters.

The loss of energy at each transfer is so great that the number of feeding steps involved in a food chain must necessarily be limited. Animals at the top of a food chain end up with a tiny energy base on which to support themselves.

Water, gases, and minerals also cycle through ecosystems. However, unlike energy, gases such as carbon dioxide or nitrogen, and minerals such as potassium or calcium do not leave the system as fast as they arrive.

The U.S. environmentalist Aldo Leopold conveys the nature of mineral recycling well in his narrative *Odyssey*. This is a tale of X, an atom of phosphorus laid down in an ancient sea in the skeleton of some microscopic organism. The organism is converted over millions of years into limestone, and that limestone is eventually made part of dry land. Eventually an oak root helps the limestone decay, and brings it back into the world of living things.

Its life continues first as a catkin, which becomes an acorn, which fattens a deer, which feeds an Indian, all in a single year. In later incarnations X becomes a plant, a mold, an oat, a buffalo, a buffalo chip, a plant, a rabbit, an owl, a fungus, and so on, and so on. And all this time X is being washed downhill a little, into a stream and a river, and then into the sea where the story can begin again its cycle of perpetual return.

What applies to phosphorus applies equally to the other nutrients, such as carbon, calcium, or nitrogen. There is no waste as we know it; everything that is left over by one organism is broken down and used by another. It is critical that it should be so, because a build-up of carrion or feces or fallen trees would soon render an environment uninhabitable. The organisms that cause the decay of dead wood, for example, are helping to return its ingredients to the soil, where they can be used by the next generation of trees.

Humans are increasingly global creatures, which means that we are involved in practically

every ecosystem, large and small, throughout the world. For one thing, we make use of many of the world's natural ecosystems as sources of food or raw materials. Nutrients, as Leopold put it, have become itinerants. Previously, they may have passed millions of years in one narrow geographic location, endlessly recycled within the local flora and fauna. But now they may be taken to the other end of the Earth as a teak table from Indonesia, a pineapple from the Philippines, or a hamburger from Costa Rica.

The ecosystems themselves are also transformed by plowing and planting, or building and mining, or by their use as sites on which we discard whatever we have no use for. The deliberate or accidental introduction of new plants or animals, whether they end up as crops or pests and weeds, affects relationships within ecosystems. We may now be changing the composition of the atmosphere and hence the pattern of climates, which is likely to affect ecosystems further.

One of the reasons that humans have become such global creatures is that we are, as far as our ecological "function" is concerned, without a specific niche in life. Most animals inhabit a very narrow niche – they eat a narrow range of food, and are in turn prey to a narrow range of predators. Others, like rats, are generalists, and so are humans. We have no particular ecological role of our own to play.

This trait became evident about ten million years ago as humans started a progressive shift from the tropical forest to the dry savanna grasslands. We used our versatility to survive in one of the most arduous environments imaginable. We learned to catch the gazelle when the cheetah was elsewhere, but were happy to share the vulture's portion of the carcass. We ate the tortoise's green leaf, but we also ate the tortoise. By learning to get by in this most challenging of environments we prepared ourselves to dominate virtually every other.

A regular 12 hours of intense sunshine a day, temperatures always between 70° and 80°F (21° and 27°C), and at least 65 in (1,650 mm) of rain throughout the year make the tropical rainforest the world's most luxuriant, diverse, and interdependent environment.

Forest clearance means soil erosion, flooding, local and perhaps global climate change, and the destruction of indigenous peoples. Species extinction entails the loss of numbers of potentially useful organisms.

Humans Emergent

According to the United Nations, the world's population reached five billion on Saturday, July 11, 1987. This number seems set to double to ten billion sometime in the next century. In fact, nobody can be sure exactly how many people there are, or will be, because population counts and projections are notoriously unreliable even in the best researched countries. In some of the countries where population growth is generally thought to be fastest – such as those in the arid fringes of the Sahara – no comprehensive censuses have ever been conducted.

And if determining the present population is difficult, estimating how many people there were in the past is even more uncertain. But whatever the doubts that might remain about exact numbers, what is certain is that the greatest increase in numbers has occurred since the Second World War.

Although our common ancestral group is probably around 100,000 years old, we have only been present in any great numbers around the world for the last few thousand years. When the last major ice age ended about 10,000 years ago, and the dome of ice that had covered large parts of the northern continents receded, there were probably no more than 20 million people worldwide, and probably a lot less than the current number of inhabitants of Mexico City.

Neither have populations expanded continuously in the ensuing millennia. Growth has been slow and painful, with frequent catastrophic declines, due to the vagaries of climate, the depredations of invading armies, or the visitations of plague.

The development of agriculture in scattered parts of the world from about 8000 BC onward initiated the first spurt of population growth, but even in the first century AD the world's population probably did not exceed 300 million. It reached one billion around 1800, doubled to two billion in the late 1930s, and doubled once more by around 1970. During this century, the average annual increase in people – the number born minus the number dying – has been around 50 million a year, more than one a second.

Some people see a cause for great anxiety in this increase. They point to a spectrum of current world problems, from the prevalence of famine and disease to the mutilation and depletion of the environment. Yet others point to our extraordinary success as a species in the past and look forward with great optimism to the future. They stress that it is not just human numbers that have increased, but also human well-being and creative and scientific development.

But whichever side is taken in this debate, our achievements as a species and the prodigious capacity of our planet to support our ambitions are undeniable. We have grown from diminutive bands of itinerant hunter-gatherers 10,000 years ago to the astonishing numbers and concentrations of settled peoples of today.

Masailand lies at the very Cradle of Mankind, in the part of East Africa where humans emerged 100,000 years ago. The Masai people's lifestyle as pastoral nomads has probably not changed appreciably in the last 10,000 years. Their traditional territory ignores both national borders and the boundaries of game reserves, and in defiance of government attempts to curtail their wandering, they and their cattle still roam the Great Rift Valley very much as they have always done.

FROM FOREST TO SAVANNA

ALTHOUGH HUMANKIND NOW EXISTS IN GREAT numbers and has spread into virtually every habitat the world offers, every modern human is closely related, sharing a common ancestral group that evolved less than 100,000 years ago. Since all other humans and near-humans have become extinct, scientific classification puts us in a single subspecies within the genus *Homo* – "human" – and rewards our success by dubbing us with the rather conceited title *Homo sapiens sapiens*, which means wise, wise human.

The attributes of *Homo sapiens sapiens* are essentially no different today from what they were when the subspecies first evolved. Although there have been some adaptations to local conditions – resistance to certain diseases, for example, and the development of racial characteristics – all evolved from the one root subspecies. If a child born today could be exchanged for one born 100,000 years ago, neither child would be out of place biologically. Neither would experience any particular advantage or disadvantage. They would grow up as normal members of their society, indistinguishable from any other.

What could account for the extraordinary geographical flexibility that enabled us to flourish wherever we went? Part of the answer lies in the fact that we have no specific niche in the economy of nature. It is our very lack of specific adaptations to particular environments, and our adaptability with regard to food, that have made us so versatile. Another part of the answer may lie in what might be called "pre-adaptation." The ancestors of *Homo sapiens sapiens* had the ability to turn physical, intellectual, and cultural attributes that fitted them in one environment to good account when they entered others.

Although all modern humans' roots go back a mere 100,000 years, many human*like* – hominid – species have evolved over the last ten million years, always, it seems, in Africa. Several of these species, such as the "ape-humans" *Dryopithecus* and *Ramapithecus*, however, ventured out of Africa as much as ten million years ago.

Finds that corroborate the facts have been reported from places as far apart as Britain, China, and India. But these ape-humans were creatures that probably rarely moved far from forests, and they must have migrated from Africa across a land bridge that connected Africa with Arabia while forests still covered large parts of southern Europe and Asia.

It is impossible to be certain why Africa should have been the crucible of human development. But the answer may lie in its uniquely interlocking habitats, chiefly of rainforest, wooded savanna, and open grassland. Most of the physical attributes that link humans most closely with apes and monkeys developed in a forest environment. But our distinctive human qualities developed in a series of species – among them *Homo habilis* and *Homo erectus* – over an extended period, several million years in duration, spent in more open environments.

IN THE FOREST

A long sojourn in the rainforests endowed ape-humans with excellent color vision and an acute sense of both smell and taste, abilities which must have been of great survival value in a dense, dark, and predominantly green environment. These abilities are no less useful today. The color and texture discriminations routinely made by artists and designers probably had their roots in a distant ancestral past in tropical forests.

The modern human's senses of smell and taste may not be as acute as they were in the forest, when they were vital to survival. These senses made it possible to distinguish between ripe and unripe fruits and berries, or poisonous and benign insects. The extraordinary feats of discrimination attained by the educated "nose" or "palate" of wine or tea tasters, chemists or perfumers, show that humans retain many latent abilities which special training can restore.

Many other sensory and physical capabilities probably date from this forest-dwelling period. For example, forward-looking eyes facilitating stereoscopic, or 3-D, vision probably developed as a response to our ancestors' habit of traveling through the forest by swinging from branch to branch. In such circumstances, any misjudgment of distance could be fatal.

This particular method of locomotion may also have provided us with the kind of versatile shoulder joint needed for throwing or hitting objects overarm with any force, whether it is a spear or stone, a baseball or a tennis ball. A diet consisting chiefly of fruits, seeds, shoots, insects, and the like doubtless also encouraged strong but delicate manipulative abilities and good hand-eye coordination.

Many of our social patterns, too, must have evolved in this context. The habit of foraging through the forest for ripe food in wandering bands was paralleled in the lives of subsequent nomadic hunter-gatherers. Such a mode of life demands making the best use of forest resources and would have placed a premium on the development of communication skills. Like modern humans, monkeys, and apes, early humans almost certainly had expressive faces and used complex gestures and vocal and other noises to make their feelings known, and nurtured their children culturally as well as physically for long periods before they attained maturity.

Some million years ago, in a cooler phase when

Much of what we know about early humanity is due to the information contained in cave and rock paintings. The earliest are more than 30,000 years old, and perhaps the best known are found in Lascaux in France – these date from about 15,000 BC.

Cave and rock art depicts mainly humans and animals, and shows features of everyday life such as hunting and cooking. Some paintings may also have had religious significance and been used as lucky charms to ensure the safety of an expedition or an abundance of food.

This painting, found on the Tanilli Plateau in Algeria, was probably executed around 5000 BC, at about the time the local people turned to farming. Hunting and the domestication of animals are predominant themes in their art, but they also depict giant buffalo, elephant, and other animals now extinct in that area.

the great Antarctic ice cap expanded and the climate of eastern and central Africa became drier, the forests in which monkeys and apes had flourished contracted, broke up, and were partly replaced by a mix of wooded savanna and open grassland. As a result of these changes, our ancestors "came down from the trees."

Biologically speaking, this move implied no radical innovation, no sudden mutation. On the contrary, their physical acquisitions in the forest environment stood our ancestors in good stead; they were already pre-adapted for life on the more lightly wooded margins of the forest. But the interrelated physical, technological, and social developments made on the savanna grasslands are the key to humankind's phenomenal geographical success story.

Among them are a fully erect stance and a two-legged gait, which meant that hands were freed to carry things. With time, hands became specialized as powerful but delicate organs of manipulation. The brain, too, was progressively enlarged and made more complex. Cooperative lifestyles were attainable, and with them came language and cultural learning.

The first humans that could begin to leave their environment of origin are known as Australopithecines, "southern apes," which evolved on the savanna about 4 million years ago.

Was there anything especially helpful to future humans about the semitropical savanna? Possibly not. In fact, it seems that the very difficulty of survival there prompted critical developments. There were certainly fewer fruits, nuts, and berries to be found on the grasslands than in the forest, and leaves, which had formed a significant part of the forest dwellers' diet, were often tough and indigestible. Seeds, roots, and tubers, however, were abundant and these foods still form our staple diet.

It is highly unlikely that humans were as yet particularly good hunters. Small animal prey, especially lizards, tortoises, snakes, and small mammals, may have been more accessible here than in high forest. But although the terrain also supported a variety of large grazing ungulates, including zebra, wildebeest, and gazelle, as well as giraffe, elephant, and hippopotamus, they cannot have been easy to catch.

In any case, there were many highly specialized killers and scavengers around, like lion, leopard, hyena, and hunting dogs, to compete (and contend) with. Fortunately, large enemies were also more visible, especially to an upright animal such as humans had become.

This new environment was in many ways far more challenging than the forests had been, and one in which there was no obvious vacant niche to fill. Yet this was the environment in which our ancestors managed to survive and, indeed, thrive. Adaptations to life here seem to have prepared us for life everywhere else.

The relative balance of population between continents and countries has been changing rapidly over the last few centuries, with important implications for patterns of settlement.

Birth rates and death rates are on the whole declining throughout the world, but differences are profound. In Europe, where death rates are low but birth rates are sometimes even lower, populations are likely to decline in the future, while Pakistan, Kenya, and Brazil, with high death rates but much higher birth rates, will continue to grow rapidly. Since many of the countries with the fastest-growing populations are also among the world's poorest – and already some of the most densely populated – such changes are likely to have dramatic effects on patterns of settlement and much else besides.

For example, in 1960 the United Kingdom had 52.6 million citizens, which made it the world's ninth largest state. By 1985 it had grown by around 3.5 million to 56.1 million, but it had slipped to 15th position overall because so many other countries were growing faster. By 2000 its population is expected to have grown by another 300,000; only enough to keep it at 22nd position. Over the same period Nigeria will have grown from 42.3 million (14th position) to 161.9 million (7th position).

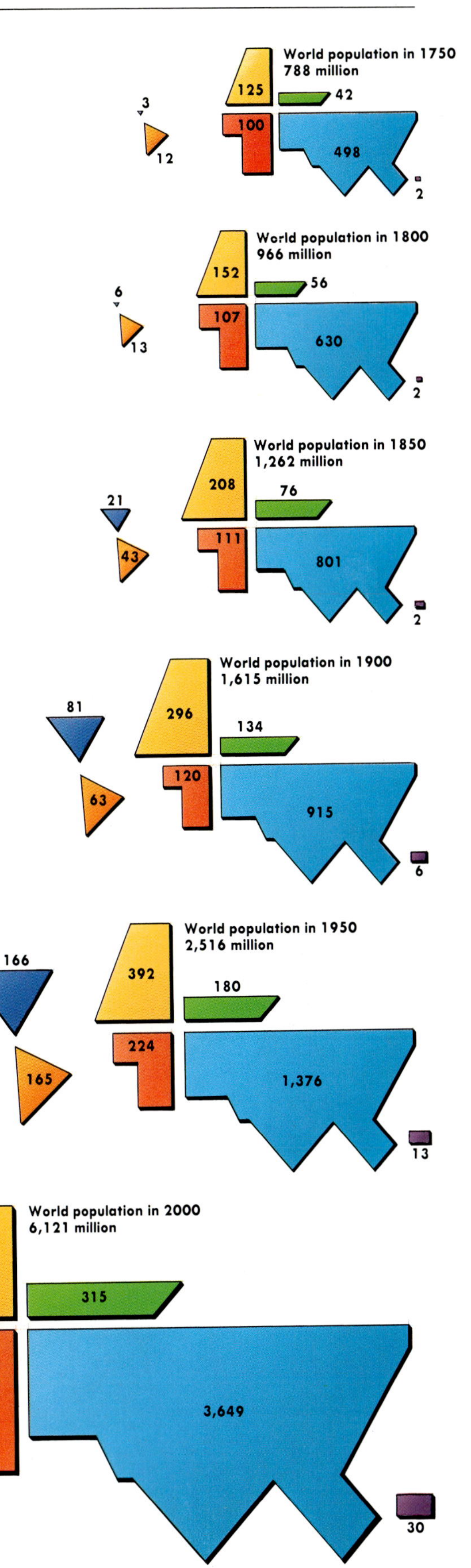

Many of the Inuit people of the American Arctic live by a form of hunting and gathering that reflects one of the earliest ways humans learned to survive. Seals, walruses, even whales, form a major part of their diet. But the demands – and potential rewards – of the 20th century are luring people away from this way of life.

The only secure employment in Greenland is in fish factories, one of the reasons that 20 percent of the 55,000 population have abandoned hunting and gathering to live in the capital, Godthab. The inhabitants of northern Canada once depended on the sale of sealskin, but this is not as fashionable – or as lucrative – as it was.

In Alaska, pictured here, oil has brought wealth, but the pollution caused by recent oil spills has destroyed much of the marine life and threatens the traditional livelihood of the area.

Out of Africa

For 99 percent of our history, humans have lived in tiny nomadic bands, exploiting only wild food resources. It was in such hunter-gatherer groups that *Homo sapiens sapiens* colonized much of the globe and adapted to its environments and climates. Migration must generally have been slow, taking perhaps several thousand years to cover a thousand miles.

Life in cooler northern lands, with less game to hunt and a more seasonal supply of plant food, demanded a profound understanding of the habits of animals and the properties of plants, the control of fire, and the invention of a complex array of tools. Rather like a modern surgeon or mechanic, our ancestors developed specialized equipment finely adapted to suit local conditions and needs. For example, tool kits may have contained equipment for hunting or trapping fish, birds, or mammals, for butchering carcasses, and for tailoring skins into clothing and bedding.

If modern hunter-gatherers are anything to go by, our ancestors would have been expert botanists, too. They would have had a detailed knowledge of the whereabouts and specific properties of hundreds of plants. They would have used them not only for food, but also as dye, poison, medicine, and materials for the construction of temporary homes and the manufacture of tools, clothing, twine, and rope.

Such skills and expertise must have taken time to develop in any new environment, but human powers of intelligence and innovation, and the capacity to pass on experience through cultural learning, cannot have been any less then than now. It seems likely that the spread of humans

across the planet may have quickened and slowed at different times and in different places, but it was inexorable all the same.

FIRST FOOTING

The spread from Africa to southern Europe and central, southern, and southeastern Asia was probably achieved quite quickly from about 90,000 years ago onward. Since this was a glacial period, so much water was held in the polar ice caps that sea levels were considerably lower, exposing shallows and throwing up various land bridges and coastal corridors. For example, Japan could have been reached from Korea, probably around 30,000 years ago.

The peopling of the Americas, however, is still obscure. No sites have been dated there before about 15,000 years ago, but it seems likely that they were reached at least 30,000 years earlier by people who had traveled across Beringia, a land bridge that joined Siberia to Alaska. While some of these early colonists may have crossed the Rocky Mountains to the dry plains to the east, many may well have traveled southward down the coastal corridor. The journey to the southernmost tip of South America, Tierra del Fuego, was completed by about 12,000 years ago.

The most recent colonization seems to be of the Pacific Islands, chiefly by people who had already acquired the ability to grow crops. In 1947, Thor Heyerdahl tested his theory that Pacific islands could have been reached from South America by people sailing in balsa wood boats traveling westward on equatorial currents. Although he succeeded in establishing that such a journey may have been possible, linguistic, botanical, and other evidence suggests that the islands were colonized not from the west, but from the eastern Pacific.

Certainly the people of the eastern Pacific are formidable sailors, undertaking journeys of many hundreds of miles using only their intimate knowledge of birds, currents, winds, and stars to guide them. Islands colonized in this way include Tahiti and Easter Island, the most remote place on Earth, around 5,000 years ago. The last large Pacific island to be colonized was New Zealand, a mere 1,000 years ago.

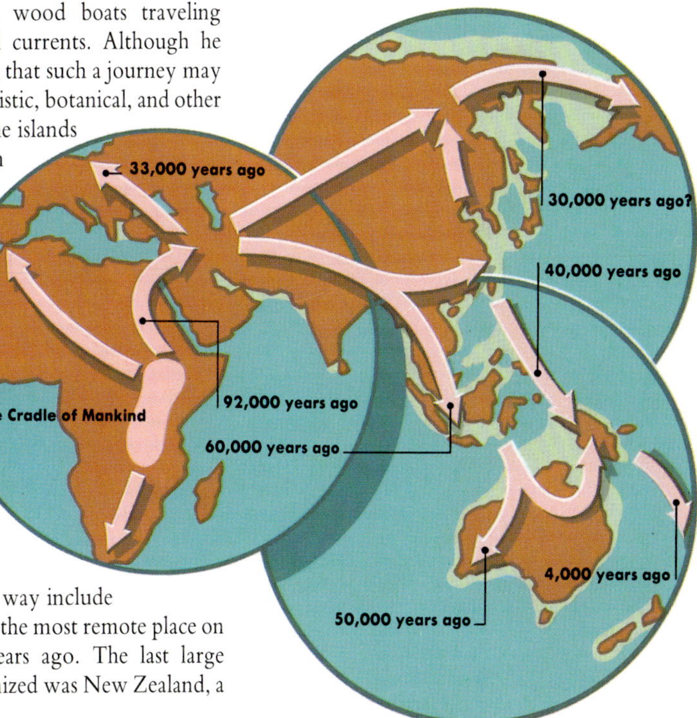

Thousands of years ago, the continents of the world were linked by land bridges that are now deep beneath the oceans (shown in pale green on the illustration). Britain was attached to Europe, Japan to eastern Asia, and the islands that now form Indonesia were connected to mainland Southeast Asia. Human expansion was primarily restricted by the extensive ice sheets that covered much of northern Europe and Canada.

As these ice sheets melted, they exposed areas of tundra that soon became rich in wildlife and were ripe for settlement by hunters. The development of warm clothing and weatherproof shelters followed, enabling humans to survive in what had previously been uninhabitable areas.

THE ORIGINAL AFFLUENT SOCIETY

A LARGE PART OF OUR KNOWLEDGE OF WHAT LIFE was like for early hunter-gatherer peoples comes from archeological finds – human skeletons, the bones of butchered animals, arrowheads, the charred remnants of campfires. But most of our knowledge about the significance of these material artifacts must be drawn by inference from the behavior that can still be observed among the small and rapidly diminishing fraction of the world's population that still lives by hunting, fishing, and gathering.

Today, hunters and gatherers survive in some of the world's harshest environments, mostly those that are too cold, wet, or arid for farming. In some cases, they have been pushed to these remote and often hostile areas by more powerful competitors, which probably means that their lives are harder

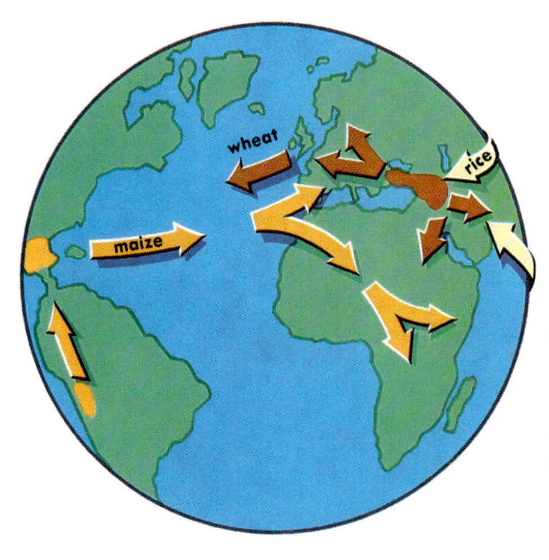

The details of the origins of agriculture are not clear, but it certainly emerged independently in a number of places in America, the Middle East, and Asia as early as 5000 BC. Staple crops in many areas remain largely unchanged – rice in India and the Far East, maize in the Americas, and wheat in the Middle East and Europe. But the discovery of the Americas, and increasing world trade since, has meant that crops have been carried around the globe and introduced into other areas. Australia is now an important exporter of wheat, while the U.S. is the world's largest producer. Maize has been introduced into both West and East Africa, where it is grown alongside cassava as a major food crop.

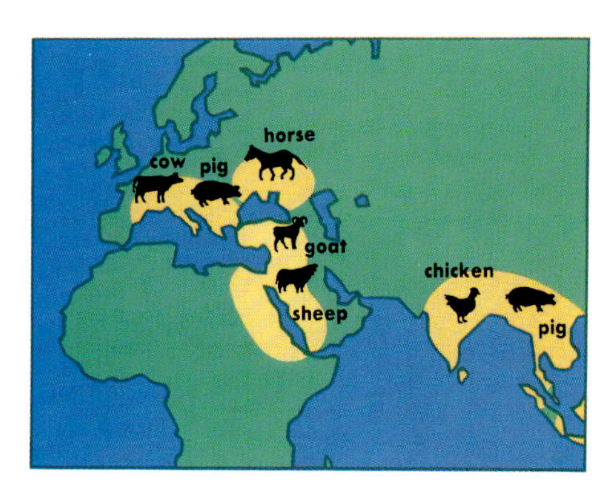

The domestication of animals was a logical extension of the development of agriculture. Animals had always been hunted for meat, but now they became a source of milk or wool. Many pastoral nomads of East Africa still rely largely on their cattle or camels for the milk and blood that is the basis of their diet.

Animals also became beasts of burden. Greater distances could be covered and opportunities for trade increased. With the development of agriculture came permanent settlements. Food surpluses – a first step on the way to the growth of cities – meant that people could make a living other than by growing their own food, and complex cultures emerged.

than they would formerly have been. Modern-day hunter-gatherers, like our ancestors, are completely at the mercy of the environment. They depend on a wide variety of naturally occurring seeds, fruits, nuts, roots, game, and fish, each of which may only be available in certain seasons and in particular places.

Most groups do not store food, although the Inuit people freeze meat in winter, and some groups accumulate seeds and nuts for periods when game is in short supply or drought threatens. Acquiring food is therefore an almost daily activity in all but the richest areas, and hunters and gatherers have to respond to the availability of food by migrating. Even so, periods of abundance alternate with periods of scarcity.

All family members are involved in acquiring food. Men are generally the hunters, while women and children are the gatherers, and everybody fishes. Food is also shared by everyone: a hunter does not keep the animal he catches for himself and his family alone. Since the availability of food is often unpredictable, food sharing is important for both immediate and long-term survival. It also promotes good relations and encourages the group to stay together.

However, where sources of wild food are plentiful, such as among the !Kung people of the Kalahari desert in Botswana, groups spend only two or three days a week acquiring food, and enjoy abundant leisure time. Some anthropologists have even described them as "the original affluent society."

Almost all hunting, gathering, and fishing societies are small scale. Nature unaided rarely provides enough food to support large populations, and the technology these populations possess, although sometimes quite effective like the harpoon, is unable to harness much energy from the environment. As a result, hunter-gatherer groups rarely exceed 100 people and are normally much smaller – around 30. In areas where resources are particularly sparse, such as the desert regions of Australia, there may be less than one person per square mile (2.5 square km). Larger numbers would exhaust food supplies and threaten the survival of the group.

Not only are resources generally too meager to support large populations, they cannot support any population for long periods of time. Being a hunter and gatherer means being a nomad. Where sources of wild food are abundant – when the salmon are running, or the trees are full of fruits and nuts – groups may stay in one place for several months. But where they are scarce, as they are for the shellfish gatherers of the Chilean archipelago, the group may be forced to move almost every day in order to survive.

The nomadic existence of hunters and gatherers has many different effects on their culture. Since they are constantly on the move, their homes take the form of temporary shelters. Caves may be used, but more commonly shelters are constructed from perishable materials such as wood, leaves, and skins, or in the case of the Inuit, snow and ice. A nomadic way of life also means that people have to limit their possessions to necessities such as tools and cooking utensils. Because of the difficulty of carrying infants, various methods of population control – including spacing of children, and even infanticide – are practiced in order to make sure that another child is not born before the previous one can walk.

Given that such nomadic peoples have to depend for their long-term survival on the continued productivity of the land, they must live in some kind of balance with nature. For one thing, they do not possess the technology to alter the environment significantly, at least compared to farmers and urban dwellers. Nevertheless, in the past they may have had a significant impact both on the landscape and on animal life.

There is a strong circumstantial case that human predation caused many extinctions among the larger mammals such as mammoths, cave bears, and saber-toothed cats shortly after the last ice age. More recently, the settlement of New Zealand by Maoris about 1,000 years ago probably accounts for the extinction of more than 20 species of flightless birds, including the 11-foot-high (3.3-m) giant moa.

Probably the most visible legacy of hunter-gatherers has come from their mastery of fire. Among the Plains Indians of North America, for example, fire was widely used to drive buffalo toward cliffs, blind valleys, or waiting hunters. In Australia, Aborigines have for millennia systematically burned off tough, invasive plant-cover to encourage the growth of new, more edible shoots. In these ways, over thousands of years, hunters and gatherers have gradually transformed many former wooded areas into vast open grasslands and semideserts.

From Foraging to Farming

It is not possible to be sure when agriculture was first practiced, or even where, but it probably required no flash of inspiration, no sudden breakthrough or remarkable "invention" to bring it about. On the contrary, hunter-gatherers must for millennia have possessed more than enough intimate biological and geographical knowledge to cultivate animals and plants had they wanted to. The issue is perhaps not of the "first agricultural revolution" but a question of why people had not wanted to practice agriculture.

Farming as a way of life is in fact deeply

abhorrent to many hunter-gatherers. They regard it well within their powers to tend plants and rear animals, but simply disdain it. In their view, agriculture is much too arduous, too time-consuming, and, of course, too sedentary. For hunter-gatherers, wild nature is an ever-full larder; it provides for all needs with minimum effort and leaves plenty of time for social and cultural pursuits such as conversation, play, music, and personal adornment.

It is perhaps no accident that the many fine cave paintings that have been discovered around the world – in Altamira in Spain or Lascaux in France, for example, or in the midst of the Sahara, or those which are still being painted today by Australian Aborigines – are always the product of hunter-gatherer societies, never farmers.

The Triumph of Agriculture

In spite of any reluctance on the part of many hunter-gatherers, a variety of grain and root crops was nevertheless brought into cultivation from about 10,000 years ago, shortly after the last ice age. Rather than rely on what nature had to offer, some humans began to experiment with a narrow range of crops and animals for food, clothing, spices, dyes, and fibers. Just how few is apparent from the fact that, even today, only about 150 of the world's 80,000 or so edible plants have been domesticated, and a mere 20 of them provide about 90 percent of the world's food. More than half the world's population relies on just three grasses: rice, maize, and wheat.

The earliest crop domestication probably occurred on the grassy mountain slopes of the Middle East, where wild wheat was already an established source of food, but there seem to have been many other sites or "hearths" of agricultural development, including central America (maize) and the Andes (potatoes). Other crops that soon followed included dates, olives, and figs. Animals, too, were domesticated. Some, such as dogs, chickens, pigs, and donkeys, ran free or were tethered or enclosed close to the settlement, foraging as well as being fed on surplus crops and the leftovers of meals. Others, including goats, sheep, cattle, and camels, were eventually herded between scattered seasonal pastures.

Selective breeding changed plants and animals both in physical form and in genetic makeup, so that they are now quite different from their wild ancestors. The present-day corncob is vastly superior as a source of food to its wild precursor of maize, which was less than 1 in (2.5 cm) long. Inbreeding has made sheep fatten faster, given them longer wool, and caused the "stupidity" that makes them much more manageable than their ancestors. The fleshy part of the banana has been developed to such an extent that the fruit can no longer reproduce from its own seed. Wheat seed cannot now disperse itself from its parent plant, but has been bred to stay behind in the head to await harvesting.

The type of food crop people grew made an important difference to their lifestyle. Cereals contain a balance of protein, carbohydrate, and fat, whereas root crops, such as manioc or yams, lack protein. As a result, people whose livelihood is based on root crops still have to hunt, fish, or collect protein-rich seeds in order to achieve a

The development of agriculture was spearheaded by the invention of the plow. A heavy iron version seems to have emerged independently in Egypt and what is now Iraq before 2000 BC, and had reached western Europe before the rise of the Roman Empire.

But the lightweight wooden plow was still sufficiently familiar in France in the 15th century for it to feature in this illuminated manuscript, Les Très Riches Heures du Duc de Berry.

balanced diet. This limits the number of people that can survive in any area – hence most urban civilizations are based on the cultivation of cereals rather than root crops. The early American empires are associated with maize, southern and eastern Asian empires with rice, and the Middle East and Europe with wheat and barley. They remain the world's most important food crops.

Wheat and barley were first cultivated in the Near East – in present-day Iraq, Turkey, and Jordan – between 8,000 and 9,000 years ago. About the same time, plants were being raised in the New World, although maize itself did not become a food crop until about 5000 BC. Rice was domesticated later, probably in India, and there is evidence of it being cultivated in China about 5,000 years ago.

We cannot be sure what prompted people to become farmers. One theory is that they may have been forced to develop agriculture as a result of the climatic changes at the end of the last ice age, which altered vegetation and contributed to the

decline of large mammals on which many hunters had traditionally depended.

Another theory is that since hunting and gathering can provide a balanced and adequate diet for a little labor, people did not begin experimenting with plants and animals until human populations began to increase or wild foods became scarce. A third theory suggests that the first farmers already lived at a comfortable margin above subsistence level, since people who are constantly threatened with famine do not have the means or time for the slow, experimental steps necessary to develop a better food supply.

There is also considerable controversy about how many times agriculture was "invented." Some archeologists have argued that it originated in Southeast Asia and diffused from there into other parts of the world; others believe that it may have evolved separately in as many as eight or more regions.

What seems clear is that agriculture developed independently in the Old and New worlds. Before Europeans colonized the Americas, few crops were common in both areas, and some of those, such as the coconut and the bottle gourd, were probably carried by ocean currents without human help. Furthermore, agricultural systems in the two worlds were quite different. In the Old World, cereals were sown by scattering rather than planting individual seeds, and farmers there also possessed the plow, the wheel, and large numbers of domesticated animals, none of which were in use in the New World.

TAKING CONTROL OF THE LAND

WHEN HUMANS FIRST TURNED TO FARMING, THEY raised crops and animals in small gardens near their homes. But as agriculture developed, they cleared larger and larger stretches of land – using fire or axes – for cultivation or grazing and developed techniques that enabled them to farm new areas.

Shifting cultivation was one of the first forms of agriculture. It entailed growing a variety of crops, many of them root crops such as manioc and sweet potatoes, until the soil nutrients became exhausted and yields declined. The farmers then cleared a new plot, abandoning the old one and allowing the forest to reestablish itself.

Once the old plot had rested for 10 to 20 years, it was ready to be cultivated once again. This form of agriculture, with its high yields, is extremely productive and ecologically balanced, but at any one time large stretches of land are "resting" and unavailable for cultivation, so it can only support small populations.

Although 50 years ago Europeans regarded shifting cultivation as a primitive and wasteful system, it is now recognized as a successful means of utilizing tropical environments where soils are generally less fertile. It is unsuccessful only in areas where the pressures of rapid population growth mean fallow periods have to be reduced, causing yields to decline. Today, shifting cultivation is found in parts of Latin America and Southeast Asia, and it is still the major system of cultivation in tropical Africa.

Permanent cultivation can support more people

Because the metal plow could cope with heavier soil than its lighter, wooden counterpart, more land became available for food production, and farming became the means of livelihood for most of the world's population.

Abandoning the nomadic lifestyle – in other words, settling down to become farmers – meant that for the first time humans were attempting to shape their environment. They were no longer completely at the mercy of the seasons. In times of plenty, they could produce more food than they could eat and store it for use in times of need. They could live in permanent settlements and accumulate possessions.

Agriculture has, of course, continued to develop, and it is now possible to plow an acre (0.4 hectares) of land in just over 11 minutes. Very few people in the developed world depend on subsistence farming for their livelihood. In many parts of the developing world, governments are encouraging farmers to concentrate on cash crops for export. But in the province of Baluchistan in Pakistan, this farmer and millions like him across Asia and Africa till the land with the help of a single ox and a lightweight plow that their ancestors would have recognized.

Throughout history, as populations have grown, pressure to produce more food has increased. Where arable and fertile land have been in short supply, various forms of intensification have been introduced. One of the oldest and most familiar of these is terracing, practiced in eastern Asia, parts of Africa, and the Andes for centuries.

*The rice terraces of Banaue in northern Luzon, principal island of the Philippines (*opposite*), were built at least 2,000 years ago, and in addition to maximizing the rice yield of this hilly region, they are sufficiently spectacular to have become a tourist attraction in their own right. Irrigated by man-made waterfalls, their length is estimated at 14,000 miles (22,500 km). This increases every year as farmers continue to hack away bits of the mountainside, using the rock to build retaining walls around small areas of rice beds. The rice itself is a delicate shade of pale pink and is highly regarded by rice connoisseurs.*

since it does away with the need for extended fallow periods, but it can only prosper if soil fertility is constantly restored. This fertilization may occur naturally, as when river banks and flood plains are fertilized by nutrient-rich flood waters. But more often, large quantities of fertilizers are needed – possible sources being human and animal dung, household waste, seaweed, bird guano, fish remains, and lime.

Food production can also be increased by extending the area of cultivation. Mountainous areas, for example, can be cultivated using a range of techniques including contour hoeing and, more commonly, terracing. Some of the world's most dramatic terraces are found in the Andes and Southeast Asia, often accompanied by irrigation systems designed to channel melted snow from mountain regions to lower altitudes. This combination of techniques not only creates cultivable land on steep slopes, it also controls soil erosion and water runoff, and adds to soil fertility.

Terraces were probably first developed in the Near East about 9,000 years ago, but they have reached the peak of their complexity in the wet rice-growing areas of Asia. As a result, land can be highly productive over long periods and can support dense populations. The disadvantage of terracing and irrigation is that the construction and maintenance are extremely labor-intensive. When the Inca population of Peru collapsed under Spanish rule, the remaining Indians could not sustain the intensive forms of agricultural production the Incas had used. Today a number of rural development projects in Peru involve the rehabilitation of these Inca irrigation systems.

Irrigation is more commonly associated with dry lands, of course. Along the Nile in Egypt and the Chang Yiang (Yangtse) and Huang He (Yellow) rivers in China, for example, canals and aqueducts date back 4,000 or 5,000 years. In addition to allowing plants to grow in areas where rainfall is sparse, irrigation also has a fertilizing effect, thanks to the nutrients in the water. But large-scale irrigation is a very sophisticated technique, requiring skillful water management and the mass mobilization of labor if channels are to be kept free of sediment and weeds.

Poorly managed irrigation systems can cause many more problems than they solve, killing crops and rendering the land unworkable. Waterlogging and subsequent evaporation may serve to bring salts to the surface of the soil, while excessive pumping from wells may cause seawater to penetrate the groundwater or cause land subsidence.

Food production can also be increased by cultivating poorly drained or swampy regions often found along rivers or lake shores. Although silt deposits make river banks and mouths fertile, and they can form some of the world's richest farmlands, they cannot generally be made productive as long as they are liable to be flooded.

One well-known way of protecting these vulnerable lands is to build dikes to hold water back, but another technique is to make raised fields. They are constructed by digging canals or ditches, and heaping the mud dredged up into ridges to form a rich soil in which to grow crops. Each year, the ditches are cleared and more mud is added to the soil, which maintains its fertility. This system was common in central Mexico, where the raised fields, or *chinampas*, of the Valley of Mexico supported the Aztec civilization. Other types of raised fields exist in wet highland areas where ridges are placed along the steepest part of the slope to encourage runoff.

HERDING

When people became farmers, they generally formed permanent settlements, but some subsequently became herders of domesticated animals and returned to a nomadic existence. Herding appears to have developed in Asia about 3,000 or 3,500 years ago as an offshoot of mixed farming, once horses and camels – essential means of transportation to the pastoral nomads – had been domesticated. Pastoral nomads are specialized herders who move their animals from pasture to pasture, water hole to water hole. Some depend exclusively on one animal, such as the camel in northern Arabia or the reindeer in Siberia; but in central Asia, herds contain a variety of animals – including horses, goats, and sheep – adjusted to local climatic and grazing conditions.

Within 8,000 years of the development of agriculture, most of the world's population had become farmers. Agriculture brought with it many profound and long-term changes. Agriculture was not without its costs, of course, as hunter-gatherers had realized. Although farming may support larger and more dense societies, predators, parasites, and diseases also seem to thrive wherever crops and people are more highly concentrated.

Yet agriculture increasingly "humanized" the wilderness and made it habitable for far greater numbers of people than could be supported by hunting and gathering. In many cases, the environments it created have been esthetically satisfying and ecologically sound, as well as economically rewarding and favorable to the continued growth of civilization.

There may be problems associated with the "agricultural revolutions" of 10,000 years ago, but it was this ability to modify and control natural environments through agriculture and permanent settlement that led *Homo sapiens sapiens* to the superior position it currently enjoys. If people's lives are not dominated by the constant quest for food, they can abandon a wholly nomadic existence and settle in permanent villages. Ultimately, more constant food supplies provided the basis for the development of wealth, the growth of cities, and the emergence of states and empires.

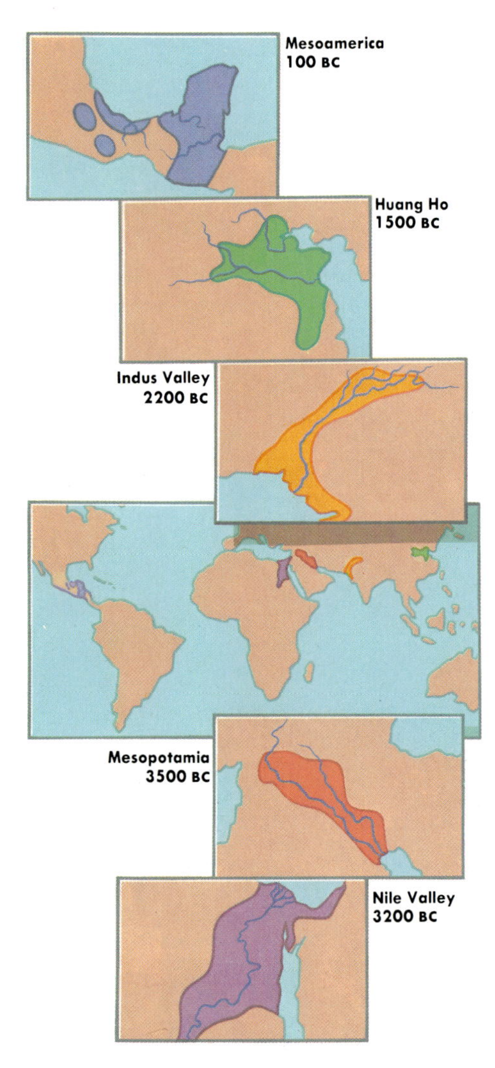

The first cities emerged along river valleys, where natural fertility ensured that food was plentiful. Only then could a community afford to concentrate on other, less obviously productive aspects of life. In Mesopotamia the Sumerians developed a form of writing around 3000 BC. This ability to transmit knowledge is a feature of all early civilizations, with its best-known form being the hieroglyphics of Ancient Egypt.

The principal cities of the Indus valley housed many skilled workers. Relics of one region found in the other show that they conducted a thriving trade with Mesopotamia.

In China the Shang dynasty supported a luxurious court with revenues from its extensive territory. Their use of bronze for ceremonial objects means that many relics of their religious and courtly life have survived.

The Mayan civilization in Mesoamerica emerged much later, and there is evidence of a complex system of religion and of great emphasis on ornamentation and jewelry – luxuries that less sophisticated societies could not afford.

THE RISE OF EUROPE

WHEN THE 13TH-CENTURY VENETIAN EXPLORER Marco Polo returned from his travels across Asia to the Orient, he brought back stories of the seemingly fabulous wealth accumulated by the emperors, emirs, and sultans he had met on his journeys. No European monarch could match the power, splendor, or technological sophistication of the Chinese or Arab rulers of the day.

Gunpowder, the mariner's compass, printing, paper – all of them destined in their different ways to transform the European world – originated in China. So, too, did more mundane but no less important inventions, such as the horse-harness and water-mill mechanism.

With the help of these discoveries, humans could utilize the power of nature and culture more effectively than ever before. Islamic technology and scholarship were no less impressive, with their repository of classical, mathematical, astronomical, and navigational knowledge. On practically any measure, Europe was very much their poor relation at this time.

Yet 200 years later, in the 15th century, it was Europe and not the eastern empires of China or Islam that emerged as the dominant force in world history, and remained so for several centuries to come. It was Europe that unified the globe, that brought every part of the whole world into regular contact with one another. What happened to bring about such a change?

Europe is geologically and ecologically varied, with well-defined areas of fertile soils widely distributed across the continent – a fact which underlies the profusion of small states, languages, and cultures that this tiny continent has spawned. A generally mild climate means that farming can flourish without constant fear of drought, flood, or devastating winter. At a time when land transportation was slow, arduous, and expensive, Europe's extensive navigable rivers and long, indented coastline encouraged both local and long-distance trade.

Although Europe lagged behind China and Islam in technical virtuosity during the Middle Ages, it was by no means a stagnant society. Indeed, in several crucial areas, it was already beginning to move ahead. Paramount among them was agriculture. Major innovations had begun to transform European farming as early as the 6th and 7th centuries.

The water mill, the heavy plow, and new field systems that allowed for crop rotation and therefore preserved soil fertility all meant that, by AD 1000, the elements of a new and highly productive system of agriculture were in place. Increasingly large and reliable food surpluses meant a growing

European explorers who thought they were taking civilization with them on their voyages of discovery found that great cities and cultures had existed in China, the Indus valley, Mesopotamia, and the New World centuries before.

The once-mighty Inca Empire was weakened by civil war by the time the Spaniards arrived, and they colonized it comparatively easily. But the great Inca city of Machu Picchu remained undiscovered until the 20th century, despite the thoroughness with which the colonists visited and despoiled every known village and town.

Perched high up in the Andes, on a mountainous crag that drops steeply away on every side, Machu Picchu, home of the Inca rulers, must have been unassailable. The hillsides that produced food for several hundred people were also engineered to prevent soil erosion. The skill that went into the construction of the buildings was such that stone walls survive intact 500 years after the city was presumably abandoned. Both its exact origins and its fate remain a mystery.

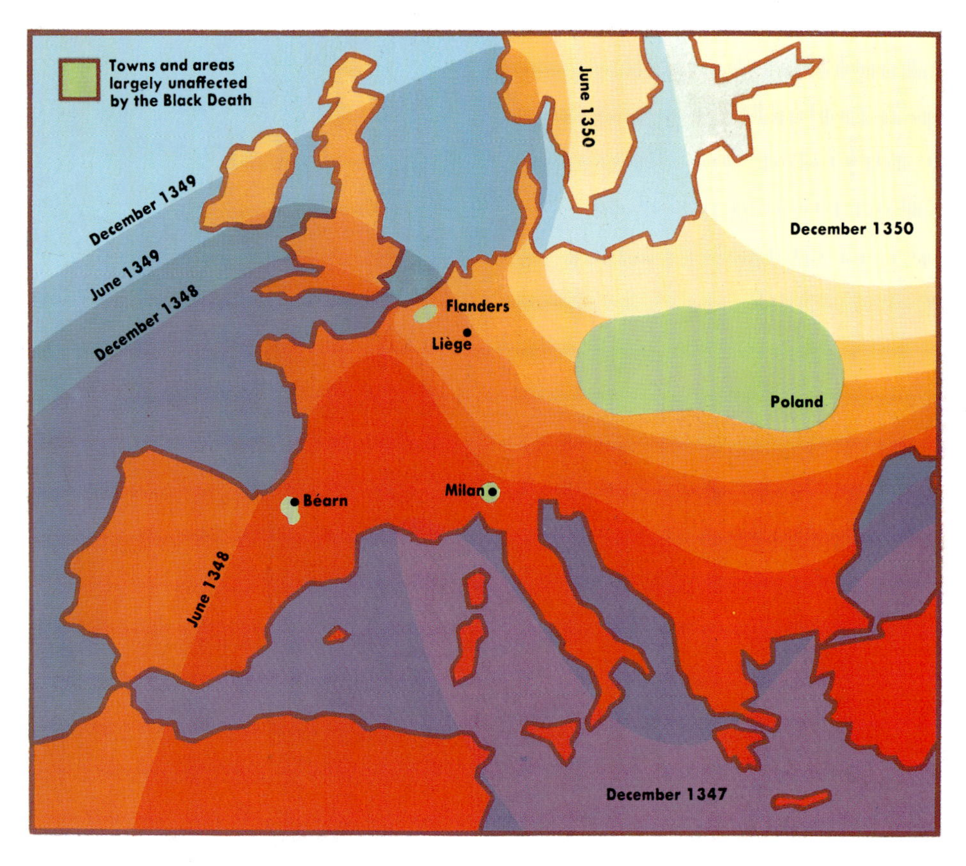

The plague known as the Black Death reached Europe in 1347, and moved in waves across the continent. The map shows the approximate spread of the disease over the next three years.

No previous epidemic or war had ever killed so many people. Disease spread most rapidly in cities, and whole villages were wiped out. Close-knit communities such as monasteries were particularly hard hit. But the devastation was not evenly spread – it is not clear why the city of Milan or a huge tract of eastern Europe should have been largely spared.

population, economic development, a proliferation of towns and cities, and an increasingly elaborate religious and secular society.

Unlike China, which had been a unified empire with a single ruler and bureaucracy from as early as AD 618 with the Tang dynasty, Europe's loose patchwork of relatively independent states meant that no one nation was ever able to monopolize power and impose its sovereign rule over the rest of the continent. Competition between these rival states in large part pushed Europe to the forefront of military, economic, political, and technological developments.

But mutual conflict is not the whole story. Although competition may have been fierce, most areas of Europe were also linked together by a common cultural heritage – that of Christianity (together with vestiges of an earlier Roman imperial past).

In times of peace, Christianity's common use of Latin overcame the profusion of European languages and allowed knowledge to diffuse across national and linguistic boundaries. Transnational monastic orders and itinerant priests became the conduits through which not only religious but also classical culture and novel technical developments were disseminated throughout the continent.

In wartime, too, the unifying force of Christianity fostered Europe's rise to power. Between the 11th and the 13th centuries, many European countries united in a series of Crusades

against Islamic domination in the Middle East. By wresting control of the Mediterranean from the Muslims, albeit briefly, the Crusades also laid the foundation for the emergence of great city states such as Venice and Genoa.

A shared culture, a fragmented political and economic structure, and a benign geographical environment conducive to the growth of agriculture and trade all laid the foundation for the flowering of medieval European society. Even

The origins of the Great Wall of China can be traced back to the 5th century BC, but most of its present form dates from the early Ming dynasty period, the late 14th and early 15th centuries. At this time, the Chinese had expelled the invading Mongol hordes, and the main reason for strengthening the wall was to keep them out.

Although the Chinese were pioneers in shipbuilding and the development of navigational techniques, they failed to compete with Europe when it came to conquering the oceans. This may have been partly because they had plenty of land of their own – if they needed more, they expanded inward. But frequent conflict with neighbors, and the Mongols in particular, may have exhausted resources that might otherwise have been spent on seafaring.

The Mongols were a warrior people whose empire had at one time – under their legendary leader Genghis Khan – not only included China but stretched as far west as Poland and Hungary. A Mongol army is said to have introduced the Black Death to Europe by catapulting plague-infested corpses into the Crimean town they were besieging.

continent-wide disasters were only temporary setbacks. When the Black Death appeared in an army of Mongols in the Crimea in 1347, the black rat – a creature adept at climbing ropes and therefore widely found in ships – carried the flea that spread the plague all over Europe. It arrived in Italy late that year. By the end of the following year, 1348, it had arrived in the south of England. By year's end, it was in Norway and Poland.

Over the next five years, the Black Death destroyed one-third or perhaps even one-half of the entire population of Europe. When Pope Clement VI ordered a count of the deaths that had occurred, the figure given was 42,836,486. Although it is unlikely to have been very accurate, the fact that he could call on priests the entire length and breadth of Christendom to collect figures is testament to the degree of social integration already reached at this time.

Yet, in spite of such catastrophic losses, Europe

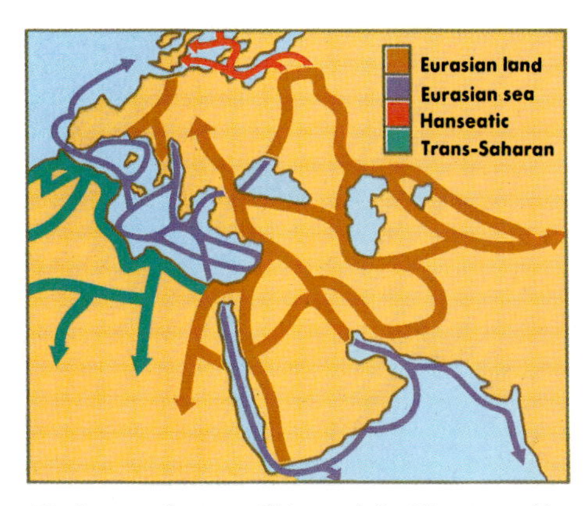

*Trade routes between China and the West – notably the Silk Road – were well established by the time Marco Polo set off on his travels. From the 13th to the 15th centuries, trade in northern Europe was dominated by a group of German merchants known as the Hanseatic League. Routes across the Sahara were later exploited in the Middle Ages by Europeans in search of gold. The illustration (*above*) outlines these land and sea routes.*

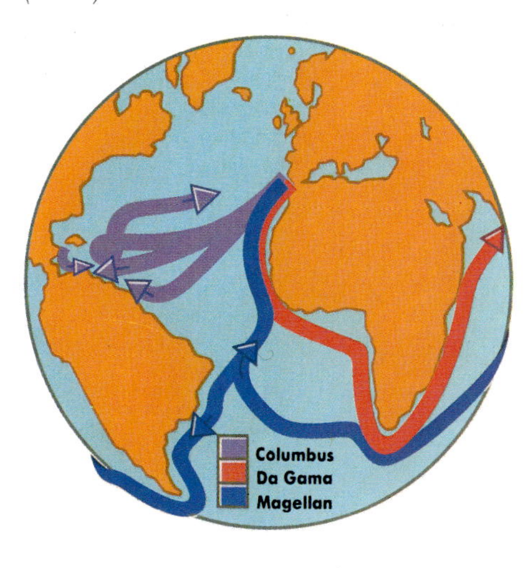

From about 1492 to 1522, Europe's achievements in exploration enhanced our knowledge of the world beyond recognition. Christopher Columbus set off to find a trade route to the east by heading westward – and discovered America. Vasco da Gama found a sea route to India in 1498, traveling eastward around the Cape of Good Hope. And an expedition led by Ferdinand Magellan was the first to circumnavigate the globe, completing the journey in 1522. In addition to proving definitively that the Earth was round, the expedition clarified our understanding of the size of the globe.

was remarkably buoyant. It soon began to recover its former numbers, internal expansion resumed, and Europe continued along its dynamic upward trajectory. Within a hundred years of the Black Death, Europe was in a position to take hold of the whole world and transform the course of history.

EUROPE CONQUERS THE OCEANS

IN THE 15TH CENTURY, THREE GROUPS OF PEOPLE – the Chinese, the Arabs, and the Europeans – had the sailing vessels and the navigational techniques necessary for overseas expansion. The Chinese had a profound knowledge of the stars and had been using a lodestone – a kind of primitive compass – since the 11th century. From the 14th century onward, their enormous square-rigged junks explored the Indian Ocean, reaching Jeddah on the Red Sea in 1432.

At that point, they stopped, turned back, and as far as China is concerned, maritime exploration all but ceased. The reasons are not entirely clear. One explanation is undoubtedly the fact that China had little need of the rest of the world. It was self-reliant in the necessities of life, and the luxury items that provided the incentive for other nations to roam the world were available locally.

Nor did China have much need to colonize lands across the seas. Whenever it wanted to expand its territories, it simply pressed farther and farther inland, although it was frequently preoccupied with the Mongol threat to its northern borders, or with the steady encroachments of Islam to the west. As a result, the Chinese came to look inward rather than overseas – the Ming dynasty (1368–1644) went so far as to prohibit any further boat-building.

The Arabs, too, were gifted navigators who regularly voyaged out of the Red Sea and the Persian Gulf in search of trade. But their ships, although effective enough for their purposes, also proved to be restrictive. Their maneuverability was limited, and in the main they could only run with the wind. Nevertheless, they had a technique that made splendid use of the peculiarities of local geography – they hitched rides on the Indian Ocean's powerful monsoon winds. With its regular reversal – the word itself comes from the Arabic for "seasonal wind" – the Arab traders were carried out on clockwise winds in the summer, and back again for the rest of the year.

In this way, Arabs established firm trading links with India, Southeast Asia, and China, and as far east as the Spice Islands of Indonesia. They also established trading stations all the way down the east coast of Africa, not only on the mainland, but also on islands such as Mombasa and Zanzibar – a fact which is still reflected today in the wide distribution of their religion, Islam, and the frequency with which Arabic names are encountered throughout the region, as in Dar es Salaam ("Gate of Peace") in Tanzania.

Why, then, were Europeans able to conquer the Atlantic and Pacific oceans, and link the Old and New worlds? Part of the explanation lies in geographical necessity. With southern Spain in the hands of Moors from North Africa until 1492, and Constantinople (modern Istanbul) falling to the Turks in 1453, the Mediterranean had become, in effect, an "Islamic lake," and overland trade routes to the east were blocked. As a result, city states like Venice and Genoa, whose trade depended upon these routes, suffered. Countries which bordered the Atlantic, however, prospered.

With a long Atlantic coastline and difficult winds, tides, and currents to contend with, first Portugal and then Spain developed boat-building skills and seamanship superior to anywhere in the world. Above all, they turned the wind to their advantage. Earlier seafarers in the Mediterranean, such as the Greeks, Phoenicians, and Romans, had used chiefly human-powered galleys, which, although slow, were tolerable for their tideless, enclosed sea. The Portuguese, however, Europe's westernmost people, developed small but astonishingly maneuverable craft, known as caravels. With a shallow draft no more than a few feet deep, these ships were capable not only of ocean travel, but also river exploration.

It was in craft such as these that the Portuguese set off after about 1400. Although they also voyaged to the north Atlantic in search of fish, and were sometimes blown off course to Brazil, their principal intention was to travel south. They reached Madeira in 1419, the Azores in 1439, and rounded the Cape of Good Hope in 1488. Within a few years, the Portuguese had established trading stations all the way around the Indian Ocean to Malacca (1511). Shortly afterward, a ship captained by Ferdinand Magellan completed the first circumnavigation of the globe (1519–22), and a trading post was established in Japan (1542).

Exploration may have opened the way to trade, but there were many problems that had to be overcome if the Europeans were to take control of the routes. Above all, they had to displace existing traders, notably Arabs, who were found at trading posts throughout the Indian Ocean. In fact, the Portuguese already enjoyed an advantage: by now Europeans led the world in gunmaking.

Although gunpowder had been invented in China, it had not been developed as a tool of war, but as a plaything. Europeans had learned of the material from Marco Polo and immediately set to work adapting it for use as an explosive to hurl projectiles. In this task, they had a curious head start. Unlike in Islamic countries, where *muezzin* call the faithful to prayer from minaret towers

high above the city, in churches and cathedrals throughout Christendom, bells were ubiquitous. As a result, Europeans had for centuries been casting massive bronze bells, an expertise that could easily be turned to new uses. After all, a bell and a cannon have similar shapes.

Bronze cannon were already in use in Florence in 1326, and soon became common throughout the continent. Although gunmakers initially strove to build super-weapons to batter down city walls, firing monstrous stone balls up to 20 in (50 cm) in diameter and weighing more than 6 tons, they later developed small cannon and hand-held guns, ideal for use on ships.

In spite of their sophistication, neither the Arabs nor the Chinese could match European maritime prowess. The Chinese had effectively removed themselves from the race by confining their horizons to the immense landmass they already controlled. The Arabs were no better off. Too politically diffuse a people to make unity possible, their lateen-rigged vessels were no match for the full-sailed, high-prowed European ships armed with cannon and shot. Nor could the Arabs switch designs easily: the whole of the Middle East was by this time desperately short of timber for boat-building. Outgunned and outmaneuvered, the Arabs were eventually driven from the seas.

Meanwhile, in 1492, Christopher Columbus, a Genoese sailor sponsored by the king and queen of Spain, stumbled upon America while looking for a westward route to India.

THE WORLD IN 1492

THE BEGINNING OF EUROPEAN EXPANSION OVERSEAS has often been regarded as a critical period in history. This expansion started in 1415 when Portuguese sailors established a fortified outpost at Ceuta in northern Africa and began their relentless journeying, which would lead them to circumnavigate the entire globe within a hundred years. At this point, the world's innumerable societies ceased to live separate lives and were bound together in a single world system of trade and cultural exchange. In fact, the only areas that might justifiably be described as being isolated from one another were the Old and New worlds.

Notre Dame de Chartres, the second cathedral and sixth church to stand on its site, is one of the finest examples of the wave of Christian building that swept across Europe as the Crusaders returned from the Holy Land in the 12th century. Its Gothic architecture and glorious stained glass windows bear witness to the wealth and confidence that characterized Europe in the early Middle Ages.

The decline of the Muslim Empire, symbolized by the recapture of Jerusalem in 1099, strengthened Europe's position with regard to the rest of the world, unifying the Christian church against the "infidel." But the Crusades also brought an upsurge in Christian fanaticism. Jews who had established themselves as merchants and traders throughout Europe now found themselves rejected – even exiled – from the Christian countries in which they lived. The Catholic church passed legislation restricting their rights and ordering them to wear distinctive dress. It was the beginning of a discrimination that has lasted until modern times.

When Columbus set off from Spain, in 1492, the world's population stood at around 350 million – only one-fifteenth of what it is today – but it was rising fast in many areas. Although only a small proportion of people lived by hunting and gathering, they covered great regions of the Earth's surface. As now, most such societies flourished in remote areas where farming was not possible because the climate was too cold or arid, the land swampy, or the soil unworkable with simple tools. Hunter-gatherers inhabited large forested and semigrassland areas of North and South America, as well as eastern Siberia, the Kalahari desert of southern Africa, and the whole of Australia. Specialized fisher-hunters could be found in the Arctic circumpolar belt and along the northern Pacific coast.

Nomadic herders occurred primarily in the Old World, dominating a swathe of arid lands extending from South and East Africa (Bantu, Masai) to North Africa (Berbers), through the Middle East (Bedouin) to central Asia (Kazaks, Kirghiz, and Mongols), as well as the icy tundra regions of northern Europe and Asia (Lapps, Chukchi).

Although most of the world's inhabitants could, by 1492, be classed as farmers, they lived in a great variety of social groupings based on very diverse forms of agriculture. Tribal societies, mainly small-scale and practicing shifting cultivation, flourished throughout the Amazon Basin, North America, sub-Saharan Africa, the Pacific Islands, and in the hill country of Southeast Asia.

More complex, hierarchical societies, with strong centralized governments, priesthoods, bureaucrats, specialist craftworkers, traders, soldiers, and other elites, were generally associated with those major river valleys where people had

In the 15th century, European knowledge of world geography still depended largely on the work of Ptolemy, the great astronomer and mathematician of the ancient world. His eight-volume Guide to Geography *had been published in both Latin and Italian, and contributed to the growing interest in mapmaking and exploration. But his map of the world (*above*), although remarkably accurate in many ways, seriously underestimated the size of the Earth. He showed Europe and Asia occupying more than half the circumference of the globe. In fact, they cover little more than a third. It was partly due to this error that Columbus calculated he had reached Asia when he was less than a third of the way there.*

The printing press, introduced into Europe in the 1450s, made possible the first – and perhaps the greatest – information revolution in history. The acquisition of knowledge no longer depended on word of mouth or memory. Literacy and literature blossomed, printed maps became more common, and geographical knowledge more widespread.

*Libraries appeared all over Europe: for the first time learning was not the preserve of the very rich, the universities, and the church. The monastery and palace of El Escorial (*opposite*), outside of Madrid, were built by King Philip II of Spain in the 1560s, and the library he founded there contains thousands of books and illuminated manuscripts.*

If the Spaniards originally went to the New World in search of "God, gold, and glory," gold at least has lost none of its attraction in the 20th century. It was discovered in northern Brazil in 1980. Within two weeks, 10,000 garimpeiros *or gold-diggers had arrived, and now 50,000 men struggle through the mud of this open-cast mine, digging for gold and carrying away huge quantities of unwanted soil.*

This massive incursion of prospectors may be fatal to the native Yanomami Indians. A free-ranging people unused to the concept of landownership, they now find that their territory has been subdivided into unconnected pockets, surrounded by areas given over to gold mining.

It is not only the land that is under threat. The people themselves have fallen victim to a number of previously unknown diseases, the most pernicious of which is malaria. With an estimated population of less than 20,000 remaining, the Yanomami's future looks bleak.

successfully mastered highly productive and labor-intensive techniques of large-scale irrigation and flood control.

Cities, civilizations, and empires had begun to emerge in the Old World around 5000 BC, particularly within the fertile river valleys of the Middle East (Nile, Tigris, and Euphrates), northern India (Indus and Ganges), and China (Huang He and Chang Yiang). Although early empires thrived for a while and then died, these original hearthlands of urban civilization continued to spawn a succession of states and empires, and were still highly populated in the 15th century. Elsewhere in the Old World, civilizations had also developed that were not based on water-control, notably those of Greece and Rome.

In 1492, China and Europe were the most densely populated regions of the world, each accounting for roughly one-fourth of the world's numbers. Other concentrations occurred in the Ottoman Empire, which dominated the Middle East from the mid-15th century, and in northern parts of the Indian subcontinent, where the Mughal empire was supreme. In northern Africa, a predominantly Muslim population was chiefly located in the Nile valley and the coastal strip.

Other areas where societies of great complexity, artistic ability, and technical sophistication also thrived included a band just south of the Sahara stretching from the Atlantic in the west to Ethiopia in the east. This region supported a number of distinct societies, including the Songhay and Benin. Great civilizations had also developed in Japan, Korea, and Southeast Asia, for example the Khmer people (modern Cambodia) associated with the great temple complex of Angkor Wat.

Smaller numbers of people, perhaps 60 million, existed in the Americas, but they were unevenly distributed, with about half of the population living in Mexico (Aztec) and Peru (Inca). The earliest New World states had emerged during the first millennium BC. Between AD 300 and AD 900, the Maya civilization flourished in the lowlands of Guatemala, the Teotihuacán state in central Mexico rose and fell, and independent civilizations appeared in Peru, Bolivia, and other parts of South and Central America. Of these, most is known about the empires of the Aztecs and Incas, through the records of their own historians and the accounts written by European conquerors in the 16th century.

These two states were markedly different in organization. The Inca state of the northwestern Andes was highly centralized and authoritarian, closely regulating every aspect of life. As it conquered neighboring peoples, the state integrated them into its administrative and economic organization, imposing on them a common religion and a common language, Quechua, which is still widely spoken by Andean Indians. Despite its high level of social organization, and its highly developed agricultural systems, the Inca empire had few impressive cities; the population remained essentially rural.

Strictly speaking, Mexico's Aztec empire was not an empire at all. It consisted of a number of small city states, rather like those of classical Greece, which constantly jostled for power. A confederation of three city states controlled the region, not attempting to establish political unity or to impose their way of life on the peoples who came under their aegis.

In spite of these peoples' great abilities and brilliance, many cultural attributes and commodities usually associated with civilization in the Old World were unknown in these states. Although they used gold and silver lavishly, they had no iron. They had not developed a system of coinage, alphabetic writing, printing, true glass, ceramic glazing, or gunpowder, and had not invented the arch, a true plow, or the wheel.

The Columbian Exchange

The year 1492 is a critical date in human history. Christopher Columbus's first exploration of the Americas brought together two worlds that had effectively been separated for at least 10,000 years. Within the next few centuries, the greatest biological, cultural, and technological exchange in history had occurred. Millions of people moved across the Atlantic and Pacific oceans, and with them countless crops, livestock, weeds, and diseases. "Races" were intermingled, entire cultural traditions transplanted, absorbed, and remodeled – among them languages, religions, and technologies. As a result, practically every one of the world's societies and environments has been radically transformed, and in many cases utterly overwhelmed.

In the 500 years since European conquest, tens of millions of people from various parts of the Old World have found their way to the Americas. By contrast, very few Amerindians have ever settled in the Old World. The asymmetric character of this migration reflects, in part, the unequal devastation caused by epidemics. In 1492, the native population of the New World numbered about 60 million. Some 150 years later, fewer than five million remained. Other factors – both military and economic – had a part in this decline, but Old World diseases played the major role.

In addition to a number of universally dangerous diseases introduced at this time, like yellow fever and malaria, commonplace childhood diseases such as influenza, measles, and mumps became first-rank killers. In the early colonial

*Amsterdam emerged in the 17th century as the center of an international trading empire. Note the large number of ships in port on the map (*opposite*). It grew to be the focus of the world's banking, lending money to foreign powers and exerting considerable political influence. At the same time, it could claim to be the intellectual and cultural capital of Europe.*

Perhaps because it was at the peak of its prosperity at the time of the industrial revolution, Amsterdam did not develop as a manufacturing center. Although its population is now nearly three-quarters of a million, and diamond-cutting and tourism have replaced banking and trading as its major industries, the canals, bridges, and wharfs laid out in the 17th century remain a prominent part of the city.

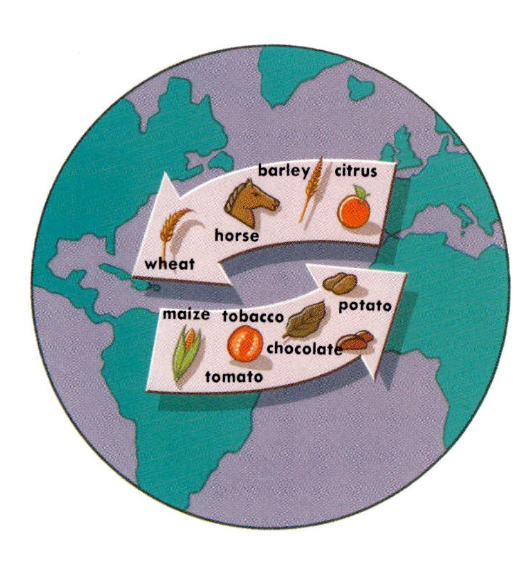

Before Columbus, people, animals, and crops were largely confined to their own continents. Entirely within the last 500 years, trade in these commodities has resulted in coffee becoming an essential part of American life, and tomatoes and potatoes becoming staples in Europe. The exchange has, of course, not been entirely beneficial – European diseases decimated the American Indian population, while the destruction of habitat, combined with the introduction of hunting for sport, led to the near-extinction of the North American bison.

period, epidemics of smallpox, plague, or typhus could appear in Indian communities every five or ten years, each time carrying off as many as half the population. Few diseases made the return journey, with the probable exception of syphilis, which seems to have appeared suddenly in Europe in the 1490s and is traditionally associated with the return of Columbus's ships to Spain.

Declining Indian populations had a profound effect on the social geography of the New World. The relative absence of indigenous peoples created the illusion of unoccupied lands waiting for colonization. It also led to severe labor shortages. As a result, the first mass migration to the Americas involved the forcible movement of West Africans, about 15 million of whom were shipped there, the majority to work on tropical plantations. Very few returned.

The collapse in numbers of the native population and the transatlantic – and later, transpacific – migrations fundamentally altered the racial character of the Americas. In many parts of the New World, such as the Caribbean islands and coastal Brazil, the native population became extinct and was replaced chiefly by people of African ancestry. New physical types also emerged, partly because the white and black men who went to the New World outnumbered women. People of mixed blood – mestizos, mulattos, zambos – became the rule rather than the exception. Fortunately, the resulting offspring of intermarriage gave the indigenous population greater immunity to Old World diseases.

With the opening up of trade and migration came crops and livestock. Many European settlers wanted to grow the crops and raise the animals that were familiar to them from home. Others made profits by establishing plantations of exotic crops for export to the home market.

The main areas of production today are often far removed from the places where crops and animals were first domesticated. For example, coffee was initially cultivated in Yemen, but the major coffee-producing areas are now in Brazil, Colombia, the Ivory Coast, and Indonesia. Sometimes crops have been so thoroughly adopted by new regions that it is hard to imagine what the local diets were like without them. Both the potato and the tomato were domesticated in the Andes, but who could imagine the Irish without potatoes or the Italians without tomatoes?

Europeans took wheat, barley, sugar, and citrus fruits to the New World, and brought back the potato, the tomato, maize, cacao, and tobacco. In the 19th and early 20th centuries, African diets were slowly modified by the introduction of maize and manioc, while in parts of Asia, sweet potatoes and peanuts have become important supplements to rice and wheat.

Botanical gardens played a key role in the intercontinental exchange of plants. The first botanical gardens were established by apothecaries for raising chiefly local medicinal herbs, but they grew in importance in the 18th and 19th centuries as scientific expeditions brought back plants from remote parts of the world. London, Amsterdam, Paris, and Madrid all had botanical gardens where plants and crops were experimented with and shipped to new colonies. Botanical gardens were also established in the colonies to collect indigenous species and disseminate previously unknown plants as speedily as possible.

A case in point is rubber, a latex material exuded from trees found in the Brazilian rainforest. When seeds of the rubber tree were smuggled out of Brazil in 1876, they were shipped first to Kew Gardens in London, and then transported as seedlings to gardens in Colombo, in modern Sri Lanka, for distribution to specially created plantations throughout Southeast Asia. The entire world rubber trade was thereby relocated and control of production and trade placed in the hands of the British. This came about partly because of the greater efficiency of the plantations, but also because disease greatly reduced Brazil's indigenous industry.

Although the exchange of crops was relatively balanced, this was not true of the exchange of livestock. The New World had few domesticated animals to offer; it possessed only the turkey, Muscovy duck, guinea pig, llama, and alpaca. Europeans, however, brought with them horses, cattle, sheep, pigs, goats, and chickens.

It is hard to imagine North American Indians without the horse, but in fact they had no experience of these animals until the 16th century. The horse, and other introduced animals, greatly enhanced the supply of food, clothing, and energy, but they also caused ecological damage and profoundly disrupted North American Indian social relations and subsistence patterns.

Colonial Expansion and Influence

Up to the 18th century, European settlement in the Old World was limited mainly to numerous tiny fortified trading posts and supply stations strung out along thousands of miles of coastline. Long sea voyages were dangerous enough, and few dared to venture very far inland. Through fear of attack away from the protection of shipborne cannon, or of contracting tropical diseases, colonization amounted to little more than a huddle of shops, inns, and lodging houses, with surrounding gardens to grow food for sailors and more permanent inhabitants. The Gold Coast of West

Africa was particularly lethal. With two out of three Europeans dying within a year of their arrival, it fully justified its reputation as the "white man's graveyard."

For a century or more, Portuguese sailors controlled virtually all of the settlements along the African coast and into the Indian Ocean, some of which had previously been established by Arab traders. By the 1540s, Portugal had managed to set up a string of ports from the Cape of Good Hope at the southern tip of Africa, to Hormuz in Persia, Goa on the west coast of India, Macao in China, and Nagasaki in Japan.

After 1700, the Dutch, British, French, Danes, and other Europeans founded their own ports, but also displaced each other. Some ports, like Cape Town, changed hands many times in the course of a century. As a result of these developments, European powers wove a net of routes between Africa, Asia, and Europe, and in the case of the slave trade, with the Americas, too.

Initially, at least, long-haul trade centered on Europe's demand for luxury items – gold, silks, and spices. Although these commodities came from Asia, European merchants became so dominant that they eventually destroyed Muslim

The West African republic of Mali, formerly French Sudan, has at various times over the last thousand years been a center for trade in gold, slaves, ivory, civet, and gum arabic. Two of its principal cities, Djenné and Timbuktu, were havens of Islamic scholarship. Having for centuries formed part of a succession of African empires, the area was overrun by the Moroccans in 1591. Trading routes and centers of learning were destroyed in the chaos that ensued.

French colonization in the 19th century went some way toward pacifying the region, and French is now its official language. But the waves of colonization are reflected in the composition of the people of Mali – Berbers and Moors from the days of Moroccan occupation, and many African groups descended from the ancient empires of Ghana, Mali, and Songhay.

The great mosque at San (above) is a relic of the zenith of the Islamic culture.

operations in the Indian Ocean and controlled the trade in many key commodities, especially cotton textiles, spices, and even porcelain.

In the Caribbean and Americas, the story is rather different. The Spanish and Portuguese were the only imperial powers in the 15th and 16th centuries to settle extensively in their colonies, a fact which only partly reflects the relative ease with which they could penetrate a continent virtually depopulated by disease.

It is often said that the Spanish and Portuguese went to the Americas in search of "God, gold, and glory." Certainly, churches and mission stations proliferated throughout the continent. The conversion of the American Indians to Catholicism and their instruction in "civilized" ways of life were taken seriously by the Iberian monarchs, if not always by their subjects. Compared with the Spanish and Portuguese, the interests of the British, French, and Dutch were almost entirely economic; cultural domination was a consequence of, rather than a motive for, expansion.

Gold and silver were the chief prized resources, much of it brought back to Europe as bullion or minted coins, but immense quantities were also transported across the Pacific, via the Philippines, to finance trade with China and Japan. Although the conquistadors found a considerable amount of precious metals already purified and in the form of jewelry and other items, tales of an "El Dorado" – a City of Gold – proved groundless. Even glory proved hard to earn. Once the Spanish had plundered all the treasures of the native American empires, they were forced to extract the silver and gold ore for themselves. Some of these mines became associated with large-scale and permanent planned settlements that have outlived the resource on which they were based.

As European numbers rose, traders increasingly turned from the search for booty to less immediately profitable sources of wealth, and established large estates to produce crops such as sugar, cacao, and indigo for export to the European market. In the absence of local labor, hundreds of thousands of Africans were forcibly brought to work on these plantations as slaves.

In North America, however, colonization was still impractical. In the last half of the 16th century, the French in Florida and the British in what is now North Carolina attempted to establish footholds on the continent, but crop failure and disputes with native Americans meant their settlements were soon abandoned.

A new and more permanent phase of settlement did not start until the next century, and then became established only very slowly. Philadelphia, for example, North America's largest city until the 19th century, was first settled by Swedes in 1643, but even in 1700 had a population of only 10,000.

The level of European colonization only changed when the supply of particular goods proved insufficient to satisfy demand, or when the control of productive areas was threatened by local resistance or rival powers, typically other European states. Then the armies and administrators moved in to establish formal colonial rule and set up plantations. The British possessed trading colonies in Bombay, Madras, and Bengal, but did not attempt to rule India effectively until the 18th century, when native empires began to break up and other Europeans started to threaten their domination of trade.

On the Brink of the Modern Age

While European expansion overseas was uniting the world in a complex system of trading and political relations, at home separate national identities were increasingly asserting themselves. Based on shared language and a common history, large unified nation-states increasingly displaced the city states and small principalities of former times. In particular, western and northern states bordering on the Atlantic rose to prominence while the Mediterranean city states waned. Collectively, these momentous and interrelated changes transformed the face of Europe, and the course of world history.

Intercontinental trade boosted more local sources of wealth arising from agriculture and industry. In the 15th and 16th centuries, precious metals and spices were the chief sources of wealth; but over the next 200 years, the range of exports and imports widened, with coffee, copper, cotton, furs, sugar, tea, tobacco – and slaves – among the most lucrative goods to be traded.

Not all countries benefited in equal measure from the continuing economic growth. Although Spain and Portugal pioneered the voyages of discovery and grew fabulously rich as a result, they failed to develop any large-scale industries. When the supply of gold and silver from the New World began to dwindle in the 17th century, their primary source of wealth dried up and their economies withered.

In time, the focus of power and prosperity moved north, to the three arch-rivals – France, the Netherlands, and Great Britain – who remained locked in seemingly interminable conflict for over 200 years. By the middle of the 18th century, thanks largely to its domination of North American trade, Britain began to pull ahead of its rivals and was poised to control world trade.

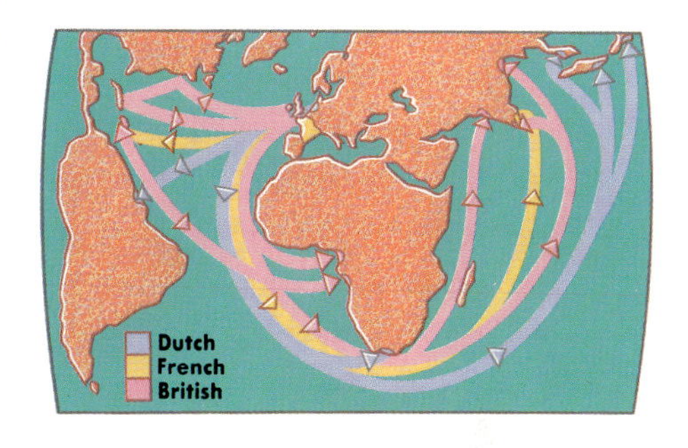

Although they dominated the conquest of South America, the Spaniards had never really explored Asia, and the Portuguese also failed to exploit the economic potential of their colonies there. This meant that large areas of India and the East Indies remained available for the British, French, and Dutch to set up trading posts. By the beginning of the 17th century, both the Dutch and the British had established East India companies and within a hundred years there was also a thriving French/Chinese trading company.

Politically and financially, the 17th century was a difficult period for Spain, and its control of overseas territories was undermined by "pirates," who often operated with the connivance of local French or British governors. Although a few major colonies were wrested from Spain, the Dutch, French, and British all acquired small but important settlements in the Caribbean.

Conflict in Europe always led to conflict in the colonies. England won territory from Spain as a result of the War of the Spanish Succession in the opening years of the 18th century. Wars over trading rights and territory continued, despite the fact that every war ended with a treaty designed to resolve the grievances. But political stability at home gave the British, French, and Dutch a strong basis on which to build their trading empires.

NE
Y. S. LINE
Y. S. LINE
Y. S. LINE

Globe Inc

The Growth of the World System

The Industrial Revolution brought social and economic change on a scale the world had never seen before. Britain was at the forefront of the industrializing nations, but over the next century or so others emerged to join an ever-changing economic core. By the beginning of the 20th century the U.S. was a world power; after the Second World War Japan became a global force, and its position is now being challenged by a new batch of "developed" nations.

Putting the World to Work

As industry, agriculture, transportation, and communications developed, so did a worldwide system of production and marketing. Vast sums of money now flash around the world in seconds, while commodities as varied as wheat and automobiles can be seen as part of the global economy. The third world supplies cheap labor and – it once seemed – inexhaustible resources.

Crises and Consequences

The repercussions of the developed world's exploitation of less prosperous countries are felt all over the globe. Tracts of forest have been destroyed; tourism brings valuable foreign exchange but may damage local environment and culture beyond repair; and the demands of the world market mean that cocaine is a more profitable crop than coffee for South American farmers.

The Growth of the World System

When did the world's economies truly become one global system? The development of printing in the 15th century allowed words and thoughts to be circulated far beyond the range of those who heard them spoken aloud. Explorers and traders opened up new lands. But the real origins of what we call Globe Inc can be found in the Industrial Revolution, which changed economic patterns forever by drawing individual strands together and weaving them into an intricate worldwide network of interdependent economies.

At the heart of the Industrial Revolution lay James Watt's steam engine, patented in 1769, and Eli Whitney's cotton gin, patented in 1793, which brought limitless power to the textile mills of England and unimaginable speed to the processing of cotton in the British colonies. Pouring into the country from plantations in India and America's Deep South, cotton was king for decades in Manchester.

But Britain would never have become the leader of the industrialized nations without Whitney's invention, which allowed the short fibers of the cotton boll to be separated quickly and cleanly from its seed. Slaves, who had previously performed this tedious and painstaking job by hand, could now clean 50 lb (22.5 kg) of cotton a day instead of 1 lb (0.5 kg).

All over the world the invention of new machines changed the work people did and the way they did it. Slowly but surely, the combination of natural resources, topography, and aggressively competitive attitudes transformed four neighboring countries – Britain, Germany, France, and Belgium – into the first core group of industrial manufacturers.

Germany's economic power surpassed Britain's when the need for machinery, both large and small, exceeded the need for almost anything else. Now steel became king. As well as being virtually tireless, machines were stronger than humans and infinitely more adaptable. Fortunes were to be made in manufacturing weapons. Factories and mills needed steel, and so did the expanding network of railroads.

As if anticipating industry's requirements, nature had lined the fertile valley of the Ruhr River with rich seams of coal and iron ore, and threaded the land with navigable waterways that could transport even the heaviest manufactured goods. The Ruhr yielded water power to the manufacturers, too, while Germany's huge population provided ready buyers for their goods and a steady flow of workers. Lured by the burgeoning towns, people forsook the farms for the factories.

In early America, industry triumphed despite the lay of the land, the absence of an indigenous work force, and the scarcity of skilled laborers among the immigrant population. While southern landowners grew rich on the cash crops of their plantations, the homesteaders had to push westward for the chance to expand their land holdings and their incomes. The first manufacturers in the Northeast produced cheap, durable domestic tools and agricultural implements. The immigrant workers here had little interest in settling down in the factory towns and little desire to master any mechanical skills. So, to employ them profitably without sacrificing productivity, the manufacturers constantly refined, simplified, and standardized their production methods.

The north–south corridors of the country, at least as far west as the Mississippi, were easily navigated by paddle steamer. Connecting the trading centers of the East Coast with the agricultural cornucopia of the Midwest, however, posed enormous problems that the railroads could only solve by defying nature. They eventually reached Chicago and pressed on, across the desert and toward the gold that was first found in California in 1848. The territory of the U.S. expanded as foreign nations sold and ceded their territories to the increasingly important central government, and the nation became a manageable unit in 1869 with the completion of the transcontinental railroad at Ogden, Utah.

American industry served its own growing population first and still found itself with plenty of grain, beef, textiles, metal ore, steel, and manufactured goods to export. Once again, as in Europe, the manufacturers held the strings of economic power and determined, by the numbers they employed and the wages they paid, the density of urban population and the standard of living many people enjoyed.

Deeply enmeshed in the global network of trade by 1900, America's financial and industrial corporations faltered during the Depression, but their strength was consolidated during the Second World War and remained strong for 30 years. But rising labor costs, exhausted facilities, and a collapsing infrastructure in the cities, along with steadily increasing competition from abroad, eventually took their toll. In 1971, as the world's manufacturing moved elsewhere, the U.S. suffered its first trade deficit in 80 years.

The working population left the industrial cities of the Northeast for service industries all over the country. Since mines were closing and farms could no longer pay their way, the land ceased to occupy the work force. Fast-food restaurants, hospitals, and microchip factories sprang up everywhere, and in the new postindustrial economy information emerged as the most valuable commodity.

High-tech industries like computing, finance, and communications demanded highly qualified personnel, and they, in turn, demanded comfortable working conditions and competitive salaries. So the country's business moved south, to the Sunbelt, where there was room to expand and the living was easy.

Few people would have predicted that a cluster of islands on the other side of the Pacific would one day rival America in the production of sophisticated electronic technology and the acquisition of a vast share of the world's economic wealth. Relatively small in area and geographically isolated, Japan was too mountainous to be an agricultural power and lay at the mercy of floods, hurricanes, and volcanic eruptions.

When Japan finally rejected feudalism and opened its doors to foreign trade in 1854, it threw itself into modernization with the same determination it had previously employed to keep the rest of the world at bay. By encouraging commercial agriculture and developing technology, it had caught up with the West by the time the Second World War began.

Japan's economic progress after the war blossomed from two root factors, methods and money. Japanese industry evolved a system of flexible production that allowed specialized manufacturers to supply specialized components where and when they were needed. Technological innovation flourished as these subcontractors helped the industrial giants turn out everything from ships and cars to hairdryers and pocket computers. Money helped, too. While other nations amassed huge defense budgets, Japan's defense spending after the war was limited to 1 percent of its GNP, which meant it could funnel its soaring wealth into industry instead.

Vertical integration proved to be the system by which the corporations could best control all aspects of a product's cost and manufacture and the efficiency of its worldwide circulation. How better to dominate the market for, say, computer software or cassette tapes than by designing the item, building and running the factories that produce it, arranging its distribution, and also providing the machinery that brings the consumer and the product together?

Having established its corporate presence in America through mergers, multimillion-dollar acquisitions, and new subsidiaries, Japan has strengthened its links in the intricate net of the global economy. By moving into western Europe and, more recently, into eastern Europe, it hopes to become a part of the economic landscape there before the new import tariffs of the European Community (EC) cut into its profits.

But Japan is no longer the only Asian nation trading on an international scale. New core groups of nations, or newly industrializing countries (NICs), are snapping at the heels of its economic supremacy. Four countries, South Korea, Taiwan, Singapore, and Hong Kong – the Four Dragons – pose the most serious threats, with the Little Dragons – Malaysia, Thailand, Indonesia, and the Philippines – close behind. In all these areas labor is cheap and productivity high.

Isolated from the outside world by its own choice until the 19th century, South Korea has surged ahead of the others in its manufacturing output. It began its industrial history as a Japanese colony, supplying raw materials and manpower to the more advanced economy, but since 1964 it has competed successfully for a share of the overseas market for textiles and industrial machinery. Ever more ambitious, it is finding cheaper and quicker ways to produce computer equipment and software, and has already stolen much of the U.S. market for pocket calculators and video recorders from the Japanese.

As the Western world slides toward recession and the third world suffers a steady decline in its living standards, the dragons on the Pacific Rim ride the prevailing tides of global economics, pulling the rest of the world in their wake.

TIED BY A SINGLE THREAD

How Manchester clothed the world

In the first 100 years of the Industrial Revolution, Manchester, the center of Britain's cotton industry, came to embody and symbolize the awesome power of the new global economic order. Into its vast steam-powered factories came raw cotton from India, Egypt, and the U.S. to be spun, woven, bleached, dyed, and printed by a new army of industrial workers – most of them women and children – using the most sophisticated machinery of the day, finally to emerge as the finished fabrics that clothed the world. The consumption of raw cotton grew at a phenomenal pace. Eight thousand tons in 1760 became 25,000 tons by 1800, and 300,000 tons by 1850. The first factory chimney appeared in 1781. By 1800 there were 50 or more.

Until 1750, most textiles woven in England were made from wool. But as the population grew, increased demand for food displaced sheep from their pastures so that more land could be brought into cultivation. Yet demand for fabrics – for clothing, furnishings, and so on – was also increasing. As the price of wool began to rise, manufacturers were forced to look elsewhere for cheaper alternatives.

At first, small amounts of raw cotton were imported from the East Indies, but supplies were erratic. A more stable source would have been the Caribbean and the southern U.S., where English plantation owners had been using slave labor to grow cotton since the early 18th century. Yet production had remained small. Away from the eastern coast, the climate, especially in the north, was simply too cold. A restricted growing season meant the cotton fibers or "staples" remained short, making it extremely difficult to separate them from the seeds in the tangled boll. It took a slave a whole day to pick and separate a single pound (0.5 kg) of cotton.

But with the invention of the cotton gin in 1793, which could easily process short-stapled cotton, American plantations boomed. Once restricted to a few areas on the eastern coast, plantations – and slavery – spread westward across the Mississippi. Production soared from 3,000 bales – a single bale weighs 500 lb (225 kg) – in 1790 to 4.5 million bales in 1860. Most of it was destined for Manchester.

Manchester's mills became so dependent on American cotton that when the U.S. Civil War broke out in 1861 and cotton imports were halted, factories closed and thousands of workers were laid off. Faced with a collapse of trade, merchants and mill owners turned in desperation to other sources of supply.

As a colony, India had already been drawn into Great Britain's vast administrative and trading system. Although its cotton was of considerably lower quality than either Egyptian or American cotton, it had become a more reliable source of supply.

Events in India illustrate clearly how Britain came to dominate all aspects of world cotton production. India once had a flourishing home-based textile industry. Indeed, for hundreds of years its fine printed cloths or "calicoes" (named for Calicut in southwest India) had been imported into every part of Europe. But by 1700 opposition from British wool, linen, and silk manufacturers had led to bans on their import. Thus protected from foreign competition, British-made cotton textiles, which were far inferior in quality, received a powerful boost.

Having successfully excluded Indian fabrics from the home market, Britain further undermined India's textile industry by flooding it with cheaper factory-produced cloth. As India's industry declined, even more raw cotton was made available for export – to Manchester.

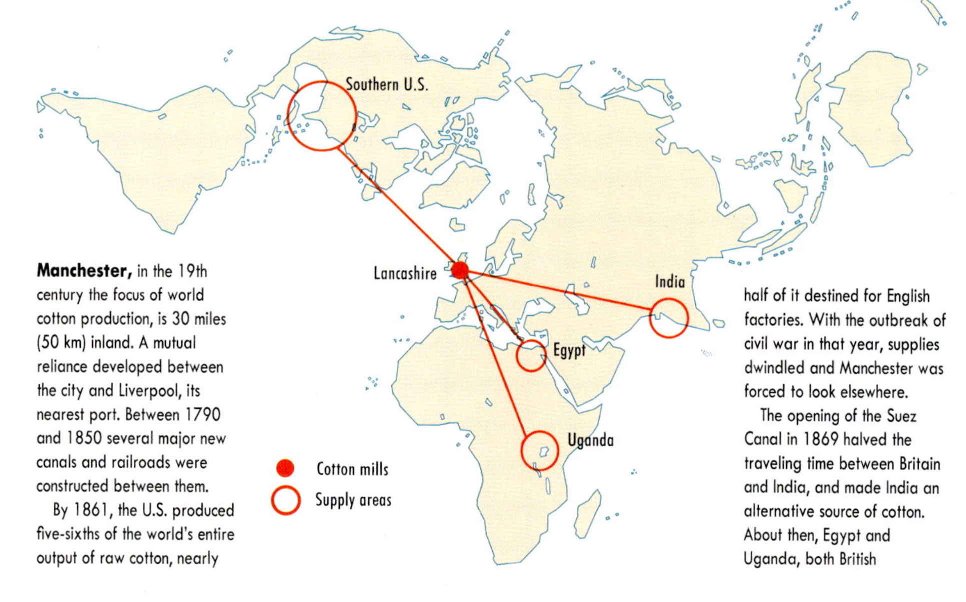

Manchester, in the 19th century the focus of world cotton production, is 30 miles (50 km) inland. A mutual reliance developed between the city and Liverpool, its nearest port. Between 1790 and 1850 several major new canals and railroads were constructed between them.

By 1861, the U.S. produced five-sixths of the world's entire output of raw cotton, nearly half of it destined for English factories. With the outbreak of civil war in that year, supplies dwindled and Manchester was forced to look elsewhere.

The opening of the Suez Canal in 1869 halved the traveling time between Britain and India, and made India an alternative source of cotton. About then, Egypt and Uganda, both British protectorates, were becoming important cotton producers in their own right, thus ensuring that British manufacturers would never again be dependent on a single supplier.

In the middle of the 19th century, most of Britain's textiles were exported: to Europe; the U.S.; and Latin America, especially Brazil and Argentina; but also to its colonies, including both Africa and India.

By 1800 Manchester's cotton factories, or mills, were beginning to turn over from handpower or waterpower to steam-driven machines. Bands carried the power from the engine room throughout the mill to each of the spinning or weaving devices.

Although many craftworkers were displaced by the new machinery, the demand for Manchester-made goods was so great that hundreds of thousands of new machine minders or "operatives" – most of them young women and children – were drawn into the area. In 1773 Manchester had only 23,000 inhabitants. By 1851 the figure had risen more than tenfold to 250,000.

Conditions in the mills were hot, humid, and dangerous, and hours were long. Most operatives lived in densely packed hovels close to the mills amid scenes that appalled many of their more affluent contemporaries.

For some, the factory system and the industrial city were visions of hell on earth dominated by the grime and squalor of unrestrained industrial capitalism. Others considered that the regularity of factory production and the wealth generated by manufacturing and trade produced enormous benefits to owners and workers alike.

KING COAL

The industrialization of the Ruhr

If cotton was king during the first phase of industrialization, the second phase was ruled by coal and iron. Thanks to its head start in the race to industrialization and to its lavish coal and iron ore deposits, Britain remained the world's leading industrial nation well into the 20th century. But from the mid-19th century it was joined by other countries in Europe, notably Germany, France, and Belgium, as well as by the U.S. and, a little later, by Japan. In this way an economic core was formed that came to dominate the economies – and indeed, through its military might, the political systems – of most of the rest of the world.

What part did location play in this rise to dominance? Europe was exceedingly well placed to take a leading role in industrialization. It has a varied but generally temperate climate, large areas of fertile land, and numerous navigable rivers. Equally vital was its large population, which formed both a labor force and a domestic market. Between 1750 and 1900 the population tripled to over 400 million – one-fourth of the world's people.

Europe was a rich market. Already by 1750 items that had once been luxuries, such as books, furniture, tea, coffee, and sugar, were in daily use. Even Europe's multiplicity of nation-states stimulated economic growth. On the one hand, it encouraged economic competition between states, and on the other, it ensured that no single country dominated the continent militarily.

But none of these advantages could have been realized without liberal supplies of coal and iron ore. These were not distributed evenly over the continent, but in pockets and belts, often associated with river valleys that cut through to the seams deep in the ground. It is this distribution that helps to explain the subsequent development of Europe.

Those regions of Britain, France, Belgium, and Germany blessed with ample supplies of coal and iron ore were the centers of explosive industrial growth. Marginal areas of Europe, however – the Iberian Peninsula, southern Italy, Ireland, northern Scandinavia, the Balkans, and east-central Europe – remained outside the emerging industrial core. Indeed, even now they are behind in the chase for economic growth.

The Ruhr region of Germany was one of the greatest centers of industrial expansion. It was in the mid-19th century, when demand for steel soared, that this region came into its own. Blessed with abundant supplies of both coal and iron ore, it also had navigable rivers ideally suited for the transport of bulky commodities.

Initially, the region's steel manufacturers had to import iron because the local ore contained too much phosphorus to make good steel. Then, in the 1870s, advances in processing made it possible to cope with locally mined ore. From that moment on, progress was phenomenal. By 1875, the region was producing half of the world's steel.

A great range of interrelated industries emerged in the area, including textiles and heavy machinery. Coal was soon utilized not only as a fuel but also as a raw material for the most advanced chemical industry, producing dyes, acids, and agricultural chemicals.

Despite devastation during the Second World War, the Ruhr's industries soon revived. In the 1970s, its fortunes were undermined again by the collapse of heavy industry throughout the industrialized world. Coal production plummeted because of competition from oil as an energy source and a raw material for chemicals. As a result, there was a steep decline in the number of pits, together with intense modernization and job losses.

Yet unlike most other "rustbelt areas," the Ruhr has successfully reindustrialized and the region is now enjoying a renaissance. It has remained one of the greatest steel-producing areas, largely because the local economy switched to high-quality consumer durables – notably "white goods" such as refrigerators and washing machines.

As industrialization took off around 1750, people moved in their millions to the main centers of production, most of them sited on coal and iron ore fields. Within decades, former villages and market towns were transformed beyond recognition. Although death rates were high in these dangerously polluted industrial cities, birth rates were even higher, so natural increase reached levels without precedent in human history.

The world's first such cities emerged in Britain in the late 18th century, led by Manchester, Glasgow, and Birmingham. In the following century, they were joined by cities throughout Europe, many situated along a swath of coalfields that crosses northern France, Belgium, Germany, and Poland.

Early industrial growth often took place in rural landscapes. But trees and fields gave way to pithead gear, smelters, factory chimneys, warehouses, railroad stations, and mass housing. Overnight, agricultural laborers became factory workers.

This 1875 engraving shows the Krupp's armaments factory, which was the cornerstone of the growth of Essen. At the turn of the century the population stood at 200,000, about one-fifth of whom were employed at Krupp's works. The name was changed to Krup in 1945.

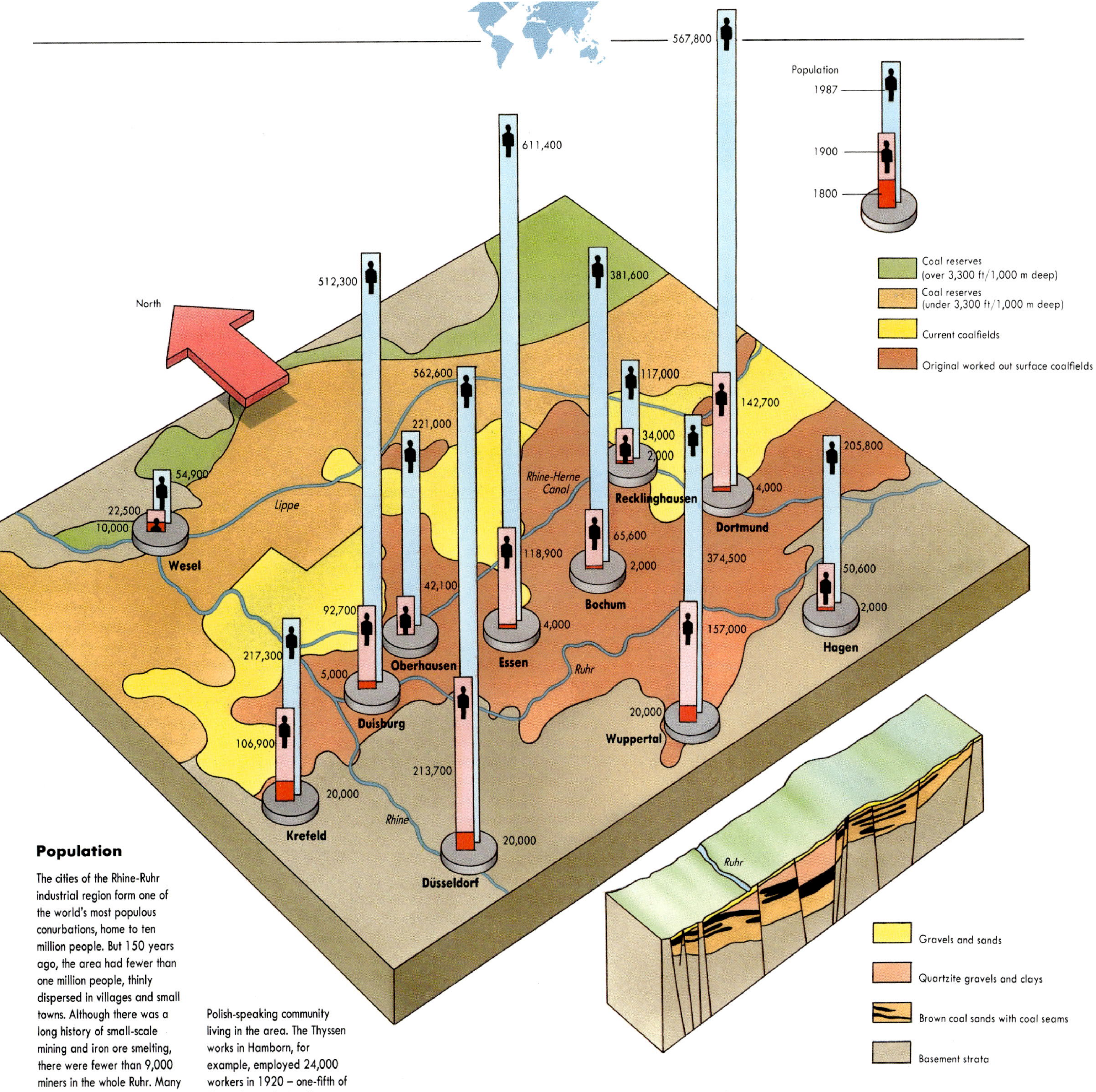

Population

The cities of the Rhine-Ruhr industrial region form one of the world's most populous conurbations, home to ten million people. But 150 years ago, the area had fewer than one million people, thinly dispersed in villages and small towns. Although there was a long history of small-scale mining and iron ore smelting, there were fewer than 9,000 miners in the whole Ruhr. Many worked only part-time in tiny pits. By 1900 there were nearly 30 times that number working long hours in new deep pits.

In the latter half of the 19th century, when local coal started to be "coked" for making iron, and steel industries began using local ore, the regional economy blossomed. New factories attracted a huge workforce, many from eastern Europe. There is still a substantial Polish-speaking community living in the area. The Thyssen works in Hamborn, for example, employed 24,000 workers in 1920 – one-fifth of the city's population. Twenty-five years earlier, Hamborn as a whole had a total population of only 5,000.

By 1910 the population of the Ruhr had grown to over 3.5 million. Duisburg, at the confluence of the Ruhr and northern Europe's main river artery, the Rhine, had expanded to become the largest inland port in the world.

Since then, the population has continued to grow, in part thanks to a massive influx of Greek and Turkish workers, who form a majority in some areas. Like other regions of Europe and North America whose economies were dominated by mining and heavy industry, the Ruhr region faced total collapse in the 1960s and 1970s. However, systematic investment has transformed the industrial and commercial base.

Geology

The coal seams of the Ruhr region were among the thickest and most extensive in the world. At first, coal was mined mainly in the south of the region, where the seams come close to the surface and were therefore easy to mine with primitive equipment. Flooding, gas explosions, and suffocation were a perpetual threat until the development of high-powered steam pumps. With ventilation and drainage more secure, access to seams at greater depths was made possible.

The sinking of the first deep shaft in 1837 revealed immensely thick coal seams of excellent quality up to 3,300 ft (1,000 m) below the surface, which have yet to be mined. Gradually mining settlements moved northward, probing the limits of the coalfield.

THE GIANT AWAKES

American expansion west

SINCE THE UNITED STATES ACHIEVED independence from Britain in 1776 its economic power has multiplied many times over, bringing incomparable military strength, political importance, and cultural influence. Home to just 5 percent of the world's population, it produces 22 percent of the world's cars, and consumes 25 percent of the world's coal and energy, as well as 30 percent of its crude oil. Without U.S. grain exports many countries, including the Soviet Union, would run short of food. Half the world's largest corporations are American. How did this giant of the world scene emerge?

It was an inauspicious start. To early European navigators aiming for the Orient, North America was simply in the way. For more than a century, until 1607, every attempt at permanent settlement on the eastern seaboard failed. Extremes of climate, pests, wild animals, mountains, forests, swamps, deserts – the very geography of the place seemed deliberately built to deny Europeans a toehold.

Even in 1776 the U.S. was little more than a narrow strip along the East Coast. It had a tiny manufacturing base, and little more than three million inhabitants of European or African origin. Only its maritime capacity was considerable: its shipbuilding industry was able to draw on seemingly infinite supplies of timber. (Europe had largely exhausted its own forests.) Most manufactured goods were brought from Europe in exchange for the produce of plantations in the semitropical South, especially sugar, rice, indigo, tobacco, and cotton.

But while slave-based agriculture flourished in the South, the Northeast initiated a crash program of industrialization. This divergence, based on geographic differences, was to prove fateful. Within a century, it plunged the country into bloody civil war.

One momentous development in the North was the construction in 1792 of the world's first industrial park – Paterson, New Jersey. Initially powered by waterfalls, its factories produced cotton goods, then steam locomotives, and later the repeating handgun and rifle (Samuel Colt set up here in 1836). Elsewhere, steel nails, axe heads, plowshares, barbed wire, wind pumps, sewing machines, labor-saving reapers and threshers – all items vital to the opening up of the interior – were mass produced by highly mechanized, highly standardized methods. Goods had to be serviceable but cheap; pioneers had no surplus cash.

Moreover, skilled labor was scarce. Although millions of immigrants were arriving through the gateway industrial cities of the Northeast, most had been farm workers. This meant that production methods in the new factories had to be greatly simplified. In any case, the lure of the West made these new Americans reluctant to settle for protracted apprenticeships. As a result, the U.S. established an early lead in mass production techniques, which it maintained until Japan took over in the 1960s.

Charlie Chaplin in *The Immigrants* (1917). Europeans flooded into the U.S. in the 19th and early 20th centuries, both to escape hardship and in search of new opportunities. Although many stopped in the industrial gateway cities of the East, others headed west. Chaplin himself left England in 1910, and made his first film in Hollywood in 1913.

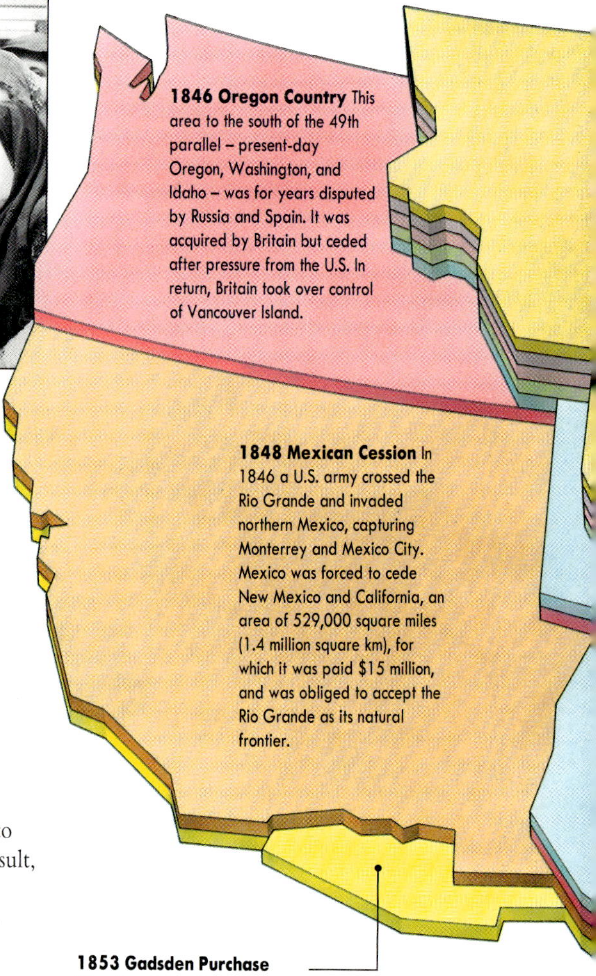

From around 1800, pioneers squeezed through natural breaks in the Appalachian Mountains and headed west, coming into conflict not only with native Americans, but also with earlier Spanish settlers. New transportation routes pried open vast and hitherto inaccessible regions. Flat-bottomed paddle steamers encouraged north–south trade along the shallow rivers of the interior, like the Mississippi, the Illinois, and the Ohio.

But then came the steam locomotive. In outright defiance of the geography, most railroads went east–west, putting the industrial and trading cities on the Atlantic coast in touch with the rich agricultural domains of the Great Plains. Wheat and beef flooded eastward, manufactured goods westward. Chicago became the hub of the nation's transportation system, home to the world's greatest stockyards, slaughterhouses, and meat-packing factories.

Meanwhile, in 1848, gold was discovered in California. The slow trickle of people moving to the Pacific coast became a torrent. In 1869, at Promontory Point, Utah, a railroad out of San Francisco finally met with one that ran all the way back to the Atlantic. It was a critical point in world history.

1846 Oregon Country This area to the south of the 49th parallel – present-day Oregon, Washington, and Idaho – was for years disputed by Russia and Spain. It was acquired by Britain but ceded after pressure from the U.S. In return, Britain took over control of Vancouver Island.

1848 Mexican Cession In 1846 a U.S. army crossed the Rio Grande and invaded northern Mexico, capturing Monterrey and Mexico City. Mexico was forced to cede New Mexico and California, an area of 529,000 square miles (1.4 million square km), for which it was paid $15 million, and was obliged to accept the Rio Grande as its natural frontier.

1853 Gadsden Purchase This relatively small area of 30,000 square miles (77,700 square km), now part of Arizona and New Mexico, was bought from Mexico to make way for a railroad through Tucson to join California with the Gulf of Mexico.

Stepping westward

The painful, piece-by-piece extension of the U.S. from the Atlantic across to the Pacific was completed a mere 70 years after independence in 1776.

In the process, native Americans were removed from all but a few scattered and largely unproductive patches of their former land, and former colonial powers were ejected. Having already lost control of the East, Britain soon forfeited parts of Canada and the Pacific coast. Meanwhile France sold off a large tract of land in the center of the continent, Spain abandoned Florida, and newly independent Mexico was forced to withdraw beyond the Rio Grande, thereby losing fully half its territory. Nothing could stand in the way of the new diverse and phenomenally dynamic American people.

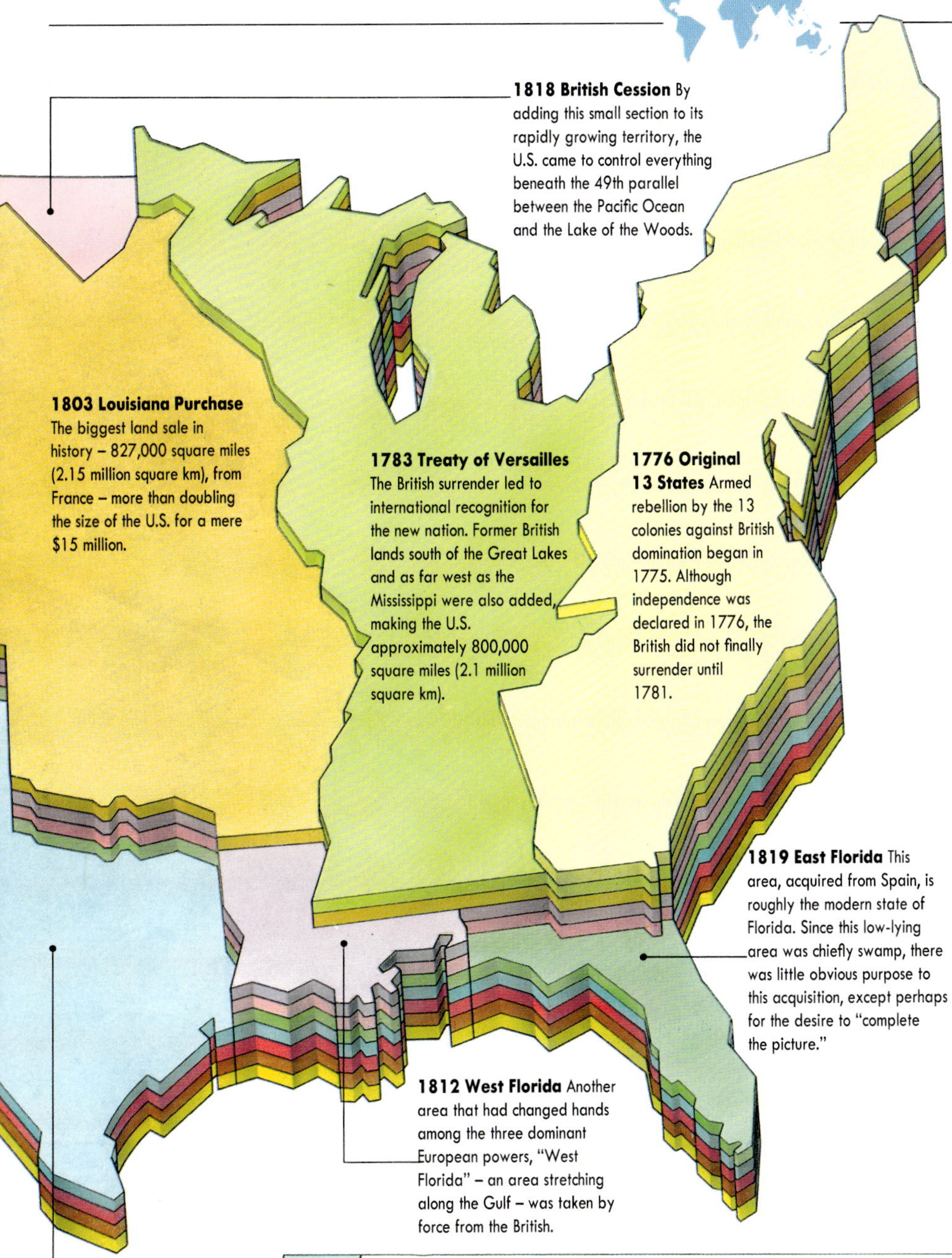

1818 British Cession By adding this small section to its rapidly growing territory, the U.S. came to control everything beneath the 49th parallel between the Pacific Ocean and the Lake of the Woods.

1803 Louisiana Purchase The biggest land sale in history – 827,000 square miles (2.15 million square km), from France – more than doubling the size of the U.S. for a mere $15 million.

1783 Treaty of Versailles The British surrender led to international recognition for the new nation. Former British lands south of the Great Lakes and as far west as the Mississippi were also added, making the U.S. approximately 800,000 square miles (2.1 million square km).

1776 Original 13 States Armed rebellion by the 13 colonies against British domination began in 1775. Although independence was declared in 1776, the British did not finally surrender until 1781.

1819 East Florida This area, acquired from Spain, is roughly the modern state of Florida. Since this low-lying area was chiefly swamp, there was little obvious purpose to this acquisition, except perhaps for the desire to "complete the picture."

1812 West Florida Another area that had changed hands among the three dominant European powers, "West Florida" – an area stretching along the Gulf – was taken by force from the British.

1845 Texas Republic Larger than any European country with the exception of Russia, Texas, at 390,000 square miles (10.1 million square km), was part of Spain's colonial empire until Mexico became independent in 1821. Having declared independence and having won victory over the Mexican army in 1836, Texan-Americans successfully sought annexation by the U.S.

THE PITTSBURGH STORY

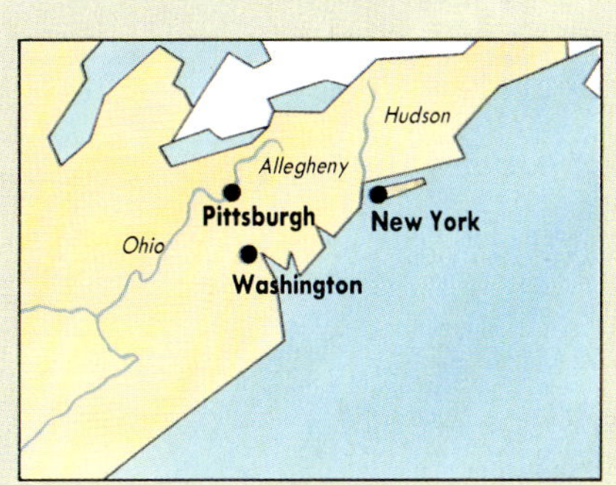

The changing fortunes of Pittsburgh, Pennsylvania, are due largely to its geography. On the site of a former Indian village and fur-trading post, it had a defensible position in a river fork that made it an important military installation. The French built the first fort here in 1753, but lost it to the British within five years.

When the frontier moved west, Pittsburgh would surely have declined but for the discovery of rich coal and iron seams. With great demand for railroad track and locomotives, Pittsburgh turned itself over to steel. By 1900, the city was the fifth largest in the U.S., producing half the nation's iron and steel. Some of the largest corporations of the U.S. were born here, including US Steel, Westinghouse, and the canned food giant H.J. Heinz.

After floods in 1936 and economic decline in the 1970s, Pittsburgh is buoyant again. Recently voted America's favorite city by corporations seeking new head offices, Pittsburgh has lived down its metal-bashing, smoke-choked image. It is capitalizing on all its geographical advantages, especially its fine river fronts, congenial rural surroundings, and a central location in the northeast manufacturing belt.

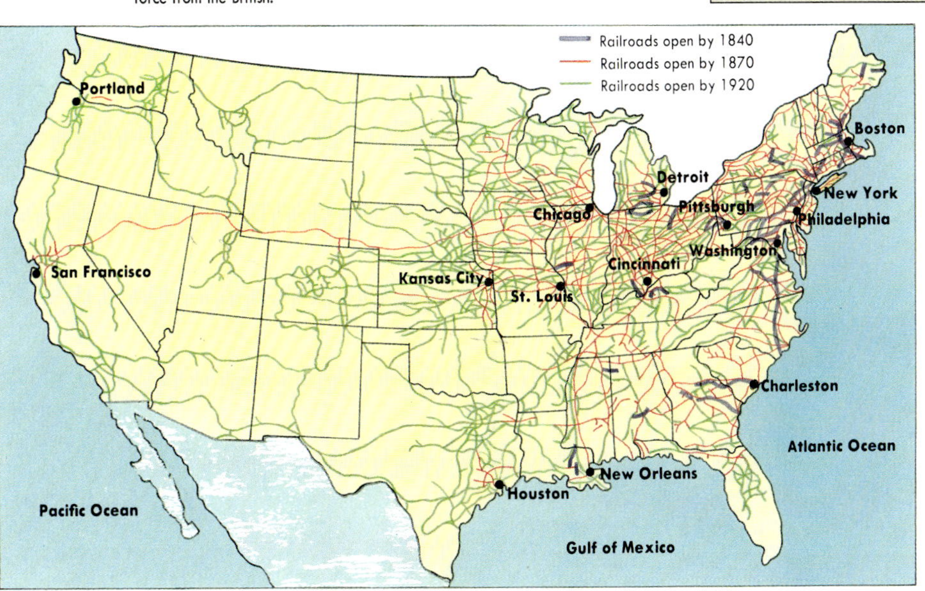

By 1890, the U.S. had more railroad track than all of Europe. Always densest in the industrially dynamic Northeast, the network left the South poorly connected.

The railroad pried the continent open. Freight costs and travel times plummeted, traffic flows multiplied.

Chicago, one-time French fur-trading station and U.S. fort, became the hub of the nation's transportation system, and home to the world's greatest stockyards. Thanks to the locomotive, the young Chicagoan Joseph McCoy brought Texas beef all the way "back east" in prime condition – he could now guarantee delivery of "the real McCoy."

FROM SNOWBELT TO SUNBELT

The U.S. changes its center of gravity

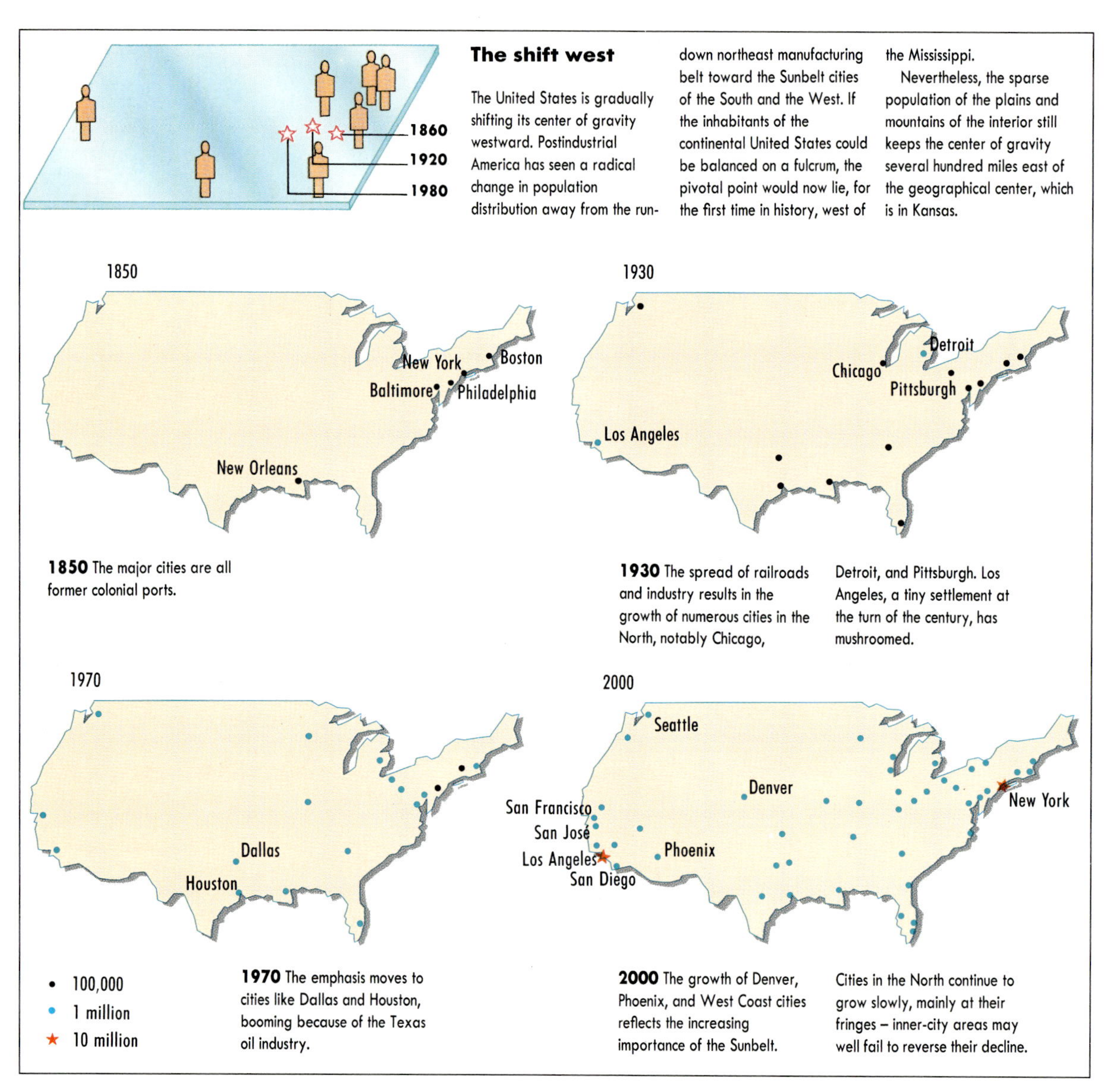

The shift west

The United States is gradually shifting its center of gravity westward. Postindustrial America has seen a radical change in population distribution away from the run-down northeast manufacturing belt toward the Sunbelt cities of the South and the West. If the inhabitants of the continental United States could be balanced on a fulcrum, the pivotal point would now lie, for the first time in history, west of the Mississippi.

Nevertheless, the sparse population of the plains and mountains of the interior still keeps the center of gravity several hundred miles east of the geographical center, which is in Kansas.

1850 The major cities are all former colonial ports.

1930 The spread of railroads and industry results in the growth of numerous cities in the North, notably Chicago, Detroit, and Pittsburgh. Los Angeles, a tiny settlement at the turn of the century, has mushroomed.

1970 The emphasis moves to cities like Dallas and Houston, booming because of the Texas oil industry.

2000 The growth of Denver, Phoenix, and West Coast cities reflects the increasing importance of the Sunbelt. Cities in the North continue to grow slowly, mainly at their fringes – inner-city areas may well fail to reverse their decline.

By 1900 the United States had already become a leading economic force on the world stage. Its huge internal market had fostered the early growth of now giant industrial and financial corporations, among them the Ford Motor Company and Exxon's parent, Standard Oil. A deeply ingrained entrepreneurial spirit and spontaneous enthusiasm for mechanical and electrical devices spurred economic progress. The path to prosperity was not always smooth, of course. Economic depression during the 1930s robbed millions of factory workers and farmers of their livelihoods. But the onset of war in Europe in 1939, and American involvement from 1941, marked an end to the downturn.

Untouched by the widespread destruction that befell its major competitors – Britain, Germany, Japan, and the Soviet Union – the U.S. emerged from the Second World War economically and militarily dominant. Manufacturing and consumer spending surged. Two million automobiles were made in 1946, and four times as many in 1955. In 1946 only 17,000 television sets existed countrywide, but by 1950 they were being sold in their millions.

But the boom could not last indefinitely. High costs of labor and old plants eroded profitability. Rather than modernize at home, many corporations began to transfer their operations to low-cost countries overseas, particularly to Southeast Asia and South America. During the 1970s General Electric, for example, expanded its total work force by 5,000, but numbers employed in the U.S. fell by 30,000. In the face of stiffening international competition, manufacturing at home went into decline. In 1971 the U.S. imported more goods than it produced, resulting in the first trade deficit since the 1890s.

The changes have had a dramatic impact on the geography of production in the U.S. The traditional manufacturing heartland of the Northeast has been hardest hit, the cold winds of global competition freezing the region into an economic "snowbelt." Mines and industries have closed, steel mills rusted away. In Detroit, home of the U.S. auto industry, 70 percent of manufacturing jobs have been lost. The picture is the same in agriculture. At the start of this century more than 30 million Americans – more than 2 in every 5 adults – lived and worked on farms. Now the figure is less than 5 million – fewer than 1 in 50. Farming has become big business, and is no longer a family affair.

But now a "postindustrial" era has arisen, resulting in a general shift away from manufacturing toward the provision of services. Today, more people work in McDonald's fast-food restaurants than in General Motors or US Steel. People are now more likely to be employed in an office, hospital, restaurant, or laboratory than in a factory or on a farm. And

where industry does thrive it is often high-tech, its symbol the silicon chip, not the wrench or the grain silo. As information replaces steel – or oil – as the key resource in the global economy, the economic center stage is taken by computer scientists, lawyers, and financial analysts.

These changes are transforming the social landscape. Proximity to coal and steel may be irrelevant to a modern corporation's location, but geography is still a vital concern. Repetitive assembly tasks call for cheap, nimble-fingered labor, while high-salaried expert staff expect a pleasant environment, abundant leisure facilities, low taxes, and plenty of cheap land on which to build. A third demand is for lucrative contracts (which is why high-tech industries tend to cluster around military bases). All of these can be found in abundance in the "sunbelt" states of the South and West. While manufacturing centers in the Northeast are slumping, the Sunbelt is displaying great economic vigor.

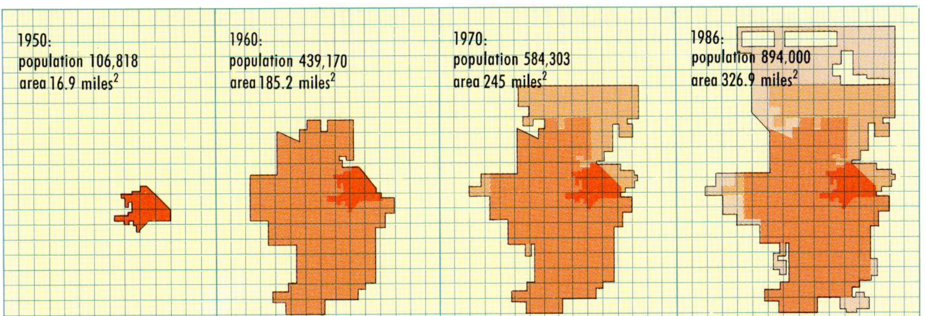

Phoenix, Arizona, the Sunbelt's fastest-growing city, owes its existence to the air conditioner. Fortunately for the inhabitants of this desert metropolis, electricity is cheap, thanks to hydropower in the Rocky Mountains.

The city's tough labor laws and heavy investment in infrastructure have helped Phoenix attract large numbers of high-tech industries – and people. Since 1950 the population has increased tenfold, the built-up area twentyfold. Migrants, most from the Northeast, continue to arrive at the rate of 3,000 a month. If current growth continues, the city will double in size within six years.

Although electricity is plentiful, water is more of a problem. Thanks to the Central Arizona Project, a 330-mile (530-km) aqueduct brings enough water every year from the Colorado River to drown the whole city to a depth of 8 ft (2.4 m). But there is a cost – the once-mighty Colorado, the river that carved the Grand Canyon, now fails to reach the ocean.

The cheapest way to distribute water around a city as sprawling as Phoenix is to lay pipes in a rectangular pattern. The uniform grid pattern is rigidly maintained, largely ignoring the substantial Phoenix Mountains northeast of the downtown area (*top*).

JAPAN INC AT HOME

The economic miracle

MADE UP OF HUNDREDS OF mountainous islands, Japan is scattered over more than 1,500 miles (2,400 km). With few natural resources, severe shortages of land, and regular earthquakes, typhoons, and volcanic eruptions, it seems ill-equipped to be one of the world's foremost industrial powers. Yet in little more than a century, Japan has been transformed from a conservative, feudal regime in self-imposed isolation from the world economic system into a highly innovative superpower, its phenomenal wealth based entirely on international trade.

Cultural contact with the West after 1854 disrupted the social order. Feudalism was swiftly dismantled. An elected government removed internal barriers to trade, gave peasants the right to buy land, and encouraged commercial agriculture – particularly silk cultivation. Above all, it promoted technical education, modern industry, and territorial expansion. Resource- and labor-rich areas such as Korea were included in its new empire. Although defeat in 1945 wiped out earlier conquests, and much of Japan's industry was devastated by bombing, economic progress scarcely missed a step.

Japan refers to the phases of its astonishing postwar progress as the three "golden ages." During the 1950s and early 1960s heavy industry soon surpassed textile production in export earnings. By 1956 Japan had overtaken Great Britain as the world's major shipbuilder.

The key products of the second golden age, from the late 1960s, were automobiles and consumer electronics. By the mid-1970s Japan's annual output exceeded 6.9 million vehicles – 21 percent of the world's production and second only to the U.S. Japan's fuel-efficient compact cars – oil has to be imported – allowed its companies to capture a large slice of the world market following oil price hikes in the 1970s.

The third and current golden age is characterized by knowledge-intensive products, particularly computers and biotechnology, coupled with the application of computer-controlled manufacturing techniques.

What explains the Japanese economic miracle? Low labor costs and high productivity certainly helped nurture industry in its early phase of development. Minimal spending on defense left more funds for investment (after the Second World War Japan's military spending was not allowed to exceed 1 percent of its GNP). The Ministry of International Trade and Industry (MITI) has provided powerful state support through subsidies and import restrictions.

Japan's geographical location and political stability meant that it could act as a bulwark against the spread of Chinese Communist influence. This factor was important in attracting Western – especially U.S. – aid and investment after the Second World War. Cultural attitudes, particularly the idea of *kazoku* (literally, family or harmony), which underlies the relationship between large firms and their employees, have prompted good labor relations and bolstered productivity.

But the structure of Japanese manufacturing is perhaps the key factor in the country's phenomenal postwar success. A large number of small but specialized firms are subcontracted to each of the major industrial giants, such as Nissan, Mitsubishi, Sony, or Toyota. This system is geared toward flexible production and "just-in-time" delivery of components. It is highly successful in promoting new products and technological innovation and in producing high-quality goods for specialized markets. Other manufacturing systems, which concentrate primarily on mass production, are unable to adapt so rapidly to changes in demand.

Japan's success has not been achieved without problems. Manufacturing and urban growth have been concentrated in fairly small areas, causing high levels of air and water pollution. The traditionally egalitarian nature of Japanese society, in which income differentials were far less than those in Europe or the U.S., is also beginning to change. The *tochi-richi* – literally land-rich, but now applied to all who make large fortunes in novel ways – have benefited from high land prices, speculative investment, and deregulation of the stock market.

On the other hand, living standards for the majority of workers have remained relatively low. The widening gulf between different income groups is leading to the breakdown of social consensus. As recent upheavals in the traditionally calm atmosphere of Japanese politics suggest, and as the rise of the Socialist Party confirms, changing economic circumstances are starting to inflict stresses and strains on the social fabric of Japan.

THE JAPANESE SUCCESS STORY

1543 The first Europeans visit Japan: Portuguese sailors introduce the musket, which becomes a major feature of Japanese warfare. Early contact with the West is largely developed through Jesuit missionaries, who greatly admire the civilized nature of Japanese life and find the people friendlier and much more receptive to Western ideas than the haughty Chinese.

1641 Japan's ruling shogunate – a military dictatorship enforcing a feudal social order – reacts violently against Western, and especially Christian, influence, which it sees as destructive to the established order. As the Dutch have visited Japan only to trade, not to convert, they are allowed to continue to send three ships a year to Nagasaki. All other Europeans are expelled. Japan will remain cut off from the outside world for 200 years.

1853 Commodore Matthew Perry (*right*) of the U.S. Navy sails into Tokyo Bay with a request for a treaty. He returns the following year with several warships. A treaty of friendship is signed, effectively reintroducing Japan to the Western world.

The Japanese are once again intrigued by Western innovations and delighted by the miniature steam locomotive Perry has brought to demonstrate to them the superiority of his country's technology.

Japan is an unlikely candidate for one of the world's most successful nations. Volcanic in origin, its hundreds of islands are still vulnerable to eruptions, earthquakes, and tidal waves or *tsunami*. It also straddles the well-established western Pacific typhoon track, which frequently brings fierce tropical hurricanes and floods to coastal areas. Risks are now exacerbated because many of the modern cities are built on land reclaimed from the sea.

Four-fifths of the country is mountainous. Forests, which are held sacred, cover more land here than in any other industrial country – 68 percent. Yet Japan is one of the most densely populated countries in the developed world, with around 124 million people, most of whom live in cities on the four main islands. The sprawling Tokyo/Yokohama megalopolis is the largest urban area in the world, with a population of over 25 million.

Japan is self-sufficient in its staple, rice, yet only 15 percent of the total land area is cultivable, and farmers average holdings of only 3 acres (1.2 hectares) subdivided into tiny pocket-handkerchief plots. Rice production is heavily subsidized – the price is fixed at four times world rates. The climate in the south is ideal for rice, which grows quickly in the warm, wet summers. A second crop – of rice, vegetables, or poultry feed – can also be grown in the mild winters. Ever since the late 19th century, wheat has been grown on the colder, northern island of Hokkaido.

Fish, too, are a staple, either locally reared in ponds, caught in coastal waters, or imported. The pressures of space mean little stock and no dairy animals are kept – the Japanese eat no milk-based products. A growing demand for beef is largely met by imports from Australia and the U.S.

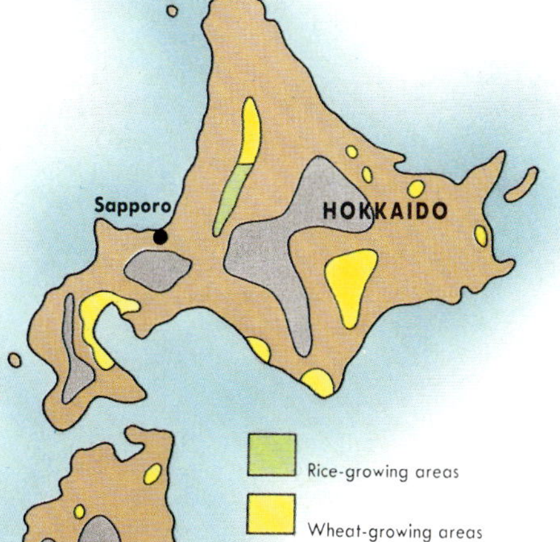

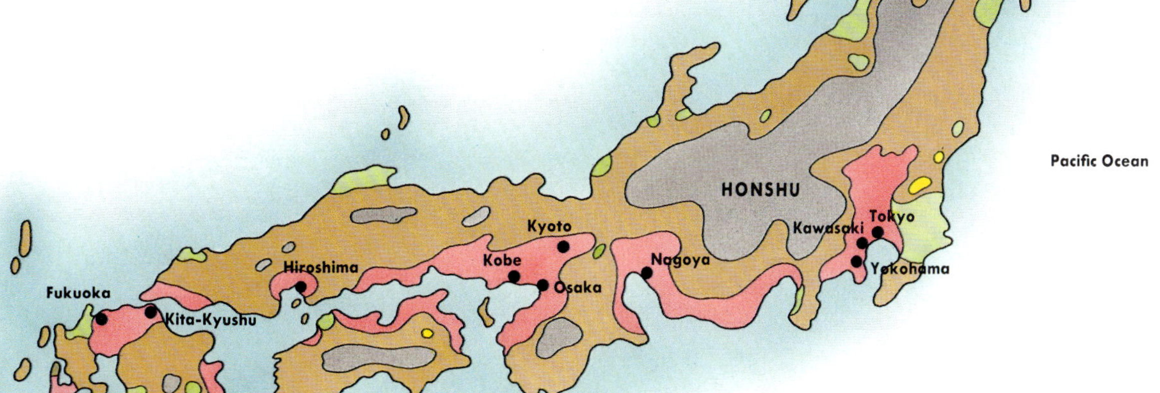

Japan's principal industrial region is a belt that stretches west from Tokyo, across the south of the main island, Honshu, and along the northern coasts of Shikoku and Kyushu.

Japan has always been relatively energy efficient. It has made careful use of its limited coal reserves, and it was quick to harness falling water for hydropower. It has even turned its resource limitations to good account. In the absence of very much iron and other minerals, it has pioneered miniaturized products, such as transistor radios, personal stereos, and compact cars.

1867 With the accession of the young emperor Meiji, feudalism gives way to an industrialized but still highly militaristic society. Yataro Iwasaki plays an important role in the modernization process: the Mitsubishi Corporation will grow from his shipping company.

Western advisers are crucial to the changes that are wrought in the Japanese political and economic systems. Education is encouraged for all, and bright young men are sent abroad to learn all they can of Western nethods. Imitation of Western habits and dress (*left*) is sometimes carried to extremes.

1894–1939 Japan's growing economic strength, but lack of resources, promotes imperial ambitions. Japan defeats China to take control of Korea, and takes Manchuria from Russia. Largely uninvolved in the First World War, it prospers at the expense of the devastated Western economies. The government continues to pursue policies of industrial development at home and aggressive expansion overseas.

1941–45 Having invaded eastern and Southeast Asia, the Japanese bomb Pearl Harbor, Hawaii, precipitating war between Japan and the U.S. Four years of conflict in the Far East and the Pacific are abruptly brought to an end when the U.S. drops atomic bombs on Hiroshima and Nagasaki (*right*). By this time, almost every major city in Japan – where prewar building was largely in wood – has suffered serious damage from incendiary bombs.

The U.S. occupying forces break up the *zaibatsu*, family firms that have been the backbone of Japanese industry. Nevertheless, firms like Mitsubishi, Sumitomo, and Mitsui remain important in the economy.

1950 onward The Japanese economic miracle begins.

JAPAN INC OVERSEAS BRANCH

The birth of a superpower

When the Rockefeller Center, one of New York's best-known landmarks, was snapped up by the Mitsubishi Corporation in 1989, the deal symbolized the growing Japanese investment in the U.S. Japan now owns more overseas companies, real estate, stocks, shares and treasury bonds, sources of raw materials, and manufacturing plants than any other country.

An estimated 10,000 Japanese companies now operate abroad, with 2,000 in the U.S. alone, and another 700 in Europe. In addition, Japan's banks dominate international finance – eight of the ten wealthiest banks in the world are Japanese, with offices in most of the major international trading centers.

Yet Japanese foreign investment is a relatively new phenomenon, the result of money pouring into the country chiefly since the 1970s. Earnings from exports of Japanese-made goods have been running at two or three times the cost of imports. Japan has increasingly invested these immense surpluses overseas.

During the early 1970s Japan's foreign investments were chiefly channeled into the low-wage economies of its Asian neighbors. The electronics giant Hitachi, for example, opened plants in Thailand, Hong Kong, Singapore, Malaysia, and the Philippines.

But from the late 1970s, and increasingly in the 1980s, attention shifted toward the U.S., Japan's biggest overseas market. Sony had built its first television plant in California in 1972, and other Japanese companies soon followed suit. Honda opened the first Japanese car factory in the U.S. in 1982, and by 1986 another four companies – Nissan, Toyota, Mazda, and Mitsubishi – had opened plants there.

Mergers, acquisitions, and the establishment of subsidiaries have all been important strategies of expansion into the U.S. When Sony took over CBS Records, a single parent company was created that controls not only the laboratories and factories that design and produce the play-back equipment, but also the record, tape, and CD manufacturing capacity, the recording studios, even the contracts with the artists.

It is a classic example of vertical integration. Similarly JVC, a branch of the giant Matsushita electronics company, has recently invested $100 million in a new Hollywood film studio. Records, films, and videos made and played in America are now increasingly the products of Japanese-owned companies.

At the same time, Japanese companies have been expanding into Europe, using the United Kingdom as a springboard. The English language, low labor costs, political support, and generous government incentives have attracted several large Japanese companies to British shores, including Nissan, Toyota, Sony, Hitachi, and Toshiba.

The establishment of plants in western Europe is to take advantage of its huge market of 320 million relatively affluent people. The recent spate of such developments is explained by the import tariffs that will arise after full economic integration of the European Community in 1992. Following the recent political upheavals, the Japanese are also moving into eastern Europe and the USSR, taking advantage of cheap labor and the proximity to western Europe.

Although it is often argued that Japanese companies have provided much-needed employment throughout Asia, and in depressed regions of Europe and the U.S., some complain that their factories are little more than assembly plants. All the important decisions are made in corporate headquarters in Tokyo. There are very few opportunities for foreign subsidiaries to carry out their own research and development, or for their local industries to provide components – these are generally brought in from Japan or another country.

As a result, bitter disputes can arise over a product's country of origin. Is an automobile assembled in England to a design generated in Japan using parts made in Taiwan British, Japanese, or Taiwanese? In 1989 France argued that unless Nissan's Bluebird cars assembled in Sunderland, England contained 80 percent European-made components, they would be counted as Japanese, and therefore subject to import restrictions.

But Japan's rise as a global investor is also causing problems for industries back home. Low-skill, labor-intensive assembly work is increasingly farmed out overseas, while Japan leads the world in the development of robotics and other techniques of factory automation. Lifelong corporation employees are now being laid off, undermining the "one big family" concept that has been the foundation of Japanese industrial success.

THE MITSUBISHI STORY

Mitsubishi is Japan's largest industrial group, a loose consortium of independent companies that grew from a family-owned combine. The 29 central companies, the *Kanyo-Kai*, act as a sort of presidential council for the rest of the group, which comprises 170 companies, in addition to a worldwide network of affiliates and subsidiaries. The Top 3 leaders, as they are known, are the Mitsubishi Bank, Mitsubishi Heavy Industries, and the Mitsubishi Corporation. The group is descended from the Mitsubishi *zaibatsu*, one of the family firms officially disbanded by the Americans after the Second World War.

The crossholding of shares is a feature of the organization of Mitsubishi and other Japanese giants. The Top 3 leaders and Tokio Marine and Fire Insurance, another company within the Mitsubishi Group, own substantial percentages of one another's shares. As well as strengthening links between the various parts of the group, this practice helps to protect the companies in question from hostile takeover bids from outside.

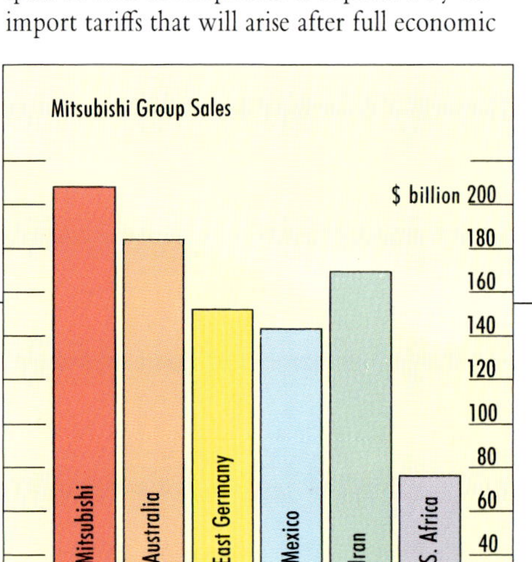

Taken as a whole, the Mitsubishi Group's annual sales exceed the gross national product (GNP) of all but 12 countries in the world. Even the sales of the Mitsubishi Corporation (the manufacturing arm of the group) are greater than the GNPs of Argentina, Saudi Arabia, or South Africa.

The graph above compares in billions of dollars the Mitsubishi Group's sales with the GNPs of several countries.

In 1987, Yasuda Fire and Marine, a Japanese insurance firm, bought Van Gogh's *Sunflowers* at a London auction for the incredible sum of £24,750,000. At the time it was the largest amount of money ever paid for a single painting. The purchase was yet another indication of the increasing importance of Japanese corporations in Western finance.

The Japanese stock exchange was revolutionized after the Second World War in an attempt to redistribute stock held by the *zaibatsu* (family firms). Now, the Nikkei Average Price Index, the Japanese stock price average, is one of the most important measures of economic stability in the world. In 1989, a troubled year for Western stock exchanges, the Nikkei average rose to record levels, barely affected by the minicrash that shook New York and London in October, and it completely recovered from the disastrous "Black Monday" of 1987.

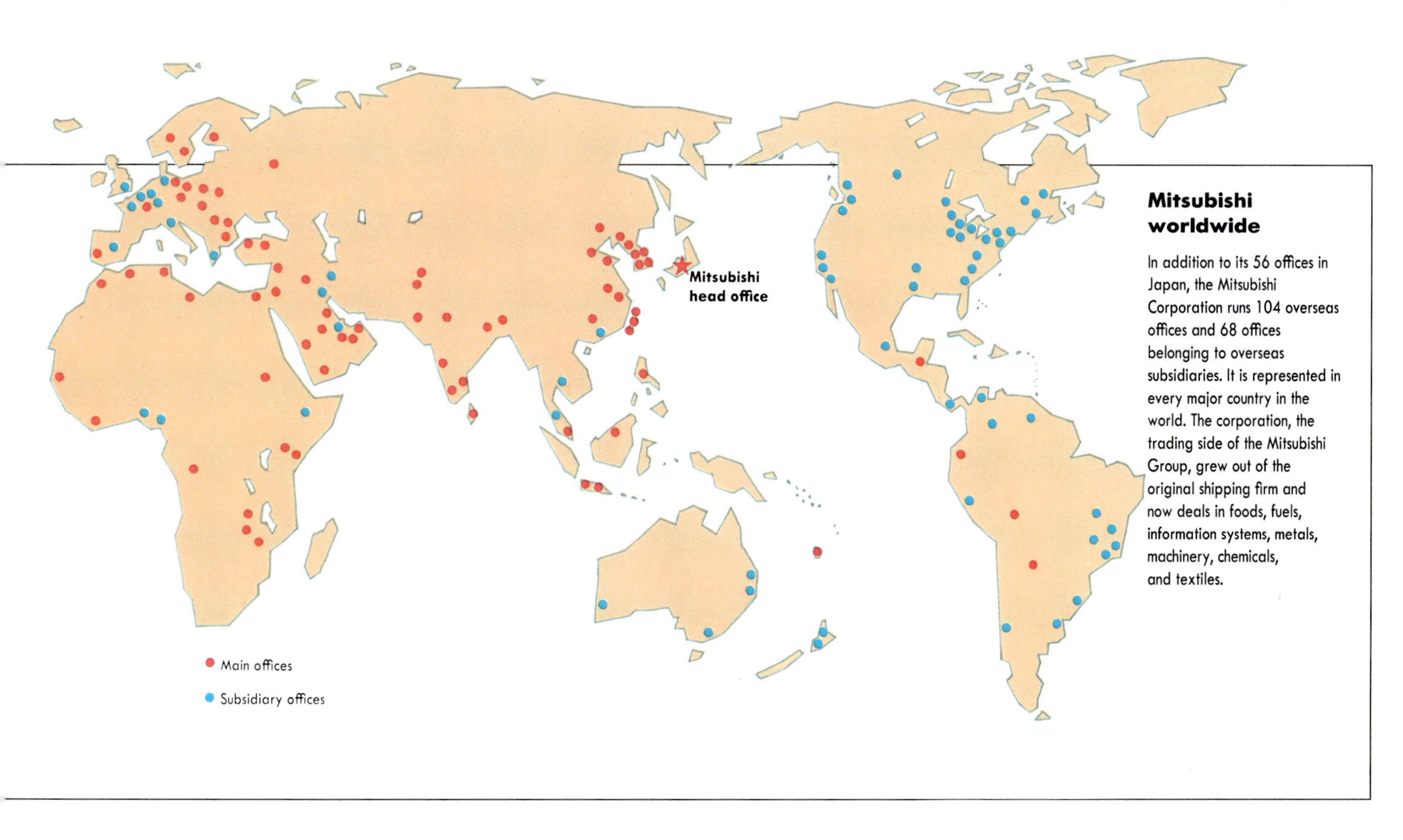

Mitsubishi worldwide

In addition to its 56 offices in Japan, the Mitsubishi Corporation runs 104 overseas offices and 68 offices belonging to overseas subsidiaries. It is represented in every major country in the world. The corporation, the trading side of the Mitsubishi Group, grew out of the original shipping firm and now deals in foods, fuels, information systems, metals, machinery, chemicals, and textiles.

CATCHING UP WITH THE CORE

Industrialization of the dragons

During the 1970s, at a time when much of the developed world was falling into recession, and low commodity prices saw standards of living in many third world countries decline, the economies of four Asian nations took off. With their low labor costs, stable (though typically repressive) political regimes, and receptive attitudes to foreign investment, the "Four Dragons" – South Korea, Taiwan, Hong Kong, and Singapore – forged ahead with rapid export-led industrialization.

And not far behind them were several other countries in the Pacific Rim with even lower labor costs, notably the "Little Dragons" of Malaysia, Thailand, Indonesia, and the Philippines. Elsewhere in the world such "Newly Industrializing Countries" – NICs – continue to emerge and alter the pattern of world economic development.

NICs are high-growth areas. Their populations are undergoing a rapid shift in employment from agriculture to industry, manufacturing output (most of it for export) is soaring, and incomes are generally rising. As their economies grow, so the gap between NICs and wealthier, developed economies is reduced. In Latin America, Brazil and Mexico have led the way, while in southern Europe, Greece, Spain, Portugal, and Yugoslavia are also growing rapidly.

But it is in eastern and southeastern Asia that the most spectacular development is taking place, led by the Four Dragons. Growth rates in this region over the last three decades have consistently been double or triple those of the U.S. or Europe, sometimes even exceeding 10 percent a year.

The most spectacular of the dragons is Japan's neighbor South Korea. In 1950 South Korea was a poor and underdeveloped country, largely dependent on rice cultivation. As late as 1965, 70 percent of Koreans were still farmers. Today, however, the same proportion live in towns and cities, and South Korea has become a major industrial power.

Once known as the "Hermit Kingdom" because it was so anxious to avoid contact with the rest of the world, in the last century Korea was pried open by Japan, the U.S., and several European powers. Unequal treaties made it easy for them to penetrate Korea's backward economy. From 1910 until 1945 Korea was a Japanese colony, exploited as a source of cheap food, minerals, and labor (during the Second World War, around two million Koreans were put to work for the Japanese war effort in near-slave conditions, a fact that still excites profound bitterness).

With the defeat of the Japanese in 1945, Korea was immediately divided along the 38th parallel of latitude into a Soviet-backed Communist North Korea and a U.S.-backed capitalist South Korea. In 1953, after a bitterly fought civil war that drew in troops from many other nations, the division was confirmed. Anxious to prevent any further spread of Communism in the region, the U.S. set about strengthening the South by providing economic aid. Within the space of ten years, South Korea's rice-based economy had been totally transformed.

Initially the government followed a policy of import substitution, placing high tariffs on imports in order to encourage manufacturers to produce goods at home. From 1964 the policy switched to promoting production for export. At first this chiefly involved textiles and clothing, but from the 1970s the emphasis shifted to heavy industry – steel, ships, chemicals, construction, and industrial machinery. South Korea's automobile industry, which is controlled by Hyundai, was virtually nonexistent before 1980. Now it has an annual output in excess of one million vehicles, half of which are exported.

South Korea is moving still farther up the technology ladder. Having captured the low end of the U.S. calculator, television, and video recorder market from the Japanese, South Korea is in the process of expanding into more sophisticated areas – computers, software, and communications equipment. Its major computer corporation, Daewoo, is growing fast, while its biggest electronics firm, Samsung, has recently become the 20th largest corporation in the world. It is even opening up factories abroad, not just in low-wage areas but also in Canada, the U.S., and southern Europe, close to its biggest markets.

The huge Hyundai dry docks – the world's largest – symbolize South Korea's drive for economic muscle. Since the late 1950s, the country has undergone spectacular industrial growth.

Once an underdeveloped Japanese colony, the country is now a world leader in products as varied as fur coats, supertankers, and modular building construction. South Korea's steel is among the cheapest in the world, and only Japan builds more ships.

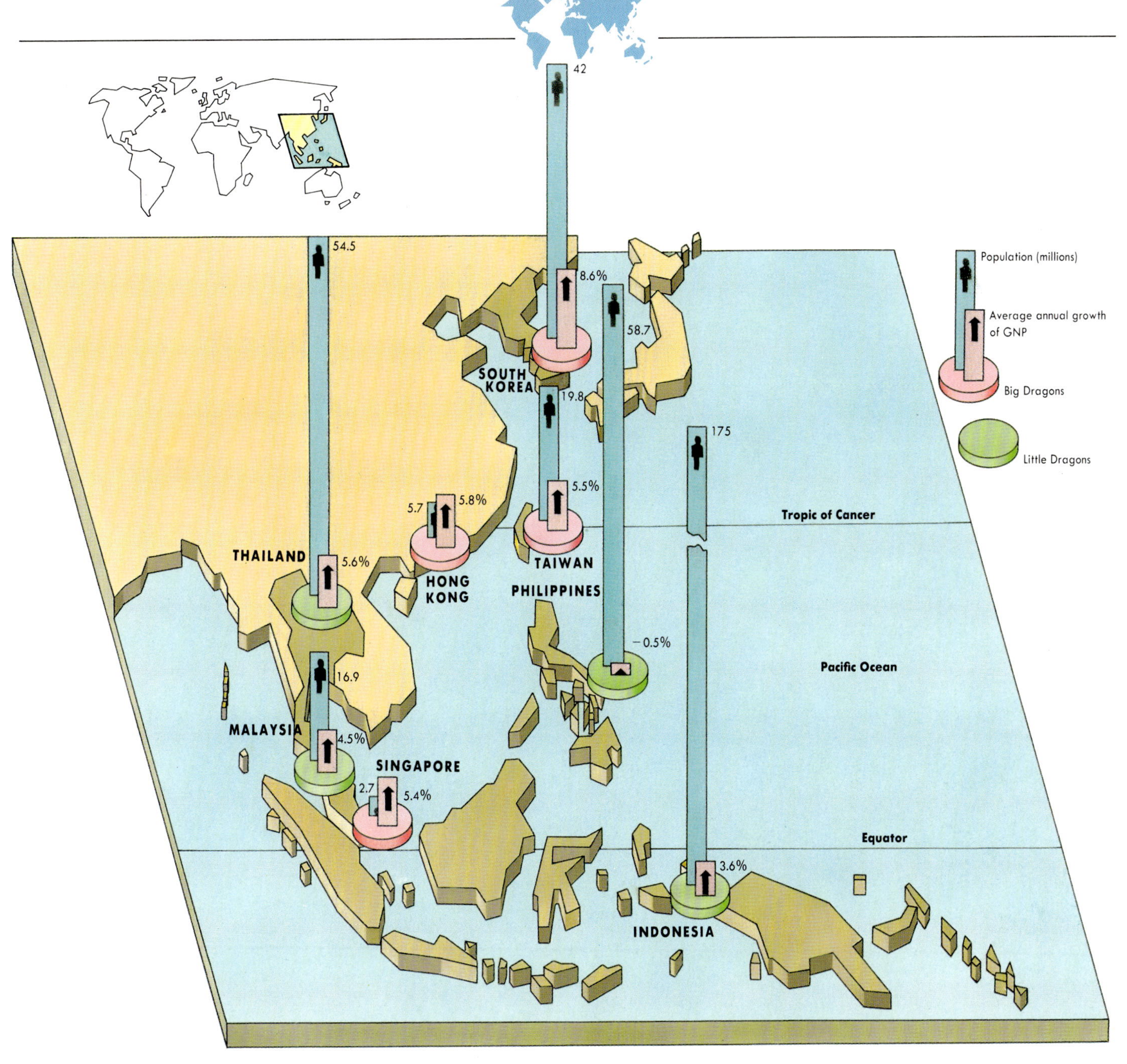

Plentiful cheap labor has been a major factor in the success of the dragons, but rapid industrial growth has led to large-scale migration from country to city. Population growth puts increased pressure on housing and services.

Rates of population growth are slowing, but they are still well ahead of the West. Indonesia is the fifth most populous country in the world, and Malaysia's population growth, at 2.5 percent, is the fastest in the region.

South Korea's population is very young, with over 55 percent being under the age of 25. The government has an active family planning policy and aims to reduce population growth to 1 percent by the year 2000. (The current rate of growth in the U.S. is 0.8 percent, and of the countries of the European Community only Ireland tops 1 percent.)

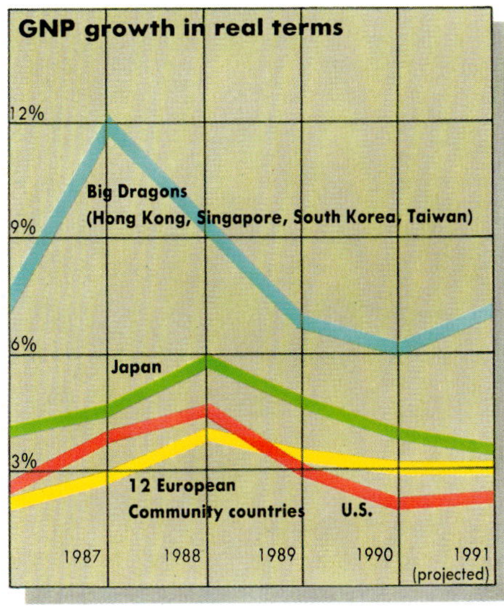

Gross national product is an internationally accepted measure of a country's wealth, related to the value of goods produced and services sold. The only non-Communist country to rival the growth of the Four Dragons in the 1980s is Thailand, with Malaysia and Indonesia not far behind. Only six other nations averaged more than 2 percent growth a year over the same period: Japan with 3.8 percent, Brazil 3.3 percent, U.S. 3.1 percent, U.K. 2.6 percent, and Italy and Spain 2.1 percent.

Putting the World to Work

Like the medieval drawings of a snake biting its own tail, a truly global economy has no beginning and no end. No country on Earth can remain entirely free of its effect. The vast geographical system that determines what goods are produced where and by whom, and where and when they are consumed, operates night and day. Relentless competition for the international markets requires equally relentless attention to them, and an unflagging application of effort is the surest route to the largest share of the economic rewards. No matter who the players are, the intricate game of work, trade, and profit is the inescapable result of technological progress.

The extent and variety of transportation methods have changed the face of the globe almost beyond recognition. The first coast-to-coast flight in America took more than 21 hours in 1922; now an ordinary journey over the same distance takes less than seven hours, and Concorde can fly the Atlantic in four.

Although it is no longer the world's principal thoroughfare for international trade, the sea carries larger vessels than ever before, and more of them. Container ships provide a cheaper, if slower, alternative to transporting goods by air, although their immense size keeps them from using formerly crucial waterways like the Panama Canal. The expanding road and rail network has penetrated the remotest areas, spawning new towns and carrying labor and goods in and out of land once considered inaccessible.

Since the end of the Second World War, the surplus labor force of relatively poor countries has flooded the core nations in search of work. Their arrival gives their host countries a decided competitive edge, since the migrants are eager even for low-paid, repetitive jobs, and as long as the workers send some of their wages home, they continue to help support the economic base they have physically abandoned.

Money moves faster around the global network than anything else. National banks have opened international branches to cope with the added business that technology has made possible, and the resulting increase in financial transactions has, in turn, demanded more sophisticated technology.

Some people mean money when they say power; others mean energy. In the form of oil, natural gas, coal, electricity, or hydropower, energy means big business the world over. Although the price of oil rose in the 1970s, it still supplies 36 percent of the world's power. Many countries are now tapping renewable sources like wind or solar power for electricity.

Cars are a ubiquitous example of the way technology has transformed global society. Luxury consumer items until Henry Ford began mass producing them in 1908, cars are now manufactured in America, Japan, and Europe, in huge plants that dominate and employ entire towns. To meet the demands of a highly competitive market while holding production costs in check, certain manufacturers, like Ford and Volkswagen, have developed intricate systems of manufacturing automobile

components in different locations – often different countries – and assembling them wherever labor is cheapest.

Countries that are not as technologically advanced as the U.S. and Japan still play an important role in the network of global trade. Their strength lies in the provision of labor rather than raw materials or sophisticated machinery, and many of them employ this strength in clearly defined industrial estates known as export processing zones. EPZs offer their multinational customers considerable financial advantages, providing tax breaks and grants along with cheap labor, power, and transport.

Over a million people, mainly women, now work in EPZs, which are most heavily concentrated in Hong Kong, South Korea, Singapore, Taiwan, and Mexico. Working conditions are often bad, safety standards may be low, and workers are dependent on the erratic demands of distant markets for their continued employment. In addition, since components come from abroad and workers receive scant training in handling them, the host country gains little useful knowledge.

Feeding the world involves the entire world, and the movement of grain around the globe reflects human needs, animal needs, and the diversity of agricultural practices. Wheat, maize, and rice supply the most basic food for more than half the people on Earth, and after oil, wheat is the most valuable commodity in the international marketplace. Coarse grains, which include barley, oats, and millet, feed livestock rather than humans.

Gradual changes in land use and farming methods mean that fewer people work on the land than ever before, but specialized machinery, careful breeding and rotation of crops, and the use of fertilizers and pesticides have nevertheless increased agricultural yield so much that farming has ceased to be a private rural pursuit and has become agribusiness.

The world is now producing enough grain to feed its entire population. The United States leads the short list of its important exporters, and the USSR, which once exported vast quantities of cereal, now imports more of it than any other country. As the populations and economies of the

Pacific Rim countries grow, then grain imports increase too. But the African nations that most need the world's surplus food can least afford it.

Of course, any country will provide for itself if it can, and some have taken extraordinary steps to make this possible. In the Netherlands people have had to pit themselves against the sea and turn it to their advantage in order to survive. In reclaiming land for both habitation and cultivation, the Dutch have developed the agricultural skills that nature forced on them and now base their global trade on the expert breeding of farm animals and flowers.

Reclamation may permit a country to overcome nature's obstacles, but it guarantees no protection against starvation, which looms as a constant danger wherever population growth exceeds food production. Scientific research over the last 30 years has brought about a Green Revolution in agricultural techniques and production that offers some hope through new high-yield crop varieties (HYVs).

Initial HYV research focused on grains because of their nutritional and market value. The new strains increased rice production so successfully that by 1985 most Asian countries could grow all the food they needed. But the effects of drought, poor soil, and invading pests have limited the success of these strains elsewhere and necessitated the widespread use of expensive agrochemicals and pesticides. The chemicals, in turn, can harm both the environment and the local farmers.

In wealthier countries, where hunger is not an immediate danger, consumers can purchase foods that were once strictly seasonal at any time of the year and drink imported beverages regularly. Wherever they live, their tea probably comes from Sri Lanka and their coffee from Brazil, and their fruit salad may contain Thai pineapple or New Zealand kiwi fruit.

The combination of cheap labor in the countries of Southeast Asia where such produce is grown, cheap rates for air freight, and massive processing and marketing systems have brought these foods within the range of the ordinary Western family budget. Greater availability has increased demand, and multinational corporations now realize substantial profits from their sophisticated global trade in produce.

Tea and coffee have achieved a similar worldwide success. Unknown in Europe until the 17th century, tea became a mainstay of British trade by the 19th. Rather than continue to pay the Chinese for the leaves, the British established tea plantations in India, where they could use the newly built railroad system for transport and the natives for cheap labor. The growing industry moved to Kenya when Indian laborers demanded better working conditions; it moved to Sri Lanka when production there could undercut costs in Kenya.

Such huge profits accrue to the fishing industry that legislation is developing to protect the resources of the oceans from being exhausted. Long-distance fleets owned and operated by the U.S., USSR, Japan, and North and South Korea trawl the seas, with advanced technological equipment, for a catch they can no longer find in their own waters. Shrimp farming supports a flourishing industry in Hong Kong and the Philippines, and the Chinese raise carp in rice paddies for home consumption and export. But in Bangladesh and Honduras such farming has destroyed agricultural land and coastal swamps that are vital to the ecology, so a place in the global industry may cost a competing country more than it earns in export revenue.

THE SHRINKING GLOBE
How transport collapses time and distance

IN 1522, THE FIRST CIRCUMNAVIGATION of the world took three years. Today, an airline passenger on scheduled flights can complete the journey in 45 hours. Information travels faster still. Computer data, TV images, telephone conversations, and faxes crisscross the world via cable or satellite. Goods, people, and information travel so far and so fast that some geographers talk of a shrinking globe – of "space–time convergence." This has manifold effects on the extent of global accessibility, interaction, and therefore also vulnerability.

Until the last century, goods, messages, and people generally traveled at much the same speed. Long-distance transport was always slow, expensive, and risky. When Ancient Rome's multi-oared galleys brought grain from Egypt, it took up to 50 days to battle back across the Mediterranean against the wind. Over land, things were worse still. It could cost as much to transport a sack of wheat 50 miles (80 km) on an ox-drawn wagon as to carry it the length of the Mediterranean by ship, partly because of the roads, but also because the oxen would be eating their load as they went along.

Even in 1800, a wooden sailing vessel carrying goods between China and Europe took up to two years for the return voyage around the Cape of Good Hope, although streamlined clipper ships had reduced the round trip to six months by mid-century. Nevertheless, further improvements were restricted before the coming of steam-powered steel ships because the use of wood and wind imposed strict limits on size – the maximum load was some 1,200 tons. Today's supertankers can exceed 500,000 tons.

Bulk transport by powerful vessel, railroad, truck, and pipeline is now so efficient over large parts of the world that transport costs no longer bar long-distance trade. Even air freight is so cheap and swift that perishable fish and out-of-season fruit and vegetables can be transported from one hemisphere to the other.

Global trade demands a global transport system. Wherever transport (including the means to keep perishable food refrigerated in transit) is unavailable, both production and consumption suffer. Many parts of Africa, South America, and Asia are still profoundly transport poor. Crops could be grown, minerals extracted, or products manufactured, but it is impossible to get them to market.

Even Europe has areas that are so inadequately served they are falling behind in the struggle to modernize. Portugal is a case in point. For more than 500 years, historical rivalry has meant that few good roads cross its frontier to Spain. Spain, on the other hand, is in the process of completing an elaborate motorway system to connect with the rest of the European Community.

U.S. airlines fly more than 300 billion passenger-miles (500 billion passenger-km) annually, most of them within the U.S. itself – over 1,200 miles (2,000 km) for every U.S. citizen.

The map shows part of the air network of a major U.S. airline. It includes cities of the continental U.S. with more than 100,000 departures a year, as well as routes between those cities that are served by more than 100 flights each week.

Sea transport may be slow, but it has the virtue of being relatively cheap. For bulky goods like oil, grain, or timber that have a low value compared with their weight, movement by ship is more economical than by air. The total size of the world's shipping fleet has changed little in postwar times, but vessels have become larger. The biggest of the supertankers now exceed 500,000 tons – more than five times the tonnage of the biggest warships. Once vital waterways, like the Panama and Suez canals, are now too shallow to accommodate them.

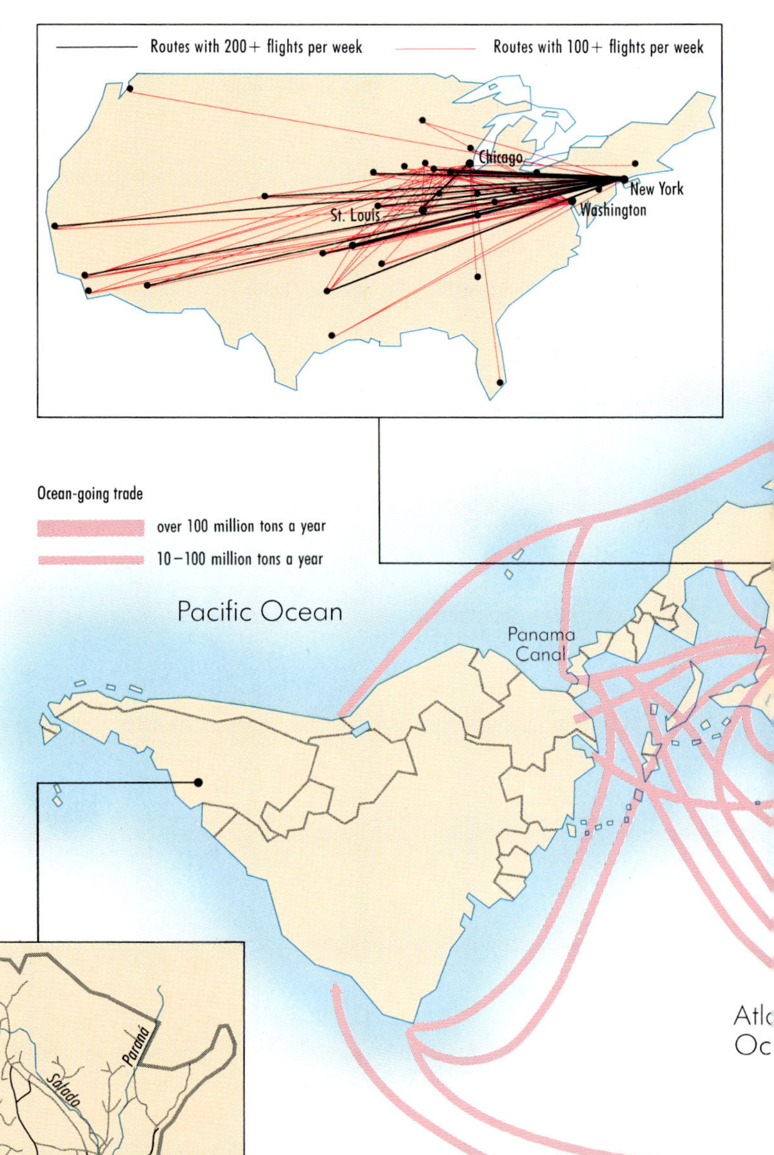

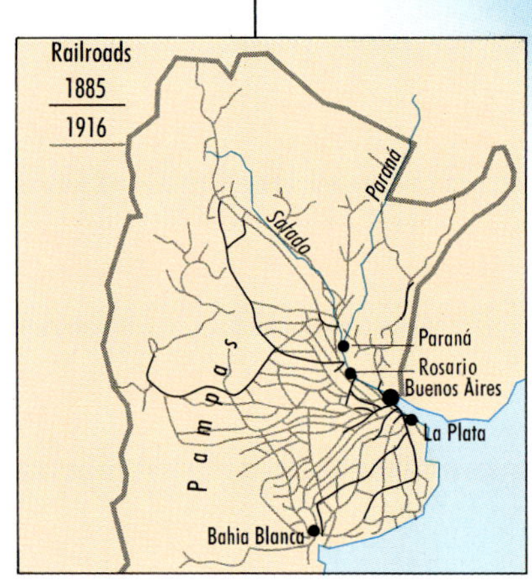

Argentina's fan-shaped railroad system, with simple linear routes running into the hinterland from Buenos Aires and the River Plate estuary, betrays its purpose: to move goods from the interior to ports for export to Europe.

As these routes increased in number, and some became more important, they, in turn, spawned small inland towns, which eventually developed their own interconnections. Finally, a dense web of railroads was created to serve the rich agricultural land of the pampas.

Until the coming of the railroads, gauchos had hunted cattle only for their hides, which were lightweight and easily transported by wagon. Carcasses were left to rot. The development of refrigerated ships and trains in the 1870s meant that Argentina could also export its beef.

DIMINISHING DISTANCES

Successive advances in transportation have cut the journey time coast to coast across the U.S.

1849 The first gold seekers leave New York on October 6, 1848, to sail to San Francisco: 146 days.

1858 Stagecoaches between St. Louis and San Francisco take 24 days, 20 hours, 35 minutes.

1869 The Union Pacific and Central Pacific railroads join up in Utah. To travel by rail from New York to San Francisco now takes eight days.

1909 First New York–Seattle motor race is won by a Ford taking 22 days.

1911 First transcontinental flight by Calbraithe P. Rogers in a Wright biplane takes 84 days, as the engine fails 6 times and Rogers crashes 12 times. Actual flying time is $3\frac{1}{2}$ days.

1922 First coast-to-coast flight in one day takes 21 hours 28 minutes.

Now A scheduled passenger flight from New York to San Francisco takes 6 hours 20 minutes.

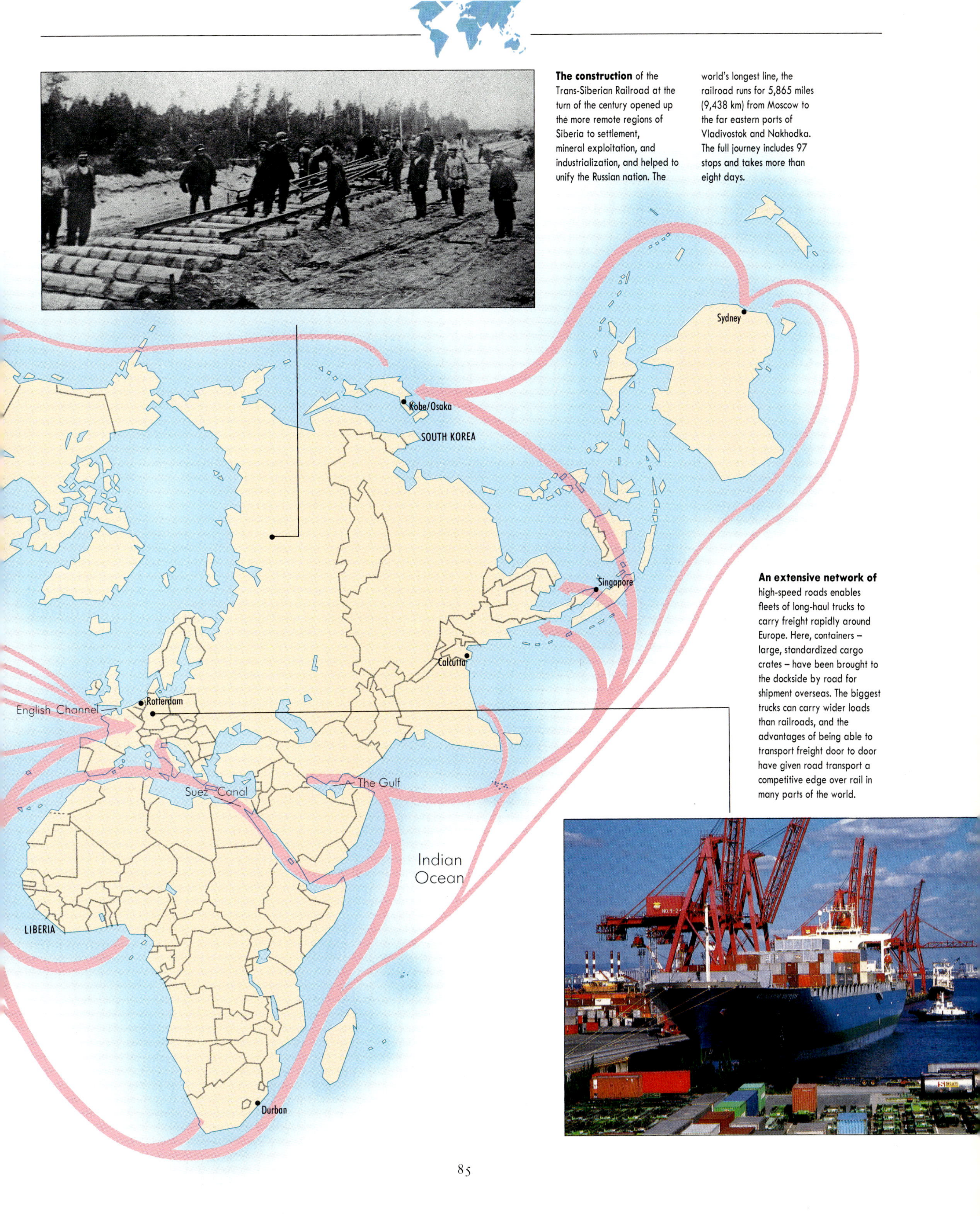

The construction of the Trans-Siberian Railroad at the turn of the century opened up the more remote regions of Siberia to settlement, mineral exploitation, and industrialization, and helped to unify the Russian nation. The world's longest line, the railroad runs for 5,865 miles (9,438 km) from Moscow to the far eastern ports of Vladivostok and Nakhodka. The full journey includes 97 stops and takes more than eight days.

An extensive network of high-speed roads enables fleets of long-haul trucks to carry freight rapidly around Europe. Here, containers – large, standardized cargo crates – have been brought to the dockside by road for shipment overseas. The biggest trucks can carry wider loads than railroads, and the advantages of being able to transport freight door to door have given road transport a competitive edge over rail in many parts of the world.

THE PEOPLE TRADE

Labor's flow in a global economy

EVERY YEAR MILLIONS OF PEOPLE around the world leave their rural homes and cross international frontiers in search of work. These "economic migrants" have been pushed from their own homelands by lack of economic opportunity and pulled to wealthier countries, sometimes many thousands of miles away, by the prospect of better jobs and higher earnings.

For some, migration will be temporary, either from choice, or because the receiving country will only grant them fixed-term work permits. For others, return to their home country will be delayed until the end of their working lives. For many, the move is permanent. They may vacation "back home," or visit in search of a marriage partner, but they will never again live in their country of birth.

As a result of this people trade, villages and towns in the economic periphery are intimately linked with the richer core countries. Mutually dependent, their fortunes come to coincide. For the core, the inflow of people provides relatively cheap labor prepared to take on the least desirable jobs, an important factor behind the competitive advantage of some core countries. For the periphery, the portion of their income that migrants send back can be vital, not only to the well-being of families and villages, but also to the nation's balance of payments.

On the other hand, deprived of its young adults – the most economically vigorous sector of its population – indigenous agriculture and manufacture in the periphery are often further atrophied, thus driving even more young people away. Fields are untended, houses deteriorate, villages have an air of dereliction about them. Meanwhile, core countries may face difficulties in absorbing migrants and their children, whose culture is often very different from that of the host country.

Until the 1970s, much labor migration was long distance and on a permanent basis. Since then, however, recruitment has increasingly become more local. Accordingly, each core region is adjoined by a "semiperiphery" from which it draws workers. This forms an accessible labor reservoir to be called upon in

Sweden 75,000
Denmark 85,000
Netherlands 380,000
U.K. 1.3 million from India, Pakistan, Bangladesh
Germany 2 million
Belgium 285,000
France 1.8 million
Austria 60,000
Switzerland 80,000
Italy 830,000
Spain 450,000
Portugal 60,000
Greece 60,000

Labor migration

From 1945 to the mid-1970s the countries of western Europe sucked in migrants from less developed parts of the world. Such flows of people in search of work typically connect former colonies with the "mother country."

Huge numbers of these immigrants have a Muslim background – there are now at least 7.5 million Muslims in western Europe (*see map, left*). Many of Germany's two million "guest workers" are Turks, while France draws much of its foreign work force from former colonies in North and West Africa. The result is an increasingly multicultural society in which contrasting beliefs and practices coexist. In the photograph (*above*), an employee at the Renault factory in Paris faces Mecca for his afternoon prayers.

Britain's "immigrant" work force has come mainly from India, Pakistan, and the Caribbean, some by an indirect route. Thousands of Indian Muslims moved to East Africa after partition in 1947. In 1972, all Ugandan Asians holding British passports were expelled by the dictator Idi Amin. Large numbers came to England, and some have since moved on to countries where opportunities are greater, especially Canada.

times of economic expansion, but to which migrants can more easily be returned when the boom finally comes to an end.

Northern Europe is mainly served by southern European countries and North Africa; the United States draws upon migrant workers from Mexico, Central America, and the Caribbean; South Africa depends on the country that it encircles, Lesotho, as well as on neighbors to the north.

Although most migrants from the periphery are poor agricultural workers with few skills, some are highly trained, and include doctors, nurses, scientists, and engineers. These expert workers, many of whom have gained their qualifications in the core, are well placed to join the "brain drain" that deprives countries of the periphery of the most educated personnel. But even within the core an equivalent process operates, causing a flow from the *relatively* poor, least dynamic countries of the core to the *relatively* rich, most dynamic areas – those that, like the U.S., can offer higher financial rewards, and better research facilities.

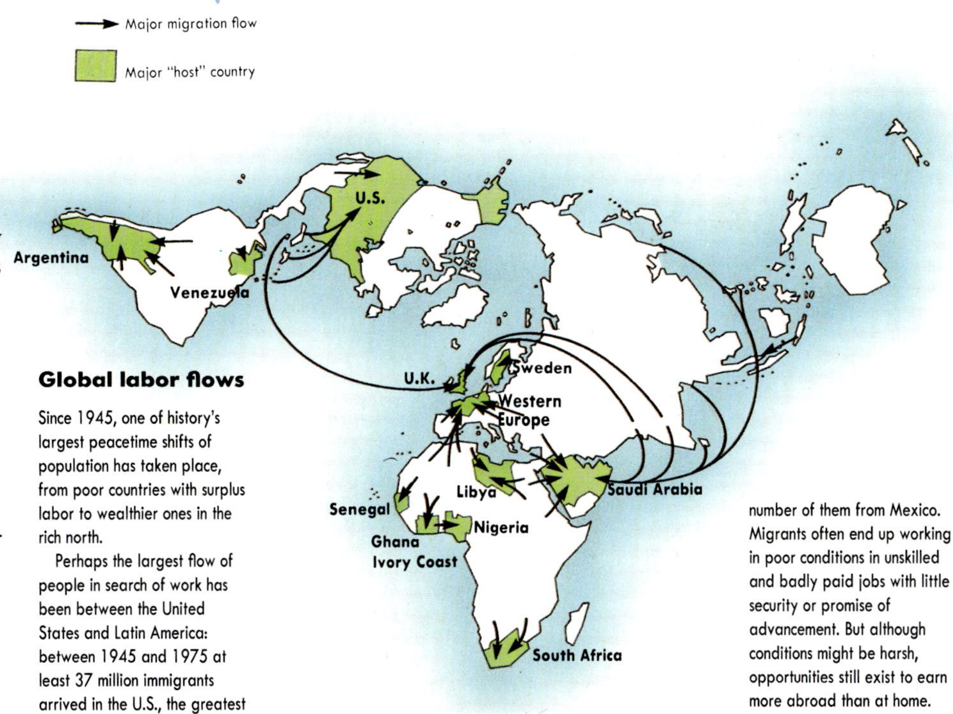

Global labor flows

Since 1945, one of history's largest peacetime shifts of population has taken place, from poor countries with surplus labor to wealthier ones in the rich north.

Perhaps the largest flow of people in search of work has been between the United States and Latin America: between 1945 and 1975 at least 37 million immigrants arrived in the U.S., the greatest number of them from Mexico. Migrants often end up working in poor conditions in unskilled and badly paid jobs with little security or promise of advancement. But although conditions might be harsh, opportunities still exist to earn more abroad than at home.

Other Arab countries 60,000
Jordan 175,000
India and Pakistan 30,000
Europe and U.S. 15,000
Saudi Arabia 775,000 foreign workers
Other Asian countries 20,000
Egypt and Sudan 130,000
Oman 15,000
North and South Yemen 330,000

Since the 1970s the oil-rich Middle Eastern countries, especially Saudi Arabia and Libya, have attracted millions of migrant workers. In 1980 Middle Eastern countries had nearly three million foreign workers, with over one million in Saudi Arabia alone. The map (*left*) shows the countries of origin of those million workers.

With rapidly growing economies based on oil exports, but without large populations, such countries were forced to rely on migrants – particularly from heavily populated countries such as Pakistan, India, and Sri Lanka – and refugees from war-torn Yemen.

Now that the oil boom has finished and the rate of economic growth in Middle Eastern countries has slowed, the demand for migrant labor has also begun to diminish.

ENERGY RICH, ENERGY POOR

Fuel consumption and production

EVERY DAY, U.S. FAMILIES THROW away as much energy in the form of junk mail as would satisfy the entire energy needs of some ten million people in the third world – easily enough to provide for Cameroon or Guatemala. Even their unopened mail (42 percent of the total) would keep all of Cambodia's industries running.

Of course, burning junk mail is a pretty inefficient way of producing useful energy, most of which is now provided by fossil fuels – coal, oil, and natural gas. In the 1960s oil overtook coal as the world's single most important energy source. At its peak in 1973, oil was providing 41 percent of the world's energy. Since then this figure has shrunk slightly, thanks to price rises, which have led to more efficient use and the substitution of other sources of electricity.

By 1990 oil accounted for 36 percent of the world's total primary energy needs (coal 25 percent, gas 17 percent, hydropower 5 percent, nuclear 4 percent, other sources 13 percent). New oil fields are being opened up, but even so, at current rates of use – around 60 million barrels a day – the world is likely to run out of oil around the year 2050.

Nations produce electricity in different ways. France, Belgium, and South Korea generate most of theirs in nuclear reactors; Denmark is investing heavily in windpower; Canada, Norway, India, and many countries in South America and Africa rely chiefly on hydropower. As yet, there are few substitutes for gasoline. Electric cars are increasing in number, notably in rich but pollution-prone areas like California and Switzerland, while Brazil is converting surplus sugar cane or corn to alcohol. This can be mixed with gasoline to form "gasohol," which is cheaper and less polluting.

Canadians lead the world in energy consumption. Averaging the nation's total energy use, each person uses the equivalent of 9 tons of coal or 45 barrels of oil a year (a barrel is 42 gal/159 l) – enough to drive three times around the world. This is just ahead of the U.S. figure of 42 barrels a person. Western Europeans and Japanese use much less, between 15 and 30 barrels. Surprisingly, at the time of unification in 1990, East Germans were using a third more energy than West Germans (33 barrels a person as opposed to 25) – largely due to inefficiencies in energy use brought about by cheap oil subsidies from the USSR.

Energy-poor countries like Nicaragua, Zaire, or India must make do with little more than an average of one barrel per person a year. And while much of the energy available in the industrialized world is versatile in form – oil, natural gas, or electricity – in the third world most energy is available only as wood for fuel, crop wastes (such as straw), and animal dung.

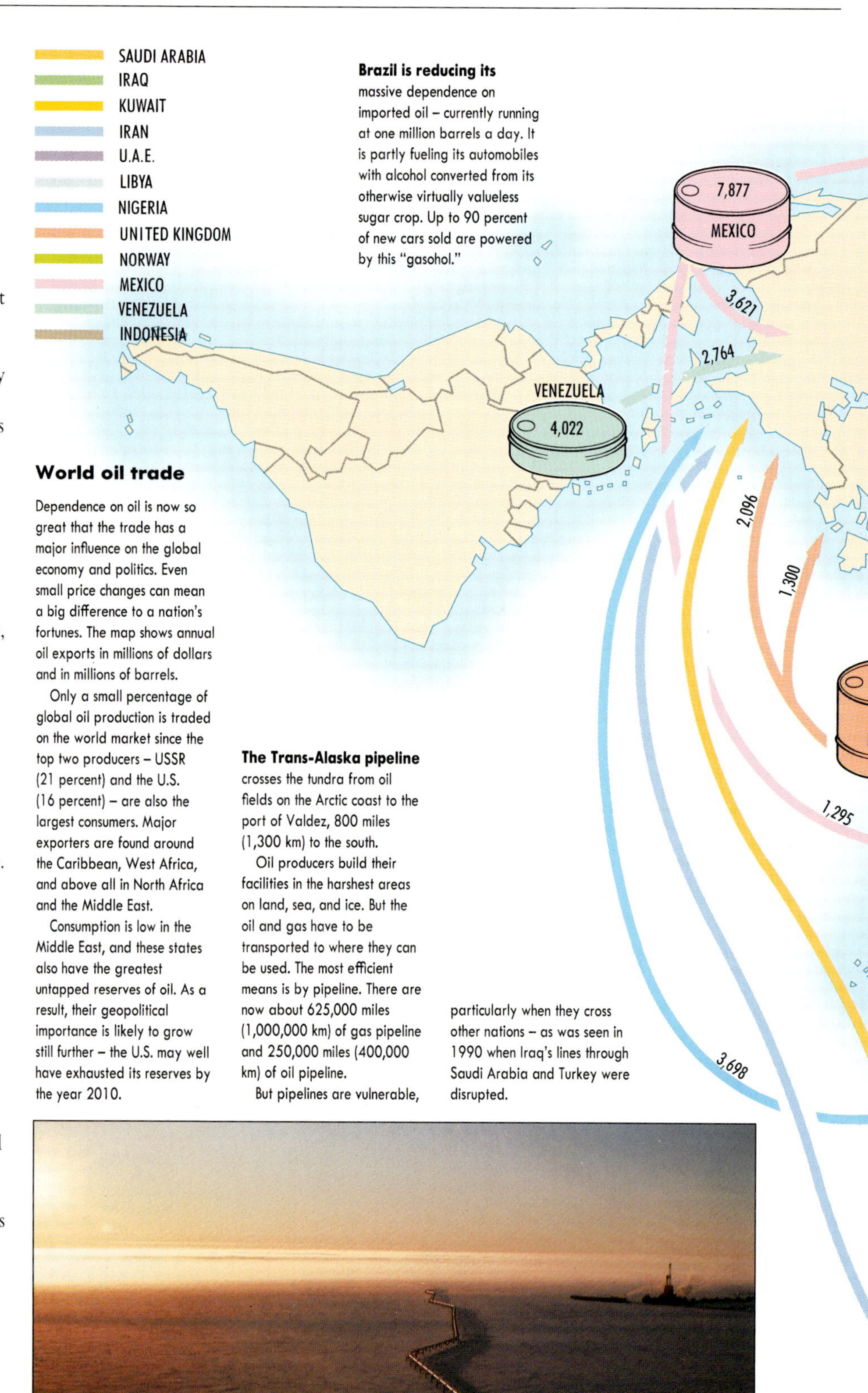

Brazil is reducing its massive dependence on imported oil – currently running at one million barrels a day. It is partly fueling its automobiles with alcohol converted from its otherwise virtually valueless sugar crop. Up to 90 percent of new cars sold are powered by this "gasohol."

World oil trade

Dependence on oil is now so great that the trade has a major influence on the global economy and politics. Even small price changes can mean a big difference to a nation's fortunes. The map shows annual oil exports in millions of dollars and in millions of barrels.

Only a small percentage of global oil production is traded on the world market since the top two producers – USSR (21 percent) and the U.S. (16 percent) – are also the largest consumers. Major exporters are found around the Caribbean, West Africa, and above all in North Africa and the Middle East.

Consumption is low in the Middle East, and these states also have the greatest untapped reserves of oil. As a result, their geopolitical importance is likely to grow still further – the U.S. may well have exhausted its reserves by the year 2010.

The Trans-Alaska pipeline crosses the tundra from oil fields on the Arctic coast to the port of Valdez, 800 miles (1,300 km) to the south.

Oil producers build their facilities in the harshest areas on land, sea, and ice. But the oil and gas have to be transported to where they can be used. The most efficient means is by pipeline. There are now about 625,000 miles (1,000,000 km) of gas pipeline and 250,000 miles (400,000 km) of oil pipeline.

But pipelines are vulnerable, particularly when they cross other nations – as was seen in 1990 when Iraq's lines through Saudi Arabia and Turkey were disrupted.

Denmark is reducing its dependence on oil and avoiding nuclear power by putting its faith in wind and other renewable sources. Some 1,400 windmills are already linked to the national electricity grid.

In New Zealand, some automobiles have been converted to run on locally abundant natural gas. Although methane gas produces 20 percent less carbon dioxide, the pressurized storage tanks it requires are too bulky for many motor vehicles.

Europe has a prodigious appetite for natural gas (*see map below*). In addition to its own sources, notably those in the North Sea, a great and increasing number of gas pipelines and specially adapted vessels bring natural gas from Asia, and under or across the Mediterranean from North Africa. Pipelines from Siberia supply 15 percent of Europe's gas demands.

Once just burned off as valueless or too expensive to transport, natural gas is an important energy source and chemical raw material in its own right. Much cheaper to recover, transport, and distribute than oil, natural gas may have great potential for supplying third world needs.

It is also much cleaner to burn than oil, and it produces less carbon dioxide. However, gas leaks may be seriously polluting to the environment.

South Africa is economically and politically isolated. With little oil of its own to exploit, it has turned to its massive coal reserves. It now leads the world in technologies that make better use of coal both for energy and as a raw material for the chemical industry.

India, with its limited oil reserves, has invested heavily in hydropower and nuclear reactors, but the social and economic costs have been great. Hopes are now being pinned largely on biogas converters for rural areas. These will use human, crop, and animal wastes to generate cooking gas and useful amounts of electricity.

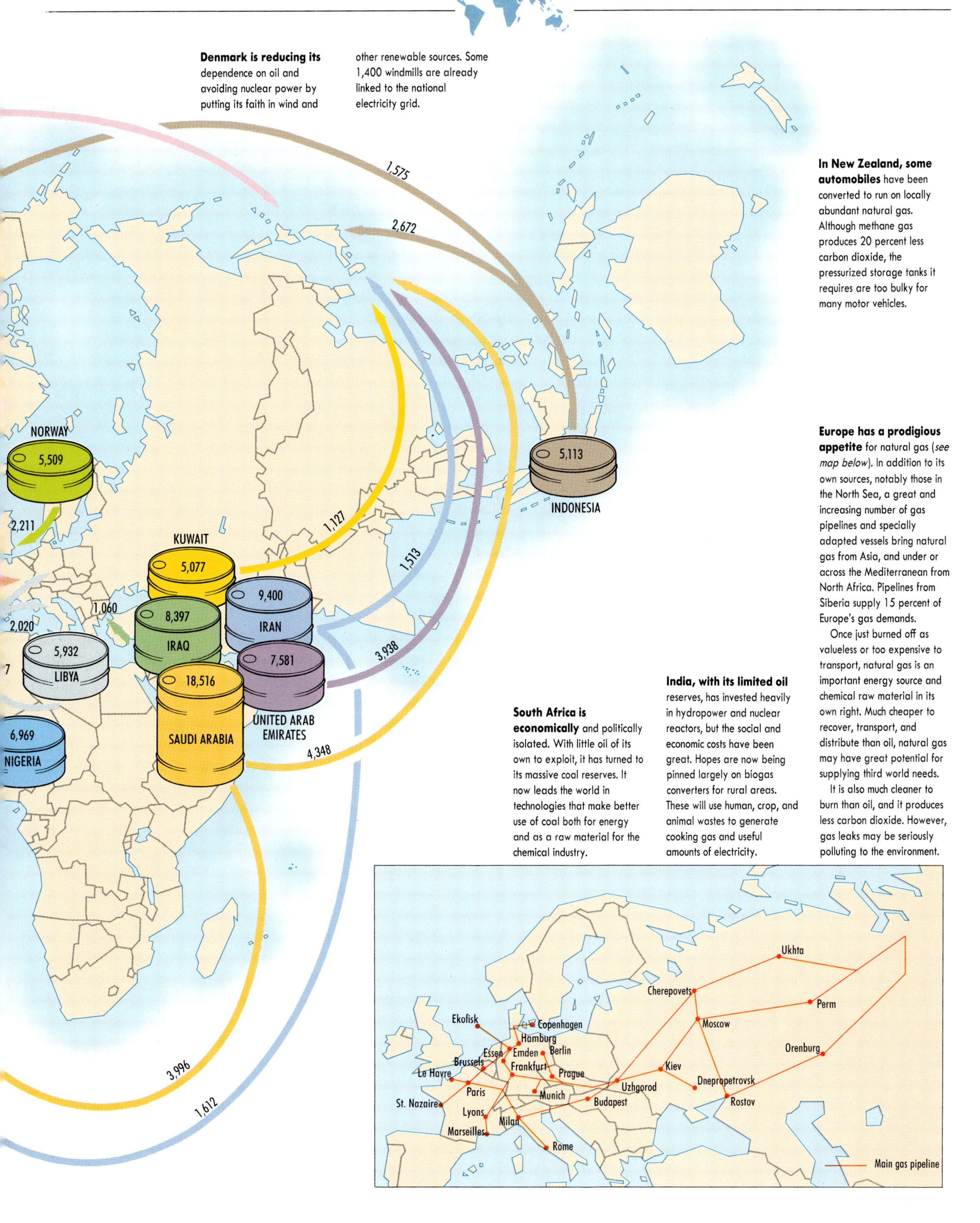

INSTANT CREDIT
The world's high-tech stock markets

TODAY, MONEY MOVES FASTER, farther, and more freely than ever before. Without leaving their homes, Finns can arrange all their banking by tapping into their account by telephone. They are answered by a voice synthesizer that is programmed to respond either in Finnish or Swedish according to their personal code. American tourists in Hong Kong or Madrid can key into an automatic teller machine and debit their account in New York. A Japanese currency dealer can purchase U.S. dollars in London and trade them within seconds for German deutschmarks or Swiss francs. From personal finance to international currency dealings, the world is now effectively a single, integrated money market.

Money may now be a commodity like any other, but it has not always been. In 1944 leaders of the major world economies met in the small New Hampshire town of Bretton Woods and agreed to tie the value of the major international currencies to the value of the U.S. dollar. This agreement provided stability in currency dealings in the period of economic uncertainty that followed the Second World War.

For nearly 30 years the Bretton Woods agreement held firm and the values of international currencies were determined no more than once a year. But in 1973, with oil price hikes devastating the world's financial and commodity markets, the agreement collapsed. With many currencies floating freely on the foreign exchanges, rates fluctuated wildly, as bouts of panic buying or selling forced the price upward or downward.

Along with the deregulation of currency dealing has come the growth of international banking. To some extent banking has always been a global affair, with European and U.S. banks in the vanguard. But since the 1970s, high and rising export earnings by east Asian economies, particularly Japan's, have prompted rapid international expansion of their banks.

The second major change in the world's money markets involves the widespread use of computers and high-speed communications links. Financial flows are now recorded and coordinated by computers linked in a network of electronic communication that spans the entire globe. New fiber-optic cables capable of carrying 40,000 calls simultaneously crisscross the Atlantic and Pacific oceans, linking Europe, North America, Asia, and Australia. By 1992 over 16 million miles (26 million km) of fiber-optic cables will be in place, sufficient to encircle the world more than 2,000 times.

Voice, data, text, or image can now be transmitted anywhere, at any time, at the speed of light. The transmission of information on prices is now instantaneous. In place of exchange rates fixed every year, screens in the world's major currency dealing centers are now updated every five seconds.

As the volatility of the world's financial markets has increased, so information about money has become enormously valuable in its

The world's money markets are largely channeled through a small handful of financial centers. London, Tokyo, and New York are by far the most important, but substantial transactions are also carried out in Frankfurt, Hong Kong, and Los Angeles.

With each of these cities in a separate time zone, trading takes place around the clock. When Tokyo dealers are closing the office and their Hong Kong counterparts are in the middle of their afternoon's trading, the European markets are just opening for business. By the time the London dealer is going out for lunch, the Frankfurt trader is coming back and the New York office has just opened.

Los Angeles opens while New York is at lunch, and by the time it closes, the Tokyo dealers are switching their computers back on to catch up with world events. In a sleepless cycle of activity, deals for huge sums of money are struck every minute of the day by financial traders around the world whose only contact with each other is by telephone or computer screen.

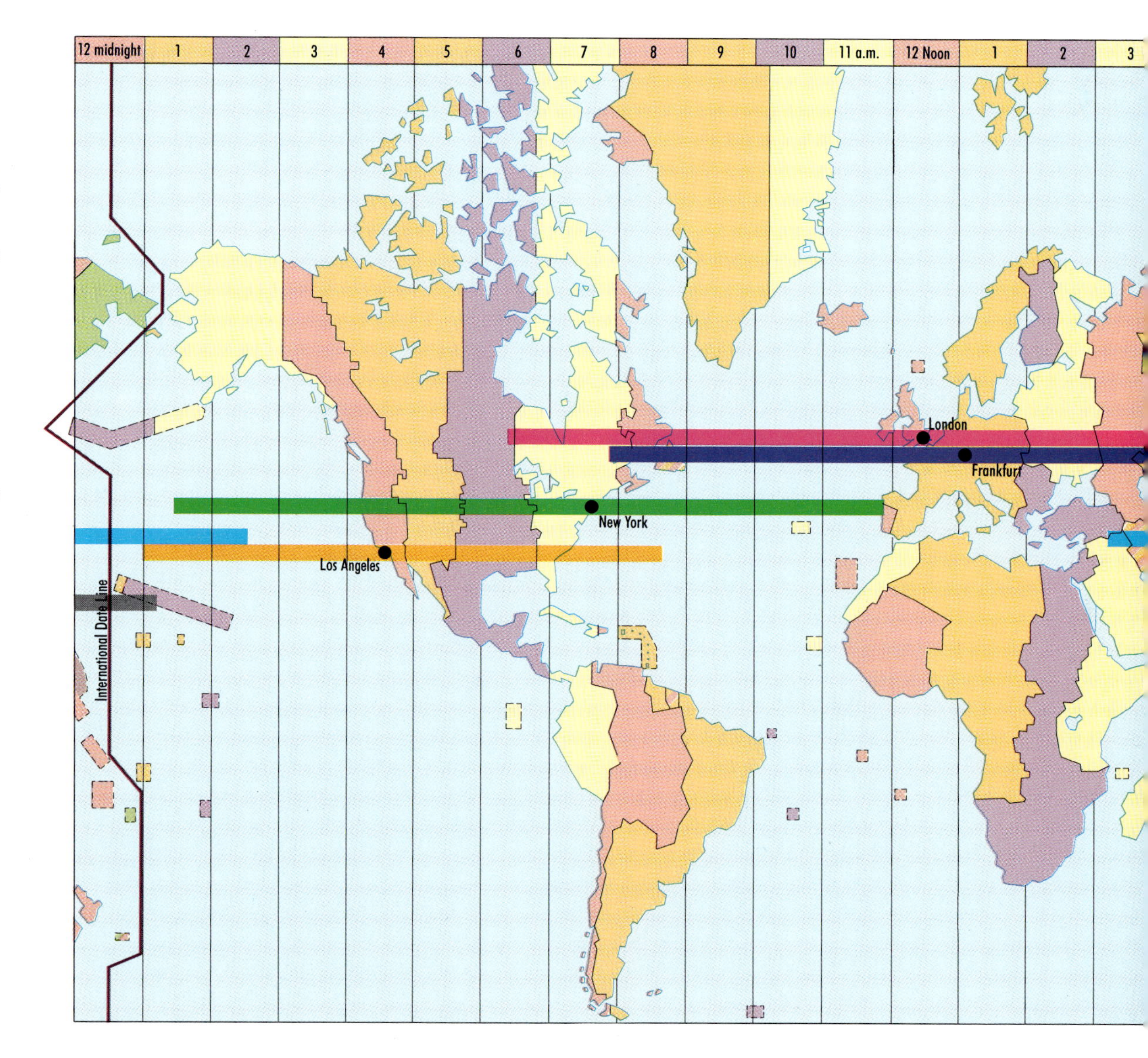

own right. Transnational manufacturers are particularly vulnerable to changes in the money markets. One night's panic buying or selling could make nonsense of all their billings, more than wiping out any profit on a long-term foreign contract.

Such enormous and growing amounts of global data need ever swifter, more capacious, and more integrated computer systems to receive, analyze, and display information as near to instantly as possible. Through the use of its highly advanced system, the New York Stock Exchange now handles well over 100 million shares a day, more than 10 times the volume traded in the late 1960s.

Also in New York, a clearing computer charts the lending transactions between all the world's largest banks – now totaling $2 trillion a year. With such huge sums being traded, the minutest decimal-point variation in prices or interest rates can make the difference between a killing or total collapse.

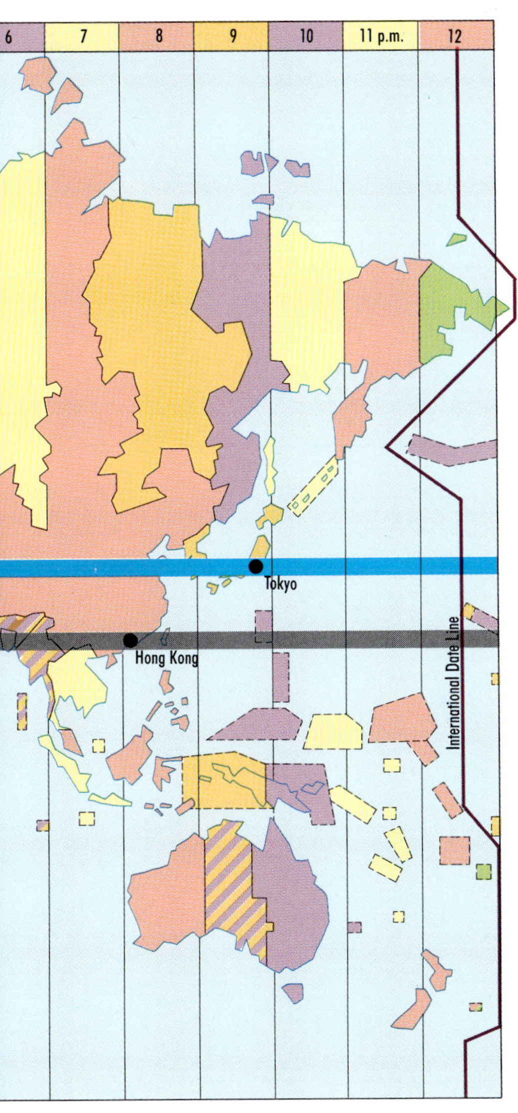

The New York Stock Exchange – Wall Street – is the largest in the world; the market value of its stocks totals three thousand trillion dollars. On the day after the worldwide stock market crash in October 1987, over 600 million shares were bought and sold on the exchange. (At the time of the Great Crash in 1929, the exchange handled 16 million shares in a day, a record that stood for nearly 40 years.) Business on Wall Street is conducted like a continuous auction, with stock in any commodity being sold to the highest bidder.

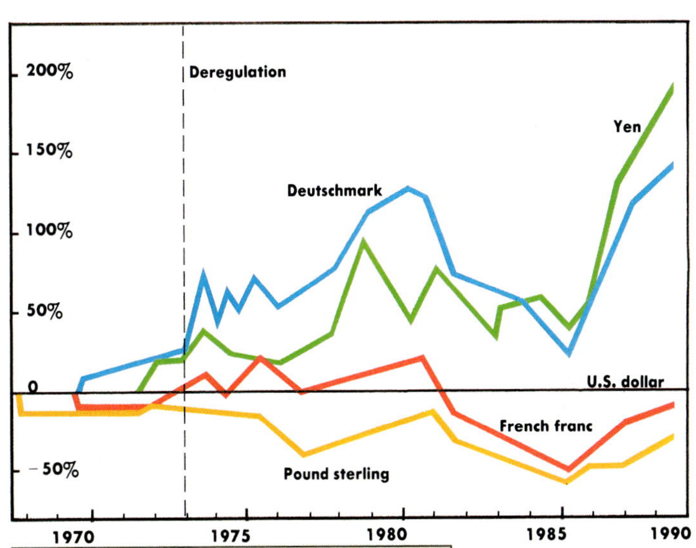

Since 1973, the world's currencies have been free to find their own level. Many plummeted in the mid-1980s, but the prosperity of the Japanese and West German economies is once more reflected in exchange rates.

The French franc was devalued four times between 1979 – when the European Exchange Rate Mechanism was introduced – and 1986, but has now entered a period of growth. It remains to be seen what effect joining the ERM and the unification of Germany will have on British sterling and the deutschmark.

The diagram illustrates the deviation of major currencies from their relationship with the dollar in 1967.

THE RISE OF JAPANESE BANKS

1969 Seven of the ten largest banks in the world are based in the U.S.

1 BankAmerica Corp. (Bank of America)
2 First National City Corp.
3 Chase Manhattan Corp.
4 Barclays Bank Group (U.K.)
5 Manufacturers Hanover Corp.
6 J.P. Morgan & Co. (Morgan Guaranty)
7 National Westminster Bank (U.K.)
8 Western Bancorp (United California Bank and 22 others)
9 Banca Nazionale del Lavoro (Italy)
10 Chemical New York Corp.

1989 Not one of the top ten from 20 years earlier is still there – the highest is Barclays, which is 17th. The former number one bank, BankAmerica Corp., has slipped to 45th place. Now eight of the top ten are Japanese.

1 Dai-Ichi Kangyo Bank
2 Sumitomo Bank
3 Fuji Bank
4 Mitsubishi Bank
5 Sanwa Bank
6 Industrial Bank of Japan
7 Norinchukin Bank
8 Crédit Agricole (France)
9 Tokai Bank
10 Banque Nationale de Paris (France)

THE GLOBAL ASSEMBLY LINE

The coming together of a world car

EACH YEAR ANOTHER 40 MILLION vehicles roll off production lines around the world, the majority made by just ten global corporations. Manufacture is often located in huge complexes, some as big as cities, where several hundred component manufacturers cluster around immense assembly areas. The best known of these is Detroit, the original motortown or "Motown," but there are now many more, including Turin in Italy, Toyota City in Japan, and Togliattigrad in the USSR.

In the most advanced of these, an integrated system of computers automatically responds to market demand, controls ordering and supply of parts, harnesses robots to perform much of the assembly (sometimes changing specifications, like color or trim, from vehicle to vehicle), and transfers part-built automobiles from one work station to another in the shortest possible time.

Early this century, there was little to suggest that the automobile industry would assume such gigantic proportions. Cars were luxury items, individually hand crafted by all-around mechanics. It was not until 1908, when Henry Ford began to mass produce the "Model T" in his Detroit factory, that the price of automobiles began to fall low enough for widespread ownership. Using a single design, standardized features ("any color you like so long as it's black"), and a conveyor belt system he had seen in slaughterhouses in Chicago, by the early 1920s Ford was producing nearly two million cars a year – 90 percent of the world's output.

In 1925 Ford's factories could make as many automobiles in one day as they had produced in a whole year under the craft system. Labor turnover was high, though. He calculated that for every 100 extra workers needed, he would have to hire 963. Many of his workers were recent European immigrants, or black migrants fleeing rural poverty in the South.

American developments soon paved the way for changes abroad. Some overseas companies such as Morris Motors in Cowley, England, eagerly adopted the Detroit-developed production lines. By 1929 Ford had established 21 foreign plants, and General Motors another 16. Most were in Europe, but some opened in Japan. Even so, in 1950 nearly 85 percent of all the world's vehicles were still U.S.-made.

But inevitably, U.S. domination of the industry was eventually challenged. In Europe, for instance, factories that had so recently been making armaments were soon converted to the production of automobiles. In 1958 the European Economic Community (EEC) was formed, allowing European firms to benefit from an enlarged internal market while at the same time restricting imports from outside. Fiat in Italy, Renault and Peugeot in France, and Volkswagen in Germany expanded rapidly. Meanwhile, Japanese companies such as Toyota, a manufacturer of textile machinery,

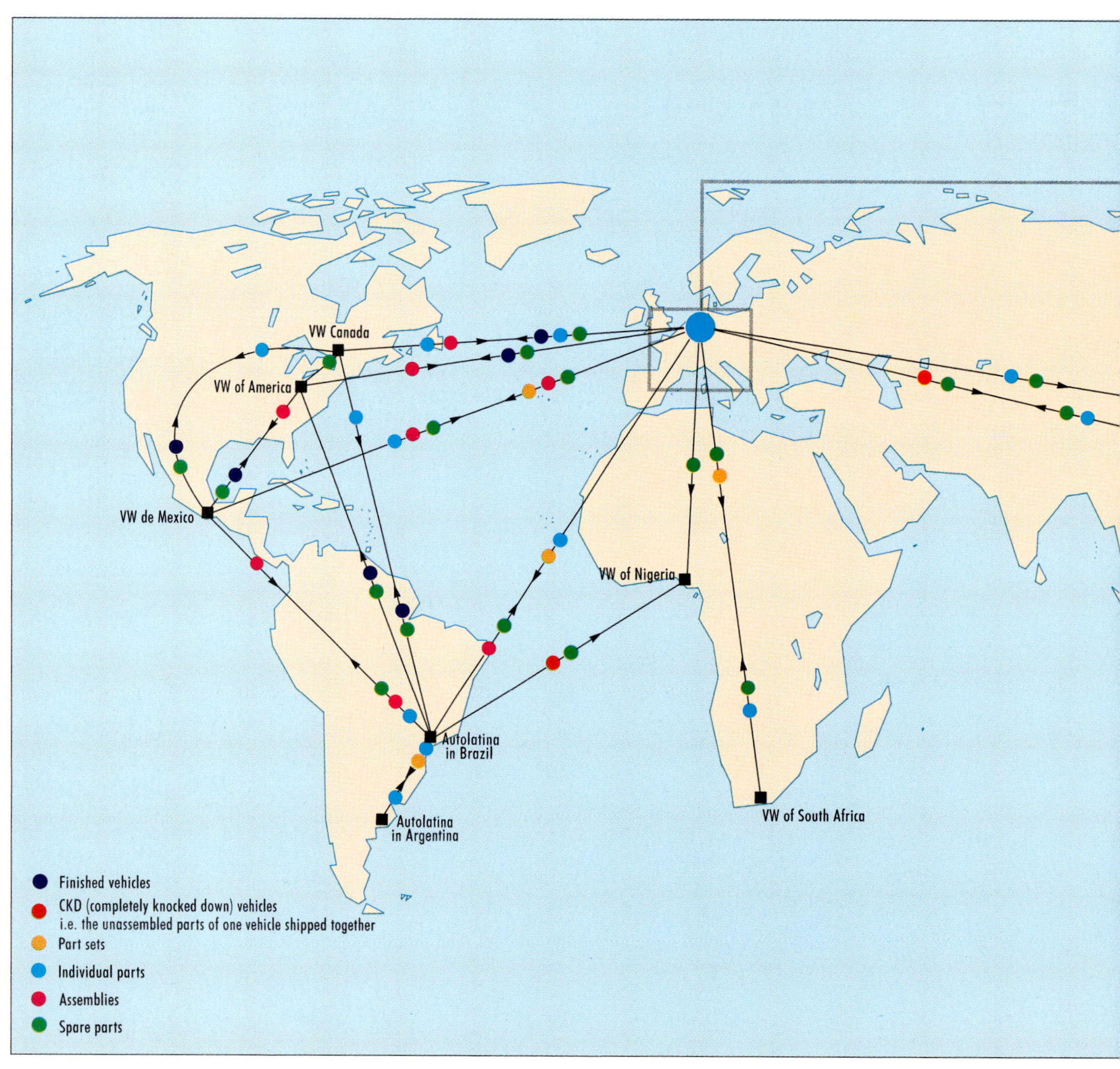

Volkswagen, like Ford, has developed a complex global network for manufacturing components and part-assembling vehicles. The map shows its headquarters in Wolfsburg, Germany, and its subsidiary companies and the flow of components between manufacturing and assembly plants around the world.

In addition to economies of scale, the advantages of this global system are that labor-intensive work can be done where labor is cheap; raw materials can be processed near their source of supply; and final assembly can be done close to the major markets, keeping freight costs down. It also means that the company is not dependent on a single source of supply for a specific component, thus reducing its vulnerability to industrial troubles and other disturbances.

VW changed the famous Beetle very little from its introduction in the 1930s until the mid-1970s, when increasing competition from other compact cars forced the company to develop other models. Its world car is known in different places as the Rabbit or the Golf.

diversified into automobile production.

Another blow to U.S. dominance came with the steep rise in oil prices from 1973. This meant that the large "gas guzzlers" typical of U.S. manufacturers became less competitive than the smaller models developed by firms in Europe and Japan. American companies found it especially difficult to compete with the advanced design, lower cost, and efficient and flexible production methods introduced by the Japanese.

Today, the U.S.-owned companies General Motors and Ford remain world leaders, but they face increasingly powerful rivals. Japanese firms such as Nissan, Honda, and Toyota have opened factories throughout Europe and in the U.S., partly to get around current or future import quotas. European car producers, like Volkswagen and Renault, now have interests in countries such as Argentina and Mexico, while independent challenges are coming from newcomers like Hyundai of South Korea.

MOTOR CITIES AROUND THE WORLD

Toyota City is one of the most highly developed motor cities. Twelve major plants cluster within a few miles of each other in a huge industrial complex 150 miles (250 km) west of Tokyo. Thousands of workers start the day with group calisthenics followed by the company song. It was Toyota that perfected the "just-in-time" system of delivery: parts arrive exactly when and where they are needed.

Flint, Michigan – "Buick City" – was a prosperous one-industry town until the 1980s, when General Motors closed its oldest plants and moved some of its work to Mexico. Half of Flint's population lost their jobs and the rate of violent crime became the highest in the U.S.

General Motors then built a new factory in Flint, half the size of the old one. Whereas in the old factory a fender was handled 22 times and traveled 8,000 ft (2,400 m) from the stamping plant to final assembly, in the new layout fenders are handled 6 times and travel only 140 ft (43 m). To ensure prompt delivery, most suppliers are located within a 100-mile (160-km) radius, and are linked via a computer network to an automatic ordering and delivery system.

In contrast, the Lada manufactured at Togliattigrad looks very like a 1960s Fiat. This is because it is made on old Turin assembly lines, sold lock, stock, and barrel to the USSR when they become obsolete for the European market.

The photograph shows the South Korean plant of Hyundai – one of the fastest-growing automobile companies with 40,000 employees.

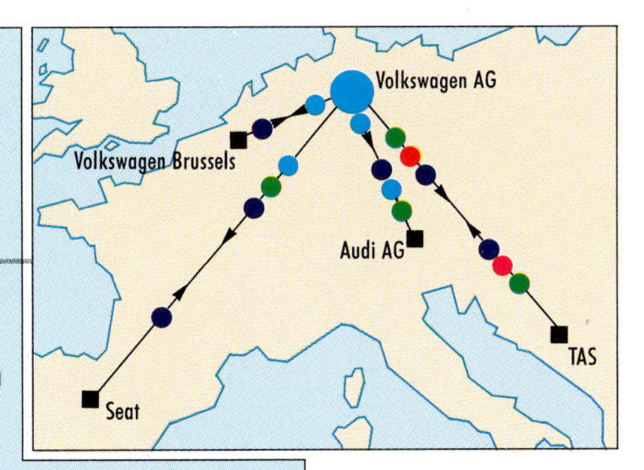

The Eurocar

In 1976, Ford introduced the Fiesta, designed for the European market and using parts manufactured in 12 different locations. Carburetors were supplied from Northern Ireland, transmissions from France, and the cars were assembled in West Germany, Spain, or England. A computer network ensured that the supply of components and cars met the demand in each country, and any shortfalls could be met by switching production between plants.

The Ford Escort followed, with components derived from no fewer than 15 countries, while General Motors adopted much the same strategy with its range of world cars, the Cavalier, Astra, and Nova.

Although the Astra is assembled in Britain, components come from as far away as Australia. Engines for many of GM's factories come from Brazil, while both GM and Ford have set up plants in Mexico, South Korea, and Taiwan.

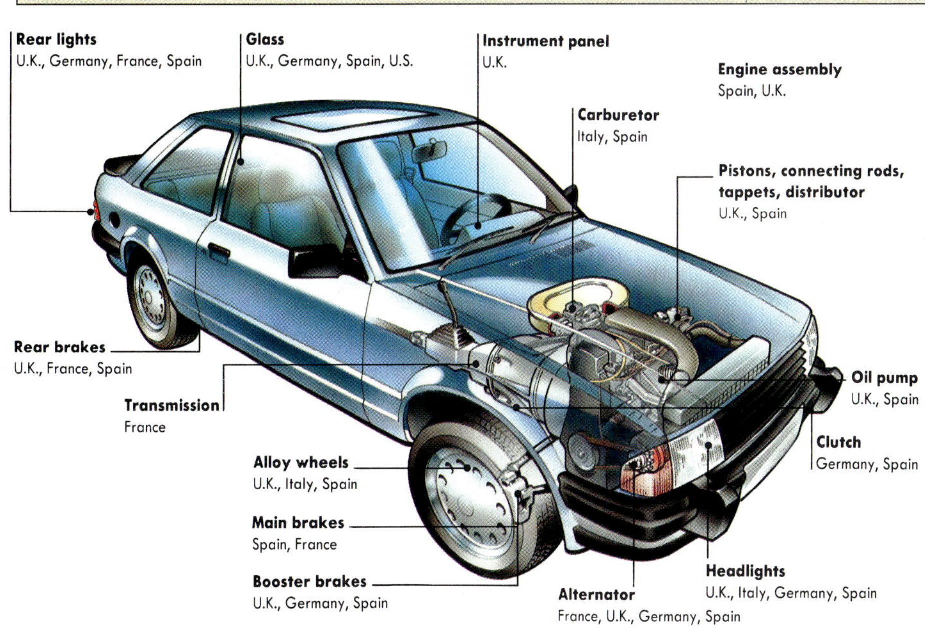

EXPORT PROCESSING ZONES

Cheap labor in the third world

An export processing zone (EPZ) is like a country within a country. Small, sharply defined industrial estates, EPZs are set up by third world governments to attract foreign transnational corporations. Low production costs encourage the companies to relocate their assembly plants there, but the raw materials and components are almost entirely imported. The finished goods – typically consumer electronics, clothing, and automobiles – are largely exported again.

What attracts transnational corporations to such sites? The principal lure is the promise of generous concessions. Host governments provide not only basic infrastructure – land, factory buildings, power, water, and transport – at cheap rates, but also a range of financial incentives including grants, tax breaks, and the removal of duties on imports and exports. Workplace health and safety standards, and pollution controls, are typically less rigorous here. Restrictions on workers' rights, coupled with an abundant supply of cheap – chiefly female – labor, add to the profitability of setting up plants within the EPZs.

The history of EPZs goes back to 1956, when the Irish government, faced with a flagging economy, designated Shannon Airport as a "free trade zone." This idea was subsequently taken up and modified by many newly industrializing countries, so that today there are about 260 EPZs in over 50 countries, employing more than 1.3 million people. Five countries account for over 60 percent of all employment in EPZs – Hong Kong, South Korea, Singapore, and Taiwan in Asia, and Mexico in Central America. Individual EPZs range in size from a handful of small factories, to giant industrial centers containing large numbers of modern manufacturing plants. Bataan in the Philippines provides work for 23,000 people.

But the long-term social and economic effects of these zones on host nations are not always beneficial. Often surrounded by fences and with armed guards patrolling the perimeter, these enclaves can be oppressive places indeed. In some, production processes verge on the perilous. Deteriorating eyesight is common in microelectronics plants, where workers have to attach hair-thin wires to integrated circuits.

Few if any components are obtained locally, so there is virtually no direct spin-off for the home economy. Moreover, since the work is predominantly unskilled, there is relatively little "trickle down" transfer of technical know-how. And because demand for products comes from abroad, third world countries in which EPZs are located become even more dependent on the vagaries of the international market. Downturns in the U.S. economy, for instance, have very serious consequences for Mexico's industries.

Many newly industrializing countries (NICs) have embarked on ambitious programs of export-led industrialization. Governments have often attempted to attract foreign investors by creating tax-free export processing zones (EPZs). The map indicates areas in the world where EPZs are found and where high-tech zones are situated. Puerto Rico, although not formally an EPZ, has an arrangement with the U.S. regarding manufacturing.

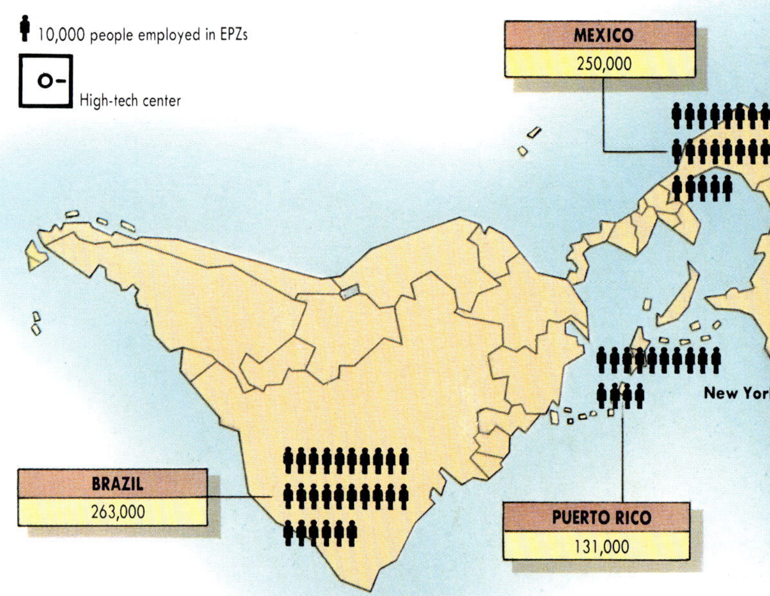

Seventy-five percent of the people working in EPZs are single women aged between 15 and 25. Many employers consider women to be more dexterous than men in handling delicate assembly work. But women are also less liable to strike and are often more willing to put up with repetitive and even hazardous work in unhealthy conditions. Perhaps more important still, women are paid lower wages than men for the same job.

In Mexico, many transnational companies have established assembly plants, known as *maquiladoras,* in EPZs just over the border from the U.S. There are now at least 850 *maquiladoras* in Mexico, employing more than 250,000 people. The companies take advantage of generous financial incentives as well as cheap labor. In 1987, the Mexican minimum wage was less than 40 cents an hour.

Vital components – silicon chips, say – are manufactured in high-tech plants in the U.S., but shipped across the border into Mexico for the relatively simple but labor-intensive job of assembly into finished products. These are then shipped back for sale in the U.S. market by the parent company.

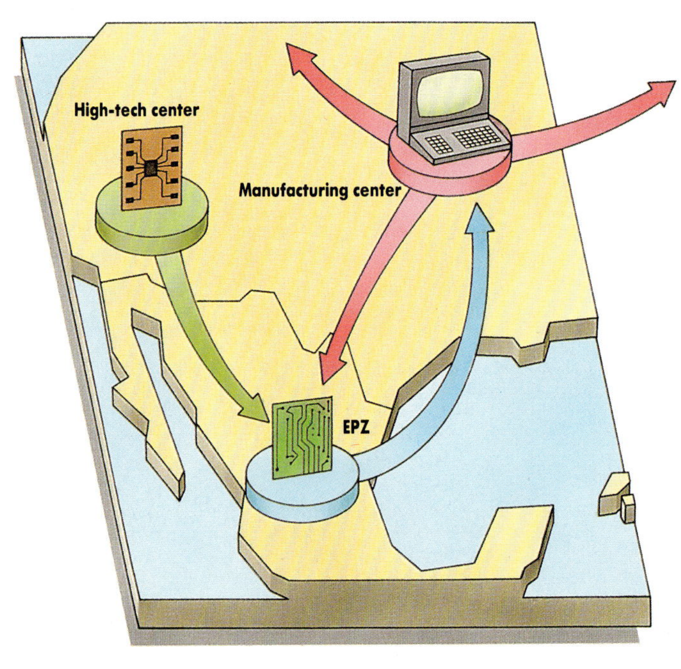

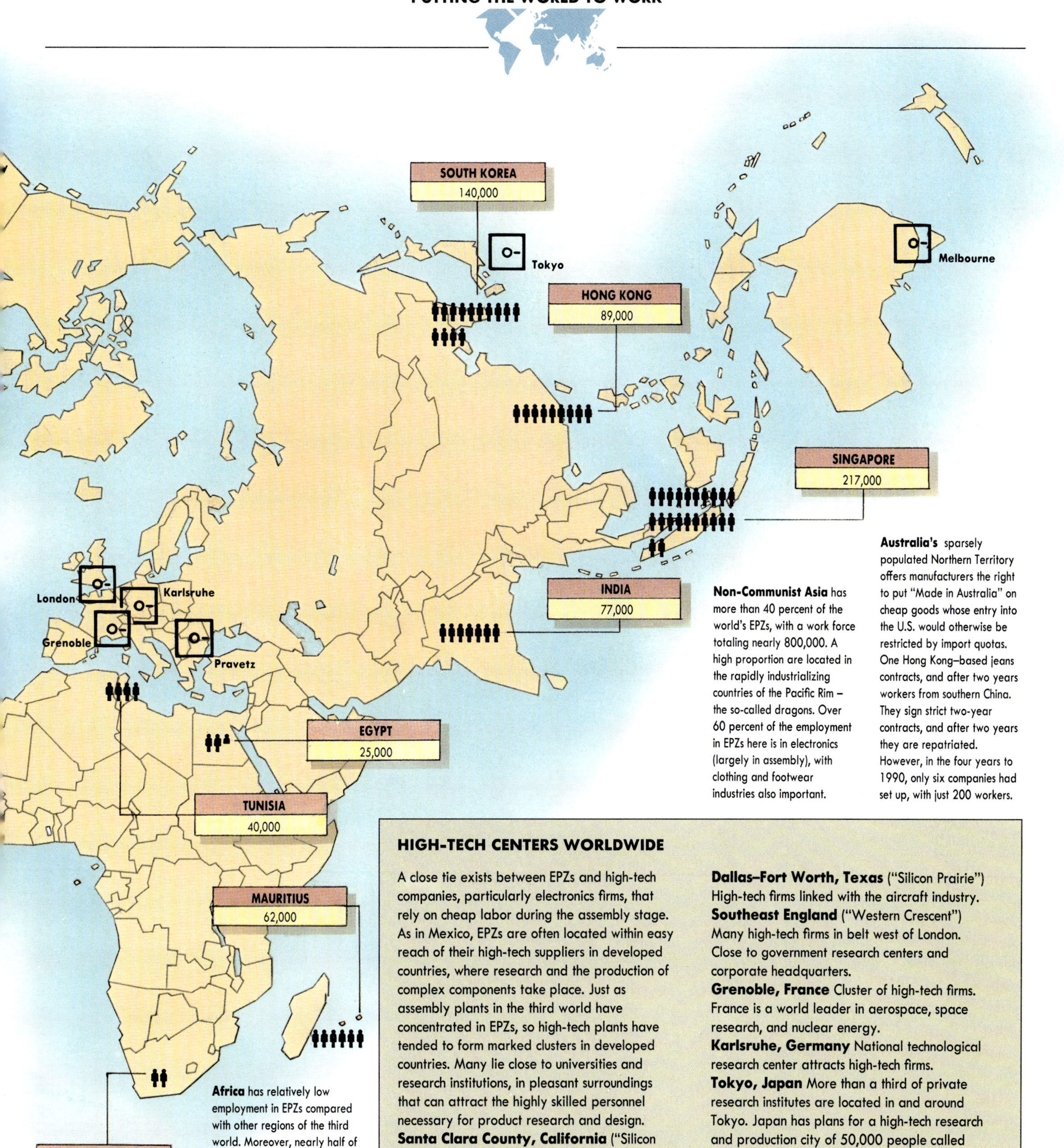

Non-Communist Asia has more than 40 percent of the world's EPZs, with a work force totaling nearly 800,000. A high proportion are located in the rapidly industrializing countries of the Pacific Rim – the so-called dragons. Over 60 percent of the employment in EPZs here is in electronics (largely in assembly), with clothing and footwear industries also important.

Australia's sparsely populated Northern Territory offers manufacturers the right to put "Made in Australia" on cheap goods whose entry into the U.S. would otherwise be restricted by import quotas. One Hong Kong–based jeans contracts, and after two years workers from southern China. They sign strict two-year contracts, and after two years they are repatriated. However, in the four years to 1990, only six companies had set up, with just 200 workers.

Africa has relatively low employment in EPZs compared with other regions of the third world. Moreover, nearly half of its total EPZ work force of 130,000 is concentrated in Mauritius, where this type of labor accounts for more than 50 percent of employment in the manufacturing industry. But the number of EPZs on the mainland is set to rise, with more than 50 coming on line in Morocco and Tunisia.

HIGH-TECH CENTERS WORLDWIDE

A close tie exists between EPZs and high-tech companies, particularly electronics firms, that rely on cheap labor during the assembly stage. As in Mexico, EPZs are often located within easy reach of their high-tech suppliers in developed countries, where research and the production of complex components take place. Just as assembly plants in the third world have concentrated in EPZs, so high-tech plants have tended to form marked clusters in developed countries. Many lie close to universities and research institutions, in pleasant surroundings that can attract the highly skilled personnel necessary for product research and design.

Santa Clara County, California ("Silicon Valley") Original center for advanced computer and telecommunications technology. By 1980 the world's most intensive complex of high-tech industry. Close to Stanford University and military bases.

Greater Boston, Massachusetts Complex of high-tech firms near Route 128. Close to Harvard University and Massachusetts Institute of Technology.

Dallas–Fort Worth, Texas ("Silicon Prairie") High-tech firms linked with the aircraft industry.

Southeast England ("Western Crescent") Many high-tech firms in belt west of London. Close to government research centers and corporate headquarters.

Grenoble, France Cluster of high-tech firms. France is a world leader in aerospace, space research, and nuclear energy.

Karlsruhe, Germany National technological research center attracts high-tech firms.

Tokyo, Japan More than a third of private research institutes are located in and around Tokyo. Japan has plans for a high-tech research and production city of 50,000 people called "Technopolis."

Melbourne, Australia Center for over 90 percent of the country's private high-tech industry.

Pravetz, Bulgaria Eastern Europe's main computer production center mostly builds U.S.-style machines under license. Chosen because it was former Communist Party boss Zhivkov's birthplace, and close to Sofia University.

THE WORLD'S BREADBASKETS

Global trade in staple foods

More than half of the world's people rely for their basic food needs on the grain of just three grasses – wheat, maize, and rice. Although these and other cereals are mainly produced for home consumption, surpluses are a vital part of world trade flows. Wheat, much the most important grain on the international markets, is now second only to oil in terms of value.

As far as exports are concerned, cereals can be divided into two groups: those bought and sold primarily for human consumption, like wheat and rice; and those used chiefly as livestock feed, such as barley, oats, maize, millet, and sorghum – the so-called coarse grains. While poorer nations tend to import mainly wheat and rice to feed their people, richer countries buy more coarse grains to feed to their animals.

Grain exports are dominated by five producers. The U.S., Canada, Argentina, the European Community (mainly France), and Australia grow about 90 percent of the world's wheat and coarse grain exports. Distribution is even more concentrated. Trade is in the hands of just five privately owned corporations. Cargill, Continental, Bunge and Born, Louis Dreyfus, and André Garnac control 85 percent of U.S. grain exports, 80 percent of Argentina's wheat exports, 90 percent of Australia's sorghum, and 90 percent of all wheat and corn from the European Community (EC).

With the exception of Cargill, which began in the U.S., all started as trading houses in 19th-century Europe. None is a household name, yet today they are international giants with interests in banks, railroads, shipping, and insurance, as well as in agricultural produce.

Of the big five exporters, the U.S. is far and away the greatest. Although other producers have increased their share of the export market – notably the EC, which has changed from being a net importer of grain to a net exporter – the U.S. is rightly called the world's breadbasket. It not only accounts for a third of all wheat exports and half of the world's trade in coarse grains, it is also the world's greatest exporter of rice. The U.S.'s chief customer for grain is the USSR. Once a leading exporter of cereals, the Soviet Union is in most years the world's largest importer.

Other major grain importers are the Asian countries on the Pacific Rim. China mainly imports grain to feed its enormous population, Japan to fatten its livestock. With rapidly growing populations and high rates of economic growth, South Korea, Indonesia, and the Philippines may soon overtake the USSR as the biggest market for the world's grain. On the whole, those African countries that are desperate to feed their populations are not big importers of the surplus grain – they cannot afford it.

Since the 1960s the world output of grains has risen faster than population growth. Thanks to new hybrids, the use of chemical fertilizers and pesticides, and the introduction of new machinery, the world now produces enough to feed everyone. Total grain available per head rose by nearly a third. Indeed, by the mid-1980s, the world seemed to be awash with grain.

Although millions in countries such as Ethiopia and Chad may have starved for lack of hard currency, elsewhere cattle feed became so cheap that U.S. burger companies, for example, could expand worldwide. But U.S. farmers became the victims of their success. Crop gluts led to low prices and bankruptcy for many. Land started to come out of production.

During the late 1980s, however, the trend reversed. Widespread droughts, succeeded by poor harvests, caused the world's total grain reserves to dip below 60 days – a critical level – and population growth began to reduce average grain weight per head. In the future, a run of bad years could plunge the world into crisis.

Changing land use

Land has always been used for food production, but the ways in which it is divided and cultivated have changed as demands have grown and technologies have developed.

3000 BC Most of southern England is thick forest with Neolithic settlements scattered in small clearings. Circular fields of around 1 acre (0.4 hectares) yield about 220 lb per acre (250 kg per hectare).

AD 1300 The medieval landscape is much more open. Crops are grown in narrow strips within great open fields. Each plowman and his whip-carrying assistant use a team of up to eight oxen to drag a heavy plow through long and thin fields. It takes a full working day of about 13 hours to plow the customary 220-yd by 22-yd (200-m by 20-m) strip.

1900 Much of the woods and common land have disappeared. The old strip fields have long been consolidated into single holdings, with hedgerows along their boundaries. Fields vary in size from a few acres up to 40 acres (16 hectares). Horses have generally replaced oxen and just two can be driven by one man to plow the standard acre in a day. Yields are around 1,650 lb per acre (1,870 kg per hectare).

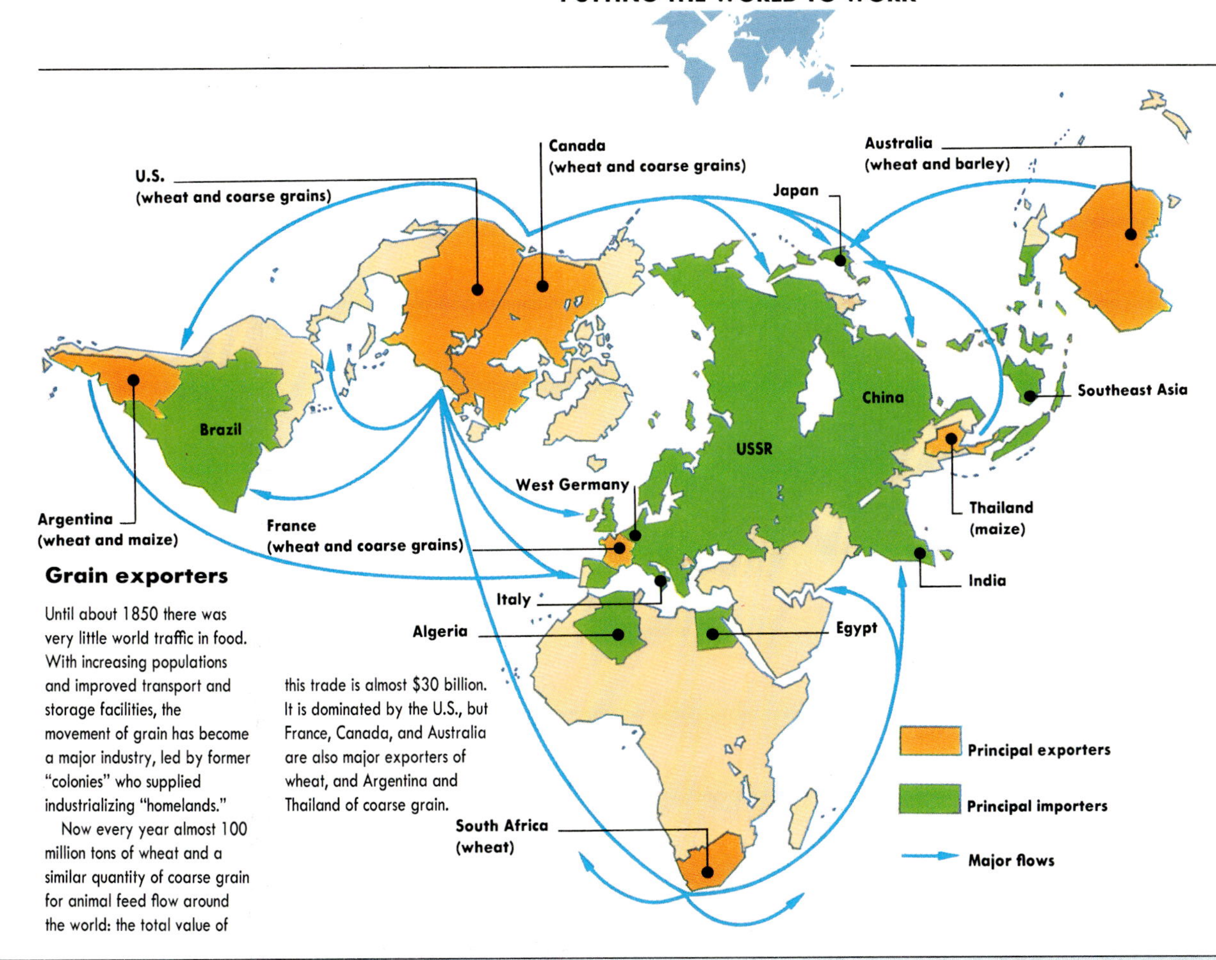

Grain exporters

Until about 1850 there was very little world traffic in food. With increasing populations and improved transport and storage facilities, the movement of grain has become a major industry, led by former "colonies" who supplied industrializing "homelands."

Now every year almost 100 million tons of wheat and a similar quantity of coarse grain for animal feed flow around the world: the total value of this trade is almost $30 billion. It is dominated by the U.S., but France, Canada, and Australia are also major exporters of wheat, and Argentina and Thailand of coarse grain.

Grain importers

The needs of grain importers fluctuate enormously. Droughts in 1988 meant that Mexico had to import much more wheat than the year before, while Brazil, usually a major importer, had a good harvest and import demand fell.

North Africa, particularly Egypt and Algeria, accounts for 10 percent of the world's imports in a bad year, but this can fall by as much as 1 percent – a difference of $300 million worth of wheat – if conditions are favorable for the domestic crop.

The two largest importers are China and the USSR, although ironically both produce more grain than the U.S. As incomes rise, meat becomes a more important feature of the diet, which means that grain is needed for animal foodstuffs as well as to feed the population.

The map shows the main importers and exporters as well as the principal flows of grain around the world.

1950 The rural landscape has changed little since 1900, but there has been a dramatic change in farming practices.

Horsepower has been replaced by tractor power, greatly increasing the area that can be covered in a single day. A small tractor with a double plowshare can plow an acre (0.4 hectares) in three or four hours. Average wheat yields are now around 1 ton per acre (2.5 tonnes per hectare).

1980 Farming has become "agribusiness." Farms of even several hundred acres can be plowed, harrowed, and sprayed by just one worker.

Tractors with 8–12 plowshares can plow an acre (0.4 hectares) in less than 40 minutes. Hedgerows are removed to create giant open fields where as much land as possible is put under crops. The use of agrochemicals and plant hybridization has doubled 1950 yields – but at the expense of scenery and wildlife.

2000 Owing to high yields and the influx of relatively cheap imported grain, much of the farming land in the south of England is now coming out of production. Some is beginning to revert to scrub, while new uses are being found for the remaining land. These include golf courses, out-of-town shopping centers, and commuter villages. Millions of city dwellers are seeking to return to their "rural roots."

CREATING LAND FROM SEA

The reclamation of Holland

"GOD CREATED THE WORLD, BUT THE Dutch made the Netherlands." By rights, much of the Netherlands (the name means "low lands") should not exist. For 2,000 years the Dutch have waged a territorial war against a formidable enemy – the sea. In some provinces half the land area has been won from the water. Overall, one in four people live below sea level.

When the Romans arrived 2,000 years ago, they were amazed to see people living on heaped-up earth mounds, or *terpen*, that became islands at high tide. Thus began centuries of reclamation. During the Middle Ages, earth dikes were built to hold back the sea, and windmills – in fact, windpumps, an idea brought back from Asia by returning Crusaders – were used to drain water from the many shallow lakes, bogs, and river margins. Nevertheless, once or twice every century a "superstorm" in the North Sea would coincide with high spring tides, overcome all defenses, and wash away up to 100,000 people.

By 1950, the Netherlands was reasonably secure behind dikes of concrete and steel, much of the land pumped dry by steam and diesel engines. Yet despite their efforts, the Dutch were still vulnerable. In 1953 a disastrous flood killed 1,800 people, made half a million homeless, and destroyed huge areas of farmland.

The Dutch have also wrested land from the sea simply to feed and house themselves – theirs is one of Europe's most densely packed populations. Reclamation continues today, albeit at a modest pace – about 12 square miles (30 square km) a year – with an emphasis on the controlled use of plant succession to restore land. After dikes, canals, and pumps have been installed, pioneer plants are sown. These species can survive in damp mud. As they grow they take moisture from the ground and help to bind the soil together. Within a few years, the soil is sufficiently dry for crops to be planted or cattle put out to graze.

The Dutch make the most of their limited space. From the 17th century Dutch growers have led the world in the intensification of agriculture. Whereas France and England were slow to accept the delights of the potato, the Netherlands was an early convert. Potatoes can support twice as many people over a given area as wheat.

While the Dutch were establishing their trading empire from the 16th century onward, they were also scouring the world for edible, medicinal, and ornamental plants. These were often brought back to their many fine botanic gardens, such as those in Leyden and Delft, for propagation and subsequent redistribution around the world.

It was from a single *arabica* seedling raised in Amsterdam in 1706 that South America's coffee plantations were started. The Dutch also introduced the tulip to Europe from Turkey. As early as the 1630s, immense areas were given

Seen from a satellite, the Netherlands still seems to be mostly water (*right*). Yet since the 12th century, the Dutch have reclaimed more than 3,000 square miles (7,750 square km) of land. (Reclaimed land or "polders" have distinctive field patterns with large amounts of blue showing through the strips.)

Much of the new land has been won from the Zuiderzee (*below*), the "southern sea" at the heart of the country. Work began with the construction of a barrier dam. The sea is now a freshwater lake, Ijsselmeer. Successive zones have been drained to form polders.

Plans are now being made to round off the country by reclaiming the Waddenzee. The West Frisian Islands will be strung together to form a continuous barrier.

New methods of reclamation are being developed. They make use of natural sand movements and dense-rooted binding plants rather than concrete and steel.

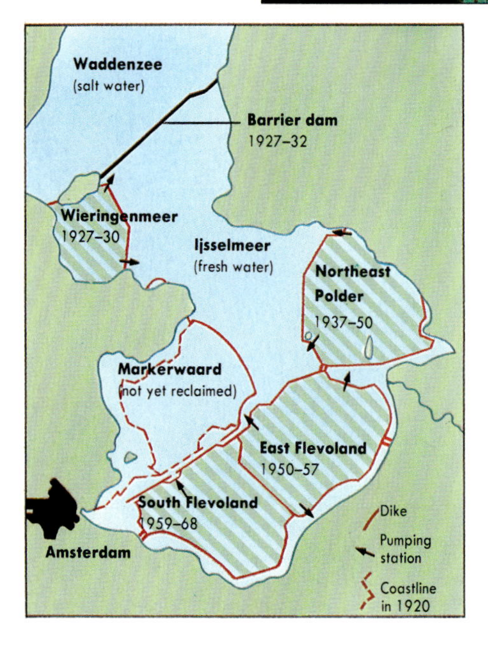

Creating a polder

Year 1 A dike is constructed around the area to be reclaimed, closing it off from the sea. Pumps are installed and canals cut to drain the water. Canals take the water to the sea, where it is pumped out or allowed to drain out naturally by opening sluice gates at low tide.

Years 2 and 3 The water level continues to fall. Mud appears. This is too saline for most plants, but rainfall gradually flushes out the salts. The mud is colonized by hardy plant pioneers that can tolerate the bare new environment. Most of these hardy plants are annuals whose light seeds are carried there naturally on the wind.

over to growing daffodils and tulips, initially as much for their edible bulbs as for their flowers.

This early lead in plant propagation is maintained today, with Dutch companies dominating world trade in exotic species. Their expertise in breeding also extends to farm animals. The black and white Holstein is now the most numerous dairy cow worldwide.

Even today, when its industrial and commercial power allows it to buy food from abroad, the Netherlands has a greater proportion of land that is given over to the production of food than any other western European country.

Most of the reclaimed land in the Netherlands is used for intensive crop and dairy farming, including horticulture under glass. But about 25 percent of the land has been set aside for housing, roads, and open spaces.

Until recently, villages were built in linear fashion along raised roads or canals. Now settlements are clustered at crossroads which makes for more efficient delivery of basic services such as fresh water, drainage, and electricity.

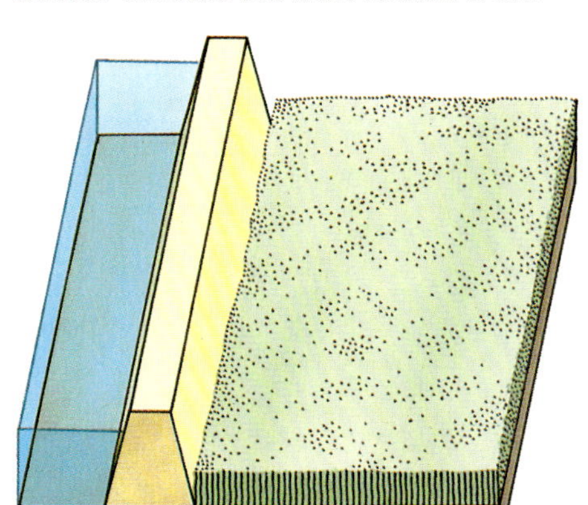

Years 4 to 6 A net or mattress of woven willow twigs is laid across the surface, and seeds of perennial rushes and reeds are sown by aircraft. The plants' underground stems and runners bind the soil and throw up new plants. They create a dense cover that draws up great quantities of water through the roots, causing the mud to dry out still further.

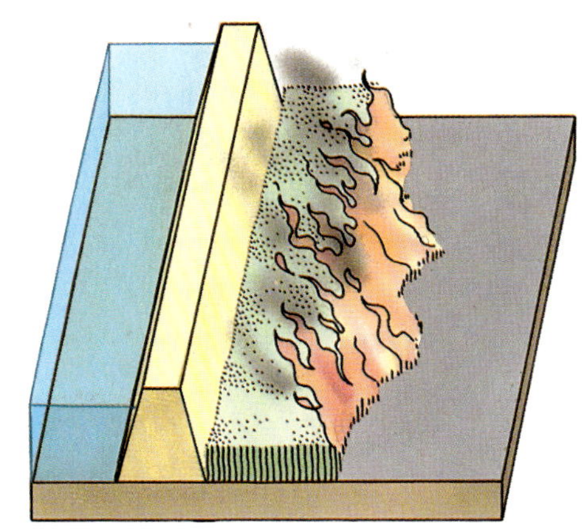

Year 7 After three years, the reedbeds are burned. The mud is now firm enough to walk on, and heavy plows are used to dig the fertile ash into the soil. Plowing dices the reeds' myriad strong roots and turns the rich, heavy earth. Plastic drainage pipes laid in the soil continue to drain groundwater away into ditches.

The polder is ready to grow crops. The first crop is usually oilseed rape. Later, wheat, root crops, and vegetables may be planted, or dairy cattle put out to graze. Within 15 years, land that may once have been 13 ft (4 m) under sea water looks as if it has been farmed forever.

THE GREEN REVOLUTION

Will it solve the world's food problems?

SINCE THE 1950S THE SPECTER OF WORLD hunger has loomed large. Populations have grown rapidly in the third world, while food production has been threatened by the drive to produce cash crops in return for foreign exchange. Major famines have indeed occurred, for example in India, China, and Ethiopia. Yet famine is nowhere near as common, or as severe, as was predicted. This is largely due to the remarkable achievements of the so-called Green Revolution.

Pioneering work by plant breeders around the world resulted in the 1960s in the development of new high-yield crop varieties (HYVs). At first, research concentrated on wheat, maize, and rice. These grains are easy to store, and it was hoped that production would increase sufficiently to provide a surplus for export.

Plant breeders were particularly interested in dwarf varieties of rice – short-stalked plants that did not waste valuable nutrients building stems and leaves rather than grains, and which were strong enough for the heavier heads not to fall to the ground and rot before they could be harvested. Research also focused on varieties that had such short maturing times that several crops could be produced a year.

Using these new varieties, Asia's rice production increased by 66 percent between 1965 and 1985, keeping well ahead of population growth and making most Asian countries self-sufficient in food. Improvements in Latin America were less spectacular and limited to particular regions, notably Mexico. Africa, however, saw little improvement. Wheat and rice are unsuited to most areas, and although there has been some experimentation with cassava and sorghum, Africa's poor soils and lack of water make prospects for major improvements gloomy.

Even where the new crop strains were successfully introduced, the Green Revolution has not been an unqualified success. For one thing, the concentration on three cereals ignored the contribution to third world diets of other grains, like millet and sorghum, or root crops, vegetables, legumes, fruits, and nuts. Nor was much attention paid to taste – some of the new hybrids proved to be less palatable than existing varieties.

There were other problems, too. Increased yields from the new crops depended on ample supplies of fertilizers. The delicate new varieties were also more vulnerable to diseases and pests, demanding the protection of pesticides. Petroleum-based agrochemicals, sometimes used without adequate instruction, damaged both the environment and the health of the farmers. They also became considerably more expensive after the oil crises of the 1970s.

Since the 1940s, grain production in many parts of the third world has risen in spectacular fashion, outstripping population growth. From 1944 to 1967, Mexico's maize output doubled and its wheat production trebled. India doubled its total cereal crop between 1950 and 1970.

Much of this increase has been achieved through new high-yield varieties developed at research centers like the International Maize and Wheat Improvement Center in Mexico and the International Rice Research Institute in the Philippines.

The new strains dramatically increased the grain harvest from the same area of land. The yield from one strain of rice developed in the 1960s – "miracle rice" – was so outstanding that for a while it seemed as if world hunger would be completely eradicated.

Countries whose yield has doubled since introduction of HYVs
Countries whose yield has increased by 50 percent since introduction of HYVs
Research Station
MEXICO
COLOMBIA
COSTA RICA
URUGUAY
Potato (Peru)
Tropical agriculture (Colombia)

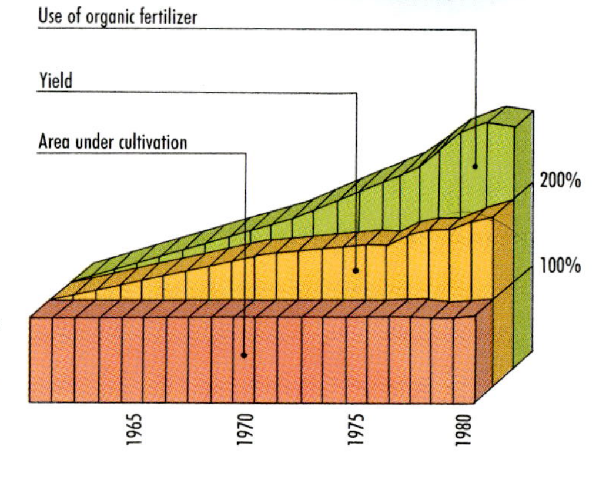

To achieve their full potential, the new high-yield varieties need a plentiful supply of nutrients – usually more than the soil can provide by natural means. Increased grain production has therefore relied on a huge increase in the global use of artificial fertilizers, which are expensive.

The diagram shows the relationship worldwide between use of fertilizers and increased yield.

Pests and diseases spread more quickly among the newly introduced crops, which were less resistant than local varieties, and grown over larger areas. Once a pest got a foothold, it appeared everywhere. This has led to a constant demand for more hybrids. The miracle rice IR8 was bred in 1966 in the Philippines; by 1968 IR8 had become vulnerable to bacterial blight and rice tungro virus. In 1971 it was replaced by IR20. By 1973 this, too, was devastated by brown plant hopper (BPH) and grassy stunt virus, and IR26 was substituted. When this succumbed to a new version of BPH in 1976, IR36 took over. IR36 is now threatened by a new variety of BPH and two new diseases.

The Green Revolution has been accused of magnifying social inequalities, causing rural landlessness and migration to the cities. The costs of new seed and chemicals, added to the expense of introducing necessary irrigation equipment and water control systems, typically put richer farmers in a better position to benefit from the improvements.

Dependence on water control systems also meant that some regions prospered more than others. The systems were widely adopted in the Philippines, Indonesia, Malaysia, and India, but were expensive to develop in Thailand, Burma, and Bangladesh. At the national level, the cost of these projects, together with the need to pay for imported fertilizers and pesticides, increased the pressure to grow cash crops for export, which in turn meant less food for the locals.

But the shortcomings of the HYVs have been recognized. Research stations are developing new varieties, particularly more drought- and pest-resistant ones, but more emphasis is now being placed on programs that are socially as well as economically beneficial. Although crop yields are not increasing as fast as in early years, progress is being maintained. So far, the Green Revolution is buying the world a little time.

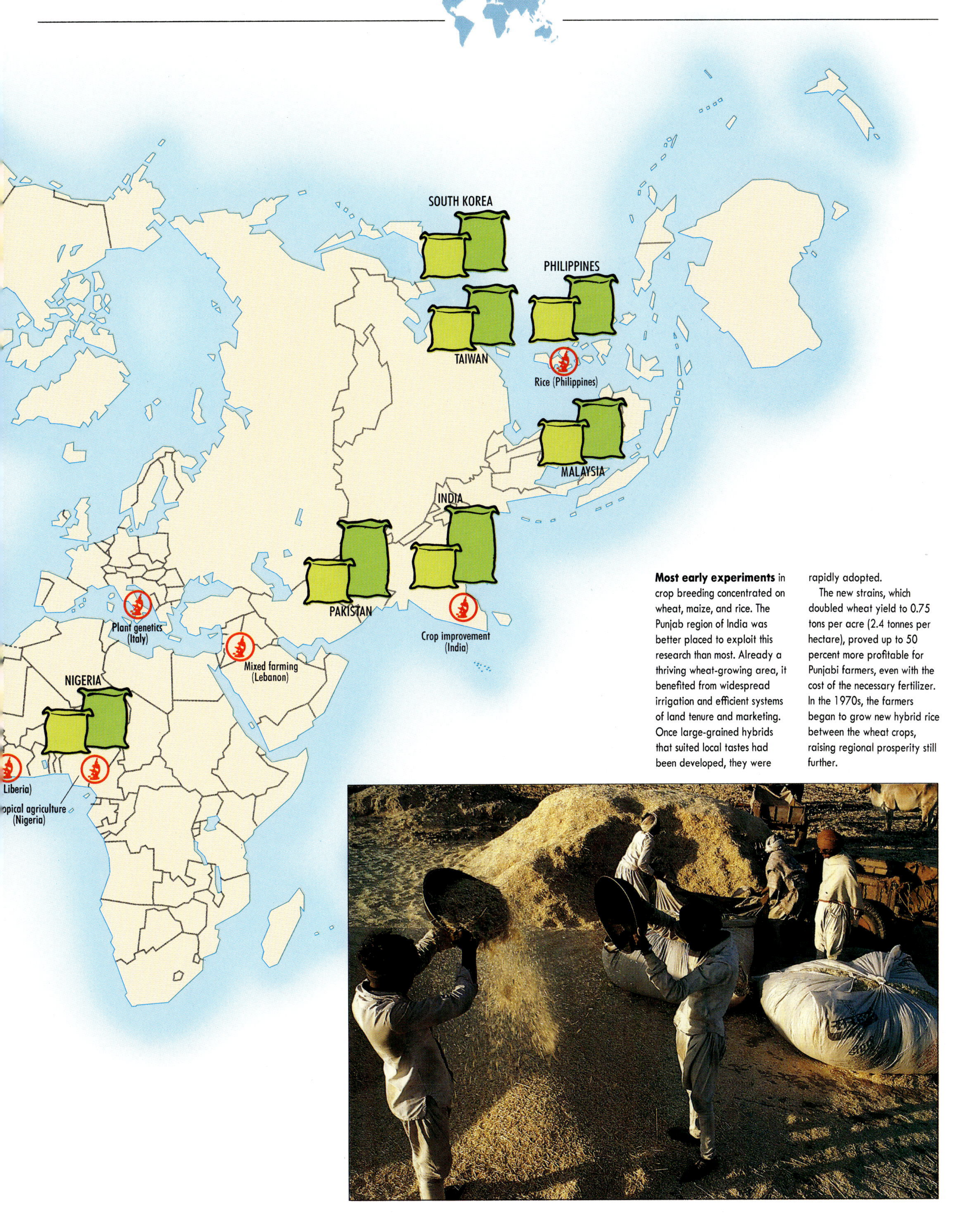

Most early experiments in crop breeding concentrated on wheat, maize, and rice. The Punjab region of India was better placed to exploit this research than most. Already a thriving wheat-growing area, it benefited from widespread irrigation and efficient systems of land tenure and marketing. Once large-grained hybrids that suited local tastes had been developed, they were rapidly adopted.

The new strains, which doubled wheat yield to 0.75 tons per acre (2.4 tonnes per hectare), proved up to 50 percent more profitable for Punjabi farmers, even with the cost of the necessary fertilizer. In the 1970s, the farmers began to grow new hybrid rice between the wheat crops, raising regional prosperity still further.

THE POWERFUL THIRST

A taste for tea, coffee, and cocoa

TODAY'S GLOBAL PATTERN OF TEA, coffee, and cocoa cultivation owes almost everything to European colonial expansion. Sri Lanka is the world's major exporter of tea, Brazil of coffee, and the Ivory Coast of cocoa, but the crops were not grown in any of these countries prior to colonization.

In ancient China tea was first used as a medicine. By the 5th century tea drinking had become a social custom there, but it was not until the beginning of the 17th century that it reached Europe. The trade in tea was originally controlled by Chinese merchants who would only accept silver and copper in exchange for the leaf. The increasing demand for tea began to drain Britain of its silver, undermining the country's balance of payments.

Britain's initial solution to the problem was to exploit China's addiction to opium. The British East India Company, which controlled trade between Britain and Asia, began cultivating Persian opium poppies in Bengal and forced Chinese merchants to trade in tea if they wished to ensure a supply of the drug.

Meanwhile, in the 1830s, the Assam Company had begun to establish tea plantations in Assam and Bengal, using seeds, plants, tools, and skilled workers smuggled in from China. Plantations were carved out of the forest, and contract workers brought from all over India were set to work in them under conditions little better than slavery. Between 1900 and 1914, Assam was supplying one-half of the tea market. Eventually, however, labor unrest provoked by the miserable working conditions forced planters to go elsewhere, many to Ceylon (now Sri Lanka) and East Africa.

Once India and Sri Lanka gained independence (in 1947 and 1948), British planters had to sell their estates, and many moved to Kenya, attracted by low taxes. Although the tea produced there is often of lower quality, it is used as a "filler" in some packaged teas and tea bags. Through aggressive advertising, these have captured a large section of the market.

In recent years, many tea plantations have been nationalized, yet the tea trade is still dominated by the major companies who handle the packing and marketing. The Anglo-Dutch transnational Unilever, through its subsidiaries Brooke Bond and Lipton, controls 35 percent of the world market.

The tea trade has come full circle. Tea plants were taken first to India to undercut production in China, then to Sri Lanka. Both these countries were then undercut by production in East Africa. East Africa now faces competition from China. Desperate for hard currency, the Chinese government has begun to promote the cultivation and export of cheap black tea as a filler in tea bags.

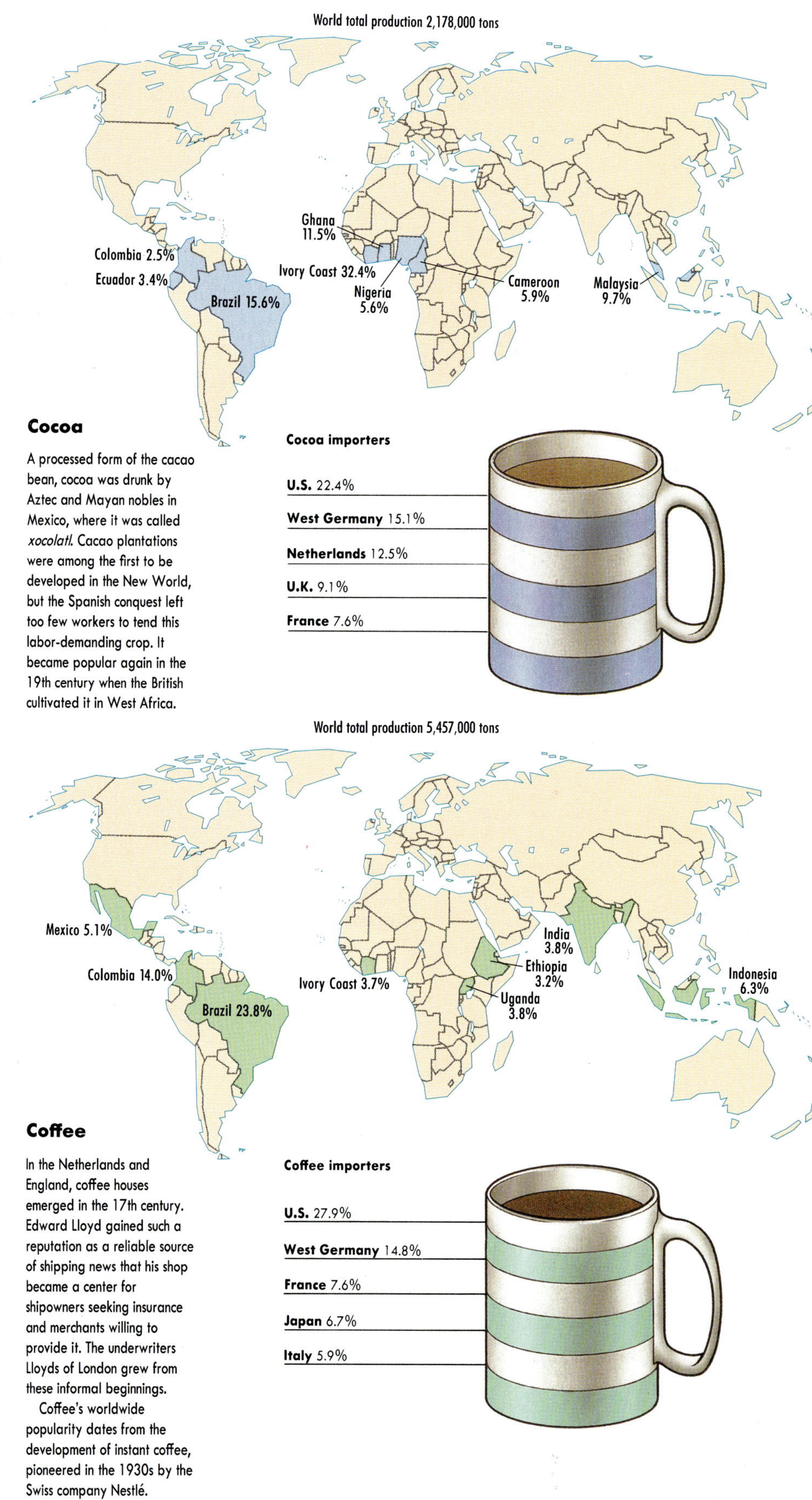

Cocoa

A processed form of the cacao bean, cocoa was drunk by Aztec and Mayan nobles in Mexico, where it was called *xocolatl*. Cacao plantations were among the first to be developed in the New World, but the Spanish conquest left too few workers to tend this labor-demanding crop. It became popular again in the 19th century when the British cultivated it in West Africa.

Coffee

In the Netherlands and England, coffee houses emerged in the 17th century. Edward Lloyd gained such a reputation as a reliable source of shipping news that his shop became a center for shipowners seeking insurance and merchants willing to provide it. The underwriters Lloyds of London grew from these informal beginnings.

Coffee's worldwide popularity dates from the development of instant coffee, pioneered in the 1930s by the Swiss company Nestlé.

Sri Lanka, now the world's largest tea exporter, was a latecomer to the trade. After Ceylon (as Sri Lanka was then called) was devastated by coffee blight in the 1870s, British planters bought up bankrupt plantations at bargain prices and began to cultivate tea.

Huge profits were made from the labor of Tamil Hindus recruited from drought-stricken southern India to work on the plantations. In part a legacy from colonial rule, the Tamils remain in Sri Lanka today, an exploited and discriminated-against religious minority.

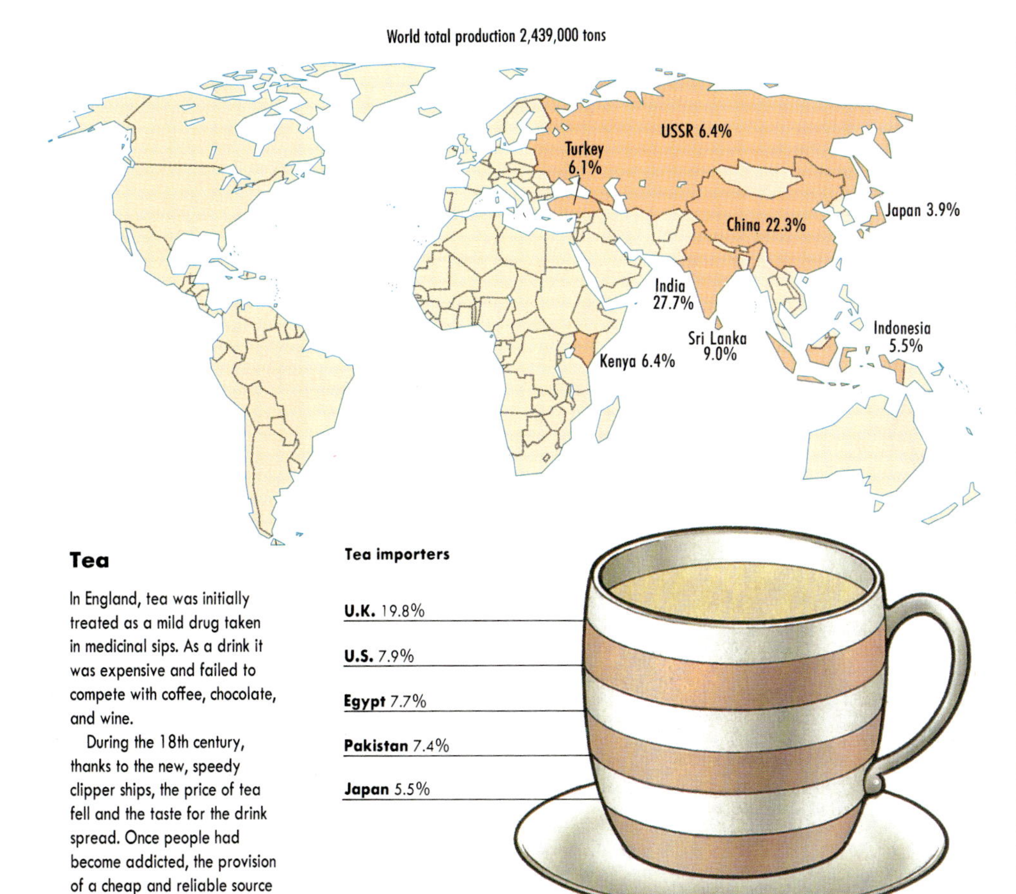

Tea

In England, tea was initially treated as a mild drug taken in medicinal sips. As a drink it was expensive and failed to compete with coffee, chocolate, and wine.

During the 18th century, thanks to the new, speedy clipper ships, the price of tea fell and the taste for the drink spread. Once people had become addicted, the provision of a cheap and reliable source of tea became a central aim of British colonial policy.

COLONIAL LEGACY

Many well-known tea brands take their names from British families involved in the Ceylon tea trade – names such as Lipton, Brooke Bond, Tetleys, and Lyons. Thomas Lipton, a Glasgow grocer, was the first to control every stage of the business from growing to marketing. Using evocative advertising slogans like "From the tea garden to the teapot," and exotic but meaningless labels such as "Orange Pekoe," Lipton's products became so popular he was able to cut the retail price of tea by almost a third and still make a fortune.

By 1914, the British had monopolized the world tea industry. All the early tea specialists were eventually bought out by larger, multinational companies, notably Unilever.

GROWING FOR GOLD

Big business in exotic foods

FIFTY YEARS AGO, PRACTICALLY THE only tropical fruit that was available fresh in the temperate latitudes was the banana. It was the perfect export crop. Cut when still green, it continued to ripen slowly on the long boat ride north from the Caribbean or West Africa.

No longer. If your fresh fruit salad recipe demands a mix of guavas, mangoes, and passion fruit, you can probably get them in your local supermarket. Thanks to revolutions in food production and transportation systems, it is possible to buy practically any fruit, vegetable, or seafood species, no matter how exotic, at any time of the year, even outside major urban centers. And it is not only tropical produce that is now available year-round. Once luxurious or seasonal crops – like strawberries and asparagus – are also often on the shelves. What is more, they scarcely count as luxuries at all; their price relative to average income has plummeted.

Today, exotic and out-of-season crops mainly come from Southeast Asia. Countries such as Taiwan, South Korea, the Philippines, Indonesia, and Thailand have not only succeeded in feeding their own rapidly growing numbers, they have also developed profitable lines in upscale food exports.

Until the 1970s, Thailand exported little other than rice, rubber, and tin. Then it became the world's biggest exporter of canned tuna and pineapple (it supplies 50 percent of U.S. imports). From this flourishing base, companies expanded into a wide range of canned and frozen fruits, vegetables, and meats. Now, thanks to cheap air freight rates, Thailand can transport fresh produce from plant to plate in less than 24 hours. Crops such as tomatoes, mushrooms, and baby corn – never previously grown in the region and not part of the traditional cuisine – are booming. Seafood is a specialty – particularly shrimp, crabs, and clams. Such a lavish product mix generates large profits

THE KIWI CONQUEST

In New Zealand, the kiwi fruit has made more millionaires than any other enterprise. Energetic marketing and a clever name switch (it started life as the Chinese gooseberry) have helped turn it into a major international fruit, on sale throughout the affluent world.

The small, wild fruit is native to China, but a handful of seeds were brought to New Zealand by plant hunters in 1904. At first it was cherished for its fine white flowers, but the greater potential of its heavy-cropping, exotic-flavored fruit was eventually recognized. Intensive breeding by selection, pruning, and grafting rapidly improved the stock.

At first, progress in exporting the crop was slow, and it was confined to luxury markets. But around 1980, kiwi fruit began to appear regularly on supermarket shelves throughout western Europe, North America, and Japan. By 1986, New Zealand was shipping more than one billion kiwi fruit to more than 30 countries.

The fruit is ideal for a country whose major markets are on the other side of the world. It crops for three months, but will last for another six if kept chilled. It can also be canned, frozen, juiced, or even turned into low-alcohol wine. Many former dairy farmers, their once guaranteed access to British markets damaged by the formation of the European Community, have turned their land over to the crop. Planted with kiwi fruit, a 100-acre (40-hectare) farm that once yielded $30,000 a year can earn more than $1 million.

But New Zealand growers are now in trouble. Having exported the vines along with the fruit, they forfeited their monopoly of production. Their market share is now declining fast. Rising production in Japan, the U.S., France, Italy, and the Channel Islands presents little threat because the cropping season is at a different time of the year. But strong competition from countries in the southern hemisphere, particularly Chile, which enjoys close links with the U.S. market, spells economic disaster for New Zealand's growers.

Pineapple processing plants are highly labor intensive. Workers remove the skin from the fruit, core out the fibrous stem, and slice the flesh into rings or chunks.

Thailand now leads the world in the production and export of canned pineapples, a plant that originated in South America. In 1987, it exported 255,000 tons with a value of $145 million – more than 34 percent of world trade. Canned or juiced pineapple is more suitable for export because the fresh fruit perishes quickly.

The main fruiting seasons for Thai pineapples are April–June and December–January, but some varieties can supply the factories all year round.

and much-prized foreign exchange.

Most food production is in the hands of transnational agribusinesses. These rely on cheap labor, sophisticated technologies, and extensive marketing systems to create a vertically integrated system linking farm to factory and factory to consumer. But rather than own and run farms themselves, many such companies employ contract farmers and truck farmers. This minimizes production risks and problems of labor relations. They are more likely to own the companies that supply the farms: seed growers and manufacturers of animal feed, fertilizers, pesticides, and farm machinery.

Although most agribusinesses are foreign owned – Japan's multifaceted Mitsubishi Corporation, for example, is heavily involved in pineapple canning and shrimp farming – some are Thai owned. The largest of these, the Charoen Pokphand Group, began as a small seed shop in Bangkok's Chinatown 50 years ago. Starting with the production of chicken feed, it moved into chicken farming for U.S. fast-food chains. Now it owns operations in 11 countries, employs 15,000 workers and 100,000 contract farmers, and has recently diversified into hog and stock raising, beer, petrochemicals, cement, and motorcycles.

The illustration shows the origins of fruit and vegetables on sale in British markets in the course of a year. The growing international trade in food produce means that, for the richer countries of the world, seasonal shortages – indeed seasons themselves – have all but disappeared. As one source of fresh food dwindles, so the produce can be brought in from elsewhere. Expanded global trade has also widened the choice of food on the shelves. Tropical fruits like avocados and mangoes, which used to be rare and expensive, are now freely available.

TRAWLING THE SEAS

International fishing business

AT LEAST TWENTY-FIVE PERCENT OF THE world's diet of animal protein is derived from fish and other seafood. While in the third world a few tiny fish sprinkled into a dish of boiled rice or stewed vegetables can mean the difference between starvation and survival, seafood is increasingly the food of choice for health-conscious Westerners.

Fish are not only economically important when they are destined for human consumption. Around one-third of the world's catch is converted into petfood, hog and chickenfeed, margarines, cooking oils, and agricultural fertilizers. It has been estimated that more than 12 million people worldwide earn their living from fishing – around 2 million of them in Java alone. Demand has been steadily rising. But can the world's supply match that demand?

During the 1950s and 1960s there was considerable optimism that the vast size and seemingly inexhaustible productivity of the waters could transform global food prospects. There was even talk of a Blue Revolution to match the Green Revolution that was occurring in grain production.

Between 1950 and 1970 global fisheries' production trebled from 20.5 million to 60 million tons, a figure well in advance of population growth. Since that time, however, catches have tailed off, and the effort required to maintain them has increased. Annual production per head, which in 1972 had doubled to reach 39 lb (17.7 kg), is steadily falling.

The resources of the waters have turned out to be no less limited than those of the land. For one thing, fish are not uniformly distributed throughout the seas. Around 90 percent of the world's fish are found in shallow water, chiefly in estuaries, and on continental shelves and slopes where the runoff of nutrients from the land assists the growth of phytoplankton, the minute plant life on which the entire marine ecosystem is based. The vast expanses of the ocean are in effect deserts as far as most marine life is concerned.

Water temperature is a critical factor, since less oxygen dissolves in warm water. Most of the world's richest grounds are found where upwellings of cold currents reach the surface, as off the coast of Peru or in the North Atlantic near Greenland. Fish stocks globally are being depleted to a dangerous extent by overfishing and ocean pollution. Estuaries and coastal waters are particularly vulnerable.

In the heavily polluted Chesapeake Bay, on the eastern coast of the U.S., oyster catches halved between 1962 and 1984, while those of the much-prized striped bass plummeted by 90 percent. A similar story could be told of California, Japan, the North Sea, or the Mediterranean. The most accessible waters are the ones that are most easily overfished.

In the last 20 years, some countries have promoted the development of long-distance superfleets able to stay at sea for months on end. Among the most powerful fleets are those of the USSR, Japan, the Koreas, and the U.S. (many of whose vessels operate under the flags of small nations such as Bermuda and Liberia). Having overfished and polluted their own waters, they scour the world's seas – often with little regard for the 200-mile (320-km) Economic Exclusion Zones of other nations.

Even the best-stocked grounds have now exceeded their limits. After 20 years of

Modern commercial fishing fleets use sophisticated navigational, radar, and sonar equipment to locate elusive fish shoals, and deploy a great variety of mechanized nets and hooked lines to secure the fish. Such methods may exhaust stocks of larger fish and go on to deplete the numbers of immature fish and devastate nursery areas. Experts believe that stocks in many areas will never recover their former numbers.

The problem of overfishing is exacerbated by the need to reach international consensus. Fish quotas and dates of close seasons are hard to agree on, and difficult to police.

Until recently, the sea was regarded as a common area open to all. Each country fished to its capacity on the assumption that if it did not, its neighbors surely would. But uncontrolled fishing led to serious disputes over territory, and countries began unilaterally to declare Economic Exclusion Zones around their coasts.

The Law of the Sea, passed at an international conference in 1983, was intended to put an end to these conflicts by giving every coastal nation control of the waters within 200 miles (320 km) of its coastline, yet disputes are increasing in number.

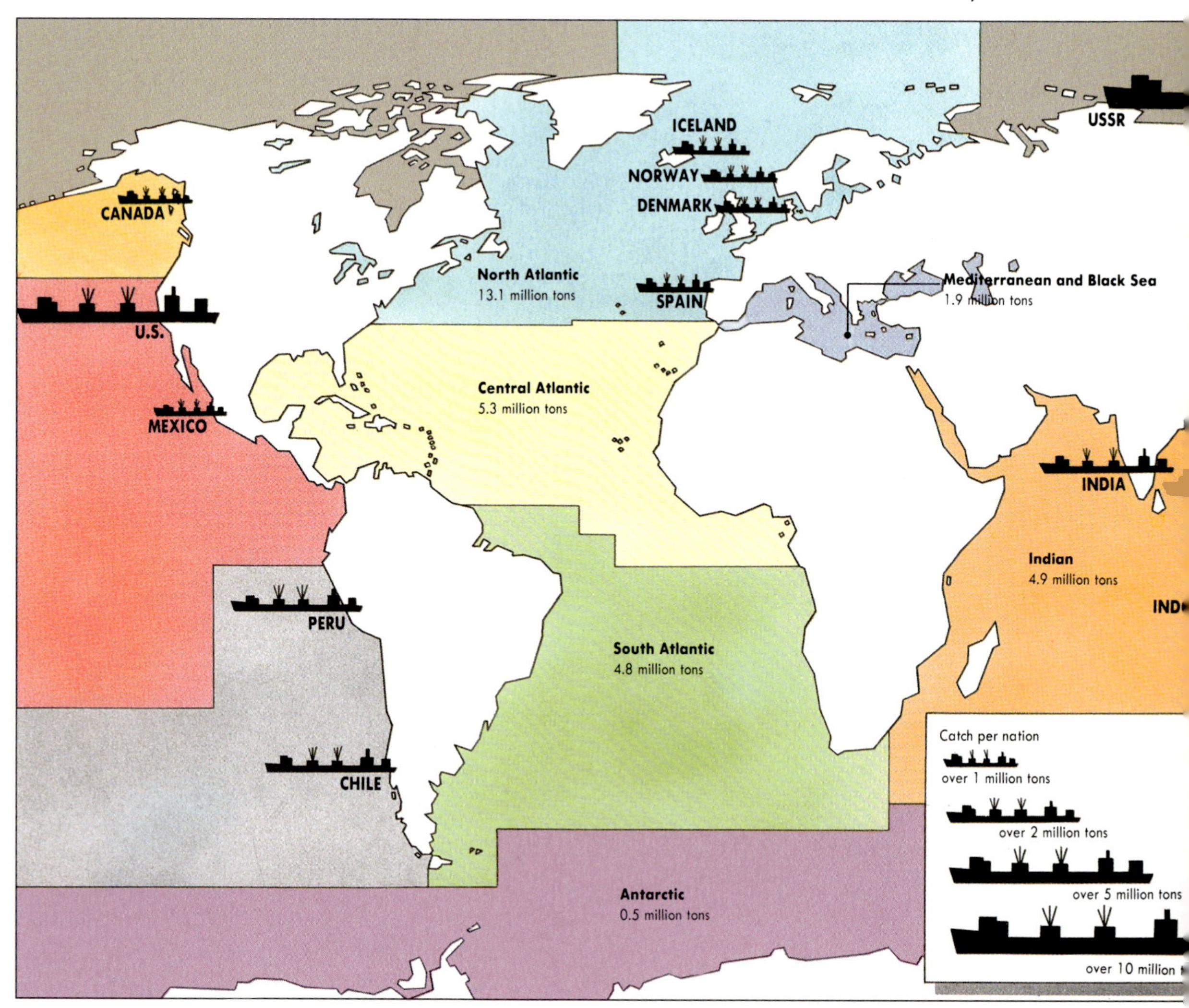

THE FISH THAT MAKES DESERTS

Today, 30 years after its introduction into Lake Victoria, the voracious Nile perch comprises 80 percent of all fish caught in the lake. Since it is prized in East African cities like Nairobi, and now Europe, new companies have been created to catch, preserve, and export this fish, which grows up to 550 lb (250 kg). But there can be no substitute for the millions of fish the Nile perch has displaced. Of the lake's 300 unique species of tiny colorful cichlids, 200 are believed to be extinct, and the vital protein they provided for local consumption is fast disappearing.

The Nile perch can never be removed, and it may be making a desert of the lands around the lake. The indigenous species could be sun-dried, but the perch must be smoked on wood fires.

continuous growth in catches of anchovetas, which by 1972 had made Peru the world's largest producer, the fish had virtually disappeared. In many areas of the North Atlantic, fish that were once common, such as cod and haddock, have become scarce, and the much-loved herring is virtually extinct. Pilchard numbers in the South Atlantic are greatly depleted, as are tuna stocks in the Pacific. The Antarctic, which was widely expected to offer immense opportunities, has already been fished out of cod.

Neither have many of the development programs been as successful as was anticipated. Such plans generally required major capital investments in fishing vessels and equipment, when even a small outboard motor, the gas to power it, and repairs can be beyond the resources of local fishing communities. Large-scale production to supply foreign markets makes even more massive capital demands, for high-powered vessels and large nets, and cooling, freezing, and canning plants. As indebtedness rises, larger operators step in and displace small-scale local producers.

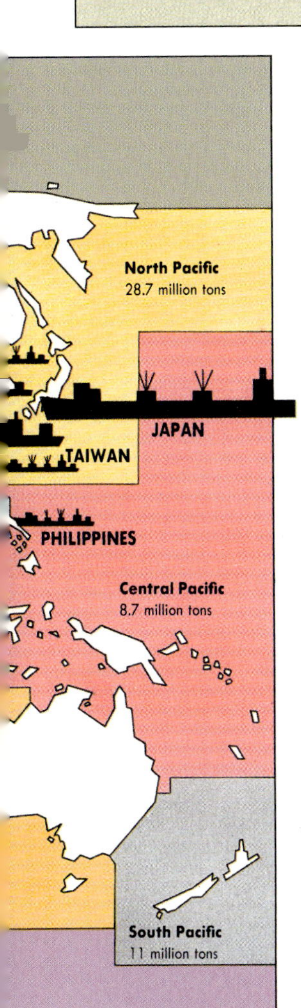

Shrimp farming has emerged as a major industry here in Hong Kong and elsewhere in Southeast Asia, turning a sought-after luxury into a relatively low-cost food for home consumption and a high-earning export. On the Philippine island of Negros, 40 percent of the sugar lands have been converted to shrimp ponds. Profits from shrimp can be four times those from sugar.

For at least 4,000 years the Chinese have added to their protein intake by rearing carp and other fish in their rice paddies and irrigation canals. Today shrimp are an important cash "crop," largely for the export market.

Shrimp production is also growing elsewhere in the world, sometimes with serious consequences. In 1989 Bangladesh's shrimp earned $150 million, but the salt water channeled inland to the shrimp pools made surrounding land unfit for rice cultivation.

In Honduras, shrimp ponds are expected to triple in area by 1995, largely at the expense of mangrove forests. These coastal swamps protect against sea erosion and provide habitat and food for marine life and forage for cattle. Any loss could lead to far greater costs than would be recouped from profits gained from shrimp.

Crises and Consequences

One of the most basic laws of physics states that there is no action anywhere on Earth that does not produce a reaction. This law applies equally to social, economic, and political behavior. The global network of business and communications has increased profits for many and facilitated new developments in the manufacture and distribution of goods. But it has also created crises of international proportions whose consequences grip the world in a relentless and potentially lethal embrace.

While industrial and technological growth have lined the coffers of large, politically stable nations, the countries of the third world have slid deeper and deeper into debt as they struggle to compete in the international marketplace. Starved of capital, they have borrowed enormous sums for development from Western nations, only to discover that the cost of servicing their debts surpasses the amounts their imports can earn.

As the price of raw materials falls and the price of manufactured goods increases, the less sophisticated debtor nations spend more and earn less. Inflation in those countries has soared – to 250 percent in Argentina and a staggering 700 percent in Brazil – and there is no sign of this spiral of debt coming to an end.

In such desperate economies, austerity measures bring a widespread loss of jobs or cut in wages. As a result, urban areas support an informal sector of workers, many of whom have adopted street trades – hawking, begging, entertaining, or even washing car windows – in order to earn a living. Domestic service or casual manual labor draw some; prostitution and petty crime claim others.

Away from the urban landscape, the problems are different. In Africa, drought, soil erosion, inadequate agricultural programs, deforestation, and population explosion have combined to produce a food crisis that threatens one nation after another. Famine is often blamed entirely on drought and the harshness of the environment, but Africa's population growth, which is more rapid than any other continent's, places greater demands on the natural resources than they can sustain.

Warfare eats into Africa's food crisis as fiercely as climate and population. Investing money in arms rather than agriculture, Sudan, Chad, Angola, and Mozambique have all waged debilitating civil wars that have destroyed lives, jobs, and crops and added hungry refugees to an already large and starving population. The gradual introduction of new seed varieties, terracing, and fertilizers may be beginning to turn the tide on Africa's famine.

A different tide is slowly swamping other parts of the third world. Tourism, once largely confined to Europe and North America, has spread in recent years, bringing Western visitors to such exotic vacation areas as Bali, Nepal, and the Seychelles. Initially the host countries welcomed this stream of visitors, since they brought foreign currency and employment for native inhabitants. As tourism grew, however, the impact on the local economy developed its negative aspects. Internal costs rose as countries struggled to improve their infrastructures and rushed to import food, drink, and furnishings that would please their guests. All over the world, indigenous customs, ceremonies, and crafts have been rescued from oblivion and crudely altered to increase their attraction and retail value.

A country that relies on tourism for its livelihood is a hostage to fashion, disease, political upheaval, and natural disaster, any of which can singlehandedly erase its appeal. Colombia survives even more precariously, basing its economy on coffee and cocaine, stimulants far more enticing than travel.

Colombia's drug problem is an offshoot of its geographical location and agricultural circumstances. It is ideally placed for trade between coca-producing countries and the cocaine markets of Europe and North America. As the price of coffee has dropped, Colombia has devoted more of its money and personnel to the illegal drug trade, using its ancient links with Spain, through which three-fourths of Europe's cocaine passes, for access to the European market. Florida's long coastline is the gateway for American trafficking, and Japan now pays even higher street prices for cocaine than the U.S. The two Colombian cartels that control the global cocaine trade realize

annual profits estimated at over $2 billion. Little of this money, laundered in foreign banks, returns to Colombia.

Nature is not always tended as carefully as Colombia tends its most lucrative crop. In the tropical regions that nourish the lush vegetation of rainforests, rampant deforestation destroys plant and animal life as well as the trees themselves. Where few fuel alternatives exist, as in Africa, timber is culled for firewood. Certain industries – like cattle ranching, which uses twice the land area it did 30 years ago in Costa Rica – cause forests to be sacrificed for the sake of profit.

Two-fifths of the world's rainforests have disappeared since the end of the Second World War, and irreparable damage has been done to the Earth's riches by logging, fire, and herbicides. Thousands of plant species are nearing extinction, and many indigenous populations have been forced to find new homes and new means of survival. Even the climate has been affected, since fewer forests result in less rain. As the global consequences of this problem mount, global awareness increases, too, and measures are slowly being introduced to counteract some of the more damaging practices.

The problems caused by agricultural pests know no boundaries. National isolation ended with the advent of colonialism, and no sooner had farm animals and settlers occupied the wilderness than alien plants and animal predators began invading the crops and the countryside.

Some were introduced deliberately. European rabbits, intended as prey for hunters in Australia, multiplied rapidly, stripping land intended for livestock. Only the viral disease myxomatosis finally checked them. Other pests, like the rats that traveled in the holds of colonial sailing ships, slipped into new countries accidentally and devastated birds and small animals as well as the harvest. Like the plagues of Ancient Egypt, worms, toads, cacti, shrubs, and flies have spread havoc around the globe, eluding the most sophisticated man-made traps and ignoring nature's own system of checks and balances.

Though we know we cannot live without water, humanity's careless abuse of this vital substance may soon leave some people high and dry. Like oil, natural fresh water is a limited resource: once the supply has been exhausted, it will be gone for good.

The Aral Sea, once one of the world's largest inland lakes, irrigates the enormous Soviet cotton fields that yield the area's principal cash crop. But the canals carrying the precious water let it drain into the soil, and the pesticides and fertilizers that protect the cotton pollute the water with chemicals. Surrounded by desert and thus bereft of inflow, this sea is shrinking steadily. The Ogallala Aquifer currently provides 20 million Americans with water, but, constantly depleted by its 150,000 wells and never replenished, it cannot last forever.

As the greenhouse effect makes summers longer and hotter, more and more water is lost to evaporation. Untreated sewage poisons the Mediterranean Sea, and over a billion gallons (4.5 billion l) of human waste daily contaminate India's sacred Ganges. Short-term planning and wanton waste of water are slowly killing our forests, our crops, and our ability to maintain any form of life on this fragile planet.

The crisis of industrial pollution advances hand in hand with the world's water crisis. The Rhine River, sometimes called "Europe's largest sewer," flows with pollutants from chemical, paper, and steel works, and with domestic and agricultural sewage. As the North Sea and the feeder rivers absorb this unhealthy mixture, they pass it on to marine life, migrating fish, irrigation channels, and drinking water. The effects are cumulative and increasingly dangerous.

No barriers can prevent the U.S.'s exhaust fumes from reaching Canada or protect a nation from the radioactive dust raised by an explosion at a nuclear power station, as happened in Chernobyl in the USSR. No profits from manufacturing or industry can erase the lasting damage of an oil spill that destroys the habitat of animals and birds and pollutes vast areas of clean water. Acid rain attacks plants and lakes, and the heat pollution from nuclear power stations steals oxygen from any neighboring water.

The industrial growth that has fed and clothed humanity since the Industrial Revolution, bound countries together in an interdependent web, and made them rich, now threatens to exact a massive and final payment from them all.

PAYING BACK WITH INTEREST

The crushing burden of world debt

NO ONE KNOWS EXACTLY HOW MUCH money third world countries owe to Western governments and banks. Some put the total figure at about $1,000 billion; others think it exceeds $1,300 billion. Several third world countries have debts larger than twice their entire gross national product (GNP), and some will never be able to repay what they have borrowed. The third world has a desperate need for development capital, yet it transfers more money to the West in debt repayments each year than it receives in loans and aid combined.

Oil price hikes in the 1970s meant OPEC countries found themselves awash with surplus funds. These were deposited in Western banks, which eagerly recycled the money by lending to capital-starved countries in eastern Europe and the third world. But when interest rates rose in the 1980s, debtor nations found themselves having to borrow even more just to cover interest payments. The debt crisis was born.

Borrowing money for development is not necessarily detrimental to economic growth. Some of the most successful industrializing countries, such as South Korea and Thailand, have borrowed massively. For them, increased exports more than cover the cost of borrowing. But in the majority of third world and eastern European countries, this has not been the case. A super-league of HICs (Highly Indebted Countries) has emerged, including Mexico, Brazil, Argentina, Poland, and Nigeria, that spend (or owe) more than 60 percent of their gross national product on debt repayments.

In some cases just paying the interest on the debt ("debt servicing"), without repaying any capital, can consume more than half of a country's export earnings. On this measure Latin American countries are the most heavily indebted; on average, the cost of servicing foreign debt accounts for 30 percent of all exports from the region.

New borrowing now comes with strings attached; the International Monetary Fund (IMF) insists on it. And the terms can be fierce, including cuts in public sector salaries and the abandonment of food subsidies and development programs. HICs like Venezuela and Brazil have experienced serious debt riots.

What accounts for the debt crisis? First, borrowing has become a more costly affair. Loans and interest payments are calculated in U.S. dollars. In the 1980s the value of the dollar rose, and debtor nations have had to pay more merely to repay the interest.

Second, while prices of industrial goods bought from the West have risen, the prices of raw materials have fallen sharply. Countries that depend on selling sugar (like Jamaica), coffee (Colombia), or copper (Zambia), for example, have found their export earnings plummeting.

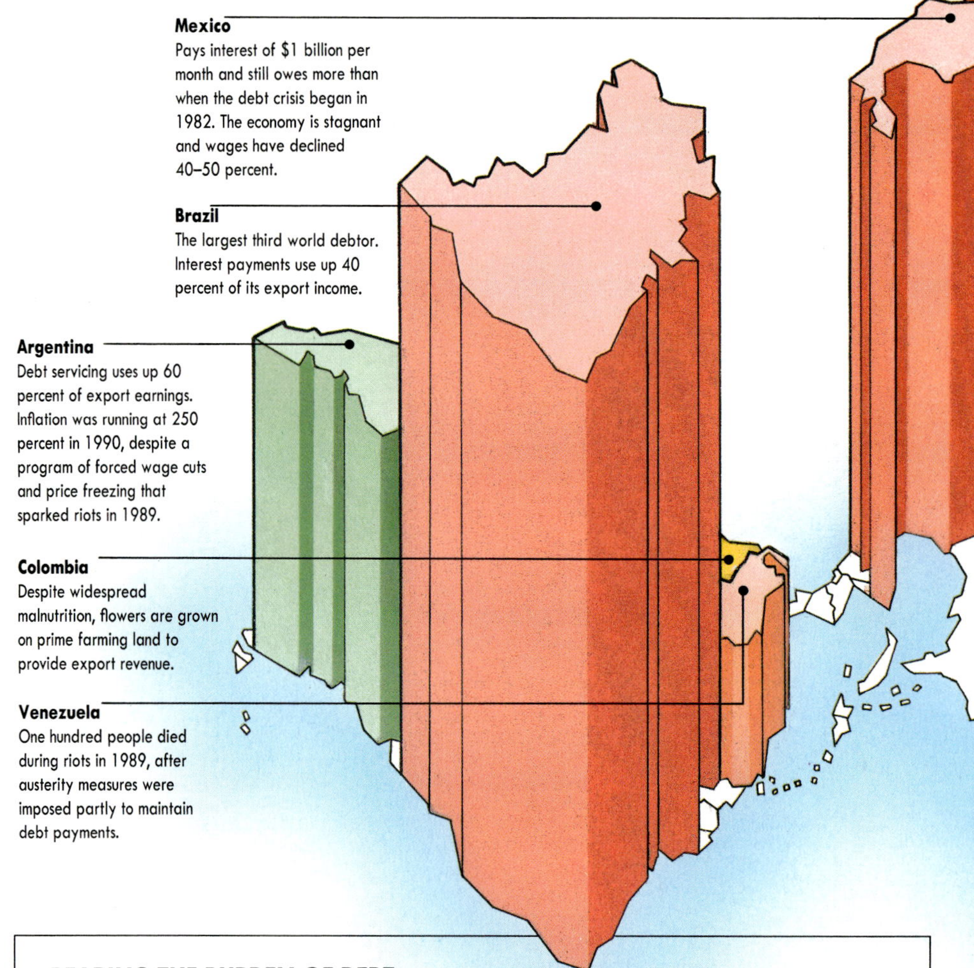

BEARING THE BURDEN OF DEBT

Brazil's massive foreign debt casts a shadow across the entire nation. Debts accumulated through the 1970s and 1980s while Brazil borrowed funds from Western banks to force the pace of development. Impressive economic growth was achieved, but it soon became clear that the debt burden had grown too large, with interest rates on loans doubling from 4 percent to 8 percent by 1985 (having soared to 22 percent at one stage), and Brazil was no longer able even to service its debt.

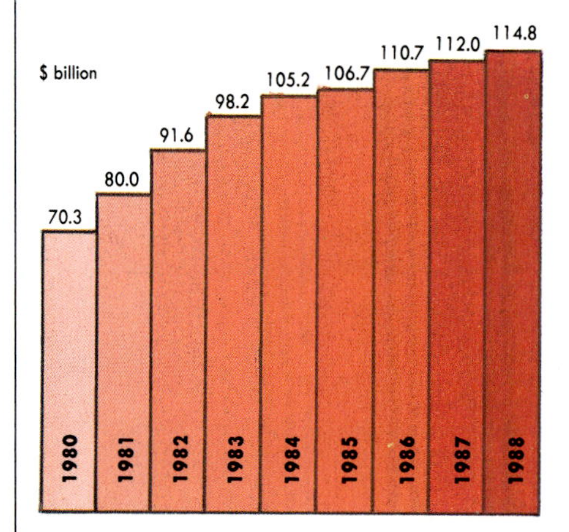

From 1983 to 1989, in an attempt to safeguard repayments, the International Monetary Fund (IMF) instructed the government to impose austerity measures. Reduced domestic spending, coupled with increased exports, was intended to curb runaway inflation and restore the balance of payments.

The measures taken included dramatic price rises – gasoline rose 60 percent and telephone charges 188 percent – but also freezes on the prices of certain essentials. At the same time, wages were cut. At $50 a month, the current minimum wage has half the purchasing power it had in 1940.

Meanwhile, the pressure to use land for export crops undermined Brazil's food supply. In 1985, the government estimated that two-thirds of the population were malnourished.

But, despite the drastic measures, the economic crisis shows few signs of abating. Annual inflation is presently a huge 700 percent, the economy is in recession, and debt looms larger than ever. The graph (*left*) shows Brazil's increasing indebtedness.

CRISES AND CONSEQUENCES

A lending and borrowing spree since the 1970s has left many developing countries with a lasting debt they can ill afford. Faced with escalating interest repayments, the world's Highly Indebted Countries (HICs) actually transfer more money now to the West than they borrow: $21.6 billion in 1987 against $38.2 billion in 1988.

Although the U.S. is in fact the world's biggest net debtor, owing $200 billion, its economy is largely cushioned by the huge volume of its exports. But for third world countries, debt is a crippling burden. At least 19 countries now owe more than the annual wealth they produce.

Several countries, like Sudan and Colombia, are forced to export food to help pay their debts, even though their populations are malnourished. UNICEF estimates that 500,000 children die each year because of the debt crisis.

The colors on the graph (*bottom right*) correspond to those on the main map, indicating the level of indebtedness in billions of dollars.

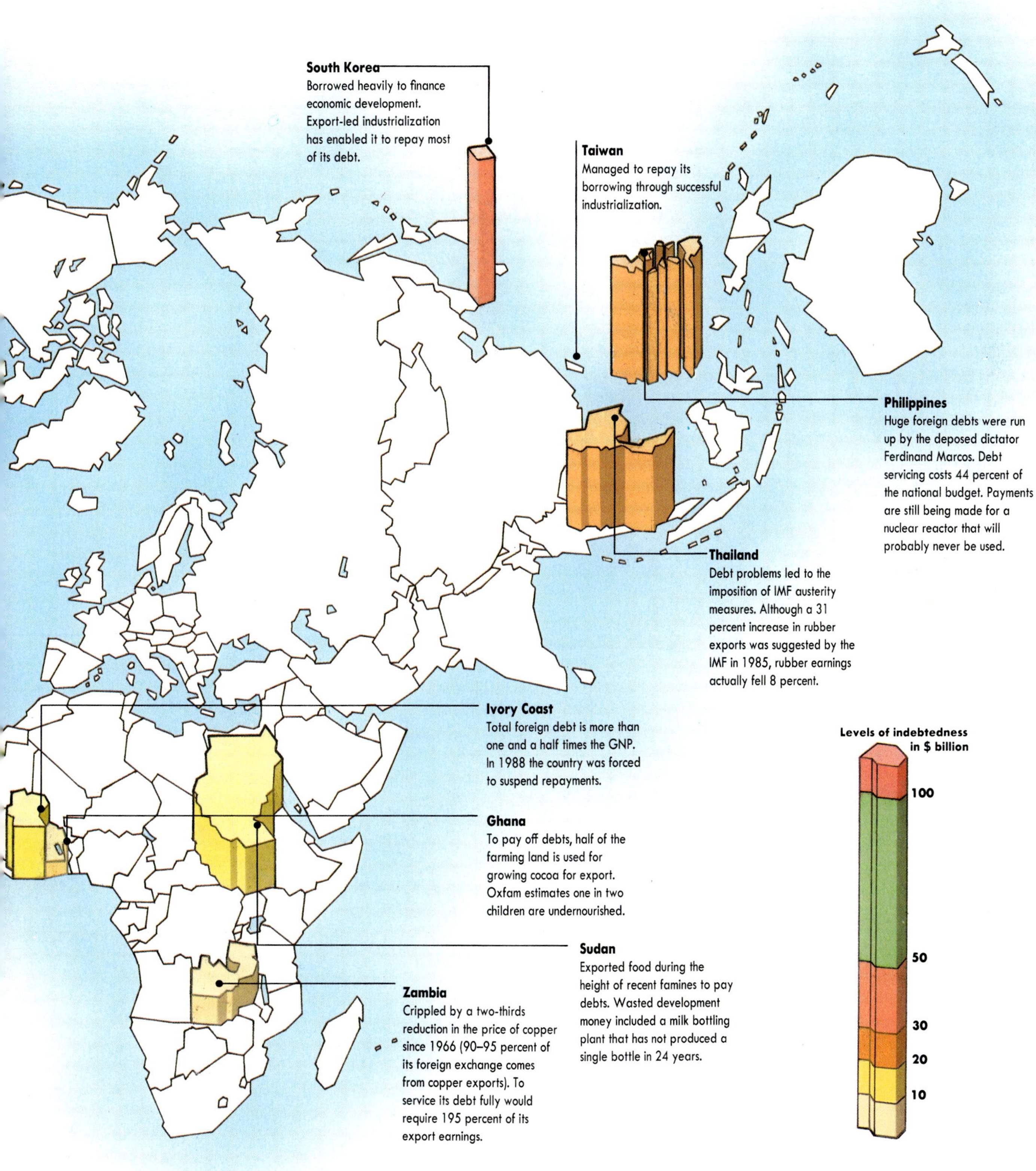

GETTING BY IN THE CITY
The informal urban economy

Among the most vivid features of the third world city is the inventiveness people apply to earning a living. Those who fail to find regular waged labor, or simply reject what it has to offer, create myriad ways of earning nickels and dimes. Washing windshields or selling newspapers to drivers halted at traffic lights, snake-charming, even waiting in line can become a profession if there are sufficient people with more money than time and patience.

This "informal sector" is very broad. Street trades are most visible, but many people also work as domestics, casual laborers, mechanics, or itinerant carpenters. Still others resort to begging, crime, or prostitution, particularly where foreign tourists are abundant.

Although such occupations may seem marginal from the point of view of the global economy, more than a billion people around the world are fed, clothed, and housed entirely from their own resources. In many third world cities, more than half of the population work in this "black" or "underground" economy.

Informal and formal sectors cannot always be clearly distinguished; one often relies on the other. And making a living in this way is not completely without the capital costs and organizational demands associated with larger enterprises. Dressmakers need a sewing machine, mechanics a tool kit; sidewalk snack-makers must buy ingredients, charcoal, and a stove.

Local and national legislation may demand that street traders acquire scarce and expensive permits, while certain favored pitches and types of activity may be jealously guarded by those already well established – stretches of pavement or even whole areas of the city may be the exclusive preserve of particular traders or shoeshiners. Even the most menial labor can be barred without the proper contacts.

In the absence of the limited means required to pursue such trades, people may earn a living from trash. No third world city is without its garbage pickers, who both restore damaged items and provide recycling plants with their raw materials.

The informal sector is often regarded as unproductive, even parasitic and demeaning. In some tourist areas, or during international events such as the Olympic Games, "national pride" sometimes dictates that all evidence of the underground economy be erased. Beggars, prostitutes, and illegal street traders may be harassed by police, and either imprisoned or

No third world city is without its garbage pickers, like these on a dump outside Manila, Philippines. Some discarded items remain serviceable; others can be renovated or modified to suit new functions. Practically everything can be turned to some advantage, particularly rags, glass bottles, aluminum cans, newspapers, and plastic. Once sorted, all can be sold to recycling plants. From the point of view of the companies, this is cheaper than producing new materials from scratch and avoids all the costs and obligations associated with formal employees.

Some street traders, lacking the means to purchase basic equipment, earn their living catching songbirds or cleaning up stray cats and dogs to sell. Those with more skills and some capital can afford to be more sophisticated: the banner behind the stall of this street dentist in Jaipur, India, promises better teeth for all.

transported to less visible locations.

In 1989, in the interests of "humanity" and "prestige," the Indonesian government banned the local three-wheeled pedicab, the *becak*, the basic means of existence for up to 100,000 urban migrants. *Becak* drivers who cannot afford the equivalent motorized cab and continue to ply their trade have their vehicles dumped at sea.

In fact, the informal sector provides a vast range of cheap goods and services that reduce the cost of living for employees in the formal sector, and hence enable employers to keep wages low. Pedicabs are cheap, nonpolluting, and make up for gross deficiencies in mass transport. Cheap lunches can be purchased at small food stalls outside factories and offices, and such services as dressmaking, carpentry, and plumbing are readily available and inexpensive.

Although low incomes mean fewer people can buy goods, this is of little concern to those companies exporting abroad. For them the informal sector provides a considerable subsidy to production, which is often eventually passed on to consumers in the developed world.

Some governments are now taking a more positive approach to the informal sector. The innovative spirit of such workers is finally being recognized and supported. Most important of all is low-cost credit. Mini-enterprise loans are rarely in excess of $100, but they are made available at subsidized rates. In the absence of security, the previous alternative was debt servitude to loan sharks. A desire for a better life for themselves and their children drives the urban poor as hard as any Western entrepreneur. They know the price of failure.

INFORMAL OCCUPATIONS

What people in the informal sector lack in educational qualifications, training, and capital, they must make up for in energy and ingenuity. The following are just some of the ways in which they may make a living.

Selling souvenirs
Selling bus tickets
Selling lottery tickets
Selling single cigarettes
Selling food and drink
Shining shoes
Washing car windshields
Street entertaining
Driving pedicabs
Guiding tours
Writing letters
Domestic service
Hairdressing
Dressmaking
Carpentry and plumbing
Dentistry
Basic health care
Waiting in line
Picking and sorting garbage
Carrying shopping
Telling fortunes
Prostitution

CAN AFRICA FEED ITSELF?

The problems of climate and population

Ethiopia's central highlands – its most densely populated area – have a pleasant, equable climate. They could be agriculturally very productive.

Although Ethiopia's population is growing rapidly, it might still be able to feed itself if soil erosion could be halted and suitable farming techniques were implemented. Planting trees would serve to stabilize the soil on denuded slopes and create windbreaks, preventing rain and wind from stripping away exposed topsoil.

But poor farming techniques are not solely to blame. Without incentives, capital to invest, or security, farmers will continue to persevere with traditional methods of subsistence agriculture.

ETHIOPIA IS A BREATHTAKINGLY beautiful and diverse country, greater in size than France and Spain combined. Its human mosaic – 50 million people from more than 50 language groups, with their unique customs, ways of life, physical appearance, and dress – is as varied as its landscape and climate.

Yet the name Ethiopia does not evoke images of diverse resources and cultures. Instead, it has become synonymous with mass starvation. Every year famine threatens to sweep through the country as it has done so often in the past. This may frequently be triggered by drought, but it reflects a more general decline in the nation's ability to feed itself.

Ethiopia's predicament captures the essence of famine across the African continent. The roots of its food crisis, like those throughout Africa, are manifold. Climate change, population pressure, deforestation and soil erosion, poor agriculture, lack of export earnings, and warfare are all important in their own right. But when they coincide, their effects are disastrous.

Famine and food shortages are often regarded too simply as a product of climate, principally of drought. But drought does not always mean disaster; even in the absence of drought, Africa would still have problems of hunger.

Population growth has a more fundamental link with famine because it continually increases pressure on land resources. Africa's population is rising faster than that of any other continent. Such increases add to the pressures on fragile environments by encouraging overgrazing and overcultivation. As a result, yields are declining, and soil is becoming uncultivable.

Irrigation is needed in many parts of Africa to overcome seasonal shortages of water. But many of the most ambitious plans involving large dams add to foreign debt, are of inappropriate scale, and cannot be properly maintained.

Ironically, agricultural development programs have often aimed to expand the production of export cash crops rather than food for local consumption. Such plans rarely benefit the rural poor. They provide little extra employment and drive subsistence farmers onto ever more marginal land.

Neither are cash crop exports a stable source of income. If international terms of trade were more favorable, Ethiopia could import food to reduce its shortages, but the price of coffee – its main export – is at its lowest since the 1920s. Yet export-led agricultural policies continue to be pursued in an attempt to support development programs, repay foreign debts, and buy arms.

Warfare has further contributed to famine in Africa. Ethiopia's civil war and conflicts with its neighbors have debilitated the country, destroyed lives and livelihoods, and disrupted the production and transport of food. By the end of the most recent disastrous period of drought and famine, in 1986, 10 out of the 13 worst-affected countries had suffered from war, civil strife, political destabilization, or a massive influx of refugees.

SWORDS AND PLOWSHARES

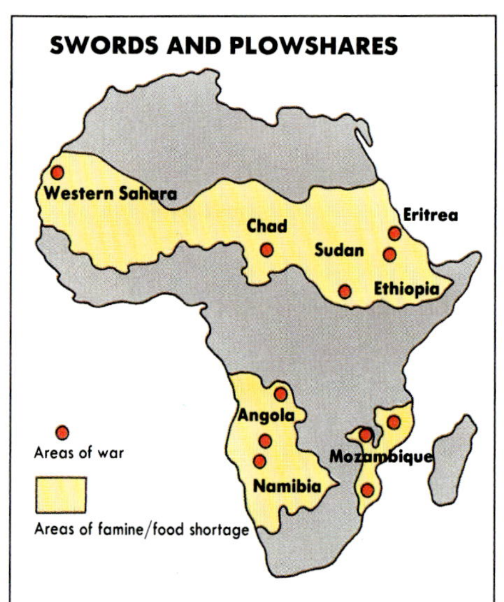

In Africa, civil war means famine. Governments that spend considerable amounts of money on arms – Ethiopia, Sudan, Chad, Angola, Mozambique – are neglecting environmental and agricultural issues, adding to the pressures that provoke internal conflict.

Refugees compound the problems of food and water shortage. War-torn Sudan has to cope with a million refugees from Chad and Ethiopia, while refugees from Djibouti and Somalia have poured into Ethiopia.

In Western Sahara, there has been war between the Sahrawis and the Moroccans since 1976. Roughly 125,000 Sahrawis – perhaps a third of the total population – are now refugees in Algeria.

The greening of Africa

During the famines of 1985, 35 million Africans suffered from acute hunger, 10 million of them abandoning their homes in search of food and water. Although this event was triggered by drought, it reflects Africa's problems with feeding itself. Since the late 1960s, food production has fallen by 2 percent a year, while populations continue to grow. Even in a "normal" year, 150 million Africans undergo some degree of hunger and malnutrition.

The pressure to increase the amount of food available for domestic consumption is enormous. All over the continent, efforts are being made to improve yields, combat localized problems of climate, and make farming more profitable for farmers. There are some projects scattered across the continent that offer the prospect of hope and self-sufficiency.

Burkina Faso
Farmers in the Yatengao province have refined a traditional technique of lining stones along the contours of their land. Acting as mini-dams, these stone lines hold back rainwater long enough for it to irrigate the land uphill, but are not so impermeable that land below dries out.

Niger
Wind is a major cause of soil erosion in Africa. Since the mid-1970s the Majjia Valley Project has planted row upon row of trees, resulting in a windbreak over 200 miles (320 km) long.

Ethiopia
Although farmland is state owned, all farmers belong to Peasant Associations, which administer the land. In contrast to the feudal system that prevailed before the 1974 revolution, this gives the farmers some hope of self-improvement.

Areas with potential for agricultural productivity

Areas of war/famine/drought

Kenya
The Green Belt Movement, started in the 1970s, has created more than 600 tree nurseries and planted more than seven million trees.

Ghana
Selective logging has degraded a number of Ghana's natural forests, but a successful rehabilitation project is running in the Subri River Forest Reserve. Land around valuable tree species is being cleared, enabling them to grow unimpeded. Felled wood is sold as timber or converted into charcoal, and the cleared land is turned over to local farmers for cropping.

Nigeria
The Green Revolution is taking off more slowly in Africa than in Asia. But an international research institute in Ibadan has developed a disease-resistant cassava whose use is spreading spontaneously as planting material is passed on from farmer to farmer.

Rwanda
The world's most densely populated country must concentrate on balancing the demands of food production against the dangers of soil erosion. Rwanda uses very little chemical fertilizer or irrigation, but protects soil against erosion by terracing in hilly areas, and by producing mainly tree or shrub crops such as bananas, coffee, and cassava.

Malawi
Malawi's trade policy is firmly protective to its farmers, with the result that the country, which imported $5.5 million worth of cereal in 1970, is now a major exporter.

Zimbabwe
The introduction of new seed varieties, seed-dressing, and increased use of fertilizer brought about a "maize miracle" in the 1980s. Although the project was initially costly for farmers, higher market prices made it attractive. Zimbabwe's maize production trebled in the seven years to 1985.

Swaziland
Increased use of fertilizer brought great economic improvement to Swaziland: cereal production grew by 50 percent in ten years.

THE PLEASURE PERIPHERY

Tourism in the third world

INCREASINGLY, THIRD WORLD COUNTRIES have become popular as exotic holiday destinations for Western tourists – so much so that the third world now accounts for around one-eighth of the industry. The countries concerned have generally been keen to promote tourism as a means of earning foreign exchange, of stimulating the local economy, and of providing employment.

But in recent years the tourist industry has come to be seen as, at best, a mixed blessing. People have begun to calculate the full costs and benefits of tourism – especially its often contradictory impacts on local cultures, the natural environment, and the economy.

Tourism certainly creates jobs. Although the work is largely seasonal, local people are employed in hotels, shops, banks, restaurants, and bars, as taxi drivers, or as manufacturers of souvenirs. But if labor is in short supply, and wage rates in tourism are relatively high, other, no less vital, sectors of the economy can be deprived of their work force. Falling local agricultural production, for example, will reduce food exports and push up imports of foreign foods.

Countries must also bear the infrastructural costs of tourism. In order to attract visitors, they may have to improve their roads, water, and electricity supplies, and build airports close to resorts. Although some of these new facilities are also of benefit to local people, many only serve tourist areas.

Dependence on tourism also makes a country particularly vulnerable to changing demand. Tourists are notoriously fickle; this year's favored resort may have gone out of fashion by next year. Political disturbances, natural disasters, and outbreaks of disease or food poisoning can devastate a local tourist industry. Fiji's military coup in 1987 caused a 70 percent drop in visitor arrivals, despite a vigorous international advertising campaign to restore consumer confidence.

The great scale and complex structure of the international tourist industry could now precipitate the complete collapse of a country's tourist trade. Tour companies can make use of any number of countries, each offering "sun, sand, and surf," and standardized facilities. Switching vacationers from one country to another presents little difficulty for the company, and may make no difference to the tourist, but is a disaster for the country that is out of favor.

The effects of tourism on local culture and environment can be contradictory. Tourism often sustains indigenous lifestyles, ceremonies, and arts and crafts that would otherwise have perished. Similarly, historic buildings and sites may be preserved, and sanctuaries created for wildlife. Yet tourism can also destroy. Indigenous cultures can be adulterated and debased, hotel developments may be unsightly, sewage and oil from boats often pollute beaches, and the breeding patterns of wild animals are disrupted. It is indeed an irony that the very qualities that most entice tourists to visit a particular country are those most threatened by their presence.

PROBLEMS CAUSED BY TOURISM

St. Lucia When tourism developed, the island's balance of payments deteriorated as workers were drawn from banana production.
Fiji The military coup in 1987 caused a sudden drop in visitors of over 70 percent. Four out of five people involved in the industry were fired. The currency was devalued by 17 percent.
Seychelles Tourism has brought much-needed foreign currency but has led to soaring land prices and the disruption of fishing and farming.
India Hotel developments in Goa divert public resources from local people. Some water mains to hotels pass through villages without piped water.
Thailand Prostitution has become the hallmark of tourism, with over half a million women involved in the sex industry.
Nepal The influx of hikers has exacerbated deforestation problems by increasing the demand for wood for fuel.
Turkey Tourist developments threaten the nesting beaches of sea turtles.
Venezuela Tourist agencies have used the powerful chemical dioxin to clear seaweed from beaches. The poison has killed millions of fish.

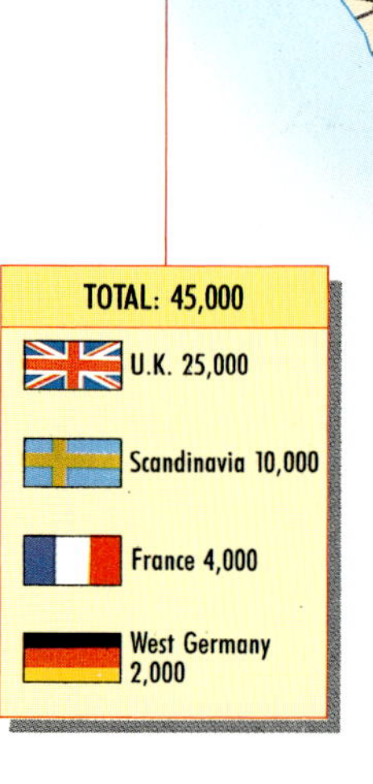

World tourism

Visits to foreign countries now total over 400 million a year, generating so much revenue that tourism is fast becoming one of the world's leading industries. Excluding airfares, each tourist now spends an average of $416, compared with $276 in 1970.

Almost 70 percent of all tourists come from just 20 of the world's 233 countries. Most still vacation in Europe and North America, but in recent years tourism has boomed elsewhere, thanks to cheap long-distance flights. Visits to third world destinations now account for one-eighth of the industry.

It is not only visitor numbers that have increased, but also the range of destinations. Well-established exotic centers – for example, those in the Caribbean and Oceania, are now facing stiff competition from other tropical islands, like Bali, the Seychelles, and the Maldives.

The map shows the numbers of visitors to eight popular tourist destinations around the world and the main countries they come from.

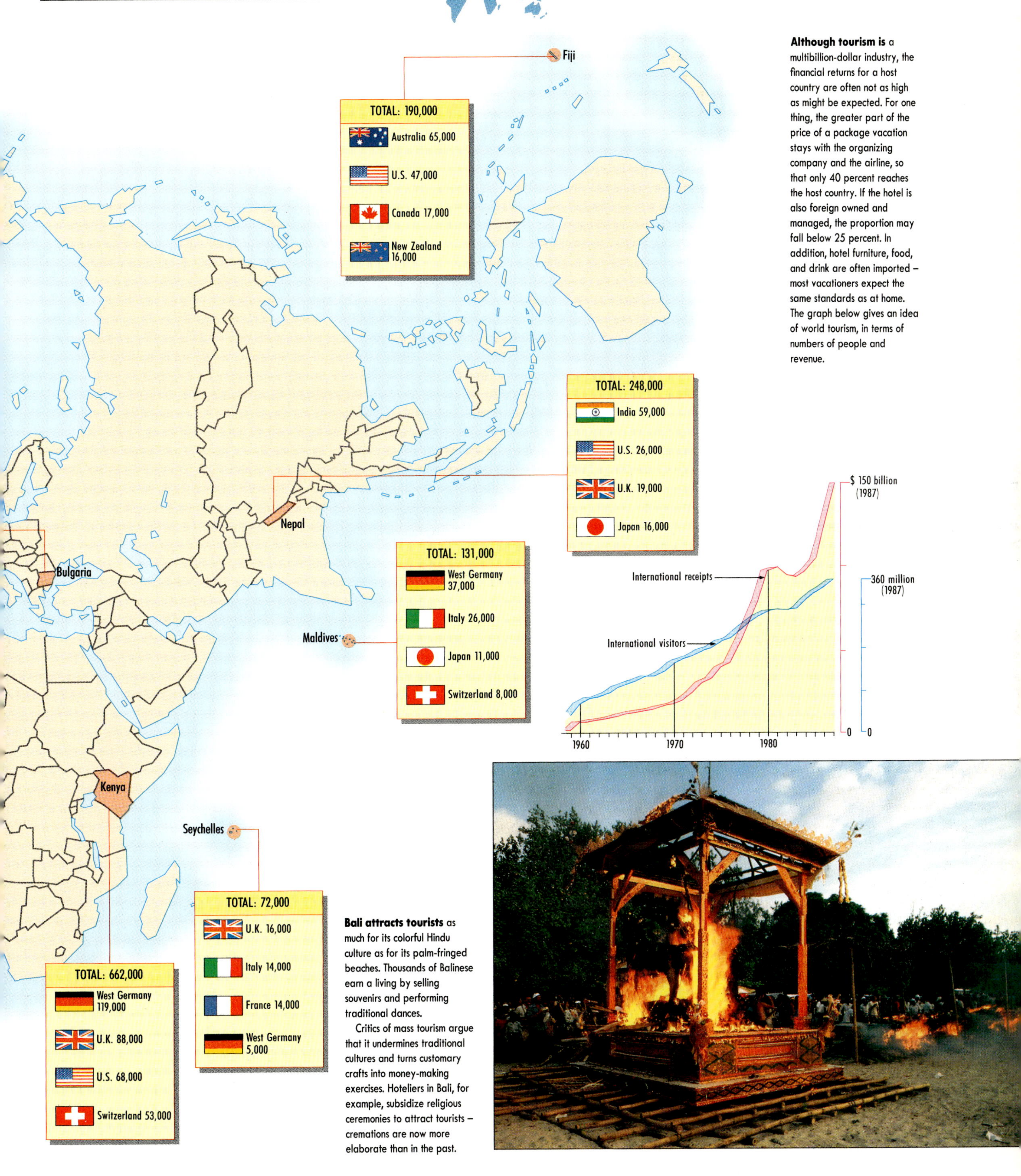

Although tourism is a multibillion-dollar industry, the financial returns for a host country are often not as high as might be expected. For one thing, the greater part of the price of a package vacation stays with the organizing company and the airline, so that only 40 percent reaches the host country. If the hotel is also foreign owned and managed, the proportion may fall below 25 percent. In addition, hotel furniture, food, and drink are often imported – most vacationers expect the same standards as at home. The graph below gives an idea of world tourism, in terms of numbers of people and revenue.

Bali attracts tourists as much for its colorful Hindu culture as for its palm-fringed beaches. Thousands of Balinese earn a living by selling souvenirs and performing traditional dances.

Critics of mass tourism argue that it undermines traditional cultures and turns customary crafts into money-making exercises. Hoteliers in Bali, for example, subsidize religious ceremonies to attract tourists – cremations are now more elaborate than in the past.

CAFFEINE – OR COCAINE?

A desperate addiction

THE ECONOMY OF COLOMBIA IS completely dependent on drugs, legal and illegal. Thanks to the country's climate and soils, it is a major producer of coffee and marijuana. And because of its unique geographical situation – it lies between producer countries, such as Peru and Bolivia, and major markets in North America and Europe – it has become the world's largest exporter of cocaine, a narcotic made from coca leaves.

Colombia exports more coffee than any other country except Brazil. A fifth of all cultivated land is devoted to its production, and coffee accounts for over half of the country's official exports. Yet it is the nature of coffee production that partly explains why Colombia has become embroiled in the risky but extremely lucrative illegal drugs business.

The fertile and well-drained slopes of the Andes at 3,500–5,500 feet (1,050–1,650 m) are particularly well suited to the cultivation of high-quality mild arabica coffee. Most is grown by small farmers, a third of whom farm less than 7 acres (3 hectares) of land. But the seasonal pattern of the industry means that while labor demands are high at harvesttime – only three months – coffee production can provide year-round employment for very few workers.

Dependence on coffee has also exposed Colombia to the vagaries of climate and world markets. Colombian coffee growers enjoyed a boom between 1976 and 1980, when frosts hit the Brazilian crop. But the price of coffee on the commodity markets has plummeted since then. As a result, many people have looked to more profitable and dependable alternatives, notably to the cultivation, processing, or transportation of marijuana and cocaine.

In the 1960s, increased demand for marijuana in the U.S. boosted Colombia's crop, and an illicit trafficking network developed. These connections enabled Colombians to take on the distribution of cocaine for Bolivia and Peru. Little cocaine originates in Colombia, since climatic conditions produce coca leaves of lower quality. It is only recently that cultivation has been stepped up as demand for cocaine has outstripped production elsewhere.

Colombia is perfectly positioned to dominate this intercontinental trade. The northernmost country in South America, with both Pacific and Caribbean coastlines, it offers a multitude of routes to its biggest markets on the West and East coasts of North America. It has also been able to exploit its cultural links with Spain to extend its network into Europe.

The trafficking of cocaine is largely controlled by two cartels – essentially integrated gangs – in the cities of Medellín and Cali. To protect their interests, these drug barons or *narcotraficantes* have allied themselves to guerrilla groups operating in the countryside. Public figures who have tried to suppress the trade have been assassinated – including three presidential candidates in 1990, as well as hundreds of judges, magistrates, and police officers.

Cocaine produces huge profits. Much of the drug barons' earnings – estimated at more than U.S. $2 billion a year – is "laundered" through foreign banks and other financial institutions. What comes back to the country does little more than fuel inflation and send local land prices soaring. Yet the production of marijuana and cocaine provides year-round employment for over half a million people, and it pays better than coffee production. Food production has suffered – and prices have risen – as peasants turn to the cultivation of drugs. Without secure alternative sources of employment, Colombia is locked into a vicious downward spiral.

The coca bush can be grown on even the steepest slopes, those unsuited to any other crop. Incas promoted its cultivation originally, and production is still concentrated in their former heartland. The leaves are chewed to provide relief from fatigue in the thin mountain air.

Coca growers harvest the leaves about ten times a year. Processors steep the leaves in acid and convert the resulting liquid into a crude cocaine base using a cocktail of chemicals. The dried product is purified in a laboratory to form a white powder, cocaine hydrochloride.

The diagram traces the diminishing volume and vastly increased price of a crop of coca as it is processed to become "street" cocaine. Even though they receive only a tiny fraction of the final value, small farmers can still double their income by cultivating coca. The big profits, though, are made by the traffickers who trade the refined product. Street prices for cocaine vary greatly with quality, supply, and demand.

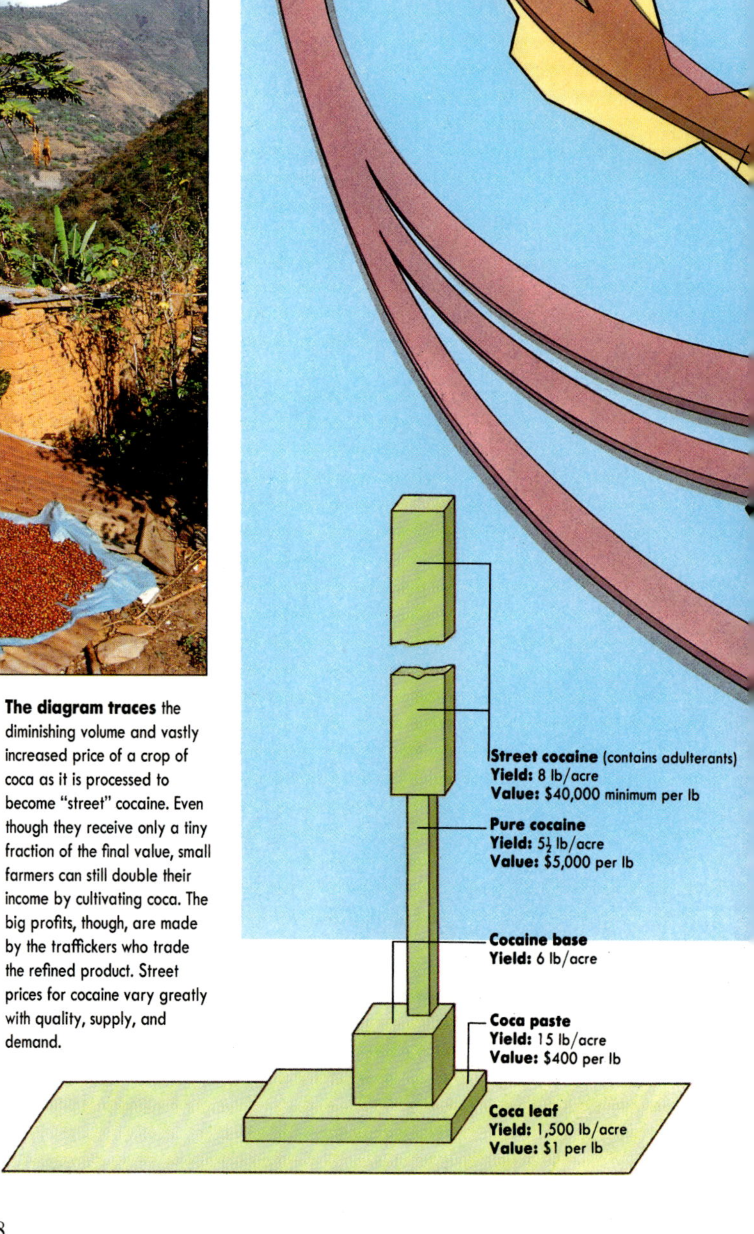

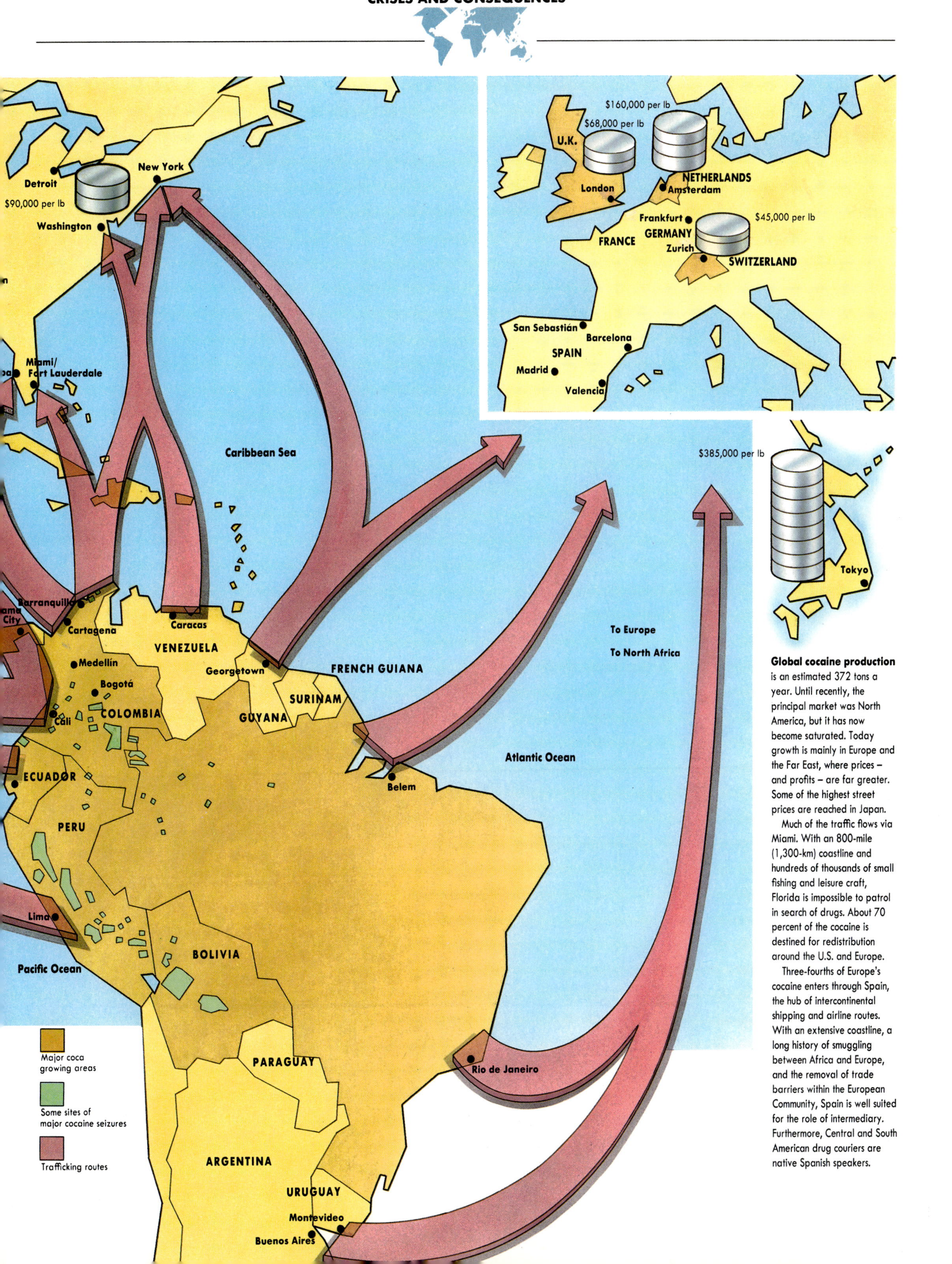

Global cocaine production is an estimated 372 tons a year. Until recently, the principal market was North America, but it has now become saturated. Today growth is mainly in Europe and the Far East, where prices – and profits – are far greater. Some of the highest street prices are reached in Japan.

Much of the traffic flows via Miami. With an 800-mile (1,300-km) coastline and hundreds of thousands of small fishing and leisure craft, Florida is impossible to patrol in search of drugs. About 70 percent of the cocaine is destined for redistribution around the U.S. and Europe.

Three-fourths of Europe's cocaine enters through Spain, the hub of intercontinental shipping and airline routes. With an extensive coastline, a long history of smuggling between Africa and Europe, and the removal of trade barriers within the European Community, Spain is well suited for the role of intermediary. Furthermore, Central and South American drug couriers are native Spanish speakers.

THE TROPICAL CHAIN-SAW MASSACRE

Forests at risk

DURING THE BURNING SEASON OF 1988, the vast rainforest of Amazonia visibly shrank. Satellite images suggest that in just a few hectic weeks the forest lost one-twentieth of its area – a fraction roughly the size of Britain.

The world's forests have been steadily shrinking since antiquity. Cut down largely to make way for fields and settlements, the woodlands of much of temperate Europe, North America, and Asia had been reduced to fragments by the turn of the century. Today, the frontier of deforestation has shifted to the tropics, especially to the luxuriant evergreen rainforests once known in the West as "jungle." These forests are being cleared at a devastating rate, causing worldwide alarm.

Loss of the forests also leads to a loss of plant and animal life. The tropical forests are home to at least two-thirds of the world's species, most of them as yet unnamed and unknown to science. According to some estimates, thousands of these are becoming extinct every year. We will never know what we have missed.

Wild plants are a vital resource. Cocoa, coffee, bananas, and sugar cane have already been improved by cross-breeding with wild species in the forest. With one in four Western medicines based on forest plants, and so much of the forest still unexplored, the potential for developing new drugs is also enormous.

Deforestation commonly results in soil deterioration and erosion. The rainforest's luxuriant vegetation may suggest that the soil is very fertile. It is not. In rainforests most nutrients are embodied in the living plants and animals, rather than held in reserve in the soil.

Each of the major rainforest regions is home to indigenous peoples. As the forests are cleared, their territories are diminishing. Some are forced into new reserves, others simply have their land appropriated and are left to survive as best they can. Their cultures are being extinguished, and many are dying from diseases like malaria, against which they have little defense. Some 500 years ago there may have been nine million Indians living in Amazonia. Today there are fewer than 200,000.

Deforestation may also be contributing to climate change. At the regional level, forest loss means less rainfall. The Panama Canal, for

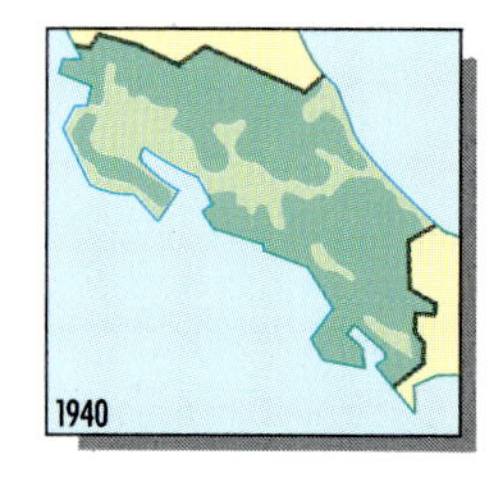

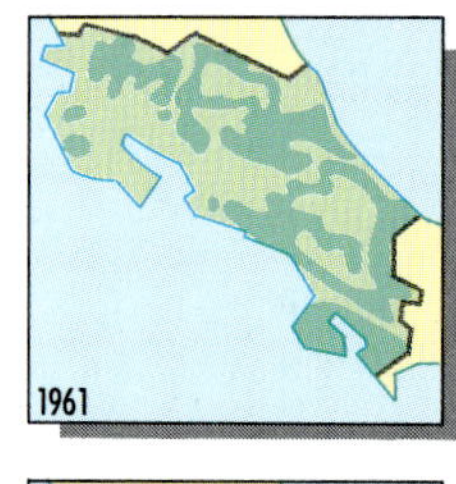

Costa Rica Its tree cover already much reduced, Costa Rica lost another third of its tropical forests between 1961 and 1987. The greatest forest loss has come through cattle ranching. Ranching is attractive because it requires little capital and labor. The total area of ranch land has doubled since 1961, and beef exports from the country have risen sevenfold.

But, more than most countries, Costa Rica is acting to conserve its remaining forests. Eight percent of its territory is already protected in reserves and national parks, and some replanting projects have commenced.

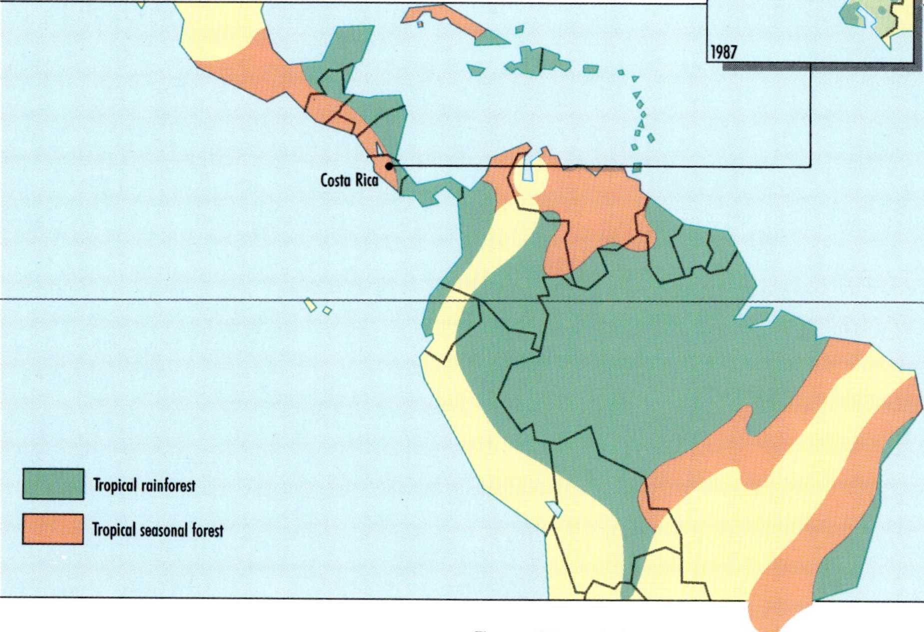

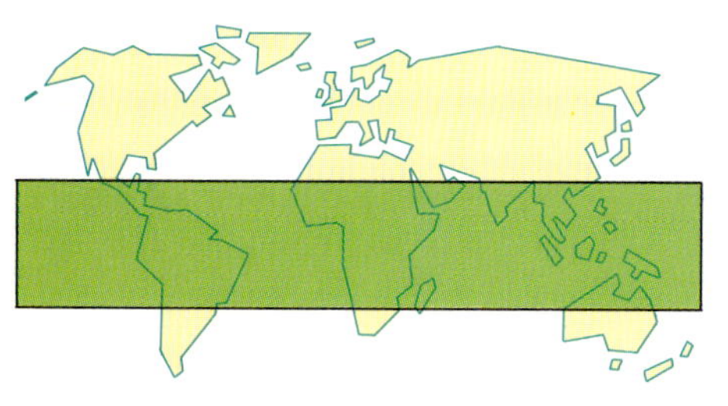

The world's tropical forests run like a girdle around the equator, covering more than 3.6 million square miles (9.3 million square km). The term tropical forest includes evergreen rainforests, mountain cloud forests, seasonal and dry forests. All are currently under threat.

Most of the seasonal forests have already disappeared or been degraded, and since 1945 two-fifths of the rainforests have also gone. Roughly 40,000 square miles (100,000 square km) of rainforest are currently disappearing every year. At present rates, Sierra Leone, Nigeria, and Thailand are likely to have few if any patches of virgin forest remaining by the year 2000. Globally, there will be virtually no rainforest left within 50 years.

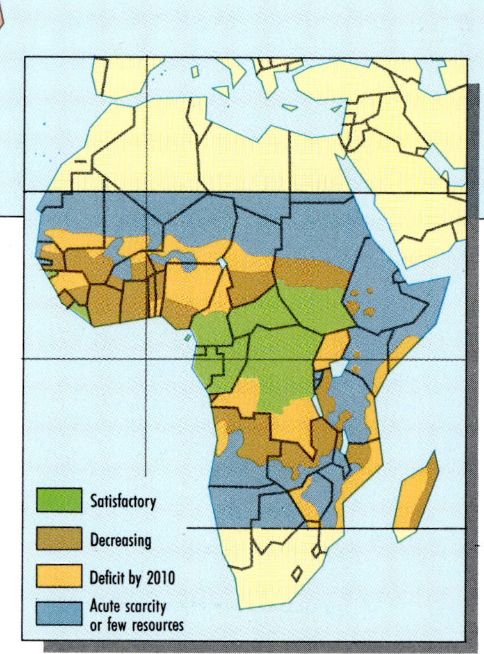

instance, is already forced to restrict shipping flows because there is too little water to fill its dams. At a planetary level, things may be even more serious since the burning and removal of trees contributes to the build-up of greenhouse gases. About one-fifth of the extra carbon dioxide currently being added to the atmosphere results from forest clearance.

Despite the gravity of the problems, there are many encouraging signs. Concern for the forests, and their peoples, has been raised worldwide. Pressure is mounting on governments and corporations that benefit from the products of current or former rainforests.

In northern India, women of the Chipko movement have denied loggers access to valuable timber by hugging the trees, a stratagem that is spreading to other threatened forests of the world. A planned dam on Brazil's Xingu River was scrapped, thanks to concerted action by local Kayapo Indians. National parks are being established, and reforestation projects primed. There is even a suggestion to offer debt relief to countries that undertake to protect their forests. But time is running short.

After fire, bulldozers, and herbicides have done their work, cleared land in Amazonia can quickly become useless. Even if abandoned after a few years of farming, the formerly rich vegetation may never have a chance to recover. Deprived of its natural cover, the soil is at the mercy of the elements. Nutrients are quickly washed out of the top soil, heavy raindrops compact the surface, and runoff can strip the earth from slopes.

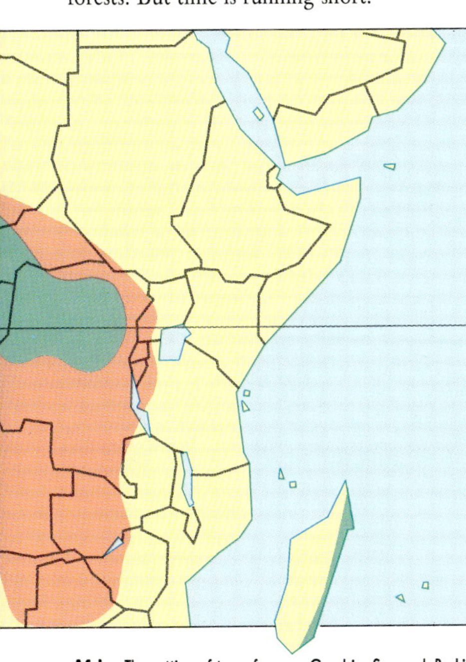

Tropic of Cancer

Equator

Tropic of Capricorn

Africa The cutting of trees for fuel adds to the strain on the world's forest resources. At least two billion people around the world use wood for cooking and heating, in the absence of alternative sources of cheap fuel. But the supply of wood for fuel cannot match the demand. Africa has some of the most severe problems. The timber in its rainforests, seasonal forests, and woodland savanna has been heavily depleted.

There is already an acute shortage of wood in much of Gambia, Senegal, Burkina Faso, Niger, Ethiopia, Kenya, Tanzania, Malawi, Zambia, Zimbabwe, Angola, and Swaziland.

Ethiopia's forests covered 40 percent of its land area in 1940 – today they cover just 4 percent. By the turn of the century, an estimated 535 million Africans will be unable to find enough wood to meet their fuel needs.

Southeast Asia Japanese demand for timber is the principal cause of deforestation throughout Southeast Asia. Japan alone accounts for one-third of the world market for tropical hardwoods. The lion's share of its supply comes from Malaysia: 440 million cubic ft (12.5 million cubic m) of logs were imported from the states of Sarawak and Sabah in 1987.

The logging companies operate on a vast scale, opening up huge sections of forest at a time. The felling of large trees brings others tumbling down with them, and their extraction disturbs the soil. Once the loggers have finished, they generally leave the degraded land and move elsewhere. Less than 0.2 percent of tropical forests are managed so as to yield a sustainable harvest of timber.

TRAVELING PESTS
Australia's unwanted guests

UNTIL 1788, NO PLOW OR HOOF HAD ever made its mark on Australian soils. There were no cereal crops, and no domesticated mammals (unless we count the half-wild dingo, introduced from India perhaps 8,000 years previously). There were also none of the common vermin that haunt the European, American, African, or Asian farmer.

European colonization terminated this splendid isolation. Spurred by the urgent demand back home for products to feed and clothe exploding populations, settlers converted vast tracts of wilderness to agriculture. In less than a century, Australia was home to over 100 million sheep and around 8 million cattle, far outpacing the number of people. Wheat crops covered huge areas in the south and southwest of the country, and great sugar cane plantations were established in Queensland.

The deserts of the interior demanded appropriate pack animals, like burros and asses. Soon there were more camels in Australia than in the whole of Arabia. And along with the foreign crops and livestock came other alien species, including ornamental plants and sporting animals. Native kangaroos were too inquisitive to offer much in the way of sport, so deer and even foxes were brought in.

Freed of nature's checks and balances, many ran amuck. The prickly pear cactus spread like wildfire, soon covering millions of acres. Rats and mice, which made their own way in ships' holds, multiplied at unprecedented rates. So did rabbits. With grass in superabundance, sandy soil for burrowing, and few natural predators, a single female rabbit can produce 25 offspring in one year. The Australian grasslands were overrun. Stripped of plant cover, precious soils were rapidly washed or blown away. Only in 1950, with the deliberate introduction from South America of the virulent disease myxomatosis, was the rabbit population finally brought under control.

For a few native species, like the larger kangaroos and the bush fly, the changed environment was a boon (the bush fly, for instance, found the proliferation of moist cow dung very much to its liking). But having evolved over millions of years in the remote solitude of their island continent, most native animals were overwhelmed.

Driven off farmland by guns, snares, and traps, forced to compete for food and shelter, and exposed to predators like rats, cats, and foxes, they retreated to remote areas. Many vanished altogether. Of the marsupials – Australia's only native mammals – 17 of the smaller species have already become extinct and another 29 are endangered.

The approaching invasion of the New World screwworm seems inevitable. The larvae of the fly hatch in open wounds on sheep and other animals, burrowing into the flesh and eating their host alive. Eradication is extremely difficult and the potential damage to Australia's sheep farming could be substantial.

But the migration of plants and animals has not been all one way. Australia, too, has been the source of transnational pests. For example, a colony of escaped wallabies has become pestilential in Hawaii, and dense mats of the Australian swamp stonecrop *Crassula helmsii*, originally introduced as an ornamental aquatic plant, are crowding out native water plants in Britain. Biologists fear there will be no way to halt the invasion. The biter has been bitten.

TWO-WAY TRAFFIC

Deliberate or accidental, the introduction of plants and animals from country to country goes on around the world. Through misjudgment or mischance, many colonists have become serious pests, with few natural predators to keep their numbers down.

California Mediterranean fruit flies have caused hundreds of millions of dollars' worth of damage to the state's fruit industry.

Galapagos The most destructive of the archipelago's unwelcome guests are goats and pigs. Goats strip leaves and bark, and compete for food with the giant tortoises and iguanas. Pigs uproot vegetation and dig for the eggs and young of birds and tortoises.

Hawaii The ornamental faya tree has escaped from gardens and is running rife on Hawaii and Maui. A rapid and vigorous grower, it crowds out native vegetation and spreads toxins through the soil.

New Zealand Rats, cats, ferrets, and dogs have had a drastic effect on New Zealand's fauna, especially its flightless birds. In just six weeks in 1987, a dog released into a forest reserve killed 500 of the park's 900 kiwis.

Southeast Asia The South American water hyacinth – a vigorous aquatic plant – has spread throughout the tropics, clogging waterways and invading paddy fields.

THE CANE TOAD

This reptile has been described as the perfect invasion machine. Introduced into Queensland, Australia, in the 1930s, it was intended to combat the cane grub that was destroying the sugar crop.

But cane toads eat practically anything. They can also reproduce in a great range of conditions, and the females can produce up to 40,000 eggs a year. As a result, they have thrived and proliferated faster than native species. Worse still, when attacked they exude a cocktail of toxic compounds so lethal that they are devastating native species and endangering household pets.

Uncontrolled by natural predators, the cane toad has now moved out of the sugar-growing area up into the unique wetlands of northern Queensland, across into the Northern Territory, and south into New South Wales. There seems no way of preventing its onward march.

Australian Pests

From the time the first Europeans settled in Australia, they began to introduce animals and plants to boost agricultural production and make their own lives more agreeable. Freed of nature's checks and balances, some of these species ran amuck.

Water buffalo
The spread of mimosa has been assisted by the escape and proliferation of the Southeast Asian water buffalo, which tramples banks into mud and reduces various lush grasses to obscurity. Its presence upsets the balance of the forest ecology, killing trees and eroding soil. Buffaloes drink too much water, which threatens the survival of other species, and they are a reservoir for bovine tuberculosis.

Mimosa
Mimosa pigra, a prickly shrub from Central America, is invading the wetlands of northern Australia. The thickets it forms are home for some endangered species, but the animals' food is still found among the indigenous plants. If these are choked out, the animals will die.

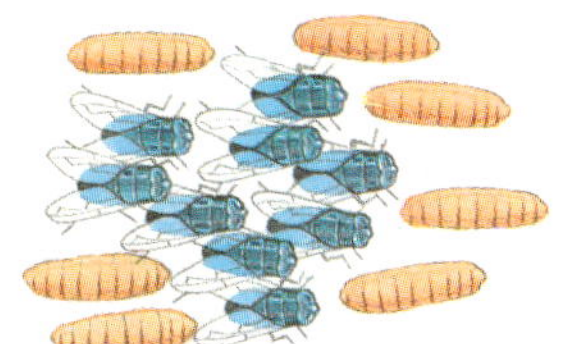

Screwworm
A native of the New World, this maggot feeds on living flesh with lethal results. It has already found its way over to New Guinea and seems likely to reach Australia in the near future.

Cane toad
A mere 102 cane toads were brought to Queensland in the 1930s. Their offspring are now wreaking havoc over large parts of Australia. Ironically, a chemical pesticide that effectively controls the cane grub that the toads were meant to eat was discovered as early as 1945. But it was too late to stop the cane toad.

Rabbit
The European rabbit has probably caused more damage in Australia than any other introduced animal in the world. It was brought to Australia in the 19th century to be hunted, and soon found that conditions over large areas of the continent were perfect. The population grew rapidly to 500 million. Grasslands were ravaged through overgrazing, and every five rabbits displaced a sheep from the savanna where they fed.

Rabbits proved impossible to control by ordinary means, so a very unpleasant disease, myxomatosis, was introduced. In 1950 it swept across Australia, killing 99 percent of the rabbits, finally bringing their numbers under control.

Today, the population is increasing again, presumably because the 1 percent that survived built up immunity to myxomatosis.

Prickly pear cactus
Introduced from the deserts of the U.S. as a garden hedging plant with a pleasing fruit, the prickly pear cactus proliferated until by 1920 it covered more than 50 million acres (20 million hectares). Houses were hemmed in by impenetrable prickly walls.

The deliberate introduction of the cinnabar moth, whose caterpillar has a prodigious appetite for prickly pears, effectively reduced its numbers.

Dromedary
Camels were introduced to Australia as pack animals because of their ability to survive in deserts, but they became unnecessary after the introduction of the automobile and the truck. They were turned loose and now run wild over much of the Australian outback.

Dingo
A domesticated variety of Indian sheepdog introduced to Australia at least 8,000 years ago, the dingo was probably responsible for the extinction of its marsupial equivalent, the native Australian wolf.

Its numbers are now enormous, and farmers attempt to control it by shooting and trapping, dropping poisoned meat from the air, and erecting dingo-proof fences hundreds of miles long.

WATER UNDER THREAT

The death of the Aral Sea

Most atlases show the Aral Sea in Soviet central Asia as the world's fourth-largest inland sea. Tables usually give its area as 26,500 square miles (68,600 square km), just a fraction smaller than Lake Superior in North America. And so, up to 1960, it was.

Ever since the 1950s, the Soviet government has been boosting production targets of the main cash crop of the region, cotton. For cotton read "white gold." One pound (just under 0.5 kg) of cotton brings more on international markets than 35 lb (16 kg) of grain. But cotton is a thirsty crop, and the area around the Aral Sea is a desert. A network of gigantic irrigation canals diverts water from the two major rivers that feed the sea.

Whatever the maps say, the latest satellite imagery tells a very different story. Since 1960, the coastline of the Aral Sea has retreated about 50 miles (80 km). Starved of inflowing water, levels are dropping by more than 3 feet (1 m) a year. At current rates, the sea will soon fragment into three residual brine lakes covering less than one-fifth of its former area.

In 1960, Muynak, then a southern fishing port, was being promoted as a future Soviet Miami Beach. Today it is not even on the coast. Where there might have been a yachting marina, a couple of dozen rusting hulks lie forever stranded amid interminable sand dunes. Only 4 out of 38 fish species can tolerate the salt levels. The fleet may be finished, but some canning factories are still working, processing frozen fish caught far out in the Atlantic and Pacific. It makes little economic sense. With half of its population already gone, Muynak has become a ghost town.

The irrigation project, too, has been a disaster. Precious water has been squandered. Smaller canals were not lined with concrete, they were simply scooped out of the soil. Half of the water drains away or evaporates before it ever reaches the crops. Nor was the water metered or charged to the farmers.

Cotton also demands prodigious quantities of fertilizers, pesticides, and defoliants. These have leached into the soil. Every drop of drinking water now has to be transported into the region by truck, as local sources are too polluted. The whole environment is saturated with salt and a cocktail of agrochemicals. Winds whip up these poisons into colossal dust storms up to 35 miles (56 km) deep and over 200 miles (320 km) long. Chemicals may be harming the health of local people: infectious diseases and cancers are proliferating, and the rates of miscarriage and fetal abnormality are higher than elsewhere in the Soviet Union.

Even the climate is changing. Much of the sea's former power to moderate temperatures has been lost. Now summers are hotter, winters colder, and the growing season is shortening. As summer temperatures rise, yet more water is lost through evaporation.

All that can be done is to limit the damage. Collective farms are now being charged for the water they use, production targets have been lowered, and work has started on lining canals and ditches. There is also a movement to establish a "green barrier" of salt-tolerant trees to suppress dust storms. But whatever is done, the Aral Sea will never regain its former status.

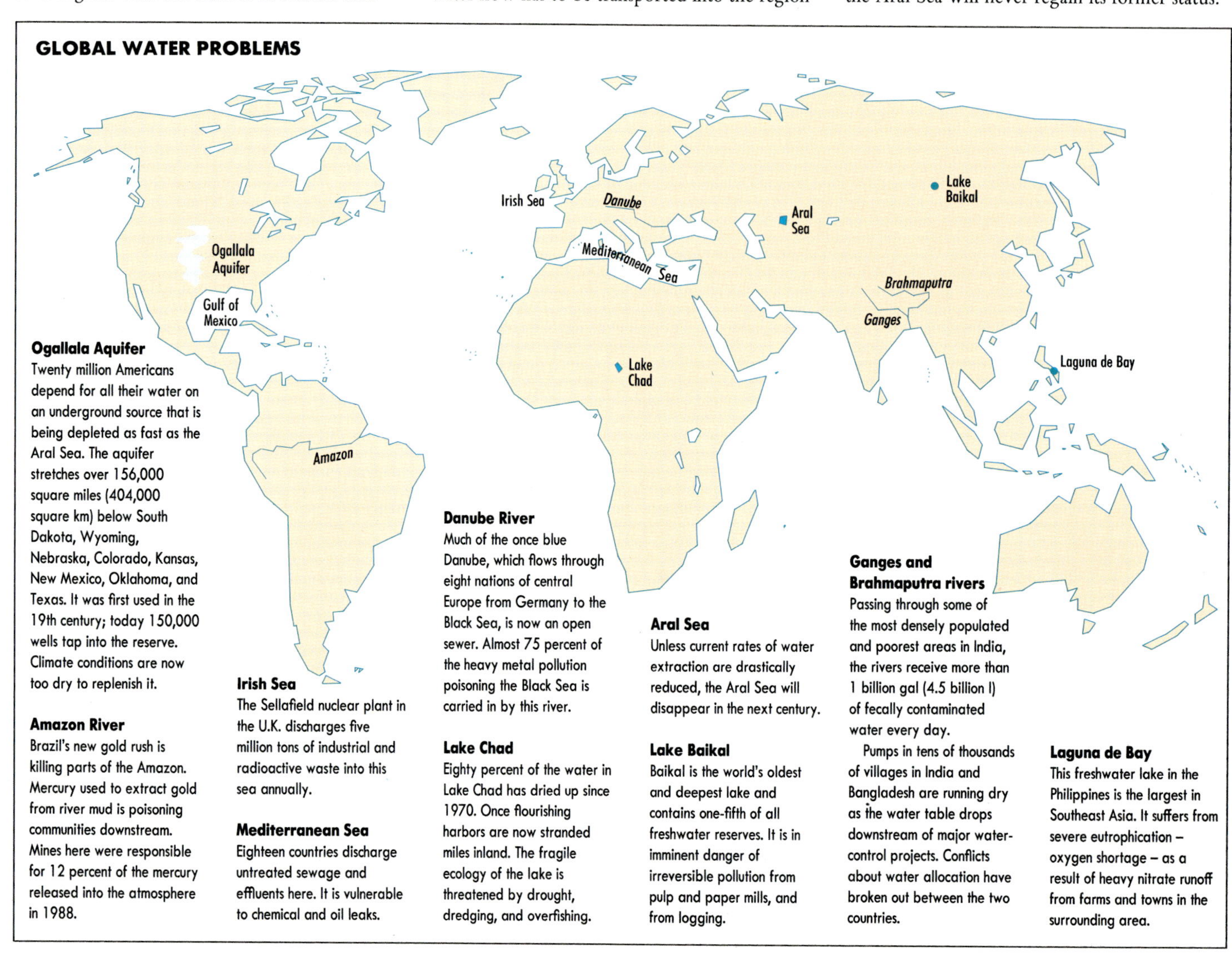

ARAL SEA DISASTER

Nature sometimes makes mistakes, asserted Soviet leader Joseph Stalin, and it has to be corrected. The tragedy of the Aral Sea (*above and right*), and the many other examples of environmental degradation in the USSR now coming to light thanks to *glasnost*, must be seen against the ideology of Soviet communism. Until recently, the view prevailed that nature – or geography – must be subdued by the will of the people. This is understandable enough, given conditions in the Soviet Union.

Geography has not been kind to the USSR. Three-fourths of the country spends much of the year in the grip of a terrible winter. It has lands that never thaw, deserts the size of western Europe, and mountains 10,000 ft (3,000 m) higher than anything found in the Alps. Yet since the 1930s, as a result of transporting millions of workers to every remote region, nature has indeed been "corrected." New canals, railroads, irrigation systems, mines, and cities have been built in seemingly impossible places. Great areas of the country have been made extraordinarily productive, though often at a terrible cost to humans and the environment.

Today, 40 years after Stalin's death, President Gorbachev admits that "ecology has got the Soviet Union by the throat."

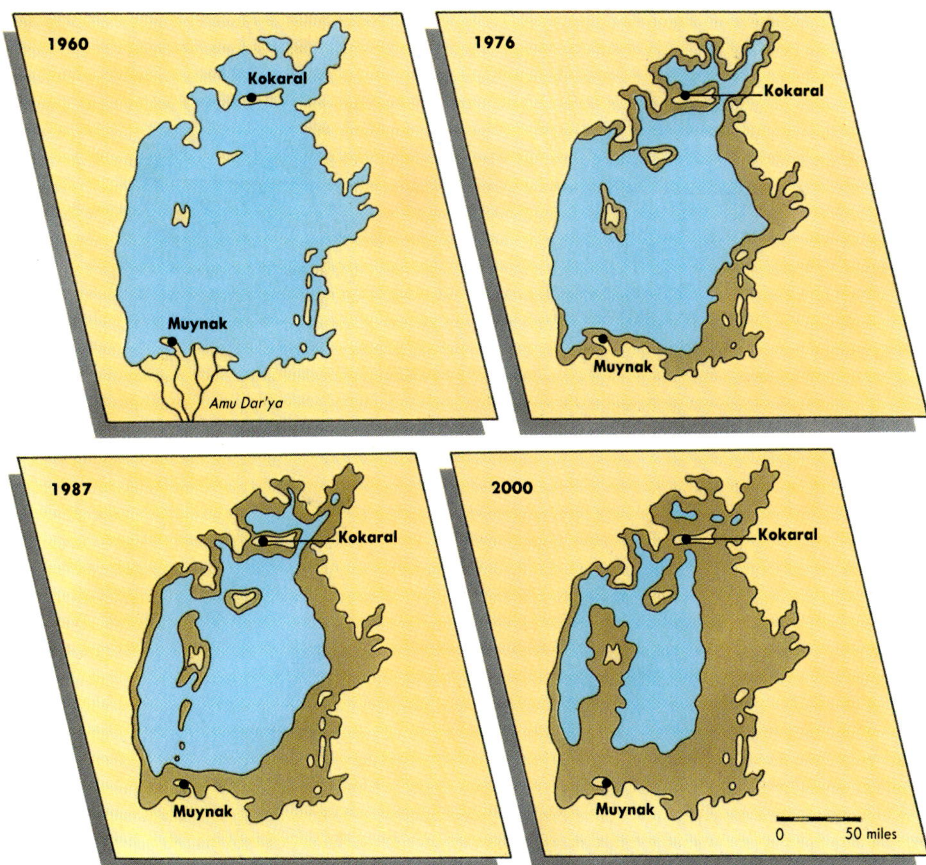

WHAT PRICE PROGRESS?

Industry muddies the waters of the Rhine

FOR OVER 200 YEARS INDUSTRIALIZATION has enriched the people of the developed world. But we are only now beginning to discover the full price that has been paid in global environmental degradation. The precipitation of acid rain, for example, demonstrates the national and international scope of industrial pollution.

Having entered the atmosphere, emissions from factory chimneys, power stations, or vehicle exhausts cannot be confined – winds recognize no international boundaries. When it takes only 18 hours for the exhaust fumes from London's rush hour to reach Switzerland's forests, or when an explosion at a nuclear power station at Chernobyl in the USSR in 1986 covered most of Europe in radioactive dust in a few days, no country can feel secure.

Water is a no less efficient means of redistributing pollution. The Rhine River, irreverently described as "Europe's largest sewer," carries away pollutants from some of the area's most heavily industrialized regions. Its daily diet includes effluents from chemical, pharmaceutical, paper-making, and steel works, as well as domestic sewage, shipping discharges, and runoff of agrochemicals and animal slurry.

Heavy metals and organic compounds, by-products of industrial processes, build up as sediments on the river bed or are carried into the North Sea where they enter the marine food chain. Every year 2,000 tons of lead, 80 tons of cadmium, and 16 tons of mercury accompany 390 tons of phenols and 315 tons of arsenic into the river. Salinity is rising as fast as salts are leached from soils, or are washed out of the potash mines of Alsace. Nuclear power stations also discharge radioactive substances.

Untold damage is caused to river life. By 1982, 9 out of 11 species of migratory fish had disappeared from the Rhine, unable to penetrate the chemical barrier separating the sea from their breeding grounds upstream. Paradoxically, the lavish supply of nutrients reaching the river from agricultural sources does not regenerate life. On the contrary, it destroys the ecological balance; algae and bacteria proliferate, robbing the water of oxygen. Along many stretches, plant and animal life have been extinguished.

Thirty million Europeans rely on the Rhine for their drinking water. But such a supply has to be heavily treated, since each liter of river water contains around 10 mg of synthetic organic substances, a large proportion of which are believed to be carcinogenic or mutagenic. Guaranteeing standards can be a real problem for water utilities. Filtration plants simply cannot cope when pollution increases after an accidental spillage or illegal discharge upstream.

It is not only domestic supplies that are vulnerable. Holland's highly productive truck farms rely on Rhine water for the irrigation of thirsty crops like tomatoes. Because of its position at the end of the waterway, the Netherlands receives water polluted by six countries over more than 800 miles (1,300 km).

In 1976, an international water authority, comprising representatives from all Rhine countries, was set up to monitor and control pollution levels. Discharges of heavy metals are beginning to drop, and oxygen levels are rising. Some fish and birds have been successfully reintroduced. Yet the accidental spillage that resulted from the 1986 Sandoz fire shows just how vulnerable the river remains.

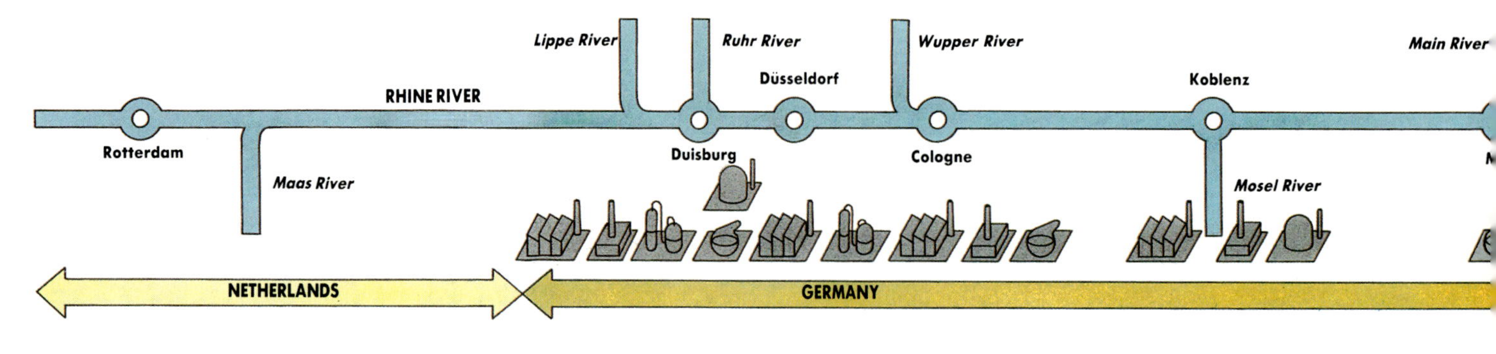

The extent of the Rhine
The Rhine River covers 825 miles (1,320 km), descending from a height of 1,300 ft (400 m) to sea level and flowing through three countries. The diagram represents the straightened-out river, and shows some of the major industrial cities which lie on or near its banks. The potential for pollution is frightening.

Rotterdam, the world's busiest port, offloads much of Europe's oil supplies. Pipelines carry the oil to refineries on the Ruhr.
Duisburg, a major port and center for steel and engineering works, lies at the confluence of the Rhine and Ruhr rivers, in the heart of the Ruhr industrial region.

Düsseldorf is home to many engineering, chemical, and paper manufacturing works, including Vega AG, Europe's second largest corporation.
Cologne lies close to Germany's main deposits of lignite, or brown coal, which is almost as important a source of energy in Germany as hard coal. Mined from huge open-cast mines – the largest artificial holes on Earth – it is used on the spot in giant power stations. Burning sulfur-rich lignite contributes to the acid rain problems of the Black Forest and northern Europe.
Koblenz lies at the junction of the Mosel and the Rhine. The Saarland, lying just up the Mosel Valley, is famous for its heavy industry.

Biblis is home to one of the world's largest nuclear power stations. When the third reactor is completed, the station will have a capacity of over 3,500 megawatts.
Mainz lies at the junction of the Rhine and the Main at the center of a major industrial agglomeration. A series of giant factories extends along the river bank, notably Opel automobiles at Russelheim.

Mannheim, at the junction of the Rhine and the Neckar, is a major engineering town. Industries in the Württemberg region of the Upper Neckar are mainly environmentally clean, but any accidental pollution here would pose a serious threat to the Rhine.

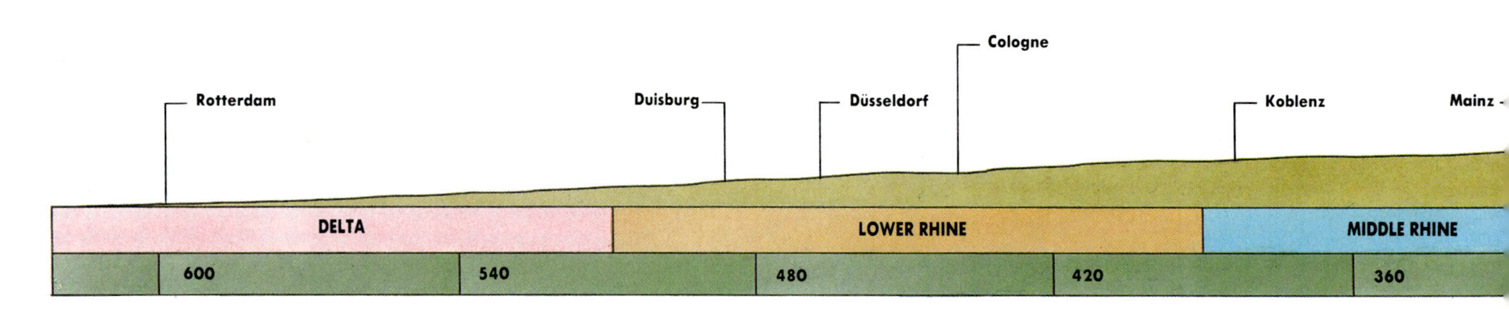

This vacation home in the Harz Mountains of Germany was surrounded by evergreen forest in 1970. Today, acid rain and disease and parasites have killed off all the firs.

It is not only plants that are damaged. Lakes become lifeless; metal, concrete, and stonework are rapidly eroded.

Rain and snow are highly acidic throughout Europe's densely populated industrial heartland, from England across Germany to Poland. Emissions of sulfur and nitrogen oxides from industrial processes, power stations, and vehicle exhausts converge on Scandinavia, blown there by the prevailing winds.

Four thousand Swedish lakes are now devoid of fish because of acid rain.

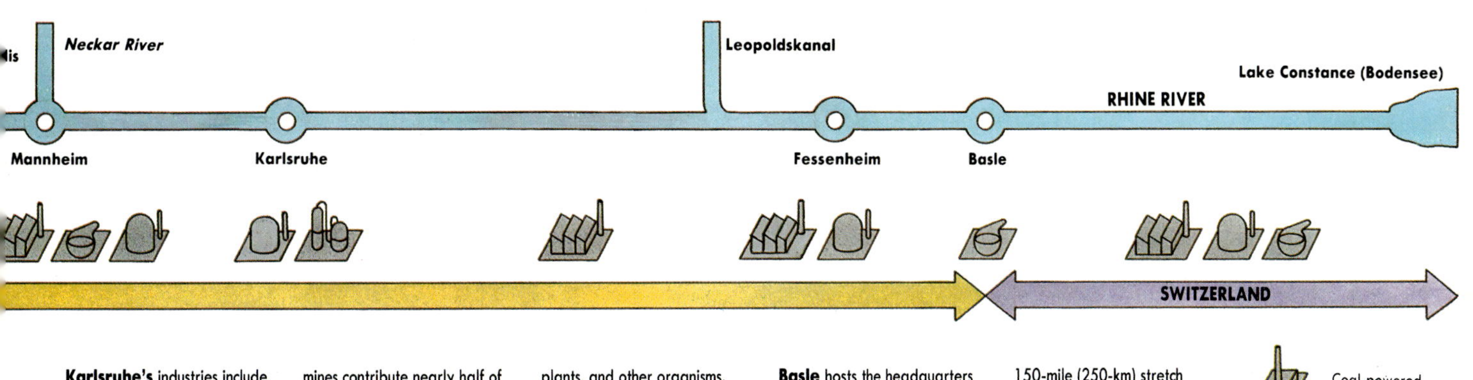

Karlsruhe's industries include oil refineries and nuclear research facilities. Major pipelines link it to Marseilles and Trieste on the Mediterranean.

Fessenheim is a village of 2,000 people situated where the Grand Canal d'Alsace meets the Rhine. Waste products of the Alsace potash mines contribute nearly half of the 56,000 tons of salts released daily into the Rhine.

Fessenheim is also the site of a nuclear power station whose heat pollution contributes to the Rhine's lack of oxygen. Since the 1970s the oxygen content of the water has been below the critical level, with detrimental effects on fish, plants, and other organisms. Although recent vigorous efforts have led to improvements in the Rhine's oxygen count, a "chemical barrier" in the Middle Rhine still prevents salmon from reaching their natural spawning grounds.

Basle hosts the headquarters of a number of large multinational chemical and pharmaceutical companies. A fire at the giant Sandoz chemical plant in 1986 resulted in an estimated 30 tons of mercury and other chemicals being washed into the Rhine. The accident killed millions of fish along a 150-mile (250-km) stretch of the river. It is estimated that the river will need at least 10 years to recover.

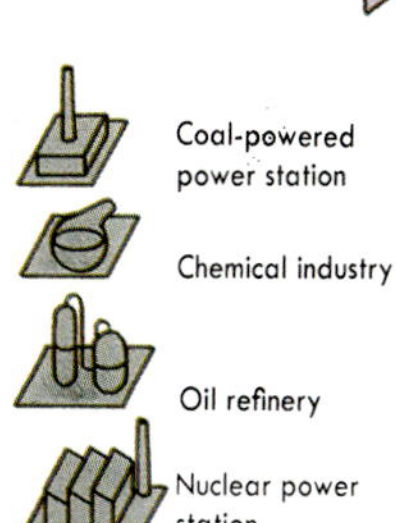

People in Place

Settling Down

Only 10,000 years ago, everyone on Earth could have been accommodated in a single modern city the size of New York or Tokyo. But they were nomads, moving continuously through the landscape in pursuit of food. Homes were temporary, simple, put up or taken down at a moment's notice.

The size, permanence, and diversity of modern homes and settlements ultimately reflects the expansion in production consequent on settled agriculture, the growth of industry, and, increasingly, worldwide trade.

When Cities Grow

The population of London – then the world's largest city – reached one million less than 200 years ago. Since then, great tides of people have surged to the cities from the surrounding countryside, or flowed across entire continents and oceans. Now there are nearly 300 "millionaire" cities. Of these, 24 contain more than five million people, and several are rapidly approaching four times that number.

The Urban Mosaic

The modern city may be a place of great size, complexity, and diversity, but it is not necessarily one of disorder. Underlying each unique street map and assortment of social groups are patterns of growth which have been a feature of cities ever since large-scale urbanization first occurred.

All cities continue to show evidence in their inbuilt environment and social structure of their origins – sometimes millennia earlier – as centers of trade and transportation, or mining and industry, or military conquest and colonial control.

Settling Down

Why do people live together? Why do we settle in more or less permanent, more or less tightly packed groupings? The ultimate answer is obscure, perhaps rooted in the evolutionary past, millions of years ago, when our ancestors banded together to secure mutual survival in the strange new worlds they were just learning to inhabit.

But we undoubtedly remain sociable animals. Even today, the need for, and the advantages of, collective life are undeniable. True, some people live as hermits in the midst of forests, or seek the solitude of a mountain fastness. But they are rare. The rest of humankind seeks out company for any number of reasons – emotional warmth, the sharing of work, the exchange of goods and services, security from external threat, sport, entertainment, and religious and political expression.

People need people, but we also need a degree of privacy – a degree that varies dramatically between different societies. Every intimacy of life in a Dayak longhouse in Borneo must be shared with up to 200 others. Yet even here each individual has his or her own space and personal possessions, albeit very few of them. Such people are frankly incredulous when they hear that there are places in the world where millions congregate. "Where do you go when you want to be alone with your ancestors?" one old man asked the Western, city-based anthropologist who had come to study his tribe.

On the other hand, many people find in the city a sense of liberation: freedom from customary constraints, or from close supervision of kin or village elders. Large numbers of people provide more opportunities for personal expression, and new kinds of association can be forged based on common interests or beliefs. Yet, despite its teeming millions, the city can also be an intensely anonymous and secretive place, home to the "lonely crowd."

The ways in which people create and maintain settlements are many and varied: from the individual home to much larger collections of buildings; from small, temporary, and relatively simple structures, like the tented encampments of the nomadic Kazaks of central Asia, to immense, durable, and complex supercities such as Tokyo, Seoul, or Mexico City. What principles guide the location and development of these different settlements? What governs their relationships, one with another? How is social life affected by the nature of villages, towns, and cities we inhabit?

Thanks to modern science and the lavish wealth of governments, corporations, and private individuals, homes and settlements can be found anywhere. Increasing numbers of people are living – sometimes luxuriously – in environments as diversely hostile as the high slopes of snow-covered mountains, the centers of hot deserts, polar ice fields, or platforms on open seas. There are even long-term settlements in space, such as the orbiting space satellite Soyuz.

These seemingly impractical places exist only because they provide the wider world with access to some exceedingly scarce and cherished resource, like a precious metal, or an energy resource, especially oil and gas. The price the world pays for the resource reflects the fact that settlements in such extreme locations are expensive to establish, and generally have to be wholly or largely serviced from outside with building materials, food, clothing, equipment, and energy. Disrupt supply lines, even for the briefest of times, and its inhabitants may die. When the resource dries up, or its value declines, the settlement itself dies.

Yet, for hundreds of thousands of years, people lived almost entirely in virtually self-sustaining communities, with next to no commerce with the rest of the world. Most were tiny, wandering groups reliant only on the natural productivity of their immediate environment. Settlements, fabricated out of materials that were readily at hand, were generally short-lived because the food resources upon which these nomads depended – wild plants, fish, and game – were rapidly used up, or were seasonal in character. As soon as supplies were exhausted, or the effort spent acquiring them outweighed the gains, the group moved on.

With the development of agriculture, a dramatic turning point was reached in human history. Farming demanded much more permanent forms of settlement than did hunting and gathering, so with the earliest food crops and domesticated animals came the first villages, towns, and cities. Settlements flourished in areas that were blessed with fertile soils and ample supplies of water.

As settlements became larger and more permanent, they developed additional functions. Markets for all sorts of produce sprang up, and merchants engaged in long-distance trade. Craftworkers, such as potters, shoemakers, and carpenters, were liberated from the need to grow food for themselves and could specialize in particular trades. Moneylenders and scribes recorded the flow of goods and collected taxes. Princes and priests grew rich on the proceeds. The village grew into a town, and in some cases, the town into a city.

For as long as agriculture remained the basic source of wealth – as it still does for most of the world's population – villages and towns depended for their livelihood directly on the produce derived from cultivation. Over time, however, the range of activities and sought-after resources has increased; and with it, the possible locations for settlements have proliferated. A good defensive site, a protected deep-water harbor, or an advantageous position astride a trade route are the most common locations, but the use of a new resource can also suddenly provide the impetus for settlement, as coal did for the Ruhr region of Germany, or as oil has done for Siberia in the 20th century.

In some cases, the nature of the "resource" that is being exploited is spiritual rather than material, as in the case of settlements that have grown around the birthplace of a religious leader, such as Mecca in Saudi Arabia, the site of a spiritual experience such as Lourdes in France, or mission stations like Los Angeles or San Francisco in California. In other cases, a site may be chosen for its symbolic significance. Capital cities like Madrid in Spain or Brasilia in Brazil were deliberately located in the center of the countries they serve.

Sometimes it is nothing more than social attitudes that determine whether a resource is valued or not, and hence whether a settlement is established. Rising affluence, additional leisure time, and cheap travel have greatly increased the demand in the developed world for new opportunities for hiking, for swimming and sunbathing, or for skiing. As a result, tourist resorts have proliferated in forests, around golden beaches, or in high mountains.

Yet for centuries, such environments have been thought of as hostile to "civilization" and hence rather repugnant. Today, however, people do not set their sights on merely vacationing in such surroundings. The attractions of more "natural" environments are now deemed so great that many people in the developed world are electing to move to refurbished or newly built "traditional" rural homes in "villages" with few, if any, agricultural functions. These people may commute to the city daily, or even "telework" from their "electronic cottages," thanks to the phone, the computer, and the fax machine.

Settlements may emerge and grow for any number of reasons, but they are also exceedingly vulnerable to disruption. In a few cases, a sudden natural disaster – a volcanic eruption or a tidal wave – has terminated a settlement's life instantly, as occurred when Vesuvius erupted in AD 79, covering Pompeii and Herculaneum in a thick layer of ash. In others, a social disaster such as an invading army or the arrival of a ship or merchant caravan carrying plague devastated the settlement.

Curiously, such cataclysmic events are rarely decisive. As long as land remains fertile, or mineral resources accessible, people generally return to the same site, or one close by, and rebuild the settlement. Many world-famous cities, including London, Moscow, and Chicago, have been razed at least once. Hiroshima in Japan, although annihilated by the explosion of an atomic bomb in August 1945, is once again a thriving commercial center with a population of more than a million

MORE OFTEN, PERMANENT ABANDONMENT IS due to a long-drawn-out change. Europe's landscape is littered with the remains of deserted villages, many of them established 1,000 or more years ago. Some were evacuated when the climate deteriorated, or when silt gradually filled their harbors, or as the result of malaria becoming established in local swamps. In others, landowners increasingly turned peasants off the land to graze sheep or create game reserves for their own sport. Now, these villages can be traced only as the archeological remains of field boundaries and ruins.

Large settlements never remain entirely dependent for their well-being on their initial resource base. Needless to say, the extent to which cities like San Francisco and Los Angeles thrive bears no relation to their former missionary activities. As settlements grow, they generally also become more complex, and hence more resilient. They support a greater variety of specialized trades and provide many more functions for the network of neighboring and complementary settlements that surround them. Even when they are faced with resource depletion, or are devastated by calamity, such towns and cities cannot be allowed to die.

CROWDS AND CONTINENTS

Where people live

THE EARTH IS A SURPRISINGLY EMPTY place. Even leaving aside the more than 70 percent that is ocean, its surface is sparsely populated. If the entire human race were spread out evenly over the land, each person could have an area the size of three football fields to himself. Standing at arm's length from our neighbors, we could all fit into an area no larger than Long Island.

In fact, much of the world is uninhabited, or nearly so, while extraordinarily dense clusters of people are scattered apparently at random across the globe. The largest of these clusters are found in the east and south of Asia, in Europe, and in northeastern North America.

It is much easier to explain where people are *not* to be found than where they are found. Icy tundra, arid desert, and inhospitable mountain ranges are all mostly without people. Disease, drought, infertile soils, geological and other local hazards discourage settlement over much of the rest of the world.

A huge proportion of the people in poorer countries are village-based peasant farmers who tend to live where the soil is rich and allows intensive food production. Villages and their surrounding fields are found wherever the land is cultivable, with occasional larger market towns interspersed among them.

In China, for example, a total population of 1.1 billion people, most of them farmers, is concentrated in the east of the country, with fingerlike extensions of the population reaching into the interior along the Chang Yiang (Yangtse) and Huang He (Yellow) rivers. The mountains and deserts of western China, Tibet, and Mongolia restrict further spread.

Similarly, the vast majority of the massive population of south Asia – India, Pakistan, Bangladesh, and Sri Lanka – live close to the coast or on the broad plain created in the east and north by the Ganges River and in the west by the Indus. Again, physical features – in this case the Himalayan mountain range and the Thar desert – clearly mark out strict limits to the spread of the rapidly expanding population.

Even so, it is difficult to explain population distribution wholly in terms of natural forces. Historical, cultural, and technological factors also play a major role. Many places in the world have fertile soil and mild climates – the areas along the banks of the Mississippi and the Nile, for example – but their respective population densities are very different thanks to their distinct settlement histories. The interplay of natural and social forces in determining population and settlement patterns can clearly

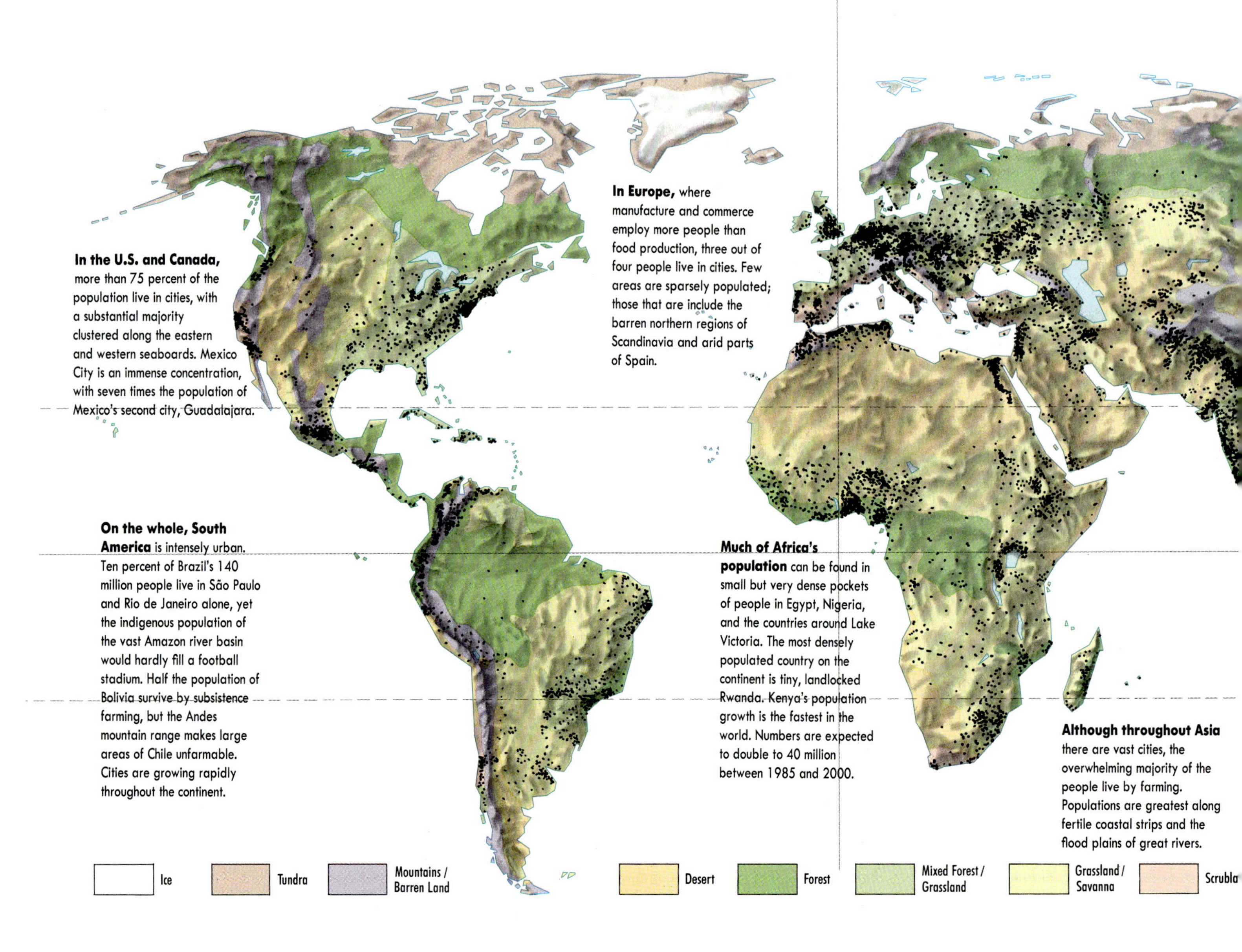

be seen in countries of similar size but with very different populations or population distributions, such as Australia and Brazil, Canada and China.

High population density is also associated with the world's most dynamic industrial and commercial regions. Food production has become much less important than manufacture and commerce as a source of livelihood in Europe, Japan, and northeastern North America, and their populations are now overwhelmingly urban. But even here natural geographical factors are important. Cities and towns that grew in Europe and North America during the 19th century owed their success to coalfields and iron ore beds. Into this century, oil fields, navigable rivers, or deep-water ports have attracted urban settlement more than favorable climates or soils for agriculture.

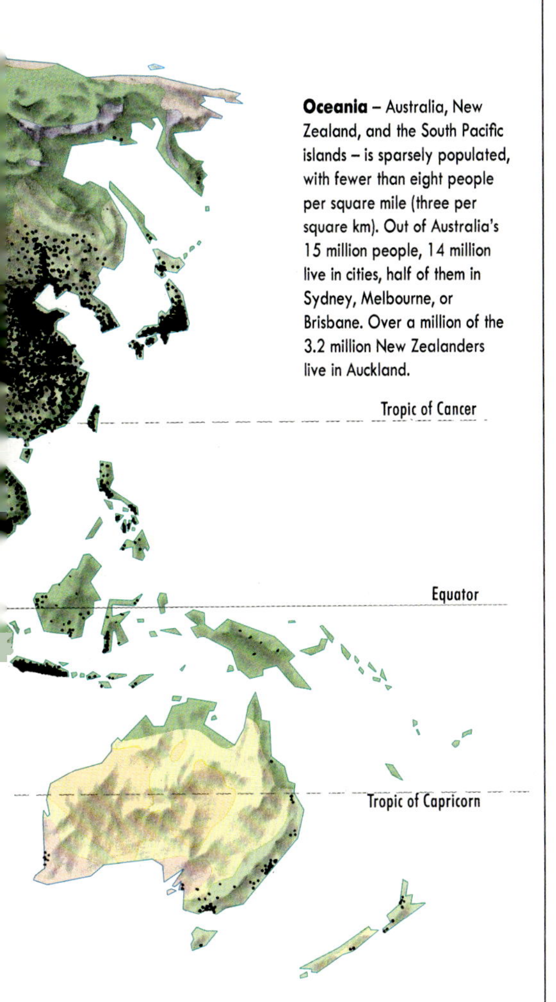

Oceania – Australia, New Zealand, and the South Pacific islands – is sparsely populated, with fewer than eight people per square mile (three per square km). Out of Australia's 15 million people, 14 million live in cities, half of them in Sydney, Melbourne, or Brisbane. Over a million of the 3.2 million New Zealanders live in Auckland.

1850–1900 Before 1850, only London and Paris had populations exceeding one million. By the end of the 19th century, another 11 cities had joined them: New York, Philadelphia, and Chicago in the U.S.; Berlin, Vienna, Moscow, and St. Petersburg (present-day Leningrad) in Europe; and Calcutta, Beijing, Shanghai, and Tokyo in Asia.

Chicago
New York
Philadelphia
London
Paris
Berlin
St. Petersburg
Moscow
Vienna
Beijing
Tokyo
Calcutta
Shanghai

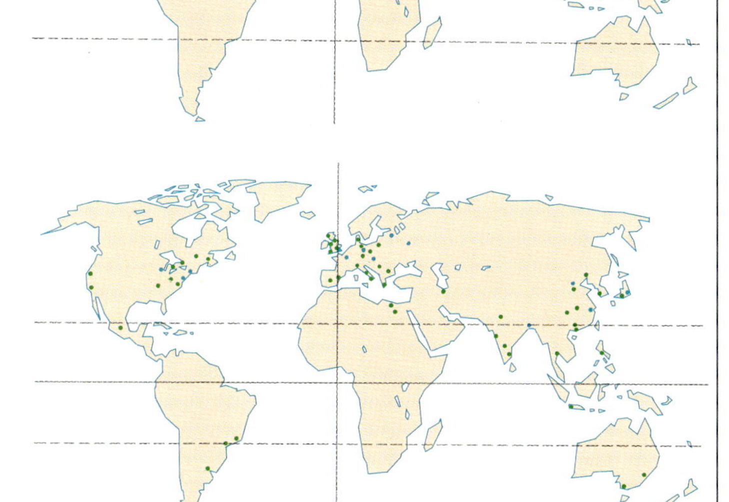

1900–1950 During the first half of the 20th century, another 54 cities rose above the one million mark. Of these, 10 were North American, 18 were European, and 17 were Asian. Latin America, Africa, and Australia gained their first millionaire cities.

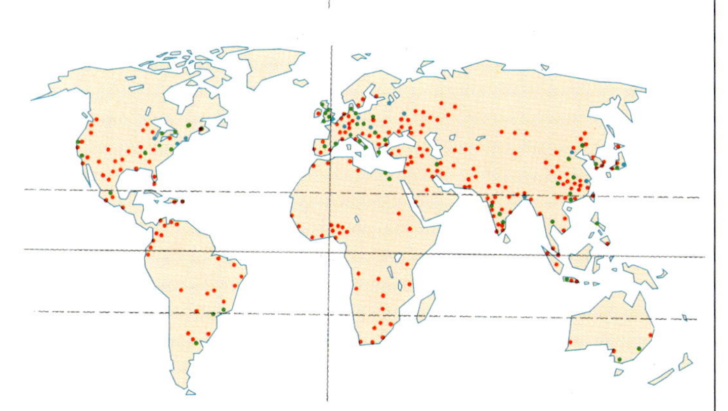

1950–2000 In the latter half of the 20th century, no fewer than 242 cities attained or are projected to reach populations of one million, with the vast majority in the developing world. By 2000, some 24 of the world's cities will have populations in excess of 10 million; 3 are likely to be larger than 20 million.

MILLIONAIRE CITIES

Few cities in history ever reached a population of one million. Most were very much smaller, rarely above even one-tenth that number. Although it is difficult to reconstruct numbers in the past, Rome probably became the world's first "millionaire city" in the first century AD. Within a few centuries, a combination of barbarian invasion, papal schism, and epidemic disease reduced numbers to around 20,000. Rome did not reach a million again until 1930.

Before the industrial revolution of the 18th and 19th centuries, only a scatter of the world's cities – Xi'an (China), Baghdad (Iraq), Byzantium (modern Istanbul, Turkey), and Edo (modern Tokyo, Japan) – may have housed more than a million people, and then probably only for a short period before they, too, suffered major population loss. The proliferation and continuous growth of urban centers seems to be a phenomenon of the last 200 years.

London became the first millionaire city of the modern period, shortly after 1800. Within the space of a hundred years its population had exploded to more than six million. In the meanwhile, it was joined by 12 other cities. Most were in Europe and North America.

London's peak size of 10.7 million – attained in 1961 – looks puny today. At that time it had already been toppled as the world's largest city by New York. Since then, however, both New York and London have started to slide down the big league table. By the year 2000 London is unlikely to figure in the listing of the top 25 cities. Although today's suburban sprawls make calculating a city's true size more a matter of where you define its limits, by the year 2000, Mexico City, São Paulo, and the greater Tokyo region are each expected to exceed 20 million.

HAVENS IN A HEARTLESS WORLD?

Variety in human habitation

THE RANGE OF PLACES THAT PEOPLE call "home" is astonishingly wide.

For one million people in Calcutta, home is the side of a road. They rarely walk through a doorway; their home life takes place in full view of thousands. For the Dayaks of Borneo, home is an almost undivided structure shared with 20 or 30 other families. Yet each member of an American suburban family may have a room of his or her own, shared use of several large rooms such as a living room, dining room, and den, not to mention a garage, lawns, and possibly a swimming pool or tennis court.

Such amenities, taken for granted in some parts of the world, are unheard of in others, where running water, electricity, and piped sewerage may be unavailable even in cities. Yet homes, be they tents or exclusive penthouses, have common and universal purposes.

Three Rs neatly summarize these functions: recuperation, reproduction, and refuge. Recuperation means body maintenance; the home provides a place for eating, washing, rest, and sleep. Reproduction refers to the home as the center where children are conceived, reared, and educated. Refuge describes the home's role as a means to survival, protecting people from weather, animal pests, and human marauders.

Until this century, most homes performed no more than these basic functions. They had few amenities, were poorly lit and heated, and offered little insulation from the outside world.

Indeed, outside the richest 20 countries in the world, the great mass of people live in such dwellings today, with only a small elite able to afford modern amenities. Most rural people still build their own homes, often on land that is occupied illegally or without title. Such houses are generally of minimal size and put up with little regard for health or safety. In the congested shantytowns of Bombay or Rio de Janeiro, they are dangerously crowded. Yet curiously, it is in these areas that "vernacular" materials and designs may have most to offer (*see p. 137*).

In the developed world, however, the modern home has extensive functions as a medium of self-expression, a status symbol, and a store of wealth. It has become the prime site for entertaining – and perhaps thereby influencing – friends and work colleagues. And, increasingly, a family's home is its most important "fixed asset," to be drawn upon on retirement or handed down to children.

Perceiving the home in this way began when its role as a workplace, in cottage industry, was undermined by the industrial revolution. As we enter the 21st century, the work function is "coming home" again, as those in the fields of information handling, product development, and research become linked by telephone, fax and computer lines to their offices.

***Tanka*, or boat people,** are a characteristic feature of Hong Kong's waterfronts (*left*). Mainly fisherfolk, their numbers are dwindling as fishing becomes less important in the Hong Kong economy. But 50,000 people still live in junks and boats as their families have done for generations.

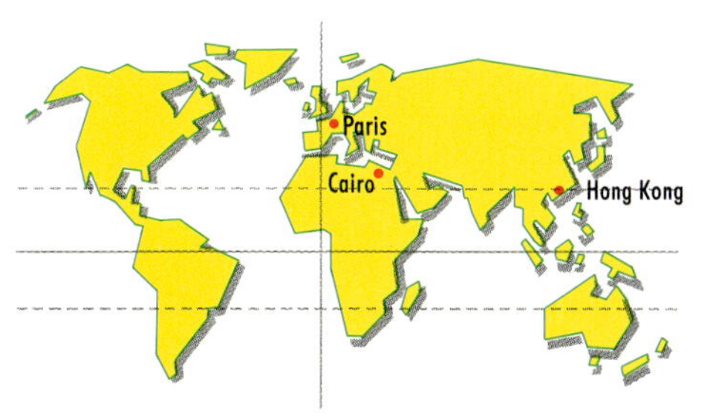

A sunny Parisian balcony (*right*) provides an outdoor haven overlooking 19th-century apartment roofs. Almost nine million people live in what is said to be a collection of 100 villages that makes up greater Paris. The presence of 1.2 million foreigners, including Portuguese, Algerians, Moroccans, and Spaniards adds a great degree of diversity to the city.

Cairo's "City of the Dead" (*below*) is a huge cemetery that houses the remains of thousands of Cairenes and perhaps a million squatters. Mosques, schools, shops, street vendors, police, and water and electricity supplies are all found in this necropolis. The squatters live in the tombs themselves, the only shelter available to them in this crowded city.

NESTBUILDERS
Homes from local materials

MAMMOTHS' TUSKS, BUFFALO DUNG, newspapers, whiskey barrels – there is seemingly no end to the materials from which people can contrive a home. If a material is durable enough to survive the local weather, is easily worked, culturally acceptable, and affordable (but preferably free), it can be pressed, shaped, and beaten into service.

When loose materials cannot be found nearby, one economical solution is to move the community to the material. Early people sometimes used or excavated caves, which can be dry, relatively warm, and easily protected. The caves of Cappadocia in Turkey are still in use, and the people of Matera in southern Italy have only recently vacated the caves so laboriously excavated by their ancestors.

Paradoxically, the more urbanized people become, the more they use space underground: parking lots, subways, nuclear weapons, and computer control centers may be built below ground, for reasons of space, security, or energy efficiency. Japan's Alice City project suggests a return to subterranean homes as an answer to Tokyo's overcrowding problem.

Most homes are built above ground, however, using whatever materials are at hand, modified as little as possible. A few trees can become a house; the trunks make the frame; the branches, twigs, and leaves are woven or thatched into walls and roofs. Packed earth, sod, sun-dried clay bricks, and stone rubble can also be turned to good account.

But "permanent residences" these are not. Plant matter rots away quickly, particularly in the tropics where rain, wood-boring insects, and fungi take their toll. Walls of untreated mud return to the earth from which they came. Sooner or later, rubble houses collapse.

Where modest, traditional buildings are recognized as typical of their region, it indicates that durable material was to be had locally – for example oak and fine stone in southern England, marble in India. Otherwise, durability was only available at the considerable expense of transporting materials, or the time-consuming processes of treating, surfacing, and reworking local products.

Building styles changed as materials were exhausted. With shipbuilders competing for supplies, oak became scarce and expensive in southern England 400 years ago. Bricks, which had not been made since the Romans left more than a thousand years before, became the substitute, their colors reflecting the chemistry of the local clay. Then, in the last century, canals and railroads allowed cheap, highly standardized bricks to be distributed around the country, along with Welsh slate for roofing. Local variation in housing was largely lost.

More recently, synthetic materials and factory

The Rendille tribe in the desert areas of northern Kenya are pastoral nomads. They plaster mud over bent sticks in order to construct their temporary shelters.

The perception of mud and clay as "primitive" and "low class" is contradicted by these elegant buildings in Morocco. Modern architects, too, are discovering the virtues of mud construction in this part of the world, while in New Mexico, adobe has become a chic building material for the luxurious homes of millionaires.

The spread of vernacular architecture

The most popular building material in the world is mud. But locally available materials – wood, brick, stone, bamboo, leaves, and bark – are used in traditional or "vernacular" architecture the world over. The emphasis is on simplicity and practicality, with traditional styles reflecting the culture of the people or region.

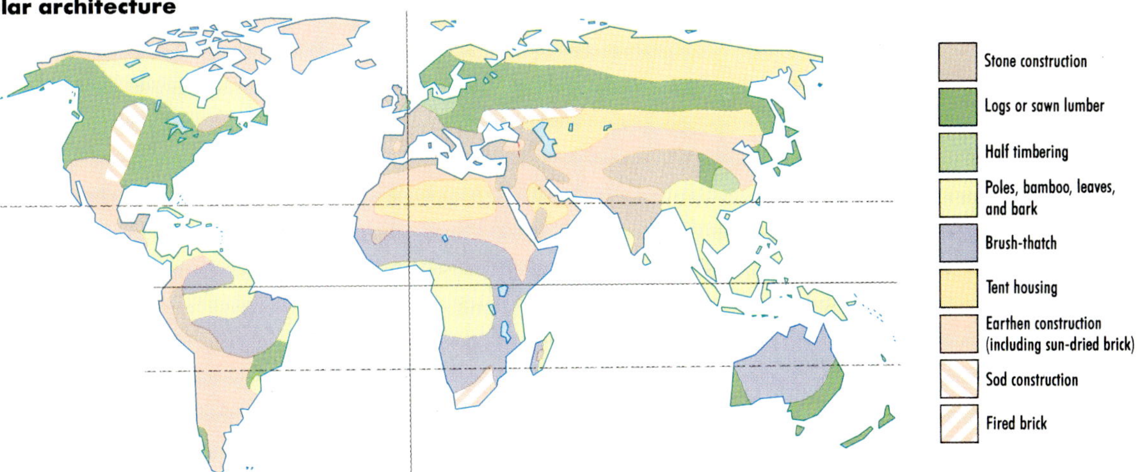

VERSATILE MUD

Mud and clay may seem unpromising materials to build with. But they can be versatile, durable, and comfortable. Many residents of "spontaneous settlements" in the third world are now returning to vernacular techniques and materials, forsaking the cardboard boxes, oil drums, and other detritus associated with shanties. The following building styles, sometimes dating back thousands of years, are still in use today.

A large part of northern China is underlain by soft and silty clay soil known as loess. As a building material it is extremely flexible, and dwellings are carved into hillsides or underground with relative ease. Early writings suggest that this "building by subtraction" may date back to the 11th century BC. The region has hot, humid summers and cold, dry winters, and the cave dwellings, with their thick walls and ceilings, are ideal for these extreme conditions.

Subterranean houses may be built around a courtyard, with living and sleeping quarters excavated on all sides. Entire villages and towns, including schools, factories, and government buildings, are built in this way.

In Yemen, and many other parts of the Arab world, substantial dwellings are constructed from mud bricks made in patterned molds, and brightly painted. Such buildings are often highly sophisticated – and surprisingly tall. A recent oil boom, and the consequent move toward urbanization, has led to an urgent need to modernize the Yemeni capital, San'a, and the introduction of more modern building technology seems unavoidable.

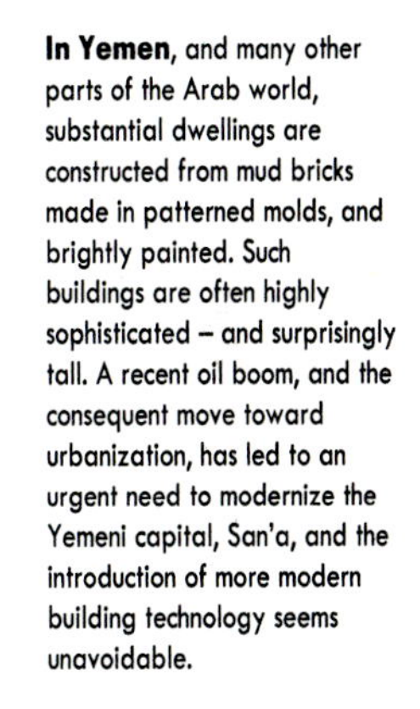

In picturesque Devon, in the southwest of England, unbaked clay is mixed with straw to produce a substance known as cob. Even today, many houses in rural Devon are made of cob, a fact that is concealed by a plaster render and paint. Traditionally, the walls were whitewashed or tarred to stop cattle licking the salt deposits exuded from the walls. Roofs extended close enough to the ground for cats to jump up them and catch mice living in the thatch.

prefabrication have continued this process of standardizing a country's – indeed, the world's – building. Iron, steel, and even glass now substitute for local timber, stone pillars, or brick walls. Softwoods, transported huge distances, chipped, and molded, are available off-the-shelf as standard components.

And since its first use in modern times, in the 1830s, concrete has become the universal building material because it is so light to transport and versatile in use. In an early display of creative marketing, its developers overcame initial sales resistance by naming it Portland cement. They claimed it was indistinguishable in appearance from the fine gray limestone mined in the Portland region of southern England and widely used in the construction of London's finest buildings.

Houses are sometimes provided, like television sets and motor cars, by specialized companies using factory methods, construction plant, and standard designs. An enormous range of materials continues to be used, from new synthetics with unpronounceably long names to traditional materials such as stone and tiles to add prestige or esthetic merit.

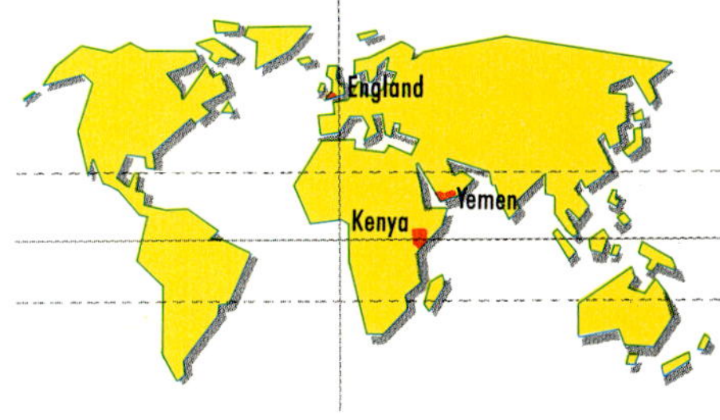

HOMES ON THE MOVE

Nomads and their settlements

For most people, home and settlement remain fixed in space. But for some they change daily, or with the seasons. A diminishing but significant number of societies still live as they have always done, hunting, fishing, and gathering, or herding animals.

Hunter-gatherer bands are generally around 30 strong. Any more, and the effort required to feed everybody is considered too great. Any fewer, and the social structure is weakened. Shifting agriculture – cultivating a patch of land until its nutrients are exhausted, then growing crops in another area while the original one lies fallow and reestablishes itself – is associated with slightly larger permanent social groups, but even here pressure on scarce resources keeps numbers low.

Yet such peoples are not "unsociable." On the contrary, periods when food is especially abundant are often taken as an opportunity to celebrate in large groups. The Ona of Tierra del Fuego, who generally spend their lives in tiny bands in a ceaseless search for shellfish, collect together whenever a whale is stranded. Likewise, the !Kung people of the Kalahari desert in southern Africa gather in groups of a hundred or more when they kill a large animal such as a wildebeest or eland. They may also collect together to share resources when these are particularly scarce, as during the winter dry period, when they congregate around the few permanent water holes.

Such gatherings are often associated with ceremonies and other collective projects – initiations, trance dancing, story-telling, healing, and the arrangement of marriages.

The movements of nomads are not random. Most are related to the seasonal and longer-term availability of food, and the recognition of the customary rights of other groups to exploit the environment. In many cases, a lifetime's travel for a nomad means covering hundreds if not thousands of square miles.

Some journeys may be undertaken regularly every year. Many pastoral nomads drive their herds between traditional spring and winter pastures – a pattern known as transhumance. Although the pastoral Kazaks of central Asia lead a wandering life for much of the year, each clan returns to its own customary territory for the winter (*see box*).

Many hunter-gatherers leave areas "fallow" for long periods, sometimes 20 or more years, before returning. If they are forced to return sooner, the subsequent supply of food is harmed because the land and its plants and animals have not had a chance to recuperate. Indigenous peoples are on the whole much less destructive of a forest's or a desert's long-term, sustainable productivity than intruders without specialist knowledge of these fragile ecosystems.

A NOMADIC YEAR

The lifestyle of the Kazaks of central Asia is largely unchanged by the 20th century. In the course of a year, a Kazak family and its herds of sheep, horses, and cattle may travel 500 miles (800 km). Summer and winter quarters may be 200 miles (350 km) apart.

Their year can be divided into six phases:

- As much as five months in winter quarters, from which they travel daily to find grazing.
- A period of constant migration, during which herds soon exhaust the early spring grasses.
- About six weeks on pastures which receive the first rains and are lush early in summer.
- Frequent migration during the hot and dry months of July and August.
- Six weeks of plentiful autumn grazing.
- A speedy return to the winter pastures when the weather turns cold.

In many parts of the world nomads are being forced for economic or political reasons to abandon traditional ways of life. In some cases lifestyles are adapted to modern circumstances. Groups of Fulani herders in Mali and northern Nigeria continue to supply beef in the area, but they have exchanged their former independence for regular waged employment. The Bedouin of the Arabian desert, one-time camel caravaners, now dominate long-haul trucking.

But for other former wandering peoples who have been forced to settle, the experience has been psychologically and culturally devastating. The social structures that once bound them together are strained to breaking point. When the Chenchu hunters of Andhra Pradesh in India were forcibly settled, the incidence of violent crime among them rose alarmingly.

Some members of one-time mobile societies are now attempting to turn back the clock to ensure their survival. Aboriginal Australians are returning to the traditional life, retracing the "song lines" established by their ancestors.

North American Plains Indians traditionally lived in tepees made of buffalo skins – later canvas – stretched over poles of pine, cedar, or spruce.

The skin-over-wooden-frame style of tent is still used by the nomadic Chukchi people of the Soviet Arctic, but they use walrus skins.

The igloos of the Inuit people are made from blocks of compacted snow, laid in an ascending spiral so that the dome stands up without scaffolding. The exit tunnel is built with a right-angled bend to keep out blasts of cold air, and the main chamber is hung with skins. If several families are camping together, individual igloos may be joined by tunnels leading to a large chamber designed as a meeting place.

The Kirghiz of central Asia – pastoral nomads related to the Kazaks (*see box*) – winter in large camps along river valleys. In this camp at Ag Djelga, the Kirghiz drive their goats and sheep into a stone enclosure for protection against the cold and wolves. In spring, most of the tribe disperses across the rich pastures; a few remain in the lower country and grow grain to tide both people and horses over the harsh winter.

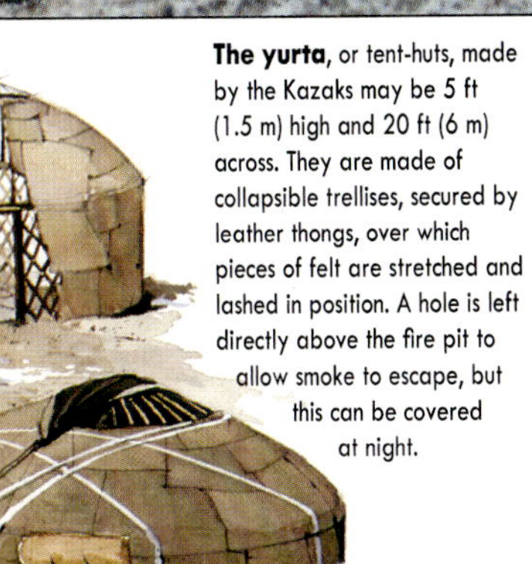

The yurta, or tent-huts, made by the Kazaks may be 5 ft (1.5 m) high and 20 ft (6 m) across. They are made of collapsible trellises, secured by leather thongs, over which pieces of felt are stretched and lashed in position. A hole is left directly above the fire pit to allow smoke to escape, but this can be covered at night.

The Bambuti are hunter-gatherers who inhabit the forest regions of central Zaire. Their homes are simple, dome-shaped huts, made from an overlapping layer of leaves laid over a framework of branches.

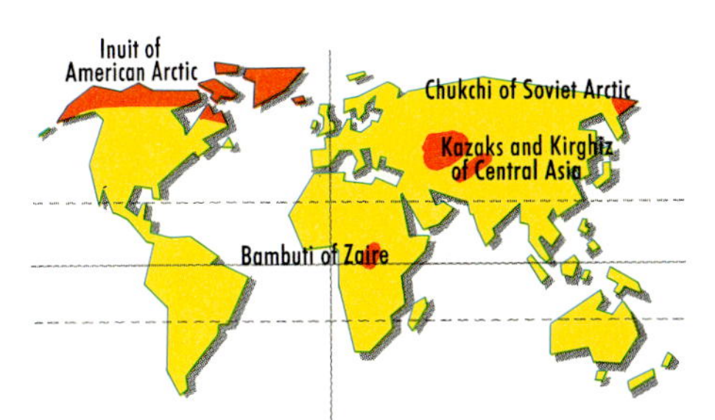

WHY VILLAGES?
The diversity of rural settlements

PEOPLE THE WORLD OVER TEND TO want to cluster together in villages and towns. In the affluent world, more people are choosing to live with their immediate family in individual homes. Even so, they generally establish themselves within a short walking distance of their neighbors; and in any case they can keep in touch with friends and relatives by means of automobiles, televisions, radios, telephones, and a postal service.

In the past, voluntary isolation was probably rarer still. Pioneers occasionally had it forced upon them when they were breaking in wilderness lands, such as northern Europe's forests a thousand or more years ago, or parts of the Americas in the last few centuries. But the norm has always been some form of more collective life. People gather in family or multifamily groups. And they have always done so: first in hamlets and villages; later in towns and, increasingly, in cities.

In general, "nucleated" or close-packed settlements affirm a human need to communicate and interact with others. We seek company to enhance our security from external threat, for entertainment and solace, for the sharing of work and the routine exchange of goods and services, and for mutual religious, artistic, and political expression.

Yet nucleated settlements have often come about through duress or constraint. People group together because they are forced to. These constraints fall into three main kinds: physical, technical, and legal.

Physical constraints involve threats to life and livelihood. Where usable building land is at a premium because of the danger of flooding or landslides, where livestock is in danger from wild animals, or when attack from other groups is likely, people tend to stick together. The need for mutual defense and support encourages solidarity between people.

An excellent example can be seen in the walled villages, or kraals, widespread in many parts of eastern and southern Africa. These consist of a number of small structures grouped within a wider circle and protected by a stone or wooden wall. The structures are similar in form, but serve separate functions as habitation, food stores, and possibly animal houses. Some kraals – the word was brought to the Americans as "corral" – have sunken inner courtyards to keep livestock from wandering.

Technical constraints mean that villages generally form when communal effort is needed. Most major landworks – rice terraces, ditches, dikes, irrigation systems – require the input of collective labor for mutual benefit.

The third form of constraint is legal or political in nature. Throughout history, people have been grouped together, whether they liked it or not, because they have been made to do so. Overlords – from the pharaohs of Ancient Egypt to Ceausescu in Romania in the 1980s – have imposed laws, tithes, and taxes on farmers, and have generally demanded, for reasons of control, that "their" farmers live in close-knit, more or less planned villages.

But duress is not the only reason for the existence of villages. Some are constructed as part of a widely held political or religious ideology, like the Israeli kibbutz and the Mormon township (the latter traditionally following the layout of the Mormons' settlement in Salt Lake City).

More generally, rural reconstruction is central to the national development plans of many third world countries. Typically these involve "villagization" programs based on the regrouping of rural population in organized settlements. The objective is partly to stem the flood of peasants to the already overburdened cities, but also to increase agricultural production and provide more people with basic services and facilities.

In Ethiopia, a government-organized villagization plan has moved up to 15 million people from scattered homesteads in the countryside to new villages, giving them better access to water supplies, schools, and clinics.

The variety of village forms

Hamlets or villages on all continents take innumerable forms, in which dwellings may be clustered (nucleated) or scattered (dispersed). The form is never haphazard: nucleated forms sometimes reflect ancient ways of working the land, or the importance of a former market, but the same pattern may apply in modern towns. Some villages surround a common space where animals were protected, or which served as a market. But 20th-century builders also provide a green for prestige.

Dual villages may be divided by a barrier such as a river, or sit at top and bottom of a cliff face. For trade, linear villages usually develop along a transport route. Most commonly, a village is made up of dwellings clustered apparently at random around a focal point.

In inhospitable terrain, a village may be merely a few isolated dwellings. Dispersed settlements may be ancient, as around estuaries in Southeast Asia, or brand new, as in recent Dutch polders.

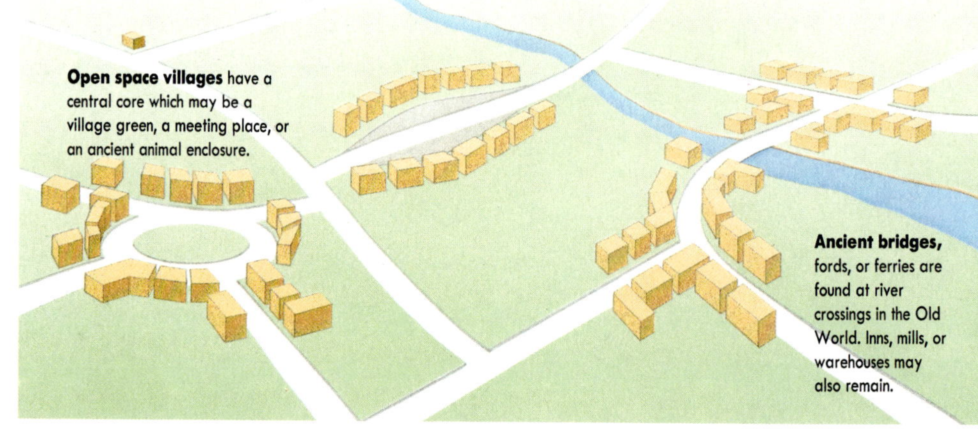

The majority of China's 1.2 billion people live in villages. Thanks to its fertile soils and temperate climate, Sichuan is China's most populous province. Rice, wheat, and a variety of vegetable crops are grown in the patchwork of tiny fields that surround each of its hundreds of thousands of villages.

Although many such settlements house several thousand people (a number that elsewhere in the world would normally qualify them for the status of towns), they are true villages – most of their inhabitants are farmers.

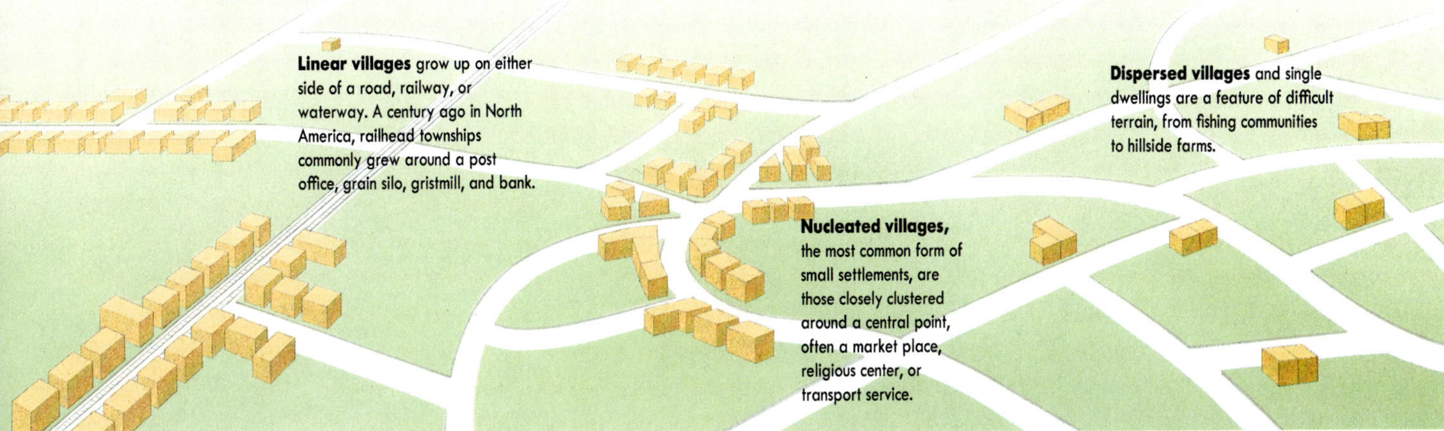

Linear villages grow up on either side of a road, railway, or waterway. A century ago in North America, railhead townships commonly grew around a post office, grain silo, gristmill, and bank.

Nucleated villages, the most common form of small settlements, are those closely clustered around a central point, often a market place, religious center, or transport service.

Dispersed villages and single dwellings are a feature of difficult terrain, from fishing communities to hillside farms.

IMPOSING PATTERNS

Field shapes and land rights

As we pass through, or fly over, a country, it is often the fields that contribute most to our experience of its "spirit of place." Fields, no less than houses, villages, or cities, are shaped by a complex combination of physical and cultural forces.

The influence of physical terrain is perhaps most evident in the tiny, irregular fields that straggle over exceptionally rugged ground. In the Greek islands or the Lake District of northern England, stone walls divide such fields: the effort involved in clearing the boulders was too great to take the rock farther away.

Fields are in any case limited by the steepness of the land. It is almost impossible to farm land more than 18 degrees from the horizontal without terracing. In steep areas with large populations, such as parts of Southeast Asia or the Mediterranean, land has been brought into production by making steps or terraces up the slope. The size and shape of the fields that result are largely influenced by the initial landforms.

Field shapes throughout the world are also influenced by water, whether naturally contained in streams or lakes, or channeled by artificial drainage or irrigation. Even the kind of equipment used to irrigate or plow can be significant in determining field shape. In dry parts of the world where rotating irrigation sprinklers are in use, as in parts of the U.S., Africa, and Australia, fields may be circular.

In some cases, cultural factors may be of overriding importance. Perhaps the most profound influence on field shapes concerns property rights and the ways in which land is surveyed and allocated.

The most impressive and formal examples include the U.S. grid (*see box*), and the slightly earlier French system of "long lots." French settlers in North America introduced lots that were long and narrow, and ran at right angles to a river. Long lots provided access to the river for transportation, and allowed people to have their own farms while being close to neighbors. They were often broken up in accordance with French law, divided between sons until parcels of land looked like beads on a string.

Subdivision of land had a profound impact elsewhere. In Ireland, until the middle of the 19th century, land was subdivided so that each son inherited a section. Fields became smaller, until some were only a few yards wide.

The only crop that could yield sufficient food in such a tiny area was the potato. So when blight struck in the 1840s, and the produce of land belonging to British landlords was denied them, there was mass starvation and disease among Ireland's peasant farmers. Around a million people died, and well over another million emigrated – most to the U.S. The derelict boundaries can still be seen.

Irregular fields of infinitely varied size and shape are characteristic of many long-settled areas. Boundaries are typically oriented to natural features, like streams and hills, or human creations, such as roads. Modern agricultural machinery usually demands larger fields, which means that ancient hedgerows, sometimes hundreds of years old, are uprooted.

Long lots in North America are of French origin; strip fields reach back from the river front. Some of the longest were laid out in Quebec, along the St. Lawrence River. Most have since been subdivided into shorter parcels, although some have also been joined to their neighbors to create wider fields.

Plowing that ignores any slopes (*above*) can lead to gullying and erosion of the earth. Contour plowing, where the plow travels around the hill at the same level, reduces runoff and conserves soil.

Circular field shapes (*below*) result when land in dry parts of the world is irrigated by rotating sprinklers.

A QUILTED LANDSCAPE

All over the U.S., at least west of the Appalachians, four-square patterns exist. With remarkably few exceptions, fields, street plans, and state boundaries are oriented north–south and east–west.

Starting work in 1785, the U.S. land survey laid down a "Geographer's Line" westward to the Pacific from the point where the border of Pennsylvania meets the Ohio River. From this base line, the rest of the U.S. could be charted and conveniently subdivided.

By the middle of the 19th century, the vast majority of the country was parceled up into rectangles a mile (1.6 km) square, imposing a geometric grid that disregarded geographical features. The purpose of the survey was not simply to map the territory, but to develop an orderly way of selling government land to pioneers, who at that time were just beginning to press toward the Pacific.

In practice, the mile-square "sections" were often bought or sold in smaller blocks – usually halves or quarters of a section. A square block of six sections by six made up a "township," the smallest administrative district. Few areas were left without a checkerboard pattern. Those that had previously been under British, French, or Spanish rule retained systems introduced by colonists. Major physical features, such as the Rocky Mountains or the swamps in Florida, also occasionally upset the military precision of the scheme.

The map (*right*) dates from 1785 and shows a section of Ohio, divided into ranges of townships by the original survey. The continuing influence of the grid can be clearly seen in the U.S. Geological Survey's map of present-day Los Angeles (*below*), where the heavier lines (running under West Hollywood and beside Beverley Hills), mark the boundaries of the 6-mile (9.6-km) square townships; and in the satellite photograph of the San Francisco area (*left*), showing field shapes defined by 200-year-old rules.

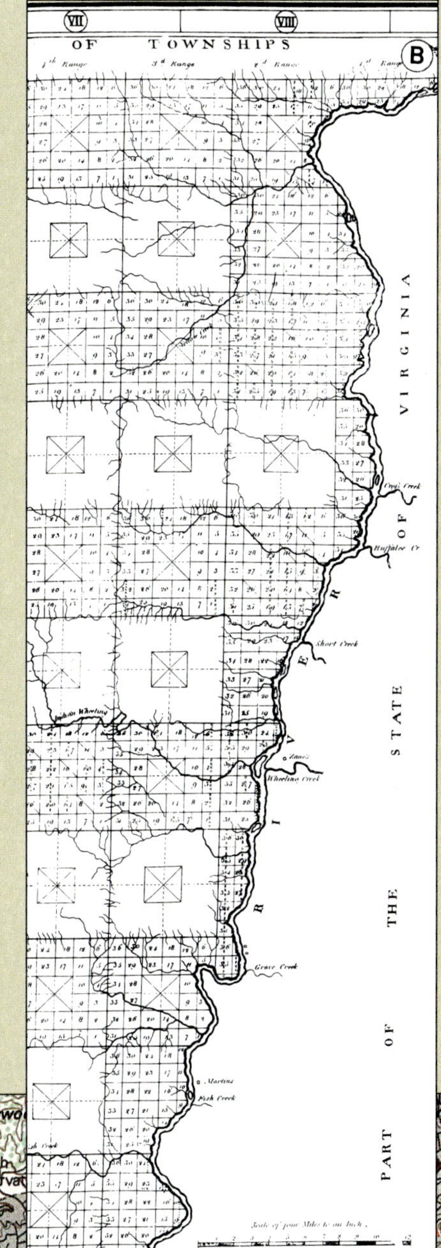

GOING TO MARKET
The hierarchy of settlements

SHOPPING MALLS AND CHAIN STORES may dominate retail trade in the richest parts of the world but elsewhere trade is still carried out in the traditional way, face to face, in markets and shops. Even in those countries where production and retailing functions have almost entirely been taken over by major international "agribusinesses" and nationwide stores which describe themselves misleadingly as "supermarkets," markets have not disappeared. In reaction to what is widely seen as overprocessed and overpackaged food, most of the produce to be found at these markets is sold loose and described as "farm fresh," "natural," and "organic."

In the Andes, in West Africa, and in Thai towns, for example, traditional markets hold sway. A sea of buyers and sellers forms. A potential customer asks a price, a figure is suggested, a spell of ritualized bargaining follows, a deal is struck. The two roles are interchangeable: a seller finds a customer for her goods and the next moment becomes a buyer at someone else's stall.

Transport is a key feature in the organization of markets. Bringing produce to market is always expensive. In the absence of navigable rivers, or good roads and motor vehicles, carriage by donkey or ox-drawn cart can double the cost of grain in less than 50 miles (80 km). As a result, most goods must be brought to, and bought from, a local market.

In most rural areas of the world, periodic markets prevail. The seasons pace the harvests, and peddlers, merchants, and carriers have to cover immense areas in a methodical way without clashes of venue. As a result, markets may be rotated in a particular order and at specified frequencies over a vast area. In parts of the northern Andes, as often in the Western world, markets are held weekly on the same day; in Yorubaland in western Nigeria they are held every fourth day; in southern China, at least until the Communist revolution of 1949, they were held three times per "week" of ten days.

As a result of periodic markets, everybody has the chance to shop as frequently as need, wealth, and availability of produce allow, while producers and traders sell sufficient produce to enable them to survive.

Some specialist markets – for carpets or livestock, say – may be less frequent, but often are correspondingly larger. Annual fairs, or markets associated with big towns and cities, may be bigger still, and may be accompanied by religious rituals and lively secular entertainment.

Cities, of course, are able to support several markets every day of the week, many specializing in a particular product – fish, perhaps, or fruit. Islamic cities are especially famed for their crowded, bustling, and diverse *suqs*: Cairo alone has for hundreds of years supported at least 35 different markets.

There have been markets for as long as there has been agriculture, and towns as long as there have been markets. Historically, the growing concentrations of people all coming together to sell and to buy at the same time and place offered scope for increasing numbers to abandon the cultivation of food and devote themselves to manufacture or the provision of services.

Market sites were often associated with early manufacturing, which concentrated on processing agricultural produce – the milling, drying, and malting of cereals; baking and brewing – and on tanning, spinning, and weaving. Gradually, a great range of craftworkers and others established themselves – woodworkers, tailors, cobblers, blacksmiths, innkeepers, doctors, lawyers, priests, and the like – and earned their living in a symbiotic relationship with the farmers from the surrounding area, and with fellow townspeople.

As a result, the market town gradually extended its influence into the surrounding countryside, multiplying the number of specialist services it could provide. Over time, some towns continued to grow in size and wealth by diversifying these functions still further.

HOW MARKETS ARE DISTRIBUTED

Markets will appear wherever there is sufficient money to be made. If there are few people in a region, and they have little money or produce, there will be few markets. If people become more numerous or more affluent, the number and frequency of markets are likely to increase.

Coming to market, whether as buyer or seller, is expensive in money and time. It is sensible to minimize journey costs by using the most accessible market. Complicated patterns arise, partly because there are many different products people wish to buy. Milk or newspapers are bought often. But few people purchase very expensive goods like vintage champagne or grand pianos, and then only rarely.

So there are different markets, shops, or outlets for different goods. Some occur frequently in any given area; others only in the largest cities. Each level of outlets has its own "catchment areas," which nestle within each other. On the whole, the larger the settlement, the bigger the area it serves, and hence the greater its distance from other large settlements.

Part of the explanation for this can be found by considering typical shopping patterns. Take a family that lives outside a village. One or another of them may go daily to the village store for bread and smaller items, but do the major shopping on weekends in the nearest town. Less frequently, they travel to a large town, where there is more choice and better facilities, to buy clothes or seek specialist medical advice. Once a year, on a birthday or wedding anniversary, they journey to a regional or national capital to celebrate.

Multiply this family's behavior by a thousand or a million, and it becomes clear how the hierarchy of settlements forms and spreads in characteristic patterns across the landscape.

When trying to understand economic behavior, economists sometimes hypothesize a person who always makes rational and efficient decisions. The geographical equivalent is the "isotropic plain" – a flat, featureless landscape where the population is evenly spread. On such a plain, the arrangement of markets and the areas they serve would be perfectly regular. This was the theory put forward in the 1930s by the German geographer Walter Christaller. He proposed orders of centers, ranging from corner shops to major cities, and drew maps of the ideal market areas, in the form of regular, adjoining hexagons.

Of course, such a perfect regularity does not exist in practice. Actual patterns may be distorted by valleys, mountain ranges, industrial belts. Neither is population evenly spread, and areas of high population attract more markets. But the retail industry has found Christaller's theory of great value in helping it to predict sales and choose locations for its outlets.

A settlement hierarchy The larger the settlement, the more goods and services it provides, and the farther it is from another settlement of comparable size.

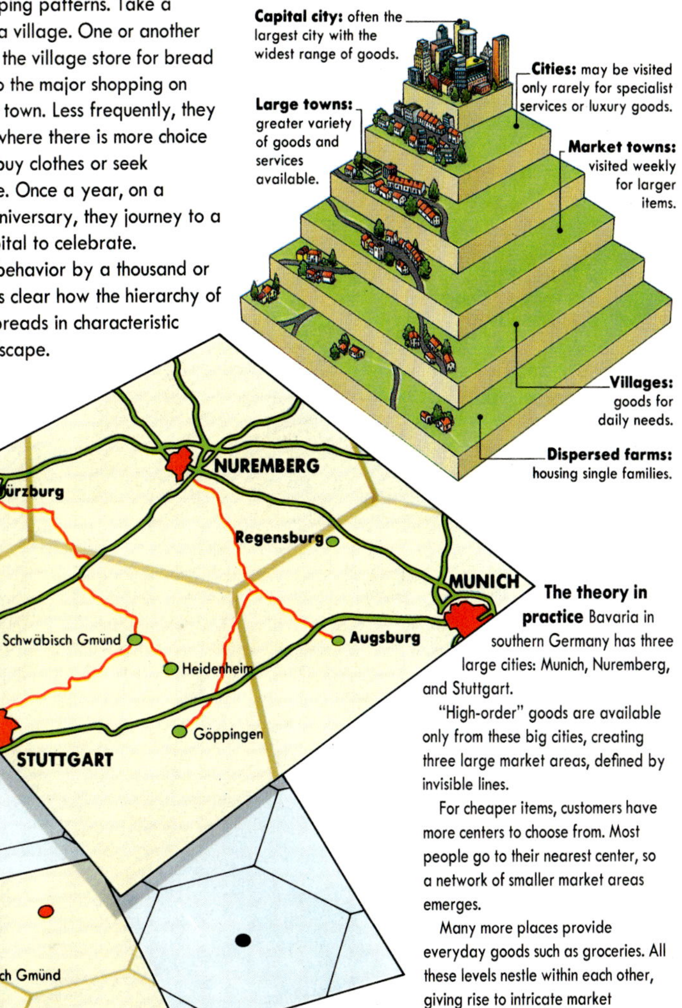

The theory in practice Bavaria in southern Germany has three large cities: Munich, Nuremberg, and Stuttgart.

"High-order" goods are available only from these big cities, creating three large market areas, defined by invisible lines.

For cheaper items, customers have more centers to choose from. Most people go to their nearest center, so a network of smaller market areas emerges.

Many more places provide everyday goods such as groceries. All these levels nestle within each other, giving rise to intricate market patterns.

Periodic markets give the greatest possible number of people the opportunity to buy and sell produce, as here in Cameroon in West Africa. One or two family members – usually female – bring the goods to market and set up their own stall.

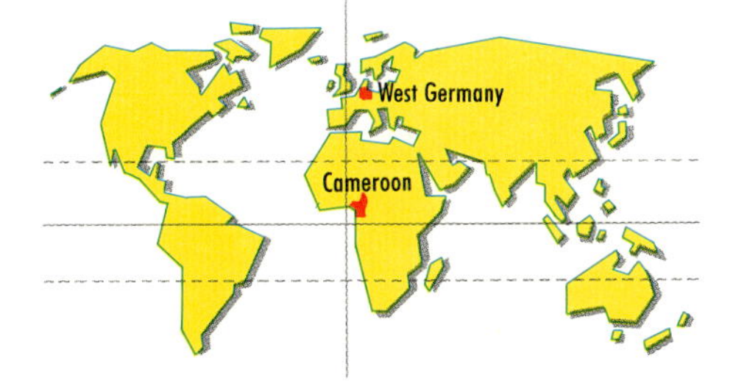

PUTTING DOWN ROOTS

Where settlements flourish

ANYBODY WHO HAS THE CHOICE OF where to live has to make certain judgments. Will this place be safe and pleasant? If not, can it be made so? Can a better living be earned from the resources available here than anywhere else? Is everything needed for survival close at hand? Is this place reasonably accessible, or is it too isolated?

No site is perfect. The ideals – among them dry, level land, a favorable aspect, protection from the elements, and abundant fresh water – rarely occur naturally in the same place, and rectifying the defects of site is often expensive and laborious.

All other things being equal, settlers are obviously likely to seek a site for their homes close to their key resources: the ones from which they earn a living, or which they need most frequently. Farmers, for example, will naturally prefer to build homes close to abundant fertile land, whereas miners will choose to be near mineral deposits.

Industries carefully calculate the costs and benefits of locations. Until the advent of electricity, energy in the form of wood to burn, falling water, or coal was often the critical consideration. Now they may decide to build their factories close to the resources from which they manufacture goods, or on transport routes that bring resources and carry finished goods away, or near the market.

An inaccessible or defensible site such as a hilltop, island, or peninsula may be another high priority. A fort, built to control a precious resource, trade route, or vulnerable site, may become the focus for the growth of a new settlement because it needs supplies of food, drink, and other goods.

A whole host of alternative natural resources are available to be tapped, and they have spawned novel settlements. A wide variety of tourist resorts center on ski slopes, beaches, and areas of "wilderness." Even the most infertile of rural areas can become desirable as a site for a commuter "village" or "science park."

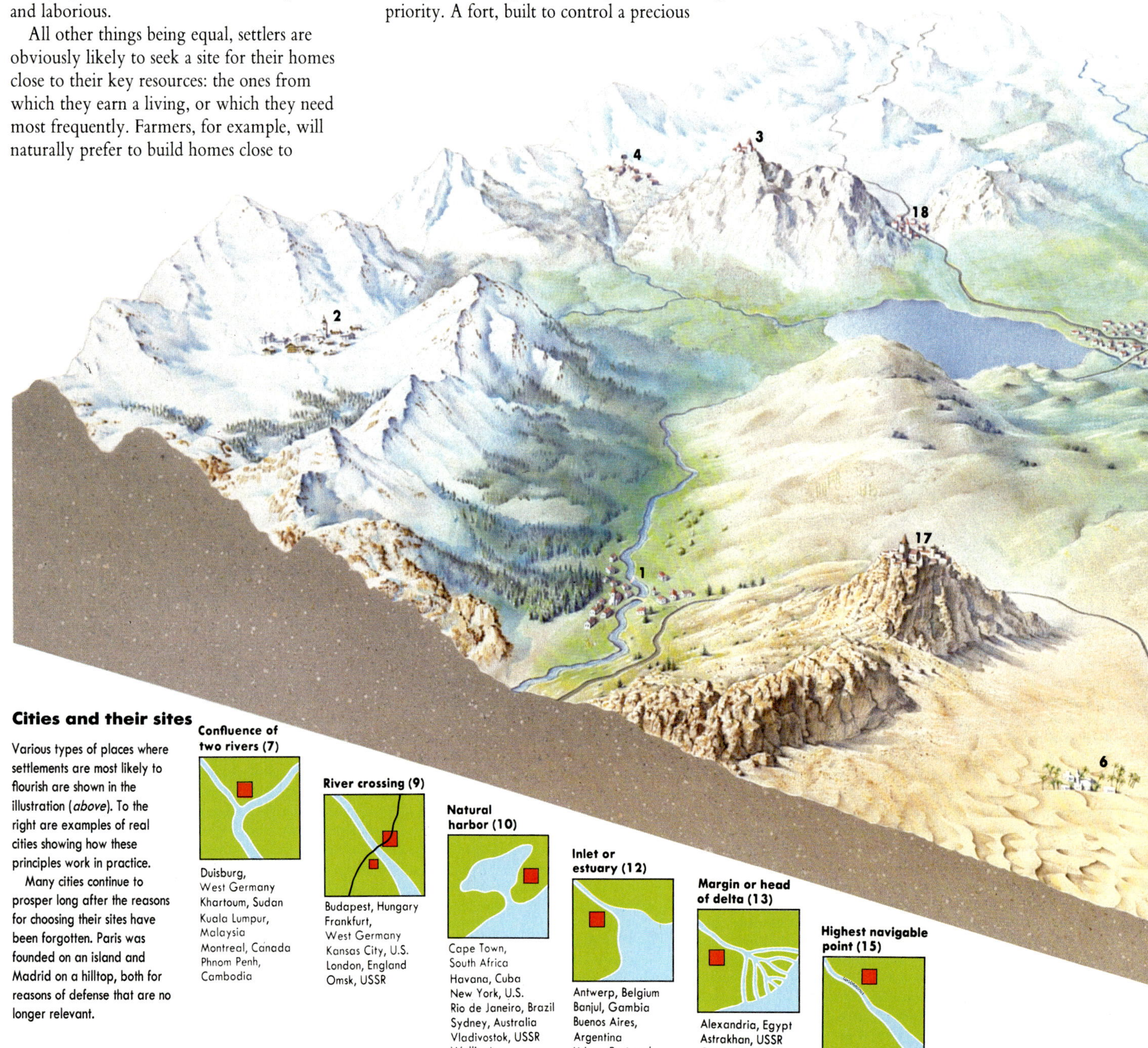

Cities and their sites

Various types of places where settlements are most likely to flourish are shown in the illustration (*above*). To the right are examples of real cities showing how these principles work in practice.

Many cities continue to prosper long after the reasons for choosing their sites have been forgotten. Paris was founded on an island and Madrid on a hilltop, both for reasons of defense that are no longer relevant.

Confluence of two rivers (7)

Duisburg, West Germany
Khartoum, Sudan
Kuala Lumpur, Malaysia
Montreal, Canada
Phnom Penh, Cambodia

River crossing (9)

Budapest, Hungary
Frankfurt, West Germany
Kansas City, U.S.
London, England
Omsk, USSR

Natural harbor (10)

Cape Town, South Africa
Havana, Cuba
New York, U.S.
Rio de Janeiro, Brazil
Sydney, Australia
Vladivostok, USSR
Wellington, New Zealand

Inlet or estuary (12)

Antwerp, Belgium
Banjul, Gambia
Buenos Aires, Argentina
Lisbon, Portugal
Montevideo, Uruguay
Oslo, Norway

Margin or head of delta (13)

Alexandria, Egypt
Astrakhan, USSR
Cairo, Egypt
Ho Chi Minh City, Vietnam
Shanghai, China

Highest navigable point (15)

Basle, Switzerland
Louisville, U.S.
Minneapolis/St. Paul, U.S.

SETTLING DOWN

A choice of location

Certain sites in the landscape are much more likely to attract settlements than others. London, for example, was established at the narrowest fordable part of the Thames, and was also on the Roman route north. The main part of the city developed along the outside meander of the river, thus allowing easy access to fresh water.

The advantages of the favored sites tend to fall into one or more of the following categories.

Topography
In some cases the physical conditions of a settlement site are of overriding importance. In a mountainous area, for example, where winds are strong and most of the land is steep and rugged, the floor of a river valley may provide the only protected level site for a village **(1)**.

However, it may be advantageous for a settlement to develop on a gentle slope **(2)** – in the northern hemisphere, sites on south-facing slopes benefit from more sunshine. (In the southern hemisphere, north-facing slopes are warmer.)

Topography may also have a symbolic significance. For example, a prominent hilltop may become the site for a religious center and its associated settlement **(3)**.

Key resources
Villages are often located within easy reach of a key resource. In such cases, the priority is to reduce the daily expenditure of time and effort. Miners and their families, for example, may choose to settle close to where mineral deposits lie **(4)**.

Once energy in the form of electricity can be brought to the industry, rather than the industry having to be located near the source of power, this has less influence on the choice of location. Exceptions arise when the demand is enormous: aluminum smelters use a colossal amount of energy, and are still usually located near hydroelectric power stations.

A dependable water supply is another vital resource, used for drinking, cooking, washing, irrigation of crops, and often also for industrial purposes. Hence, settlements may develop along a river, particularly on the outside of a meander **(5)** where the water is deeper and land less likely to be marshy. In the desert, a settlement is more likely to prosper at an oasis **(6)**.

Trade routes
The siting of settlements is influenced by trade routes whether passage is by land, sea, or air. Settlements may emerge at any point on a route, but places where routes converge, or where the flow of traffic becomes concentrated, are prime sites. These include the confluence of two rivers **(7)**, railway junctions **(8)**, and river crossings **(9)**.

Trans-shipment points – where boats must load and offload their freight – are also favored sites. Protected deep water harbors **(10)** may be found within a variety of bays, inlets, and estuaries **(11, 12)**. Ports are often located on the first solid ground on the margin or head of a delta **(13)**.

Other trans-shipment points that may spawn settlements include the mouth of a lake **(14)**, the highest navigable point on a river **(15)**, and portage points, where river freight must be transferred over land to another waterway **(16)**.

Defense
Settlements often originate at sites that are readily defensible or which provide strategic protection for resources. Many otherwise desirable sites have been impossible to build on because of the threat of invasion. In such cases, the ability to protect the site may be the first priority – immediate personal comfort is a secondary consideration.

An isolated crag or hilltop **(17)** can provide a commanding view over the surrounding plains. A site at the entrance to a mountain pass **(18)** can give strategic control over a vital cross-country route. The most easily defended sites are those partly or completely surrounded by water. Such locations include a hill around which a river meanders **(19)**, a peninsula **(20)**, an island in a river **(21)**, or an offshore island **(22)**.

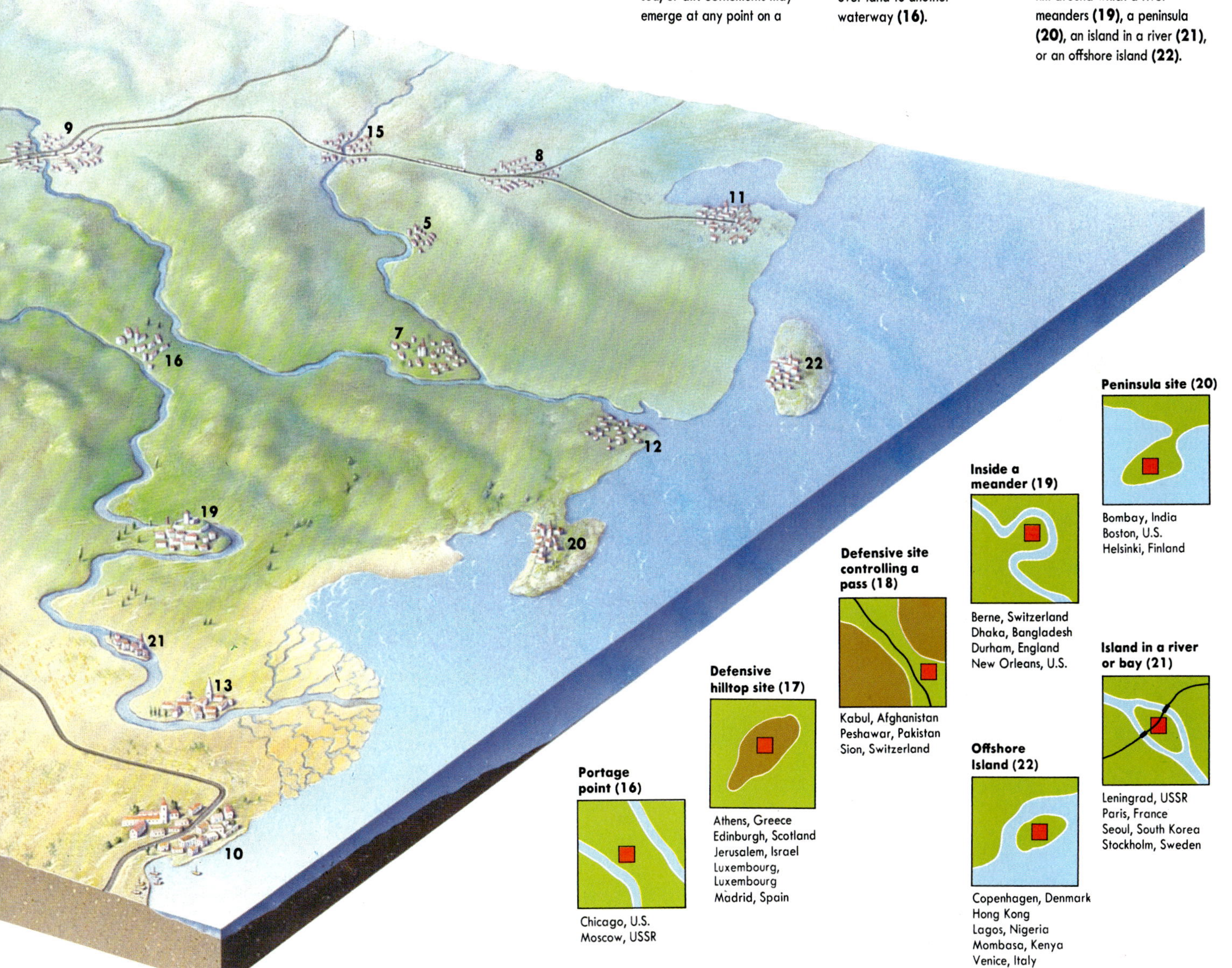

When Cities Grow

In 1850, only one country in the world – England, home of the industrial revolution – could claim to be urbanized, to house more than half its population in towns of more than 2,000 people. A century later, European countries and the U.S. had caught up with England's head start, but urban growth in the rest of the world lagged behind. Urbanization was increasing, but fewer than one in four of the world's population lived in towns and cities.

The situation now is very different. Throughout the world – and especially in developing countries – urban populations have exploded and cities overspilled their bounds. Soon, for the first time in history, more people will be living in towns and cities than in the country. As a result of this unprecedented and often uncontrollable growth, by 2020, over three-fourths of the world's population will be urban, and one in four will be living in cities of over a million people.

The location of the world's largest and fastest-growing metropolitan areas has also changed. In the 19th and early 20th centuries, cities grew most rapidly in the industrializing countries of Europe and North America. Now, the rapid rate at which urbanization is occurring in the third world is reaching alarming proportions. By the year 2000, there will be more people living in Cairo and Calcutta than in the whole of Canada, and as many in São Paulo and Mexico City as in Spain and Portugal combined.

In the 19th century, the spread of the factory system and the development of heavy industry drew millions of people away from the harsh conditions of the countryside and into large industrial cities. Women and children were needed in factories and workshops, men in mines and on construction sites. Artisans and unskilled workers alike were drawn by the lure of city life. In most cases, they went to towns close by, but from the 1850s, with the spread of railroads and steamships, some rural migrants moved thousands of miles to seek a new life in foreign cities.

None of this urban expansion and diversification would have been possible without extraordinary developments in the countryside. Until the 19th century, there was rarely enough food to sustain a sizable urban population. Failing to gather a good harvest, whether because of pests or bad weather, could spell catastrophe for rural and urban dwellers alike.

Towns and cities grew because, for the first time, food surpluses became both large and secure. A combination of new techniques and technologies – steel plows and mechanical reapers, the increased use of natural and chemical fertilizers, improved drainage, more sophisticated systems of crop rotation to ensure that soil remained fertile, and prodigious success in plant and animal breeding – produced an agricultural revolution on the same scale as those occurring in industry and urbanization.

Millions of acres of new land were also brought into production. In Europe, little other than poor-quality marsh and heathland was left uncultivated, so the possibilities were restricted. But there was plenty of scope elsewhere in the world, albeit at the expense of indigenous peoples.

Cities even had a part to play here, for they were the gateways through which European settlers first penetrated the new lands. In North and South America, Africa, Asia, Australia, and New Zealand, large areas were commandeered to supply food for Europe's burgeoning cities. American and Russian wheat, Argentinian beef, and Australian mutton became a regular feature of the European diet. Without these new sources of supply, urbanization in Europe could not have been sustained.

Greater food surpluses in Europe had another important effect on urbanization. Agricultural improvements meant that more food could be produced with less labor. Furthermore, since crops grown abroad were often cheaper than those produced at home, small-scale domestic farming could not compete. Dispossessed smallholders and redundant rural workers boosted urban populations. Indeed, without them, city numbers would have plummeted, since deaths in 19th-century towns and cities often exceeded births. In the worst areas, as many as one in three children did not live to see their first birthday. Overcrowding, grossly inadequate sanitation, and polluted water supplies provided ideal conditions for the spread of infections and outbreaks of epidemic diseases, such as cholera and typhoid.

Yet, in spite of these problems, cities continued to expand. Advances were made in health and hygiene, especially the provision of clean, piped water and efficient sewage disposal, and living standards improved. As death rates began to fall, so natural increase came to play a bigger role than rural migration in fueling city growth.

With improvements in transportation, populations began to spill over into the neighboring countryside. In 1800, London had a population of nearly one million; a century later, it had become a sprawling metropolis six times as large, swallowing

up 100 or more tiny villages in the process. Elsewhere, other urban centers were growing equally quickly, most of them, like Paris and New York, in affluent, highly industrialized societies.

In the third world – unlike Europe and North America – urbanization has preceded rather than followed industrialization and has occurred at a much faster rate. The widespread application of medical techniques such as vaccination has reduced death rates more quickly, resulting in prodigious increases in population, and rural-urban migration has occurred on an even greater scale. Migration has been stimulated by deteriorating conditions in the rural areas, derived in part from population increase, but also from unjust land tenure systems, greater use of agricultural machinery, and the sheer difficulty of eking out a living in the countryside.

At the same time, the industrialization policies of third world governments have raised expectations of better jobs in cities. Urban growth has been most dramatic in capitals, where employment opportunities are greater, particularly for women. The largest city in a third world country often contains between 10 and 25 percent of the entire population.

UNTIL RECENTLY, THE PROCESS OF URBANIZATION has been fastest in Latin America, where 60 percent of the population now live in towns and cities. Although a few countries in Africa reach this level, on average only about 30 percent of the population is urban. The same is true of most parts of Asia. Things may now be changing, though, with massive acceleration in population transfers, especially in Africa. Such large-scale urbanization creates many problems. Unbalanced growth in a tiny number of very large supercities can be detrimental to the nation's development.

In socialist countries, where the state assumes wide powers over the deployment of labor, as well as housing and industrial policy, governments have adopted a variety of measures to control urban growth. Rural industry is promoted; and migration is curbed by the imposition of a system of internal passports and travel warrants, rationing tied to a person's official home, and strict quotas on residence permits.

Attempts have also been made to relocate "surplus" urban populations. In some cases this has been forcible, as in the case of Pol Pot's evacuation of Cambodia's capital Phnom Penh after 1975. More usually, the inducement has been the prospect of higher wages and guaranteed accommodation. Some workers are directed to new towns and cities near resource centers such as oil fields, others to existing cities in peripheral areas where state control is weakest. For example, since 1950 millions of Russians have moved to resource-rich Siberia and to the Baltic states. Elsewhere, Han Chinese – the dominant ethnic group – outnumber Tibetans in their capital Lhasa.

Cities have grown to such an extent, and in the process created so many economic, social, and environmental problems that in many parts of the world their continued survival – and that of millions of their inhabitants – is now seriously at risk.

All cities share similar problems, deriving from the features they have in common: they are by definition large, dense concentrations of people who live from industry and services, and do not grow a significant part of the food they eat. Cities are therefore called upon to cater for ever-increasing food and energy needs, provide efficient transportation, guarantee a plentiful supply of clean water and unpolluted air, dispose of solid and liquid wastes, and control the spread of diseases.

Yet ironically, the very growth of the city itself makes these demands harder to satisfy. Urban sprawl covers farmland once used to feed its inhabitants. People and goods coming into the city center cause traffic jams and atmospheric pollution, increase energy consumption, and overburden public transportation systems and sewers designed for an earlier age. Local water resources are exhausted or polluted, and must be drawn from farther and farther afield, sometimes further threatening agricultural land.

The vast, flat, wholly artificial, hard and impermeable surfaces of cities can only exaggerate the effects of climate – be they extremes of heat or cold, drought or rain, or strong winds. In some cases, cities must also be prepared to cope with floods, or volcanoes, or earthquakes. These disasters would be hard enough in the best of circumstances, yet many of the most rapidly growing cities are located in poor countries that can least afford the expense.

GATEWAY CITIES

European colonies around the world

Over two-thirds of the world's largest cities are located on coasts or major rivers, the legacy of a phenomenal expansion in European trade. After 1450, Portuguese, Spanish, Dutch, and other European powers founded or developed thousands of towns – the vast majority of them ports – throughout the world to service their growing trading networks. Over the next 400 years some grew massive, chiefly those that acted as gateways for colonial expansion into the interior of the country. Cities like Sydney in Australia, Buenos Aires in Argentina, and Cape Town in South Africa expanded quickly as increasing areas of the hinterland were settled.

Forts were built to protect these ports, manufacturing started to supply the pioneers' needs, commercial and financial services developed, and formal administrations were established. These gateway cities acted like funnels through which all routes into and out of the interior passed. Through their docks the wealth of the interior, chiefly its raw materials and food produce, was shipped to Europe. Back came waves of European settlers, eager to exploit these newfound resources, and in their wake, manufactured goods for rapidly expanding markets.

Given that shipping now transports a declining proportion of global trade, and that many unused docks are being closed down, why have former gateway cities continued to grow? The answer lies partly in their early success. Having developed into major population centers, they became massive markets and commercial and industrial cities in their own right. Transport and communication routes – railroads, airlines, and so on – continue to terminate there.

African gateways

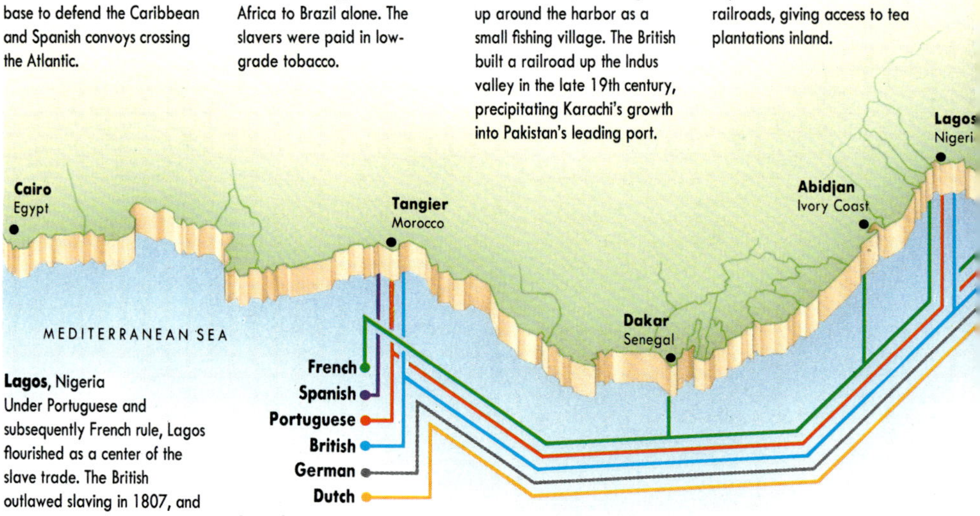

Some of the first European colonies in Africa were stopping-off points on the way to India and the Spice Islands. Later, the discovery of gold, demand for slaves, imperialist ambition, and missionary zeal all played their part in establishing European settlements.

Tangier, Morocco
Strategically placed at the entrance to the Strait of Gibraltar, Tangier has been a gateway between two continents for centuries. The Romans, Phoenicians, Greeks, and Arabs all had cities on the site. Tangier was captured by the Portuguese in 1471. Spain ruled it from 1578 to 1640, when it reverted to Portugal. In 1661, it was ceded to the British after a royal marriage, and was finally restored to the Moors in 1684.

Dakar, Senegal
One of the finest natural harbors on the Atlantic coast, Dakar came under French control in the 17th century. The first railroad built in West Africa linked Dakar to Saint-Louis, opening up the interior.

Abidjan, Ivory Coast
Although much plundered for ivory and slaves, the Ivory Coast was not colonized by a European power until the French arrived at the end of the 19th century. The French capital, Abidjan, lies on the coast and has a large, sheltered, deep-water port.

Lagos, Nigeria
Under Portuguese and subsequently French rule, Lagos flourished as a center of the slave trade. The British outlawed slaving in 1807, and oil and natural gas are now the mainstay of the Nigerian economy.

Luanda, Angola
Founded by Portuguese settlers in 1576, the capital of Angola is the oldest "European" city in sub-Saharan Africa. Until the 19th century, Luanda was the main center for the export of slaves to Brazil and the New World.

World gateways

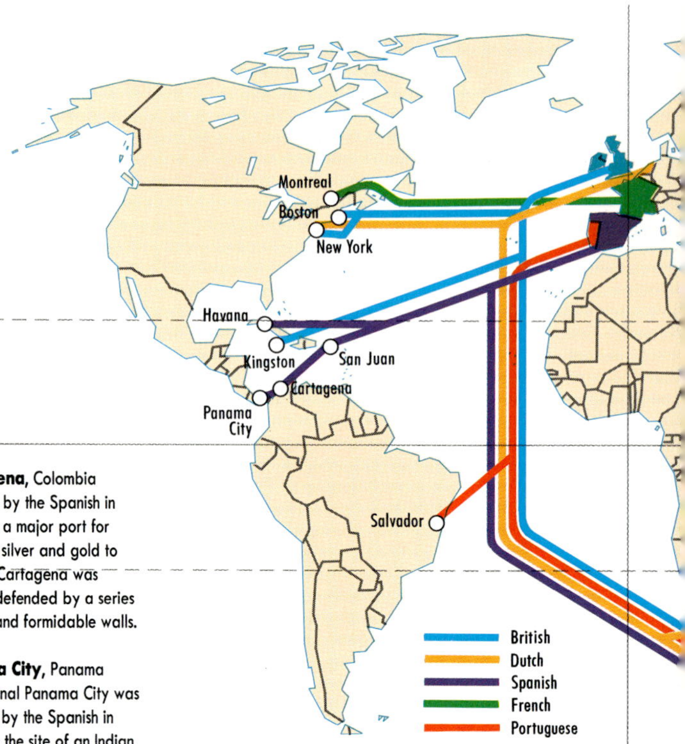

Montreal, Canada
Founded by the French in 1642, Montreal sits in an important position at the junction of the St. Lawrence and Ottawa rivers, giving access to the Great Lakes and the Canadian plains.

Boston, U.S.
The situation of Boston was attractive because of the possibilities of trade in furs and fish. As its importance as a port grew, it dealt with slavers coming from Guinea, and with ships coming from Honduras with mahogany, and from the West Indies with molasses to be distilled into rum.

New York, U.S.
Originally a Dutch fur-trading post at the mouth of the Hudson River, in the 19th and 20th centuries New York became a gateway for millions of Europeans emigrating to the New World.

Havana, Cuba
Founded by the Spanish in 1515, Havana has an excellent harbor and was the assembly point for annual convoys returning to Spain.

Kingston, Jamaica
Founded by the British, Kingston was a major port for the shipping of cane sugar to Europe. The sugar industry depended on slaves to work the plantations.

San Juan, Puerto Rico
Established by the Spanish in 1521, San Juan was used as a base to defend the Caribbean and Spanish convoys crossing the Atlantic.

Cartagena, Colombia
Founded by the Spanish in 1533 as a major port for shipping silver and gold to Europe, Cartagena was heavily defended by a series of forts and formidable walls.

Panama City, Panama
The original Panama City was founded by the Spanish in 1519 on the site of an Indian fishing port, about 4 miles (6 km) from its present site. It was linked by a trail across the isthmus to the port of Portobelo, where Spanish galleons were loaded with gold and silver on their way from South America to Spain.

Salvador, Brazil
The first European landfall in Brazil was made here by the Portuguese in 1500. They established sugar plantations and food mills that were worked by slaves from West Africa. Between 1526 and 1870, a total of 3,647,000 slaves were shipped from Africa to Brazil alone. The slavers were paid in low-grade tobacco.

Cape Town, South Africa
Founded by the Dutch in 1652 and once known as "the tavern of the Indian Ocean," Cape Town served as the last provisioning stop on the voyage from the Netherlands to Java, the main Dutch port in the East. That journey took seven months, and Table Bay provided supplies of citrus fruit to combat the scurvy that regularly cost the lives of a tenth of a ship's crew.

Karachi, Pakistan
Situated on the Arabian Sea in the Indus delta, Karachi grew up around the harbor as a small fishing village. The British built a railroad up the Indus valley in the late 19th century, precipitating Karachi's growth into Pakistan's leading port.

Bombay, India
Hot, swampy, and surrounded by cliffs, Bombay was not originally an important gateway, despite its excellent natural harbor. But once roads and railroads reached it, trade in cotton, tea, and silks commenced.

Colombo, Sri Lanka
Founded by Arab traders, and captured by the Portuguese in 1517, the Dutch in 1658, and the British in 1796, Colombo is the principal port on the west coast of Sri Lanka. It expanded after the British laid railroads, giving access to tea plantations inland.

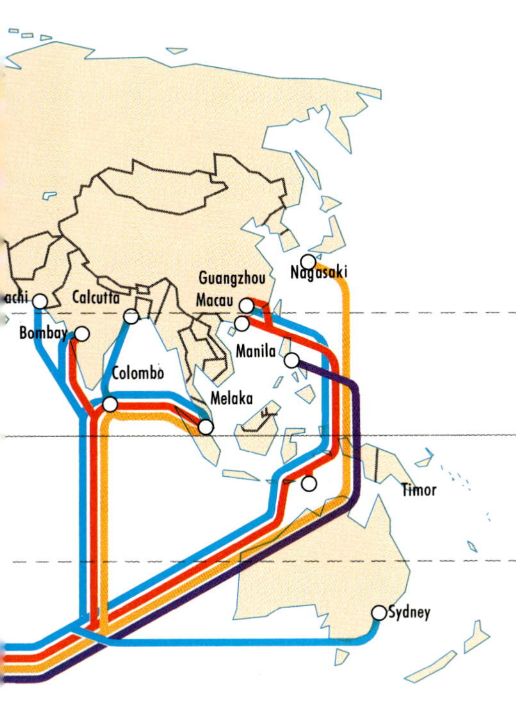

Calcutta, India
Founded by the British East India Company in 1690, Calcutta shipped jute, cottons, silks, indigo, opium, and tea to Europe.

Melaka, Malaysia (formerly Malacca)
Founded in 1400 on the strait connecting the Indian Ocean with the shallow seas of the Malay archipelago, Melaka was a major spice market and spice-shipping port under Portuguese, Dutch, and British rule. It became less important with the rise of Singapore as the principal port of the area.

Timor, Indonesia
Used by the Portuguese as an administrative center, Timor also became a collection point for sandalwood for sale in China.

Macau
Situated on a peninsula and two small islands, Macau was used by the Portuguese as a trading post and missionary base from the 16th century. It was also a collection point for Chinese silk.

Manila, Philippines
This port is situated on the eastern shores of Manila Bay, and at the mouth of the Pasig River, on a great natural harbor. It was established by the Spanish in 1574, and in a regular trade with Acapulco, silks were carried east, and silver west.

Guangzhou, China (formerly Canton)
Guangzhou lies in the fertile delta of the Zhu Jiang River, and became prominent as a center for maritime trade as early as the Han dynasty (206 BC–AD 220). Portuguese traders arrived in 1514, followed by the British in the 17th century. Eventually Guangzhou became a focus of the opium trade, but it also traded tea and silk to Europe.

Nagasaki, Japan
For two centuries the Dutch were the only foreigners permitted to trade with Japan, and Nagasaki was their only point of contact.

Sydney, Australia
Centered around a large natural harbor, Sydney was settled in 1788 by 1,487 people (759 of them convicts), and became the gateway into the newly discovered continent of Australia. From the 19th century it exported copper and wool to Europe.

WHAT IS A GATEWAY?

The first colonial settlements were, of necessity, ports. European traders could only found or expand settlements if their ships had safe harbor **(1)**.

Through these gateway ports the traders could venture into the interior of the new land. Where conditions were favorable they settled, and these settlements then became centers of trade or entrepôts for the surrounding areas **(2)**.

Industrialization at home brought a huge increase in population, many of whom emigrated to what now appeared to be "lands of opportunity." Greater productivity in Europe also meant more goods for export. The gateway cities and entrepôts grew with the expansion of trade **(3)**.

As the colonies grew, so did their manufacturing, internal trade, and lines of communication. Rail and road networks developed around the existing settlements. By the time shipping ceased to be the dominant factor in world trade, the gateways were successful enough to survive the decline of their original function **(4)**.

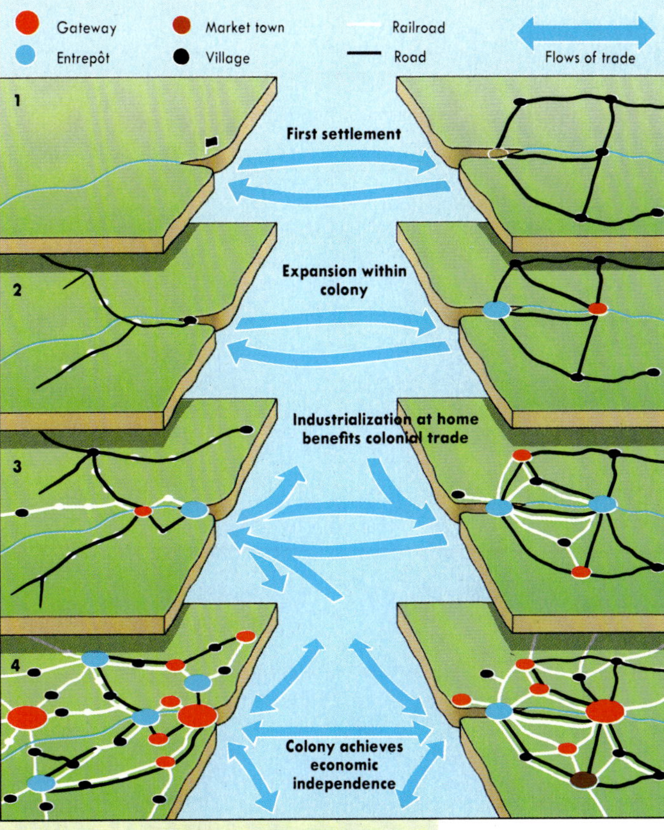

Cairo
Luanda
Angola
Cape Town
South Africa
Beira
Mozambique
Dar es Salaam
Tanzania
Mombasa
Kenya
Djibouti
Djibouti
Cairo
Egypt
RED SEA
ATLANTIC OCEAN
INDIAN OCEAN

Cape Town, South Africa
The first European station in the Cape was established by the Dutch in 1652, although the Portuguese had visited there in the 16th century.

Beira, Mozambique
Founded in 1891 as the headquarters of the Mozambique Trading Company of Portugal, on the site of an ancient Arab settlement, Beira now has rail links with Malawi, Zimbabwe, and Zambia. Together with other sea and air connections, these make it a gateway to large parts of the continent.

Dar es Salaam, Tanzania
This was a small port until the arrival of the German East Africa Company in 1885. It was the starting point for the Central East Africa railroad line, opened in 1907. Dar now handles trade for Tanzania and Zambia.

Mombasa, Kenya
Kenya's second city has been a railroad terminus since the 1890s. As a result, it is also the major port for Rwanda, Uganda, and northern Tanzania. In the past, it has been a successful trading post under Arab, Portuguese, and British control.

Djibouti, Djibouti
One of the best ports in the area, Djibouti was founded by the French. The railroad that links it to Addis Ababa, 468 miles (753 km) away, makes it the main transit point for foreign trade from Ethiopia.

Cairo, Egypt
Egypt was occupied by the Turks in 1517, but did not play a part in the European power struggle until Napoleon invaded Egypt in 1798.

STREETS OF GOLD
The third world cities

IN 1950, THERE WERE FEWER THAN 300 million people living in cities in the third world. By the turn of the century the figure is expected to rise to around 2,000 million, and double again by 2025. More than half of the world's 300-plus "millionaire" cities – those with at least a million inhabitants – are in Latin America, Africa, or Asia. In 1900, there were only 13 worldwide. Now five of the world's ten largest conurbations are in the third world – Mexico City, São Paulo, Shanghai, Buenos Aires, and Seoul. By the year 2000 only Tokyo and New York will represent the "developed" world in the top ten.

Although their growth and expansion have been comparatively recent, most of these third world cities have historic roots. Those in Latin America, the Caribbean, and parts of Africa chiefly stem from early contacts with European traders and colonialists in the 16th and 17th centuries. Elsewhere, especially in West Africa, the Middle East, India, and China, their origins may stretch back much farther, often as far as former empires.

The phenomenal growth rates of third world cities in the latter part of this century match the expansion of cities in Europe and the northeastern U.S. in the last century. Now, as then, many millions of people are being "pushed" from their rural homes by grinding poverty (and in some cases by natural disaster or political disruption), and "pulled" toward cities by the promise of work, education, and freedom from the tight constraints of family-oriented, small-scale rural society.

The journey to the city is now often much easier, cheaper, and less traumatic than in the past, thanks to new long-distance bus services. Many of these routes run direct to the largest cities in the nation. So whereas earlier generations of migrants may have tried their luck in a nearby small town first, moving crabwise from town to town in stages up the urban hierarchy, today movements to the great cities are increasingly direct.

Some estimates suggest that around 100,000 people leave the country for the city every day, although in Latin America, where over 60 percent of the population is already urbanized, flows may now be declining: city growth there

São Paulo in Brazil is probably the fastest-growing city in the world and passed the 20 million mark in 1988. The state of which it is the capital has grown rich on the profits of its coffee industry, and these have subsidized massive industrial growth. The city is now the most important manufacturing base in Latin America. Yet, across the water, an estimated two million people live in inner-city slums and squatter settlements, with water pollution and a rapidly increasing infant mortality rate among the major problems.

since the 1960s has been fueled more by natural increase than by migration. But in Africa and Asia, former peasants and farm workers continue to flood in.

Such rapid population growth, whether through migration or natural increase, can create seemingly intractable problems for the city. Whereas population movements in the developed world a century ago were more than matched by expanding employment opportunities in manufacture, this is far from true for most third world cities now.

Paid work often proves virtually impossible to find, amenities in the form of electricity, water, sewerage, health clinics, and schools are already overstretched, and the housing supply was long ago exhausted. Any work the migrant finds is likely to be insecure, arduous, and poorly paid. Home will almost certainly turn out to be a rented inner-city slum or a squatter encampment on the edge of the city.

Are the streets of third world cities paved with gold? Hardly. For many people life in a shantytown shining shoes or selling lottery tickets is still considerably better than any alternative, yet conditions in towns and cities have deteriorated in recent years. Numbers have continued to multiply, and economic recession has made it more and more difficult for people to obtain salaried jobs. At the same time, city services are reaching the point of collapse. Even so, the city still offers better prospects than the countryside. But for how much longer?

Mexico City, Mexico
Illegally occupied squatter settlements are known as *colonias paracaidistas* – "parachutists' colonies." Many people also live and work on garbage dumps, *ciudades perdidas* – "lost cities."

Bogotá, Colombia
Some 60 percent of residential building over the last 20 years has been of *urbanizaciones piratas* – "pirate urbanizations."

Rio de Janeiro, Brazil
A total of 1.5 million of Rio's five million plus population live in shantytowns known as *favelas*, a bizarre backdrop to the international resorts of Copacabana and Ipanema.

Santiago, Chile
More than a million people live in *poblaciones callampas* – "mushroom settlements."

Kingston, Jamaica
Unhealthy squatter settlements have frequently been demolished, but 25 percent of the city's 750,000 people still live in "uncontrolled" accommodation.

Casablanca, Morocco
Seventy percent of the population live in uncontrolled settlements – 40 percent in squatter settlements known as *bidonvilles*, or tin can towns. *Bidonvilles* are also found in Baghdad, Iraq, and in many parts of the former French empire.

Tunis, Tunisia
Forty-five percent of the population live in 10 percent of the area of the city. They largely inhabit the old town center or *medina*, two densely populated suburbs, and a number of squatter settlements known as *gourbivilles*.

Ankara, Turkey
Two-thirds of the population live in *gecekondus* or shanties. Of this housing, 40 percent has no more than two rooms; 60 percent no running water; 85 percent no indoor toilet, and 35 percent is shared by more than one family.

Calcutta, India
India's largest city contains three thousand separate *bustees* or squatter settlements, with some of the worst living conditions in the world. Public investment has concentrated on improving conditions rather than rehousing people, so local trades often flourish as housing becomes more permanent.

Seoul, South Korea
Although the South Korean economy continues to grow, the gap between rich and poor is increasing. Seoul is now one of the largest cities in the world, attracting vast numbers of migrants. But pressure on housing is immense, and 30 percent of the population live in "uncontrolled settlements."

Hong Kong
A huge public housing program in the 1970s did much to reduce Hong Kong's squatter problems. But large-scale immigration – not least of Vietnamese boat people – has meant that the housing available is once again inadequate.

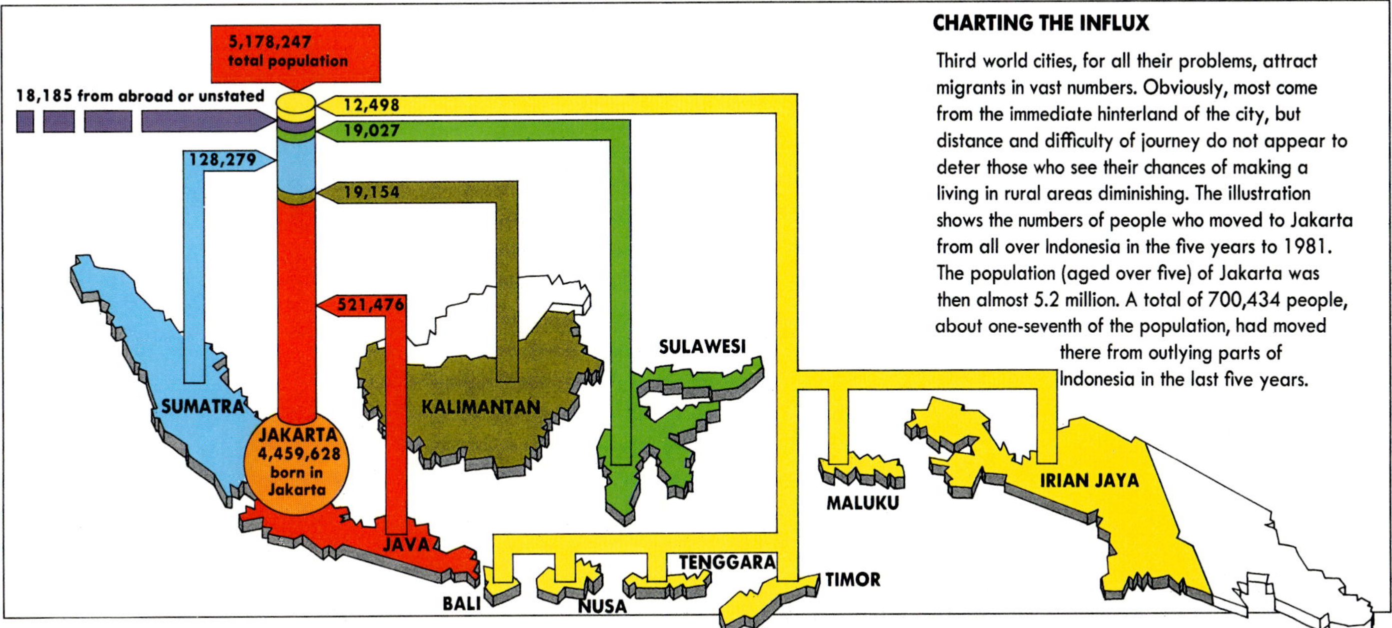

CHARTING THE INFLUX

Third world cities, for all their problems, attract migrants in vast numbers. Obviously, most come from the immediate hinterland of the city, but distance and difficulty of journey do not appear to deter those who see their chances of making a living in rural areas diminishing. The illustration shows the numbers of people who moved to Jakarta from all over Indonesia in the five years to 1981. The population (aged over five) of Jakarta was then almost 5.2 million. A total of 700,434 people, about one-seventh of the population, had moved there from outlying parts of Indonesia in the last five years.

MEGALOPOLIS
Urban sprawl and the cities of the future

About 2,000 years before Europeans set foot on American soil, the Greeks were building a new city in the Peloponnesus by bringing together the population from scattered villages and hamlets. They called this place Megalopolis – "great city."

In the early 1960s, the name was borrowed to describe the dense collection of cities along the northeastern seaboard of the United States from Boston in the north to Washington, D.C. in the south. Boswash, as this vast concentration was nicknamed, stretches 500 miles (800 km) in an almost unbroken built-up area. Its population of about 60 million people is larger than that of the United Kingdom, France, or Italy. How did the new megalopolis come about?

Cities on the northeastern seaboard first grew up as colonial ports, linking the interior with Britain's maritime trading network. Boston, New York, Philadelphia, and Baltimore were all founded in the 17th and early 18th centuries on superb natural harbors. At first their size was limited by lack of transport: until the 1850s most people got around on foot. Cities therefore had to remain small in overall area if they were to function effectively.

In the 1870s, streetcars and railroads made their appearance in cities, and fingers of development pushed out along routes leading from the center. But from the 1940s the extent of the built-up area around each city expanded as a result of the mass ownership of cars. As the cities exploded outward, so they began to coalesce into the urban sprawl that now characterizes the Boswash megalopolitan area. Rapid road, rail, and air transport have further encouraged the spread of the built-up area between the cities, until the outer suburbs of one city are more or less indistinguishable from those of the next.

What is true of Boswash is also true wherever large cities lie in close proximity to each other. Megalopolises can be found all over the world: on the west coast of the U.S. between San Francisco and San Diego, in Europe between Amsterdam and Paris, in Japan between Tokyo, Osaka, and Nagoya, and in the Brazilian Industrial Triangle comprising São Paulo, Rio de Janeiro, and Belo Horizonte.

The major cities that comprise each megalopolis are often complementary, each performing a different function. Boswash contains the federal capital (Washington, D.C.), the country's financial and commercial center (New York), and Boston, an educational and artistic center.

Some people see this vast urban sprawl as formless and featureless, with no redeeming features whatsoever to recommend it either to planners or to those who are forced to endure the boredom of endless suburbs. To others it represents the urban area of the future, a multi-centered economic region with a huge potential market and a labor force numbering tens of millions of people.

Like it or not, whether it be in the U.S., Europe, or any other part of the world, the megalopolis is here to stay. Whether it can be transformed into a pleasant place to live is a problem that confronts us all.

RISE AND FALL OF BOSWASH CITIES

Boston, New York, Philadelphia, and Baltimore all developed as "gateway" cities. Transport has been an important factor in their growth and their fluctuating fortunes. In the 19th century, Baltimore provided the fastest rail link with Chicago; but its harbor was much less suitable for ocean liners than either New York's or Philadelphia's.

Boston, the nearest to Europe, had a brief period of advantage in the early days of air travel, when flying was hugely expensive and airplane fuel capacities small. As these problems diminished, New York's airports came to dominate transatlantic passenger traffic. Only recently, with inconvenient congestion at New York, have Boston, Philadelphia, and Baltimore acquired more important international flights.

On the sea, as technology moves toward ever-larger ships, characteristics of the various city sites reassert themselves. Only New York's harbor can accommodate the most modern ships with ease.

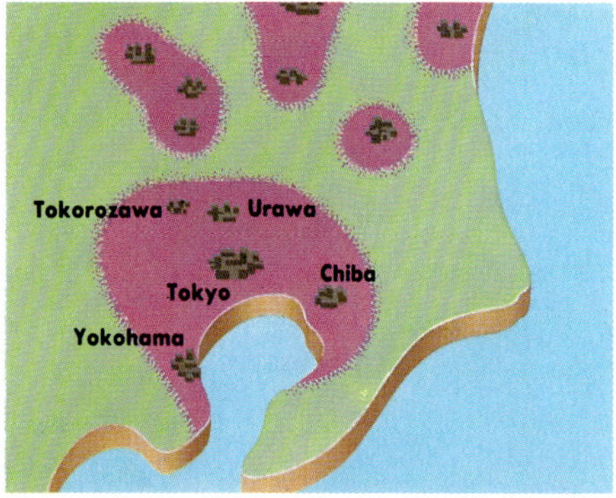

The Tokama megalopolis is only a part of the much larger Tokaido built-up area, formed by a fusion of Tokyo, Nagoya, and Osaka. It is concentrated on a narrow strip of land about 400 miles (650 km) long, following the line of the historical Tokaido Highway and the modern high speed *Shinkansen* bullet train. Over half of Japan's 120 million people live here, occupying about 1 percent of the country's land area.

Chipitts (Chicago to Pittsburgh, incorporating Toledo, Akron, and Cleveland, and reaching north to Detroit) is only one of a number of North American megalopolises: San San (San Francisco to San Diego) covers 500 miles (800 km) along the California coast, and in Canada, Windsor, Hamilton, Toronto, Ottawa, Montreal, and Quebec may be fusing into an even longer corridor, skirting Lakes Erie and Ontario.

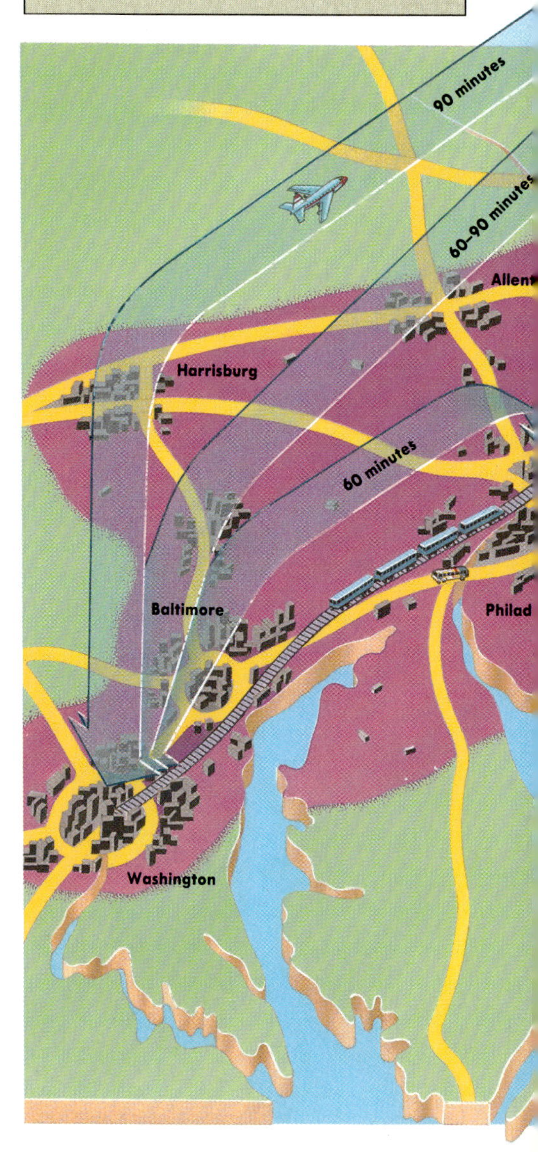

Boswash consists of more than 30 metropolitan areas, ranging in size from New York, with 15 million people, to small towns of fewer than 10,000. Medium-sized cities like Newark, Springfield, Hartford, and Arlington have their own industries, but still serve as dormitory satellites or weekend retreats for the larger centers.

Transport links are vital to the prosperity of all the cities. Supplementing a complex network of interstate highways, 650 flights a day link Boston, New York, Philadelphia, and Washington; 140 Amtrak trains and 117 Greyhound buses cover the same routes every day.

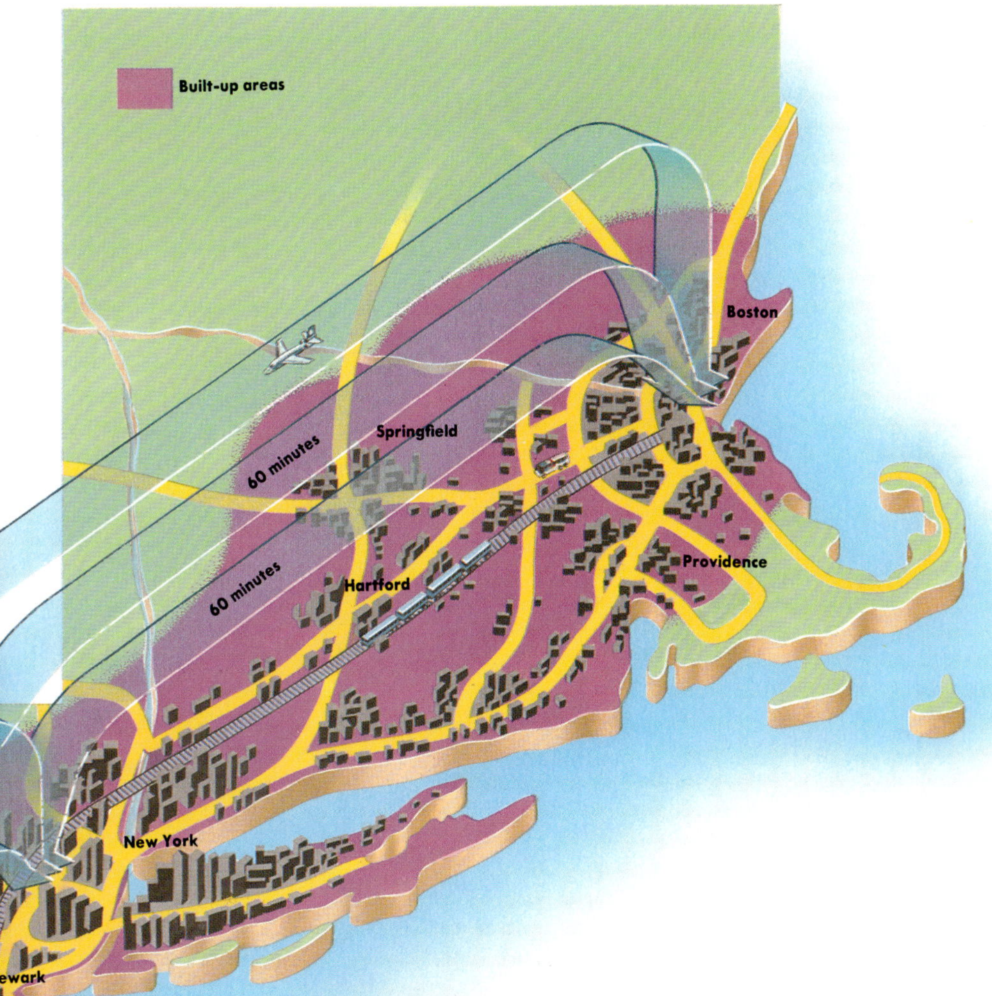

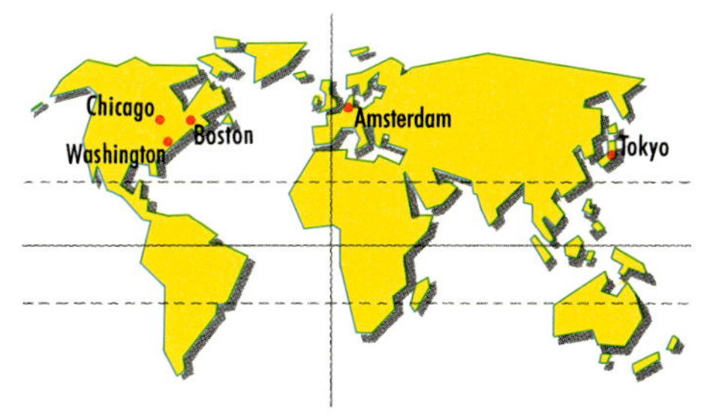

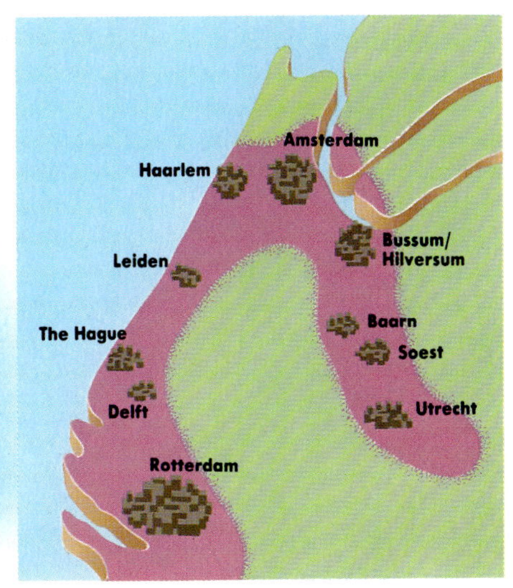

Randstad in Holland is a planned megalopolis. The name means "ring city" and the built-up area surrounds a protected "green ring." As in Boswash, the old towns play different roles. Rotterdam is a port, a center for heavy industry and trade. The Hague is the political capital; Amsterdam, a cultural and tourist center. Hilversum's wealth is based on light engineering.

Efficient transport links are as vital in the Tokaido megalopolis as they are in Boswash. *Shinkansen* is the Japanese for "new trunk line" and the first stretch of it was opened in 1964. The bullet trains which made the line famous reach speeds of 160 mph (260 km/h) and link Tokyo with Osaka and with Fukuoka on the southern island of Kyushu. The line runs about 250 trains, the largest of which can seat 1,000 people, and forms part of the busiest commuter system in the world.

LOCAL DIFFICULTIES

Coping with problems of site

HURRICANE, SNOW STORM, VOLCANO, tidal surge: the city can be exposed to any number of extreme natural torments. Some catastrophes can eradicate an entire city. In AD 79, Pompeii in Italy was overwhelmed by volcanic ash. More recently, Managua, capital of Nicaragua, was virtually abandoned after an earthquake in 1972. It remains to be seen whether it will ever be completely restored.

When natural disasters strike, one of the most important influences in terms of lives lost and destruction wrought is the city's preparedness to meet calamity. The ways in which urban dwellers cope with the problems caused by their environments probe deep into the nature of the societies themselves.

This is demonstrated by the earthquakes that struck Soviet Armenia in 1988 and San Francisco in 1989. Both registered 6.9 on the Richter scale. But whereas in Armenia at least 25,000 people died, in San Francisco there were 63 deaths.

The flimsy shacks and poorly constructed apartment buildings of Armenian towns and cities collapsed almost immediately. In San Francisco, however, quake-proofed skyscrapers resting on rubber and steel shock absorbers survived intact. The lessons for urban design are clear. But even more fundamental is the searchlight the disaster turned on Soviet society.

Site difficulties also include the nature of the terrain on which the city has been built. In some cases the physical problems associated with a site at the beginning of its settlement history may haunt a city in perpetuity, as in Venice. More often, though, the problems only become critical when populations grow and people make new demands on their site.

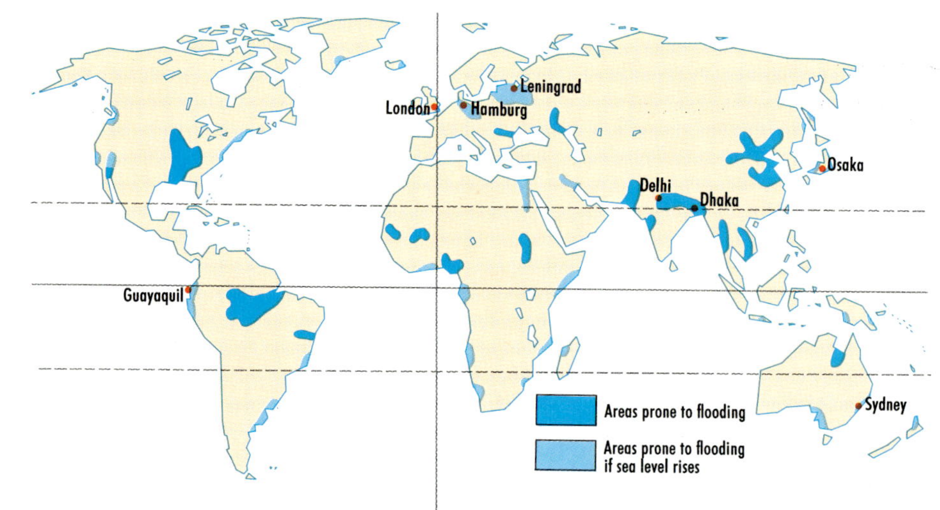

FLOOD-RISK CITIES

Flooding may be caused by a variety of phenomena: a river may overflow when its water level is raised by melting snow, or a combination of high tides and strong winds may drive sea water onto the shore. The effects of global warming may mean that low-lying areas throughout the world are inundated in the 21st century.

Guayaquil, Ecuador
In 1983 hundreds of houses were carried down hillsides because of the action of El Niño, an abnormal warming current along the coasts of Peru and Ecuador.

London, England
Southeast England is tipping into the North Sea at the rate of ½ in (1 cm) every ten years. With global warming, the Thames Barrier, completed in 1983, may not be enough to protect the city.

Hamburg, West Germany
The city of Hamburg lies 70 miles (110 km) from the mouth of the River Elbe. The river estuary acts as a funnel, which means that tidal levels are raised by up to 6½ ft (2 m) in the city. Straightening the river and dredging shipping lanes have both served to increase the mean tidal level by a further 4 in (10 cm). Already defenses in Hamburg have been raised several times in anticipation of flooding.

Leningrad, USSR
A 15-mile (25-km) barrier across the Neva estuary will soon protect Leningrad from flooding. The barrier can be raised if necessary to cope with a rise in sea levels due to global warming.

Delhi, India
A growing population has had to expand onto the flood plain of the Yamuna River. In 1980, the river breached its banks to form a lake 80 miles (130 km) square.

Dhaka, Bangladesh
The flood of 1988, one of the worst to hit Bangladesh, left 75 percent of the country under water. With rail links out of Dhaka shut down, and the airport closed, the distribution of aid to the surrounding countryside was impossible.

Osaka, Japan
Vulnerable both to storm surges and typhoons from the Pacific and to flooding from the Yodo River delta on which it is situated, Osaka's problems are made worse by subsidence caused by groundwater extraction. For the present, a series of floodgates and pumping stations protects the city.

Sydney, Australia
If global warming results in the expected rise in sea levels of 6–20 in (15–50 cm) over the next 40 years, much of Sydney, including prized resorts like Manly, built in low-lying areas, may be flooded.

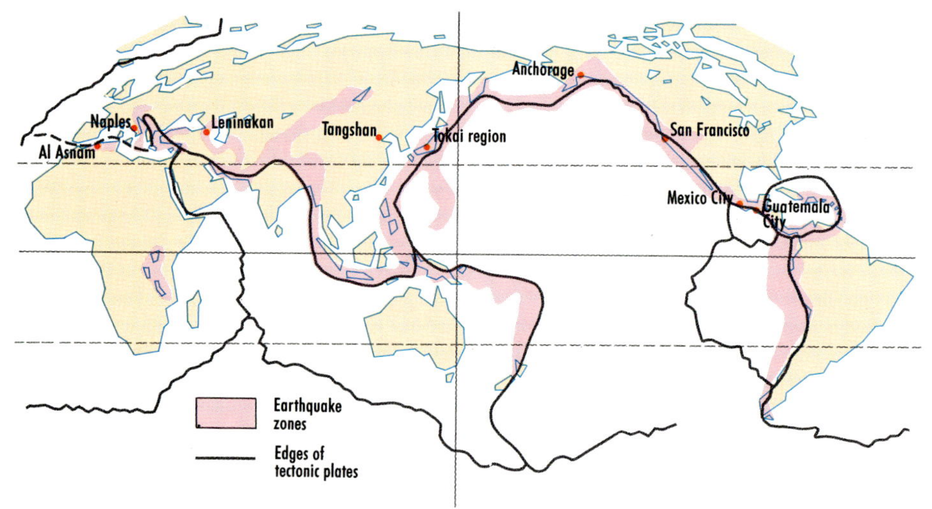

FAULTS AND INFERNOS

Earthquakes and volcanic activity threaten some of the world's most heavily populated areas, and scientists' ability to predict disaster remains vague. Although damage is obviously worst in poor countries where buildings are flimsy, death and destruction can still afflict wealthier cities that prepare for the risk.

Al Asnam, Algeria
In 1980 two earthquakes, one measuring 7.3 on the Richter scale, destroyed most of the city. The death toll reached 20,000, and over 400,000 people were made homeless.

Naples, Italy
Few cities in the world are sited as close to an active volcano as Naples. Although Vesuvius last erupted in 1944, a degree of seismic activity in the area continues.

Leninakan, USSR
One of three cities in Armenia badly damaged by an earthquake in December 1988, Leninakan saw all its buildings destroyed, and 75 percent of the city lay in ruins.

Tangshan, China
At least 240,000 people were killed in July 1976 in a freak earthquake that experts had failed to predict. The city of Tangshan was reduced to rubble in seconds.

Tokai region, Japan
The last major quake affecting Tokyo and Yokohama occurred in 1923. Known as the Great Kano quake, it destroyed 370,000 buildings, killed 59,000 people, and made 2.5 million homeless.

Anchorage, U.S.
The "Good Friday" earthquake of 1964 was the first to cause serious damage to a city. Buildings, roads, railroads, bridges, and docks were destroyed.

San Francisco, U.S.
Lying on the San Andreas Fault, the city's infamous 1906 earthquake measured 8.3 on the Richter scale. In October 1989, an earthquake measuring 6.9 caused damage worth at least $6 billion.

Mexico City, Mexico
The world's largest city was devastated in 1985 by an earthquake whose force was 1,000 times that of the atomic bomb dropped on Hiroshima.

Guatemala City, Guatemala
A series of earthquakes over a period of six weeks virtually destroyed the city in 1917 and 1918. It was badly damaged again in 1976, when 12,000 people were killed.

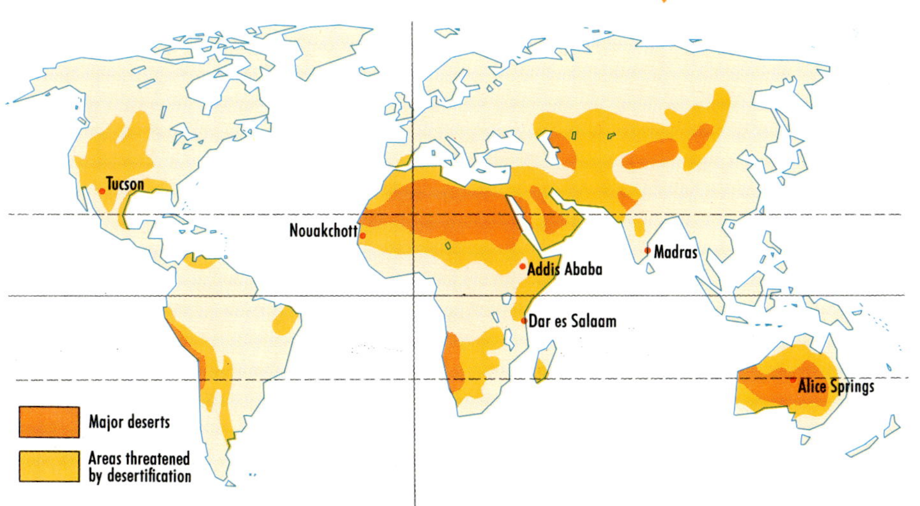

DROUGHT ZONES

Desert or semidesert covers about a third of the world's land surface: these areas regularly receive less than 12 in (300 mm) of rainfall a year. Droughts, however, can occur anywhere if rainfall is significantly less than expected. This means that the definition of a drought varies from place to place: parts of the Middle East expect no rain for six months of the year, which would be catastrophic in northern Europe.

Tucson, U.S.
The city of Tucson is so reliant on the underground Ogallala aquifer for its water supply that citizens fear the results if too much water is taken out. Since the aquifer takes thousands of years to replenish itself, land may subside into the underground reservoir, allowing salt water and pollutants to seep into the water and contaminate it.

Nouakchott, Mauritania
This northwest African country has been devastated by prolonged droughts. In the early 1970s, refugees from the droughts formed a tent city of 300,000 people on the outskirts of the capital city, Nouakchott. Today half of the city's population are refugees from the encroaching Sahel, marginal semidesert around the Sahara.

Dar es Salaam, Tanzania
The city's water system is too old to cope with growing demands. Many areas of the city are often deprived of water, sometimes for days at a time. The problem is made worse by frequent periods of drought. Water shortages have serious repercussions for Dar's textile, dyeing, and weaving industries.

Addis Ababa, Ethiopia
The existing water system is inadequate for the needs of the population, and the problems have been greatly exacerbated by recent droughts and influxes of refugees from civil wars in Tigre and Eritrea.

Madras, India
This city is the fourth largest in India and has been expanding rapidly since 1947, when India gained independence. In 1983, the monsoon failed for the third year in a row. Wells ran dry and people had to stand in line for half of every day at public taps.

Alice Springs, Australia
Situated in the middle of the Great Central desert, Alice was founded in 1872 as a staging post for overland telegraph wires. Rivers and lakes remain dry for most of the year, average annual rainfall is only 10 in (252 mm), and residents rely on water pumped up from underground.

Harnessing an adequate supply of clean water is among the earliest difficulties to confront an expanding settlement, not least because a growing population swiftly tends to foul its own supply from local streams and wells, and turns its rivers into open sewers. The explosive growth of cities in the 20th century has highlighted this problem; it is most acute in the fastest-growing cities of the third world, but highly developed countries are not immune.

Building land, too, may become scarce as a population grows and spreads out into the surrounding area. Much of this newly colonized land may need extensive modification before it is suitable for construction. Indeed, it may not be land at all. Many cities have had to create land by infilling behind sea walls, or by reclaiming swamps and marshes. Cities like Singapore and Wellington, New Zealand, have only been able to build airports that can cope with modern jets by thrusting runways out to sea. Much development in Tokyo is on reclaimed coastal marshland.

The nature of land reclamation, and the people who come to inhabit these marginal lands, illustrate the nature of the society. In Los Angeles, the very rich aspire to a home on hills and outcrops such as Bel Air. But in Rio de Janeiro, it is only the poorest and most desperate who ascend the steep slopes to build their shacks. While in California a complex of dams and drains keeps citizens safe from periodic flash floods, in Brazil hundreds have died in the mud slides that regularly follow storms.

HIGH WINDS AND RAIN

Hurricanes, cyclones, and typhoons are all names for high winds accompanied by torrential rain which occur in many parts of the tropics. They can cover an area 500 miles (800 km) wide, and may rage for several days. They regularly cut swathes of destruction across the Caribbean, the southern United States, and parts of Southeast Asia.

New Orleans, U.S.
Thousands of people were evacuated from their homes in September 1988 to avoid Hurricane Florence. Although the city is protected by dikes, it is increasingly threatened by flooding, as each year 40 square miles (100 square km) of marsh in the Mississippi delta is eroded away and becomes open water.

Miami, U.S.
Along with other parts of the southeastern U.S., Miami is severely hit by hurricanes an average of twice a year. Damage costs run into millions of dollars annually.

Kingston, Jamaica
Hurricane Gilbert, one of the most powerful hurricanes ever recorded in the western hemisphere, with winds of up to 115 mph (185 km/h), devastated Jamaica in September 1988. More than half a million people were left homeless.

Santo Domingo, Dominican Republic
Hurricane David wrecked this tiny island in August 1979. Hurricane Frederic followed six days later. In its aftermath came 24 in (600 mm) of rain.

La Paz, Bolivia
In 1976, Hurricane Lisa killed more than 500 people. There had been no adequate warning of its approach. Shanty dwellings were carried 6 miles (10 km) by the floodwaters which followed the heavy winds.

Hong Kong
Typhoons hit Hong Kong so frequently that many buildings are now specially adapted to withstand the high winds.

Manila, Philippines
These Pacific islands are often hit by typhoons. In June 1985, Typhoon Hal hit Manila, leaving 20,000 people homeless. The heavy rains that followed added to the damage.

Darwin, Australia
Virtually destroyed by Cyclone Tracy in 1974, Darwin had an efficient early warning system that allowed for evacuation in good time.

Melbourne, Australia
Dust storms – in fact millions of tons of overworked topsoil blown off the surrounding farmland – regularly choke the city. One "brown out" in 1983, which brought the city to a halt and forced all three of Melbourne's airports to close, was caused by a cloud 100 miles (160 km) long and 2 miles (3 km) high.

UNHEAVENLY CITIES
People problems

CITIES PRODUCE THE BEST AND WORST of worlds. The myriad opportunities they offer for industry, commerce, entertainment, and learning have to be balanced against the profound problems they create. These are not only crime, social conflict, graffiti, crowded public transport systems, and traffic chaos, but also less obvious threats to life and livelihood from air and water pollution, mounting volumes of solid waste, and the epidemic diseases that thrive in crowded conditions.

Wherever dense concentrations of people and industry exist, the city can turn into an unhealthy and unsightly hell. But what distinguishes cities around the world is the way problems are tackled. Air pollution, for example, shows these differences clearly.

The openness brought about by *glasnost* has exposed how pollution control in Eastern Europe and the Soviet Union has for years been subordinated to industrial output – with catastrophic results. Many third world governments, overwhelmed by debt, either abandon any thought of introducing stringent controls or turn a blind eye to those who break the few rules that do exist.

In cities of the developed world industrial change and pollution legislation have generally curbed the worst excesses of the last century's "smokestack cities." But while city air may be less laden with emissions from coal-burning fires, a new polluter in the shape of the automobile has proliferated.

The automobile produces its own weight in exhaust fumes every 10,000 miles (16,000 km), making it the single largest polluter of cities throughout the developed world. Exhausts serve up a highly toxic mix of gases that are harmful in themselves but even more lethal when cooked by sunlight to cause "photochemical smog." Los Angeles is perhaps most notorious for this, but every sunny city is highly vulnerable – particularly those enclosed by hills and mountains, like Athens, or sited at great altitude, like Mexico City.

Cities throughout the world are also struggling against the high and rising tide of liquid and solid waste they produce, together

Toronto, Canada
A fire outside the city among a dump of 14 million tires could be seen for 80 miles (130 km) in early 1990. People were evacuated, and vast amounts of oil seeped into the water table.

Bogotá, Colombia
Sewage and chemicals upriver pollute drinking water.

São Paulo, Brazil
The waste products of heavy industry lead to acid rain with a pH below 4.5 – over a thousand times more acidic than "normal" water.

Marseilles, France
Despite much investment, slum clearance, and urban renewal, Marseilles' social and racial problems are acute, and the popular image of a violent city with a high crime rate remains.

Cracow, Poland
The ornate facades of historical buildings are disintegrating as a result of acid rain caused by the toxic fumes of industry.

Athens, Greece
More damage has been done to the Acropolis in the last 25 years than in the previous 2,500, as a result of Athens's air pollution, the worst in Europe. There are plans to allow individual cars to drive in the city center only every other day, determined according to registration numbers.

Sofia, Bulgaria
The Bulgarian capital often has brown smogs caused by pollution from car exhausts and badly maintained buses.

Lagos, Nigeria
The center of the world's most congested city lies on an island connected to the mainland by two bridges across the polluted lagoon. It can take 45 minutes to cross a bridge.

Cairo, Egypt
The largest city in the Middle East, Cairo treats less than half its sewage; the rest finds its way into the Nile and lakes. Diarrhea and dysentery are common, and the risk of typhoid and cholera is high.

Bangui, Central African Republic
The city's sewerage system was built when the population was 26,000. It has never been extended or improved, and the latest population estimate is 500,000.

Addis Ababa, Ethiopia
With a population of 1.5 million, Addis is the largest city in the world with no sewerage system.

Beirut, Lebanon
Once known as the "Paris of the Orient," Beirut lies in ruins, shattered by war. Parts of "no man's land" are overgrown, and conditions in Palestinian refugee camps are appalling.

Bhopal, India
On December 2, 1984, 30 tons of methyl isocyanate escaped from Union Carbide's pesticide plant in the city. Some 200,000 people lived in the nearby slum areas. Estimates of the number of dead range from 2,352 (official) to 10,000, with up to 20,000 disabled.

Bombay, India
At least 3.5 million people live in slums, and 40 percent of housing has open sewers.

Beijing, China
A combination of solid fuel used in domestic heating and the emissions from industry means that Beijing's pollution level is 16 times that of Tokyo.

Shanghai, China
Four million tons of untreated industrial and household waste are discharged daily into the Huangpu River. In spring and summer, it produces a black, foul-smelling gas. Air and noise pollution add to Shanghai's problems.

Tokyo, Japan
Rivers in the city have such a high level of chemical pollutants that it is possible to develop photographic negatives by dipping them in the water. Acid in the air is also thought to be a major health risk.

THE WORLD'S WORST CITY?

The capital of Nigeria probably has more problems than any other city in the world. Garbage lies uncollected, traffic moves at a snail's pace, and housing is woefully inadequate for a population that is growing by 200,000 every year. The collapse of the Nigerian economy during the 1980s, as a result of a drastic fall in the value of its oil exports, has merely worsened conditions.

Lagos also has to contend with serious site difficulties. The original settlement on Lagos Island has expanded onto the mainland and now covers an area of more than 100 square miles (250 square km). The two bridges linking the old city to its suburbs are always jammed with traffic.

The rest of the site comprises sand ridges and depressions, swamps, lagoons, and creeks. The depressions are generally waterlogged and provide breeding grounds for mosquitoes. Drainage canals constructed in the 1930s have long since been built over due to pressure for housing, thereby increasing the incidence of flooding. Swamps stretch to the waterfront, and have defied reclamation efforts to make them habitable. The soil is unstable and provides only weak foundations for houses and roads. Reclamation of the lagoons has been more successful, but it is slow and expensive, and cannot cope with the rate of population growth.

Meanwhile, drainage and sewerage problems worsen as the burgeoning population of Lagos spills over onto land that is still largely a waterlogged swamp. Plans to relocate the capital to Abuja suggest that Lagos's problems may be insuperable.

Curiously, Tokyo's city center is not particularly crowded – certainly by comparison with Paris or New York. Nevertheless, space is at a premium because land prices here are by far the highest in the world. As a result, housing, public transport, and social amenities are congested. As many as 20,000 people may crowd into a public swimming pool on a holiday afternoon.

with the immense volume of runoff after every rainstorm. Concrete "jungles," unlike the real thing, are so completely impermeable that they are highly prone to flooding.

Water pipes and sewerage systems on a city-wide scale were introduced – first in New York, later in London and other major centers – in the middle of the 19th century. Today, many pipes and sewers are collapsing from a combination of age and traffic vibration, or are inadequate to meet new levels of demand. This is especially true of fast-growing cities such as Bombay, Cairo, and Jakarta that have inherited the systems built by former colonial powers.

In many other parts of the world, however, the issue is not replacement but installation of a system for the first time. Again, the expense may be so great that governments simply cannot even contemplate it. Inevitably in these cities it is children who are most at risk from intestinal infections arising from polluted supplies of water.

The average New Yorker produces 100 lb (45 kg) of garbage every week. Multiplied by the number of inhabitants, this produces a vast and ever-increasing problem of disposal. Like New York, most cities have chosen to dispose of their wastes in landfill sites outside the city. But many cities have run out of dumps, and new sites are increasingly hard to find.

Once full, dumps are filled and capped with earth or concrete. Pressure on land close to the city means that they soon become highly desirable as building sites. But there can be problems. In the short term, dumps are unsightly, verminous, and extremely smelly, but they can be even more dangerous when they are covered over. Many continue to produce toxic "leachates" that pollute local water supplies, and they may also give off methane and other poisonous and explosive gases.

Attempts to incinerate the refuse have raised problems of air pollution and the disposal of toxic ash. As a result, some cities have taken the controversial steps of dumping the ash in the ocean or shipping it to third world countries.

Paradoxically, the garbage generated in third world cities is not always such a problem. Indeed, trash itself frequently provides an important source of livelihood. In Manila, capital of the Philippines, a whole army of trash sorters, rag-pickers, and bottle collectors survive on the items they retrieve or recycle from "Smoky Mountain," a heap of city garbage.

THE IMPOSSIBLE DREAM

Venice in peril again

VENICE, THE "QUEEN OF THE Adriatic," is enthroned on 118 mud islands in the midst of a shallow, murky lagoon, its rule challenged at every high tide. The first settlers fled there from the mainland more than 1,300 years ago to evade hordes of rampaging barbarians from the north. It must have seemed an unpromising site, yet the lagoon has served Venice well.

The city proudly displays the wealth it has accumulated over centuries in the form of fine architecture and works of art, and it is more secure from flooding than at any time in its history. Yet it is under threat from a totally new quarter, a 20th-century barbarian invasion in the form of industrial and domestic pollution.

Until 1846, when a bridge carrying a railway was constructed, Venice could be reached only by boat, a fact that contributed greatly to its security and wealth. Shipborne trade throughout the Mediterranean made Venice for centuries the most important port in Europe.

Venice's successful love affair with the sea is celebrated annually in a symbolic marriage ceremony between the Doge (a dialect word for *duce* or duke) and the sea. But the relationship has always been a troubled one. Venice remains a city poised precariously at the water's edge.

In November 1966, a storm over northern Italy swelled the rivers draining into the lagoon. Winds created a storm surge over 6 feet (1.8 m) high, and the flood waters engulfed Venice, causing incalculable damage to buildings and works of art.

Since then, great efforts have been made to save Venice from drowning. A 12-mile (18-km)

A Landsat image of the eastern Italian Alps and the Lombardy plain, with Venice in its lagoon at the bottom right. The wavy line crossing the lower half of the city is the Grand Canal. The black area at the bottom left is Lake Garda; the blue-gray patches between the lake and Venice are Verona and Vicenza. The blue-white streaks across the plain show sediment brought down from the Alps by the rivers: over the years a rich alluvial plain has built up. The Lombardy plain continues to grow, extending into the Adriatic Sea.

aqueduct supplies the needs of industry, relieving pressure on the city's wells. There are sluice gates and barriers across the entrances to the lagoon, and sea defenses have been raised and strengthened. Reinforced concrete has been pumped under many of the buildings to stabilize foundations damaged by a corrosive mixture of salt water and industrial pollutants.

The physical fabric may now be secure, but other problems remain. Perhaps most important, the city is now threatened by severe pollution. Modern sanitation has been installed too slowly to protect the water of the lagoon from contamination. Furthermore, Venice's situation near the mouths of several rivers means that it suffers from both agricultural effluents and pollution from its industrial hinterland.

As a result, Venice and the nearby Italian Riviera have been hit by "killer algae" that thrive in the nutrient-rich warm waters of the lagoon. In the summer of 1988 they covered nearly 20 square miles (5,000 hectares) of the lagoon in a thick, dark mat that no sunlight could penetrate. All available oxygen in the water was consumed; dead fish were deposited in their millions on the tourist beaches; and the putrefaction of the algae made the air virtually unbreathable and filled the city with insects. Naturally, the tourist industry suffered badly as a consequence.

It is indeed an irony that having successfully survived invasion from countless human enemies and rendered itself largely immune to the threat of the sea, Venice should now be in danger of being swamped by its own waste.

Marco Polo is the most famous example of the combination of shrewdness and courage that helped Venetian merchants to dominate world trade. In the latter part of the 13th century, Polo traveled to China, where he spent many years at the court of the emperor Kublai Khan. The wealth of previously unknown geographical information he brought back was widely used during the great age of European exploration some 200 years later.

The illustration, dating from about 1400, is the frontispiece of a manuscript version of Polo's account of his experiences *The Book of the Great Khan.* It shows Marco, with his father and uncle, about to set off from Venice.

Venice could only create secure foundations on the mud flats by sinking countless wooden piles – two million alone were used in constructing the church of Santa Maria della Salute. Every major building demanded a veritable underground forest; indeed, Venice's insatiable demand for trees for ships and building piles stripped nearly every oak tree not only from northern Italy, but also from the area now called Slovenia, and much of the Dalmatian coastline besides. None of these areas has ever regained its former tree cover.

The piles were driven into a layer of clay that floats like a raft on a sea of quicksand beneath. Over time, the sheer weight of the buildings has compressed this layer, while the extraction of water for domestic and industrial use has lowered the water table and caused the whole area to subside. At the same time, the sea level has risen by as much as 8 in (20 cm) this century, threatening to swamp the lagoon. Deep channels dredged through the lagoon to allow the passage of larger ships have also added to the height of every tidal surge.

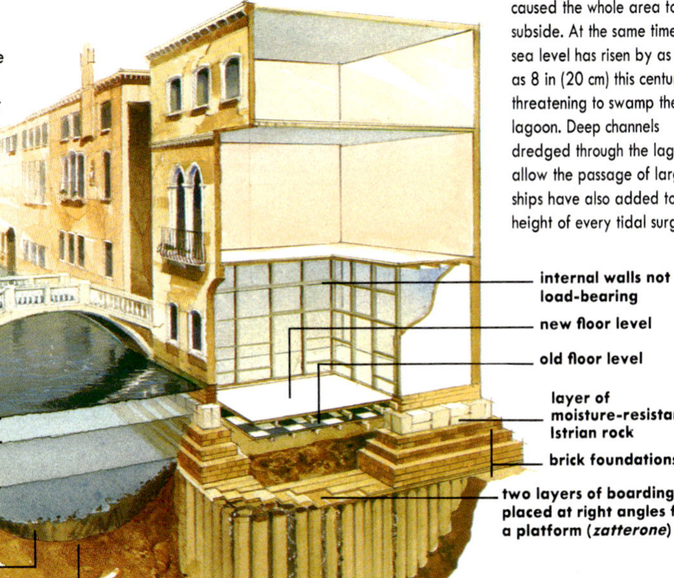

"VENICES" AROUND THE WORLD

Venice has not only inspired artists. Explorers and tourists for hundreds of years have been reminded of its beauty – and its problems – whenever they see a city dominated by water.

Amsterdam, Netherlands
One of several contenders for the title "Venice of the North," Amsterdam is built on piles in sand and mud at the mouth of the Amstel River. Three major canals developed in the early 17th century eventually encircled the old city.

Bangkok, Thailand
Sprawling along the banks of the Chao Phraya River, this "Venice of the East" is crisscrossed by numerous canals.

Fort Lauderdale, U.S.
Ideally suited to be the home of an enormous yacht marina, Fort Lauderdale's network of canals extends over 270 miles (435 km).

Leningrad, USSR
Built on 100 marshy islands in the delta of the Neva River, its separate elements linked by 700 bridges, Leningrad is vulnerable to flooding.

Manchester, England
Founded at the junction of three rivers, Manchester was a natural trading post. But its great success as a manufacturing city was assured when a network of canals built in the early 19th century made its communications even more efficient.

Stockholm, Sweden
Built on 20 islands and peninsulas, the Swedish capital's name means "island of poles" – a reference to the wooden stakes which form its foundations.

Venezuela, South America
The explorer Amerigo Vespucci named this country "Little Venice," so struck was he by the sight of coastal villages built out into the waters of the Caribbean.

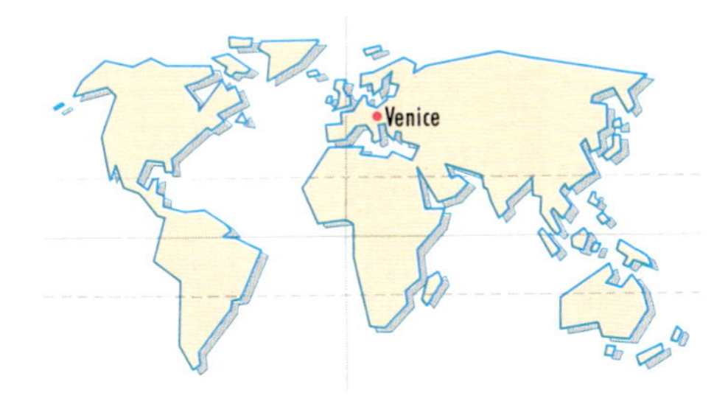

DROUGHT OR DELUGE

Los Angeles' precarious position

Los Angeles – home of angels, Hollywood film stars, and astronomical land prices – is not everybody's idea of a heavenly city. Nature has not always been generous to this semidesert basin on the fringe of the Pacific. True, it enjoys year-round sunshine, but the sun's rays also generate a poisonous smog that onshore winds and a rim of mountains hold captive in the city's "airshed."

L.A.'s ubiquitous green lawns and golf courses imply superabundant water, but virtually every drop has to be brought hundreds of miles by aqueduct from rivers and lakes throughout the region. These lifelines could easily be severed in the event of an earthquake along any of the numerous fault lines that crisscross the region.

The city is dry – its rainfall is less than 14 in (350 mm) a year, half that of Beirut or Algiers – and yet its millions of inhabitants are only defended from an annual deluge of snowmelt by the world's most costly and complicated system of flood channels and dams.

L.A. and its satellite communities stretch 60 miles (100 km) along the Pacific coast, covering virtually the entire 1,200 square miles (3,000 square km) of the basin between the San Gabriel Mountains and the Pacific. This supercity of 10 million people is the second largest in the U.S., after New York. But unlike its East Coast rival, L.A. is still growing – fast.

Yet two centuries ago, there was nothing more than a Spanish missionary settlement – Our Lady the Queen of Angels – here, at the point where an overland trail crossed an erratic stream. Gradually a small cattle-ranching center grew up, initially under Mexican rule, but captured for the U.S. in 1846.

At this time, the population was less than 6,000. Then, in the year 1882, a middle-aged government worker named Harrison Gray Otis left his job, borrowed a few thousand dollars, and bought a failed local newspaper, the Los Angeles *Times*. Otis had a dream, and the dream was a new city in southern California to rival San Francisco to the north.

But first there were difficulties to be resolved, the most pressing of which was the dearth of water. Local sources could supply at most a few hundred thousand people. After 20 years of campaigning, and largely thanks to Otis's skillful use of the power of the press he had at his disposal, local authorities voted to construct an aqueduct to secure a "never-failing source of life-giving water" from Owen's Lake, 250 miles (400 km) away to the north.

In the meanwhile, Otis and his son-in-law Harry Chandler had been assiduously buying up the entire San Fernando valley. With a superabundance of water now supplied by the aqueduct they could subdivide their land and

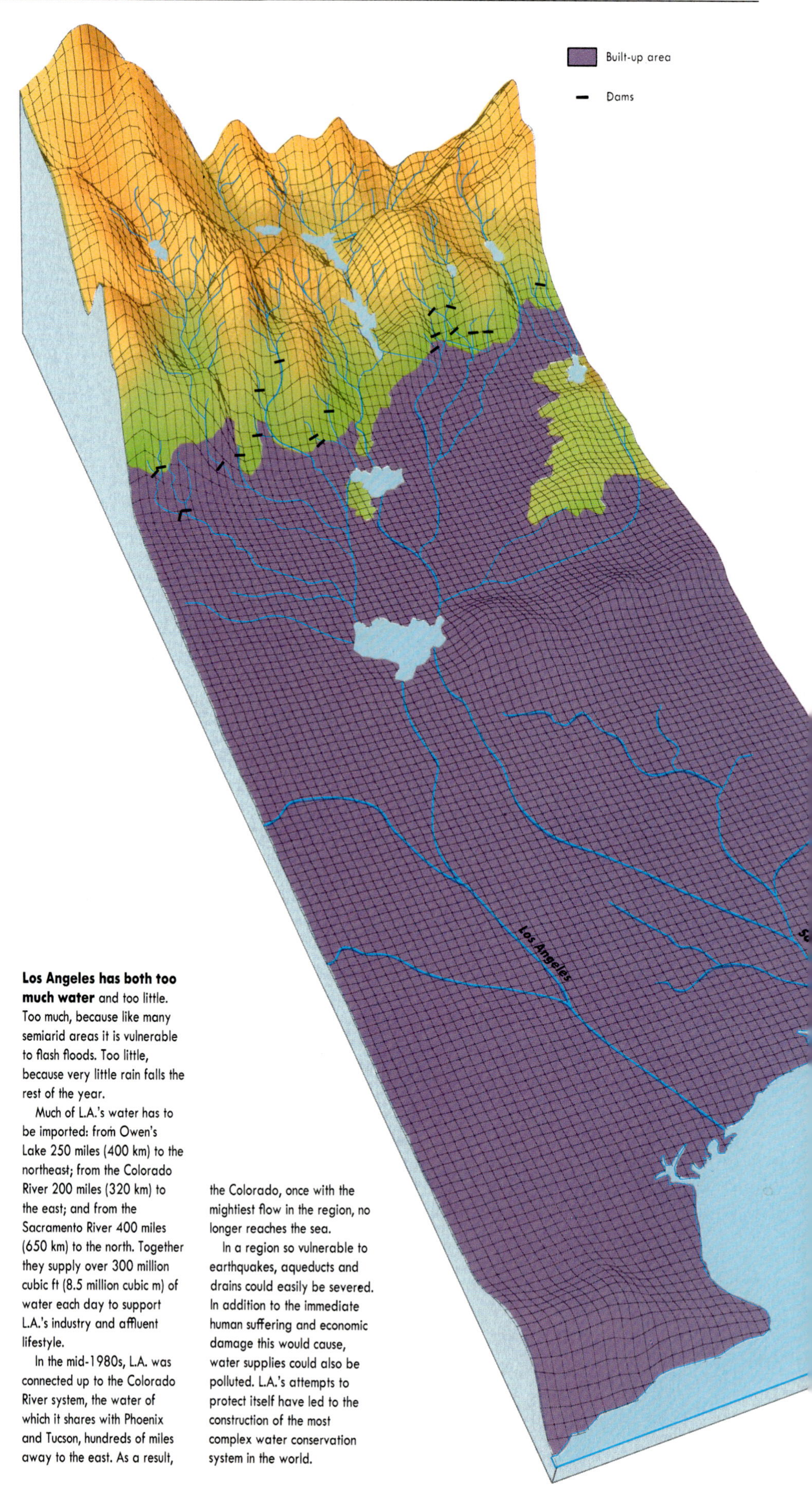

Los Angeles has both too much water and too little. Too much, because like many semiarid areas it is vulnerable to flash floods. Too little, because very little rain falls the rest of the year.

Much of L.A.'s water has to be imported: from Owen's Lake 250 miles (400 km) to the northeast; from the Colorado River 200 miles (320 km) to the east; and from the Sacramento River 400 miles (650 km) to the north. Together they supply over 300 million cubic ft (8.5 million cubic m) of water each day to support L.A.'s industry and affluent lifestyle.

In the mid-1980s, L.A. was connected up to the Colorado River system, the water of which it shares with Phoenix and Tucson, hundreds of miles away to the east. As a result, the Colorado, once with the mightiest flow in the region, no longer reaches the sea.

In a region so vulnerable to earthquakes, aqueducts and drains could easily be severed. In addition to the immediate human suffering and economic damage this would cause, water supplies could also be polluted. L.A.'s attempts to protect itself have led to the construction of the most complex water conservation system in the world.

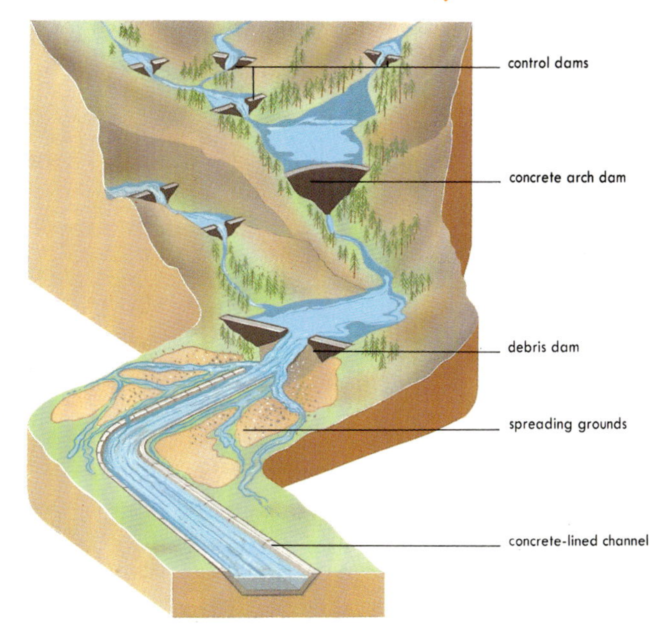

When the snow on the San Gabriel Mountains behind Los Angeles melts, water rushes down the hillsides at a great rate, producing huge amounts of debris. If all that water reached the flat L.A. basin, it would overflow the river courses and cause unimaginable damage.

To guard against this catastrophe, massive control dams have been built in the uplands to store flood water and release it slowly. Below them, debris dams trap the debris that would soon clog the spreading grounds.

These grounds allow water to permeate into the water-bearing rocks below the surface, and 500 miles (800 km) of concrete-lined river channels control the water when it reaches the lowlands.

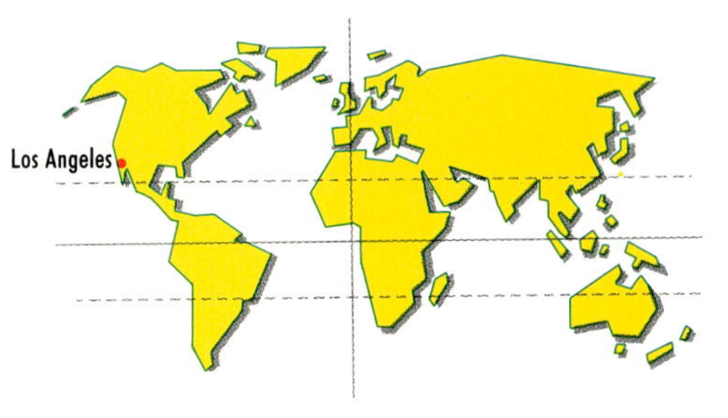

sell it as suburban plots ready for building.

In the following years, partly thanks to the glamour that surrounded the Hollywood film industry, Otis, Chandler, and other real estate speculators thrived, selling dream homes to generation after generation of immigrants. Immigration was particularly rapid during the "Dustbowl Years" of the 1920s and 1930s, when thousands arrived from Kansas, Oklahoma, and the Dakotas. Federally funded employment projects put these migrants to work on the dams that provided the irrigation, flood control, and hydroelectric power that continue to make the city viable.

Along with the migrants came industry, drawn not only by the abundant supplies of labor and the fine climate, but also by cheap electricity and water, a superb deep-water harbor and airport facilities, and the availability of limitless venture capital. By 1920, with a population in excess of half a million, L.A. had surpassed San Francisco. Within the space of 12 years it was big enough – and easily famous enough – to host the world's most lavish and prestigious sporting event, the Olympic Games.

Suburbs sprawled in every direction, laced together from the 1930s by a huge freeway system. This is a vast, complex, and disorienting cat's cradle of nonstop roads, many of which are elevated above street level. For L.A. is *the* automobile city. More space is devoted to vehicles than to people – counting the roads, showrooms, parking lots, gas stations, repair shops, motels, drive-thru eateries, and drive-in movie houses.

The need to attract the attention of the world's most mobile population – for next to nobody wants to walk in L.A., and in any case some streets have no sidewalks – makes for the most extravagant street advertising; not just gigantic billboards but also every kind of bizarre and fanciful architecture, such as diners housed in structures made to look like hot dogs or ice cream cones.

Today, with massively expanded trade around the Pacific Rim, Los Angeles and its surrounding constellation of cities, towns, air force bases, and science centers like Silicon Valley form a thriving agglomeration whose population is expected to top 11 million by the end of the century, making it the third largest urban region in the developed world.

SMOG IN LOS ANGELES

The main pollutant in Los Angeles' air is ozone, produced when hydrocarbons and nitrous oxides (emitted by the exhausts of motor vehicles) react in oxygen. The worst effects are not felt in the city but 20–40 miles (30–60 km) downwind.

The Federal Air Quality Standard for ozone is set at 0.12 parts per million (ppm) in one hour. When the ozone level exceeds this figure, a general warning is issued, as old people, those with respiratory problems, and young children may experience some difficulty breathing. In 1987 in L.A., this happened on 36 days. At Glendora, 20 miles (30 km) downwind it happened on 135 days.

When the concentration reaches 0.2 ppm, most people notice some effect. Warnings advise residents to stay inside, and to use public transport. This level was reached twice in L.A. in 1987, but 51 times in Glendora.

At 0.35 ppm, some industries in the L.A. area are told to reduce their emission levels. This level can be forecast accurately a day in advance, but it has not been reached for three years.

The levels of ozone in the atmosphere are decreasing gradually because tighter controls on emissions are being enforced. It is hoped that by 2007 the ozone level will never exceed the Federal Air Quality Standard.

SUPERCITY

Mexico's mushrooming capital

WITH A POPULATION OF ABOUT 20 million, Mexico City is the world's largest city. It has doubled in size since 1970, and by some accounts is expected almost to double again by 2025. Engulfed by shantytowns, poisoned by smog, congested by traffic and street sellers, its population density is about four times greater than London's.

It has also developed in one of the world's least promising environments. Built on an infilled lake, at a height of 7,350 feet (2,250 m), Mexico City has suffered constant problems of flooding, subsidence, and shortages of fresh water. Worse still, it is vulnerable to savage earthquakes; the last major one, in 1985, measured 8.1 on the Richter scale and caused 7,000 deaths. Yet it has ranked among the world's largest cities for at least 500 years.

In 1325 the Aztecs – or rather the Mexica, as they called themselves – chose the island of Tlateloco in muddy Lake Texcoco as a refuge and defense against their powerful neighbors. There they established a small settlement that they called Tenochtitlán, the "Place of the Prickly Pear Cactus." Within 200 years this village of simple reed huts was transformed into a lavish imperial city covering nearly 5 square miles (13 square km).

The city of Tenochtitlán was reduced to

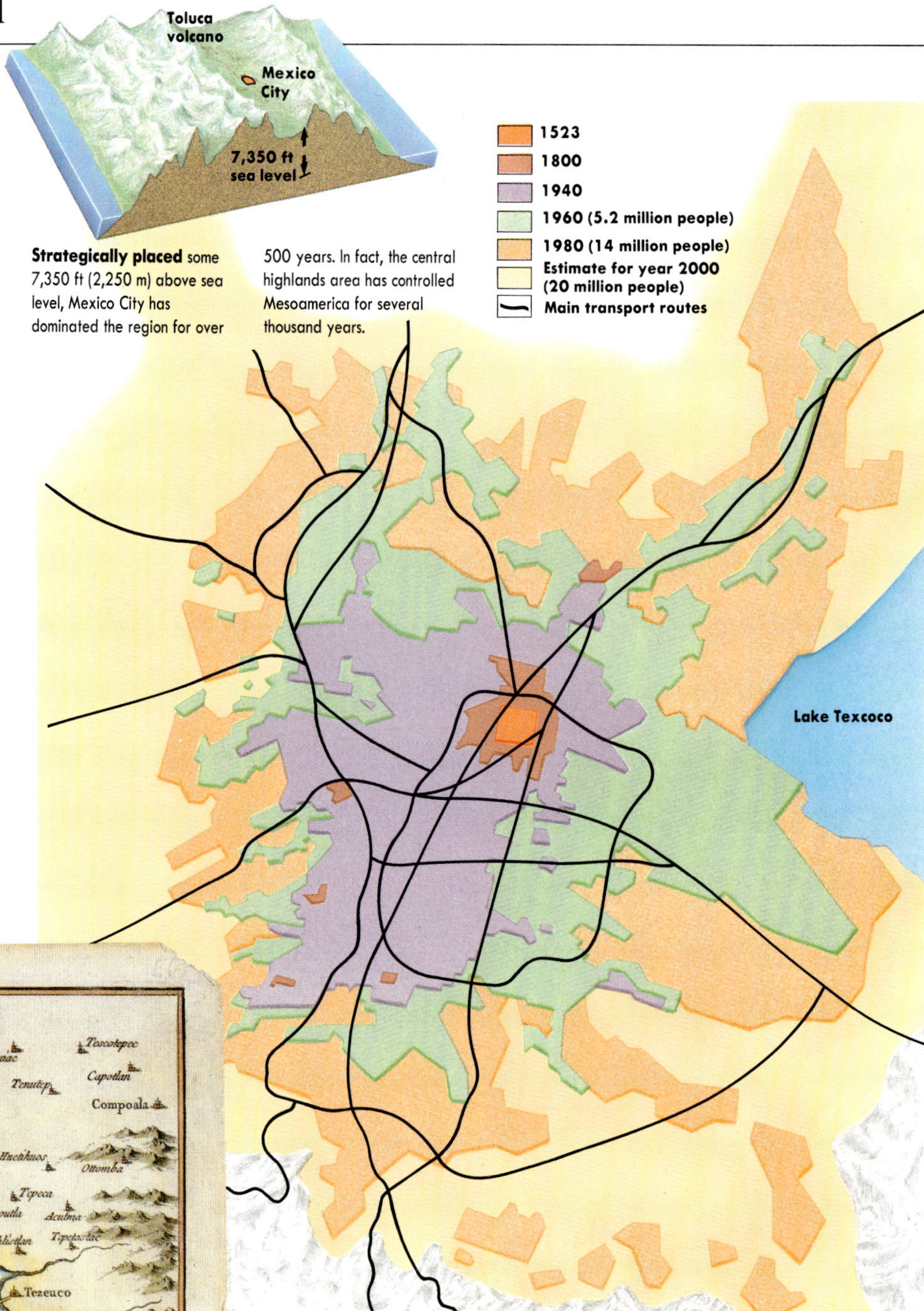

Strategically placed some 7,350 ft (2,250 m) above sea level, Mexico City has dominated the region for over 500 years. In fact, the central highlands area has controlled Mesoamerica for several thousand years.

Carte du Lac de Mexico, et de ses Environs Lors de la Conqueste des Espagnols. Pour servir a l'Histoire Generale des Voyages.

The Aztecs overcame the limitations of their island site by building three raised causeways and two aqueducts that brought fresh water from the mainland to the city. They drained swampland to create islands, or *chinampas*. These low-lying areas were constantly threatened by pollution from the salty waters of the eastern side of the lake, so to prevent this a major dike 10 miles (16 km) long was constructed.

Sealed off behind this dike, Tenochtitlán was a Mexican Venice. A network of canals and bridges divided the land into rectangular plots where its citizens grew crops. As many as 200,000 canoes provided convenient transport to most parts of the city.

The map (*left*) shows the "lake of Mexico" and its environs at the time of the Spanish conquest. The rubble from the razed Aztec capital provided the building materials for the new city, and on the foundations of a former Aztec temple arose the largest cathedral in Latin America.

On the eve of the Spanish conquest in 1519, the Aztec capital Tenochtitlán – on whose ruins Mexico City was built – already had a population of 150,000. By far the largest city in the Americas, it was also bigger than any European city of its time.

Completely destroyed and rebuilt by the Spaniards, Mexico City was again the largest city in the Americas by the end of the 18th century. Its population of 130,000 dwarfed Philadelphia's mere 40,000.

rubble in 1521 after a siege lasting 91 days. Against the majority decision of his followers, the conquistador Hernan Cortés decided to found a Spanish city on the ruins of Montezuma's Aztec capital.

Laid out on the familiar grid plan adopted wherever the Spanish settled, the city became the administrative center of the vast empire known as New Spain, which encompassed Mexico and most of Central America. Its greatness was proclaimed by its magnificent churches, monasteries, other public buildings, and fine private dwellings.

Shops of highly skilled artisans and merchants abounded, a flourishing market existed in the central square, and in the nearby park, or Alameda, high society assembled and paraded on horses and in coaches. Many travelers relate that Mexico City in the 18th century was a cleaner and more pleasant place to live than almost any city in Europe.

The start of Mexico City's urban explosion can be traced to around 1900. Initially it was due to migration from the countryside. Despite agrarian reform following the revolution in 1910 (Mexico had secured its independence from Spain in 1821), opportunities in the rural areas were limited, while development policies promoted industrialization.

The rapid growth of the city has created problems of inadequate housing, services, and transport, and caused environmental deterioration. The city experienced problems of water supply even in Aztec times, and the Lerma Aqueduct, opened in 1951, is woefully inadequate for current needs. New plans to bring water from more distant basins are in hand, but the cost of construction for a country that is wrestling with an immense international debt is enormous.

The public transport system is similarly strained and automobile ownership has led to increased traffic jams. The building of a subway system in the 1970s may have prevented the city from grinding to a complete halt, but the subway does not reach all parts of the city.

Public buses carry over 40 percent of passenger traffic, and although fares are low, the buses are overcrowded and uncomfortable. Slow-moving traffic coupled with the poor maintenance of vehicles causes air pollution, to which factories and excrement deposited in open places contribute. Smog is sufficiently severe on occasions to cause schools to be closed.

Mexico City's notoriously explosive growth has been viewed as an urban pathology, a social and environmental disaster. It remains to be seen whether the city can cope with the continuing problems posed by its ever-growing population and the peculiar conditions imposed by its site and situation.

By 1900 its population had reached 370,000, and large-scale migration from the country accounted for rapid growth in the first half of this century. Around 1,000 rural migrants still arrive there daily, but since the 1960s natural increase has accounted for most of the expansion in population. Now more than half the residents of Mexico City were born there.

United Nations estimates suggest that by 2000 Mexico City's population could be 20 million.

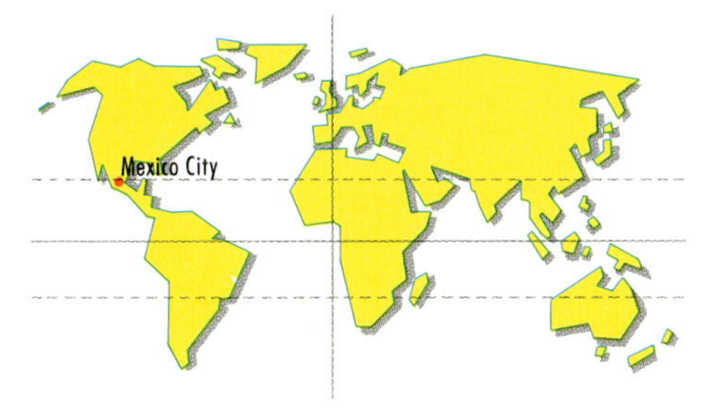

Self-built homes house some 40 percent of the people of Mexico City. Services are often lacking, and there is little prospect of improving them. Running water reaches about 80 percent of people, and the sewage system about 70 percent, but this leaves several million with no immediate hope of services.

None of this prevents residents from imposing their personalities on their surroundings, and brightly painted walls and hangings are a familiar sight.

The Urban Mosaic

A VAST AND SEEMINGLY HAPHAZARD STRUCTURE, the city is at once an economic center, a political focus, and home to many millions of people. Commuters rush daily to its offices, pilgrims flock to its shrines, migrants come in search of the good life. Ethnic groups intermingle, traffic and pedestrians jostle for space, diverse architectural styles compete for attention.

Yet, despite the apparent chaos of urban life, each group of people, each set of activities, each different type of building has its place in the urban mosaic. What principles guide the pieces into place to form a complete pattern of human activity in cities? How do these patterns emerge as cities grow?

Throughout history, the city has occupied a central place in civilization. In practical terms, it both contains people and provides a setting for their activities. But symbolically it is also a stage on which human beings have expressed their vision of the world. City layouts, major buildings, monuments, and statues are coded symbols, diagrams, or models that represent certain beliefs about the nature and structure of the universe. In earlier times, city plans, chiefly geometrical patterns based on grids, squares, and circles, were widely used to express the relationship that existed between citizens and their god, or god-emperor. In fact, these plans formed a kind of citizens' map of the cosmos.

It is no different today. The values and beliefs of societies are still reflected everywhere in the urban fabric. Middle Eastern city planners turn to the Koran for guidance. The monumental proportions of Red Square and the Kremlin in Moscow declare the authority of the centralized state. The thrusting skyscrapers in the centers of North American cities demonstrate no less clearly the awesome power of money.

In most cities, even those whose order is disrupted after generations of rebuilding, original elements still exist. Although buildings may disappear, and land use alter, some structures, street layouts, and building lots will offer greater resistance and provide historical continuity in an otherwise rapidly changing urban landscape.

The most obvious remains are physical structures. Generally only the larger ones survive for long periods, mainly those constructed for political, religious, and military purposes, such as castles and city walls, palaces, temples and churches, squares and monuments. More modest domestic buildings usually have shorter lives. But even when urban structures have been demolished, the plot of land on which they sat, and the pattern of land ownership, may remain the same. The course of city walls, for instance, can sometimes still be charted – often as roads – long after the walls themselves have been torn down. Indeed, the term "boulevard," now in almost universal use for any wide avenue, has its origins in this recycling of the past. It is the old French word for bulwark, the walkway that ran along the top of a city's walls.

The past is no less evident in modern-day street patterns. Some of these, as in Turin and Florence, follow the grid layouts established 2,000 years ago by the Romans. Even modern city planning may not wholly obscure ancient paths and trackways that predate any building. Passyunk Avenue in Philadelphia and Broadway in New York cut across those cities' formal grid plans, following the course of centuries-old Indian trails.

Cities follow social patterns as well as physical ones. But the principles that govern them often have less to do with symbolism than with political and economic factors. In capitalist economies, the competition for space among industry, commerce, and housing determines where activities take place. The result is that different parts of the city are characterized by distinct types of land use. With far more resources at their disposal, commerce and industry are normally able to outbid housing and occupy the most desirable locations in central areas. Housing is pushed into the periphery or squeezed in wherever it can find space.

WITHIN RESIDENTIAL AREAS, DIFFERENT TYPES OF housing accommodate different groups of people according to income, race, and family status. Out of these groups spring neighborhoods with individual personalities, worlds within worlds that develop their own distinctive cultural imprint – speech, form of dress, patterns of behavior. These small groups form tiny but unique pieces of the wider urban mosaic.

In fact, this mosaic is really a kaleidoscope, a pattern kept in constant motion by wider influences. Prior to large-scale industrialization, the centers of cities in the Western world were inhabited mainly by mercantile, political, and religious

elites. Here the most important buildings congregated, and from here sprang the sources of social, economic, and political power. Surrounding the center were craftworkers such as goldsmiths, weavers, and potters, who clustered in particular quarters, while the rest – the poor, those involved in noxious trades, and ethnic or religious minorities – were pushed into back alleys or out beyond the walls. The social geography of many third world cities, particularly in parts of Africa and Asia, still resembles this pattern.

In the West, all this changed with industrialization. Industry became concentrated in the central areas, the points of greatest accessibility. Along with the new wealth that resulted came congestion, pollution, and squalor. Those who could afford it generally moved to high ground, above the worst pollution, or upwind, which in cities like London and Paris meant avoiding the "east end." Many moved to suburbs, or to "exurbia," even farther afield. Those who could not were forced to remain in overcrowded, insanitary, and decaying housing.

Burgeoning urban populations have overspilled traditional city boundaries and absorbed the surrounding countryside into built-up areas. As the means of transport and communication have improved – particularly train, streetcar and bus routes in the last century; subways, increased private ownership of automobiles, and the construction of freeways in this – so urban sprawl has destroyed what remained of the once distinct break between the country and the city.

For much of the 20th century, the continuing process of inner city decay and suburban flight has transformed the pattern of North American and European cities. Elsewhere, the situation is different. In socialist countries, cities have been planned according to entirely different criteria and the social patterns differ accordingly. In the third world, the city center still attracts elite groups, while shantytowns housing the poor spring up on the periphery.

But east–west, north–south, rich or poor, to some eyes the city is universally abhorrent: every one is a vision of hell. The modern city, say critics, can never be anything other than a random, disorderly collection of strangers knowing no emotional attachments or shared beliefs other than mutual greed and self-centeredness.

Is this charge of urban chaos justified? After all, cities are the most complex human artifacts in the world. That they function at all – that food and other goods are available to buy; that water, gas, electricity, transport, and telephone lines continue to operate; and that people are able to go about their lives in the reasonable expectation that every new day will be much like the previous one – is wholly dependent on the persistent efforts of countless individuals all operating in a more or less coordinated way, albeit one which is very largely beyond conscious control.

Nowhere is the urban mosaic unchanging. Whenever political or economic circumstances alter, so, too, does the geographical makeup of the city. In the developed world, the growth of postindustrial economies is generating new forms of city structures. While manufacturing has abandoned congested city centers for greener sites on the outskirts or beyond, commercial and financial dealing are increasingly concentrated in the cores of major cities. With this new range of activities come new forms of social patterning. When housing in the heart of the city is taken over by a newly affluent "urban gentry," many longtime inhabitants are displaced to other parts of the city. Those who are left behind find themselves living, sometimes resentfully, cheek by jowl with much wealthier neighbors.

So, from the city's apparent mélange comes a rich and variegated pattern of local worlds that adds excitement and color to the urban mosaic. Distinctiveness is accentuated, cultural differences are enhanced, lifestyles endlessly transformed. Cities are not just a physical framework for domestic and economic activity. They are also the dramatic setting for the display and celebration of variety. As the point of maximum concentration of power, wealth, and cultural vitality, cities are at the very center of human creativity.

SYMBOLS IN SPACE

What do city structures mean?

AMIDST THE BUSTLE OF MODERN CITY life, a rich symbolism is embedded in the urban form. Its messages may be loud and brash, as in a monstrous statue, a gigantic triumphal arch, or the skyscraping headquarters of a transnational company; or whispered so quietly that they are barely audible, as in the subtle motifs of houses and street furniture. But the communication, although complex, is not a babble. It is largely comprehensible to anyone who takes the trouble to learn the language in which the built environment speaks to us.

Most obviously, cityscapes the world over call attention to their ruling institutions and beliefs. The city uses the infinite possibilities inherent in its fabric to declare its faith in a religion, a royal lineage, or a political order. City walls, buildings, statues, street plans – indeed any structure – can be used to promote sacred values, commemorate tragedies, celebrate triumphs, and lionize the city's most cherished sons and daughters.

Until this century, a single building – almost universally religious in nature: a cathedral, mosque, or temple – generally dominated the skyline, its symbolic power emphasized by a central location and the clever use of radial roads. At other times, royal palaces and governmental buildings have also been privileged in this way.

But over the last hundred years, such symbols of social unity have been overshadowed by a multiplicity of competing "temples to

SYMBOLISM AND SOCIAL PROTEST: TIANANMEN SQUARE, MAY–JUNE 1989

Every modern city has a symbolic area – often a square – where social protest is focused. In London, it is Trafalgar Square. In Prague, it is Wenceslas Square, where in 1968 a young Czech student, Jan Palach, burned himself to death in protest against the invasion of his country by Russia. In May and June 1989, when Chinese students in Beijing occupied Tiananmen Square, their choice of site was a gesture pregnant with symbolic defiance.

Tian-an-Men means the "Gates of Heavenly Peace," and for hundreds of years the emperor passed through these gates as he left his palace in the walled Forbidden City for the annual ritual sacrifice of the Heaven and Earth. Imperial decrees were lowered from the gate in a gilded box.

The Chinese have for thousands of years seen themselves not as a nation among other comparable nations, but as the nation that stands at the center of heaven and earth. For the last 600 years, through dynasties and republics, at the heart of that center has been the home of the emperor, the Palace City, within the Forbidden City in Beijing.

In the imperial Chinese view of things, Beijing lies at the center of a succession of zones of decreasing culture. First come directly administered royal domains, which are themselves surrounded by a zone where local rulers send tribute or taxes to the emperor. Around this lies a zone of pacification, the frontier areas where people are in the process of acquiring Chinese culture. Beyond these layers are zones of progressive barbarism and savagery. As the symbolic center of the universe, and the focus of imperial power for centuries, Tiananmen remained a focal point for the new government when the Communists came to power. Here, in 1949, Mao Zedong proclaimed China the "People's Republic."

Already vast, the square was expanded to its current 123 acres (50 hectares) and new buildings were constructed to celebrate the triumph of the Chinese revolution. On one side lies the Chinese parliament building, the Great Hall of the People, and opposite it the Museum of the Chinese Revolution. In the center of the square stands the Monument to the Heroes of the People.

Tiananmen Square also contains the massive Mao Mausoleum where the former leader's embalmed body lies in state. It is the site of the great May Day parade, and in October it is the place in which the foundation of the People's Republic of China is celebrated. It is in all respects the historical, cultural, political, and symbolic heart of China.

Its occupation in May and June 1989 by students demanding reforms was therefore an act of complete defiance that fundamentally challenged the authority of the Chinese leadership.

The largest square in the world, Tiananmen covers 123 acres (50 hectares). It is shown alongside Disneyland, California (75 acres/30 hectares); Red Square, Moscow (18 acres/7.3 hectares); Trafalgar Square, London (5.4 acres/2.2 hectares) and Piazza San Marco, Venice (a mere 3.2 acres/1.3 hectares).

San Marco
Trafalgar Square
Red Square
Disneyland
Tiananmen Square

Mammon," the towering skyscrapers that symbolize the overwhelming power of industrial and financial institutions.

Although many such buildings are designed to look ultramodern – which implies thrusting dynamism and ambition – others are deliberately made to look old. Indeed, the invocation of the power of tradition is a vital part of any cityscape. Age implies durability, which is why many government buildings, museums, art galleries, and banks are built in a "classical" style.

The Houses of Parliament in London look like a medieval cathedral. In fact, the well-known superstructure, including the Victoria Tower that houses Big Ben, only dates from the 19th century. The design was chosen because it conveyed the religious and feudal roots of this age-old mother of parliaments.

Sometimes the tradition being invoked transcends national boundaries. Washington, D.C. was planned using Greek and Roman forms. Similarly, between 1890 and the late 1930s, thousands of local "temples of government" in classical styles were constructed throughout the U.S. as seats of state and local government, courthouses, even post offices. These public places are monuments to the power of reason and democracy. Importantly, classical temples are a republican form: they are not associated with royalty or nobility.

The use of city structures as symbolic weapons in social conflict has a long history. Most usually, it means paying homage to new leaders or popular causes through the construction of new monuments and the renaming of the city, its streets, and squares. But in extreme cases, an invading group will try to destroy every trace of a city's former buildings, as happened when the Spanish destroyed the Aztec capital that stood on the site of present-day Mexico City.

Uprisings often focus on hated symbols of tyranny. After liberating the few prisoners remaining in the Bastille at the start of the French Revolution in 1789, the revolutionaries dismantled every stone and used them to surface local roads, so the people of France could trample the work of their oppressors underfoot, literally as well as symbolically. Speculators bought fragments of the prison and sold them as souvenirs, as they did when the Berlin Wall came down 200 years later.

In recent years a similar fate has befallen many statues of Lenin throughout the eastern bloc. In Bucharest, Romania, the steel from which the immense, 6.9-ton (7-tonne), 26-foot (8-m) high structure was constructed will be melted down and turned into a monument to commemorate the liberation.

Alternatively, the leaders of the new order may try not to break with the past but to coopt it. They take over buildings, monuments, or squares that possess vestiges of traditional authority, and recycle them for their own purposes (*see box*).

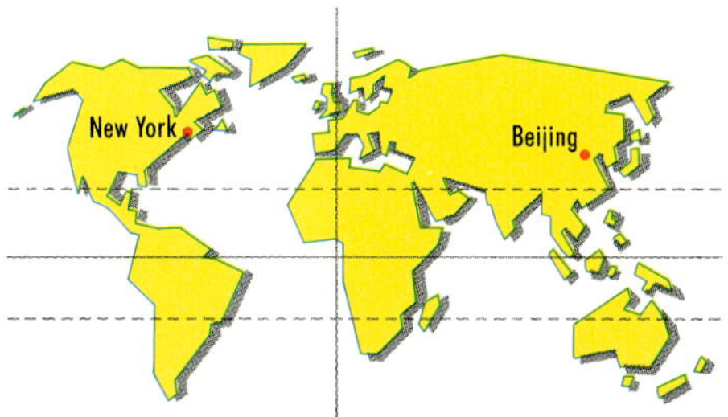

Seen here dwarfed by the twin towers of the World Trade Center, New York's coppersheet-on-steel-frame statue of "Liberty Enlightening the World" was a gift from the people of France in 1886. It was built to celebrate the victory of the French-American alliance against Britain in the War of Independence a century earlier.

For some, it may simply be the badge signifying New York. For millions of immigrants, the lamp Lady Liberty holds in her right hand has been a welcoming beacon. But for many thousands of young people demonstrating around a model of her in Tiananmen Square in 1989, she is the universal symbol of freedom from all tyranny.

CAPITAL IDEAS
Planning the perfect city

A capital city – as its name (from the Latin for head) implies – is a state's governmental, legal, and often financial nerve center. As such, it is generally the most populous city, and grand in design, with wide avenues, monumental squares, and huge public buildings proclaiming the first city's power and wealth. In many cases, as with Paris and Moscow, highways, railroads, and air routes converge on the capital, further emphasizing its national significance.

Capitals often lie in the center of their countries rather than at the edge, partly for ease of access, but also for symbolic reasons. Central capitals, like Warsaw in Poland, Madrid in Spain, or Santiago in Chile, can readily be portrayed as lying at the heart of the state, equally part of all segments of a country, and siding with none in particular. When a capital city is marginal, like London, it may be seen as partisan and out of touch with the needs of outlying regions, such as Scotland and Wales.

Federal capitals are particularly difficult to site. To avoid the charge of partisanship, they

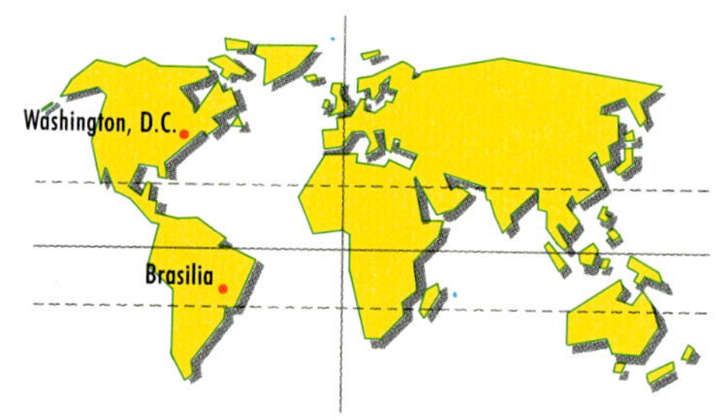

Washington, District of Columbia, is the federal capital of the United States of America, home of the Congress, the Senate, the White House, and the Supreme Court. Named after George Washington, first president of the U.S., its splendid design, with its many Greek, Roman, and French baroque elements, is typically 18th century. Yet most of its buildings are less than 100 years old.

Its layout is unlike that of any other U.S. city. The "democratic" gridiron of streets is cut diagonally by avenues named for the original states of the Union. These create impressive vistas (like the one from the Lincoln Memorial to the Washington Monument, shown here), parks, and circles reminiscent of the grand style of European autocrats. This was a deliberate design, symbolizing the supreme power of the republic.

Like Brasilia, Washington, D.C., is surrounded by areas of great poverty, with serious problems of housing and crime.

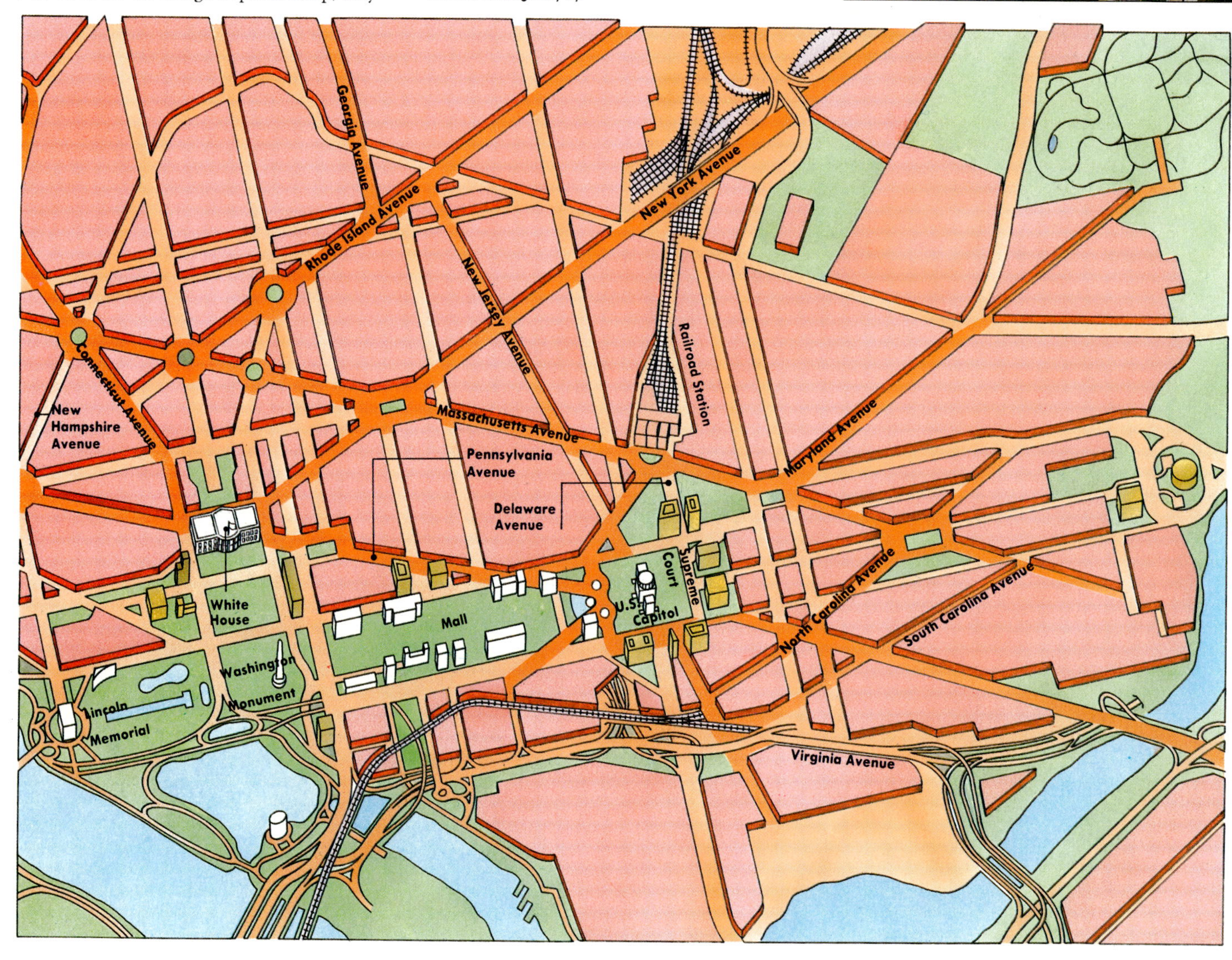

Brasilia may be grand and impressive, yet it is hard to escape the feeling that it is not really meant to be lived in.

Started in 1956, Brazil's showcase capital attempts to shift the nation's political, economic, and psychological focus away from former colonial cities on the coast toward the interior. Located high on a central plateau 600 miles (1,000 km) west of the former capital, Brasilia is planned in the shape of a cross, a bird, or an airplane.

Yet while many apartments remain unoccupied, or incomplete, thousands of people squat in *favelas* – shantytowns – that stretch for 30 miles (50 km) in all directions. As its architect Lucio Costa admits: "Of course half the people in Brasilia live in *favelas*. Brasilia was not designed to solve the problems of Brazil. It was bound to reflect them."

may have to be created afresh, with no symbolic associations with any one region, as when Canberra was created Australia's federal capital in 1927, or when Bonn was selected as the capital of West Germany in 1949.

In some cases the capital city is deliberately moved. This typically occurs not only to "center" a capital within a state, but also when a former colony gains independence and seeks to establish its national identity. Washington, D.C., for example, was created soon after the Revolutionary War, at the center of the original 13 states of the Union.

Part of the motivation for moving a capital may also be to curb the concentration of investment and population growth in the former capital. This was the case in Brazil (where Brasilia replaced Rio de Janeiro as capital in 1960), and Tanzania (Dodoma replaced Dar es Salaam in 1975).

Capitals are also frequently relocated, or at least redesigned and lavishly embellished, to reflect the absolute power of an individual or a political ideology. In the 1980s, the Romanian dictator Nicolai Ceausescu destroyed Bucharest's old city center to make way for a 3-mile (5-km) Boulevard of Socialist Victory, leading to a 7,000-room presidential palace, flanked by huge apartment blocks for his supporters, all built in the French classical style. In 1983, the Ivory Coast announced that it was moving its capital from the port city of Abidjan to the small town of Yamoussoukro, 150 miles (240 km) inland. Yamoussoukro is the birthplace of the nation's president and founding father, Félix Houphouët-Boigny.

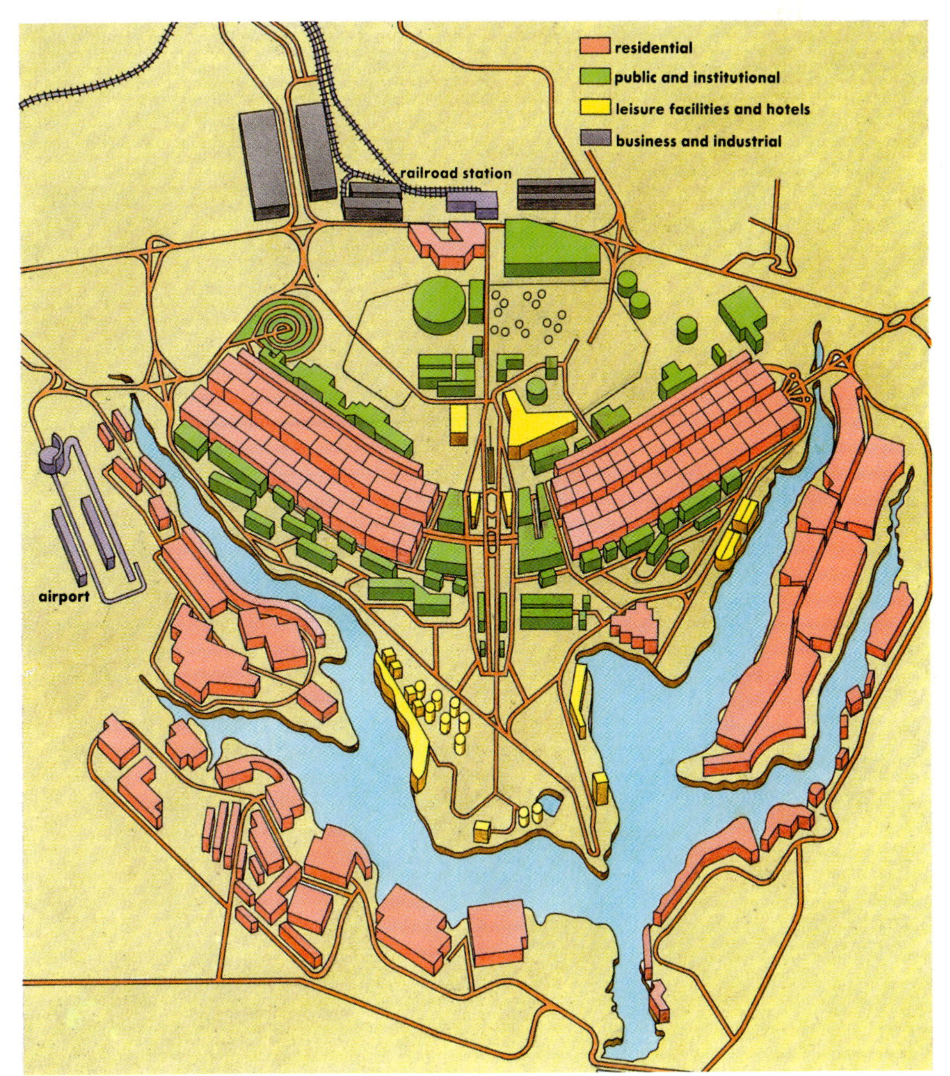

GRID CITIES

An ancient pattern in modern America

Developed to its limits in the United States in the late 20th century, the grid city has been in existence for nearly 5,000 years. Having first appeared in the Indus valley around 2150 BC, the pattern spread to the Near East, and thence to the classical world of Greece and its colonies. The same layout developed independently in many other parts of the world, including China, Japan, and the pre-Columbian Americas.

In its earliest manifestations the grid clearly had a religious function: squared-up plans, usually oriented to the sun, seem always to have symbolized the cosmos, the earth, and the proper place of people in the scheme of things.

However, it is the military version of the classical grid that has probably been the most important influence on modern town planning, at least until the last century. For reasons of discipline, symmetry, and ease of moving soldiers to vulnerable points on the ramparts, it is the perfect layout for camps and forts. It was probably these concerns that led Alexander the Great to impose grids on many of the cities he founded, including Alexandria in Egypt (laid out in 331 BC).

The Romans, too, adopted the grid as their basic unit of town planning, not only in Italy, as in Turin or Florence, but wherever they colonized new territory. Cologne in Germany was a fine example until much of the old city was destroyed during the Second World War At the heart of every Roman city is a cross of perpendicular, wide streets oriented to the sunrise on the city's inauguration day – a familiar if repetitive urban signature.

A more regular checkerboard pattern was imposed in the French fortified towns known as *bastides*, of which the classic example is Villeneuve-sur-Lot, laid out in 1153. The *bastides* played a vital part in the wars with the English throughout the Middle Ages. In fact, *bastides* also appeared wherever Europeans expanded overseas. In Latin America, Spanish towns nearly always followed the grid plan, this time based not on a cross but on a central plaza or square, around which the main church and political buildings were sited.

It was in North America, however, that the grid plan truly came of age. Virtually every new township and city from Savannah to Seattle was laid out on grid principles, usually parallel to the chief means of transport – the coast, river front, or railroad. The grid was adopted partly for military reasons – many towns started as camps and forts – and partly because most of the U.S. had been parceled into grid plots, so that land could be easily surveyed and sold. But perhaps above all, the grid symbolized equality and the infinite possibilities available in the New World.

It may have been the U.S. that adopted the grid plan most enthusiastically, but it is also here that its inherent problems have been most deeply felt. The rigid division of building plots into same-size blocks has often meant inflexibility and visual boredom. Row upon row of uniform buildings make for feelings of anonymity in urban settings, while uniform block sizes are not always easily adaptable to changes in land use. Another problem is that plots are often left undeveloped, particularly close to city centers because land prices are so much higher there than at the edge.

As an aid to the commercial exploitation of urban land, the grid was a resounding success. But the rectangular street layout is much better suited to survey and sale than to ease of transport. Most obviously, it copes poorly with the ever-increasing volumes of traffic which converge on city centers. Hold-ups at key intersections can cause the entire system to seize up – the dreaded "gridlock."

If the future of the city lies in a physical framework that can be flexible and respond to changes in society, then the days of grid cities may be numbered.

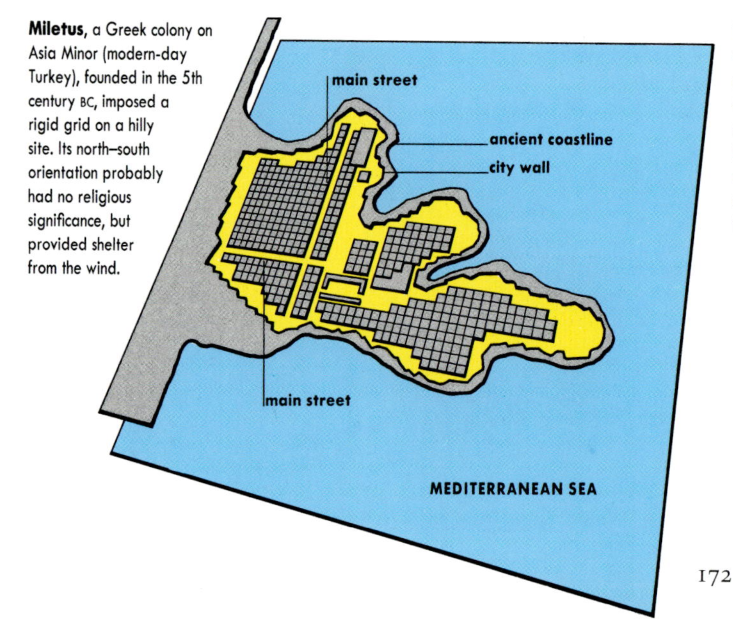

Miletus, a Greek colony on Asia Minor (modern-day Turkey), founded in the 5th century BC, imposed a rigid grid on a hilly site. Its north–south orientation probably had no religious significance, but provided shelter from the wind.

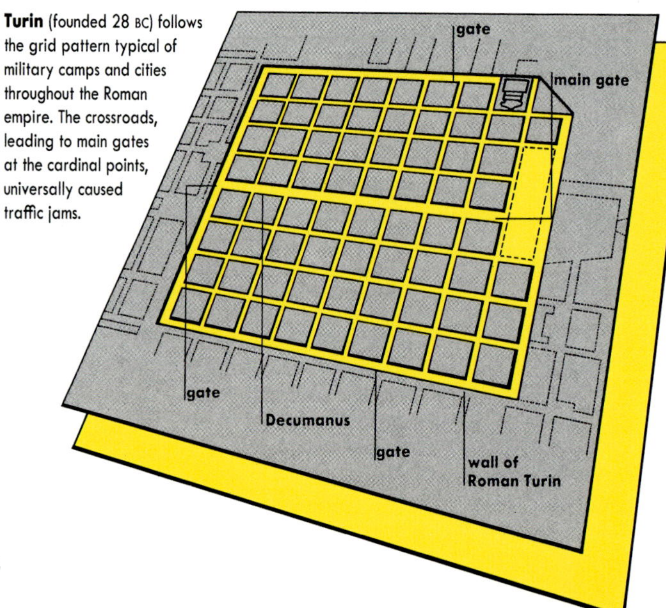

Turin (founded 28 BC) follows the grid pattern typical of military camps and cities throughout the Roman empire. The crossroads, leading to main gates at the cardinal points, universally caused traffic jams.

Manhattan, New York, perhaps the world's most famous grid city, was laid out in 1811. The name "Central" Park was a deliberate misnomer. When it was created in 1858, it lay well outside the city limits on scrubland and marsh. But its designers had faith that the regular march of the blocks would soon surround the site.

A curious anomaly in the otherwise totally formal cityscape is Broadway. New York's planners respected the Weckquaesgeck trail, a centuries-old Indian track, whose course can still be traced through the city's theatrical district.

5,000 YEARS OF GRID CITIES

There is nothing new, and nothing specifically American, about grid cities. The first known grid "city" was a temporary labor camp, situated at Kahun, Egypt, and built about 2670 BC to house workers who were building pyramids. The earliest known permanent example was at Mohenjo-Daro in the Indus valley in India, which flourished from 2150–1750 BC. It was a major trade center, advanced enough to have houses with inside lavatories and enclosed drains in the streets.

The grid flourished throughout the Roman period and once again in medieval Europe. In the 15th century, the emperor of China demanded that Beijing be established as a pattern of straight avenues around what became known as the Forbidden City. The old grid is still there, a tiny part of the vast, sprawling city.

The Spaniards took the checkerboard pattern to the New World, decreeing that their colonial cities should surround a central plaza and its adjacent government buildings.

Philadelphia (1688) was probably the first planned grid city in the U.S. William Penn's aim was to raise money from the sale of plots of land, and regular sizes may well have simplified his accounting system.

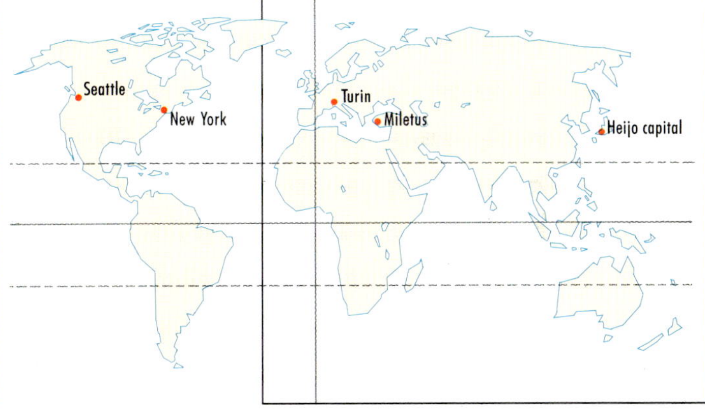

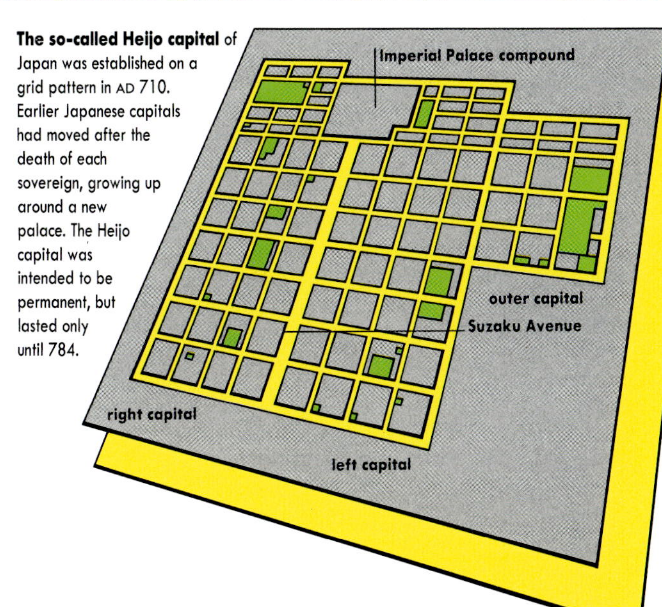

The so-called Heijo capital of Japan was established on a grid pattern in AD 710. Earlier Japanese capitals had moved after the death of each sovereign, growing up around a new palace. The Heijo capital was intended to be permanent, but lasted only until 784.

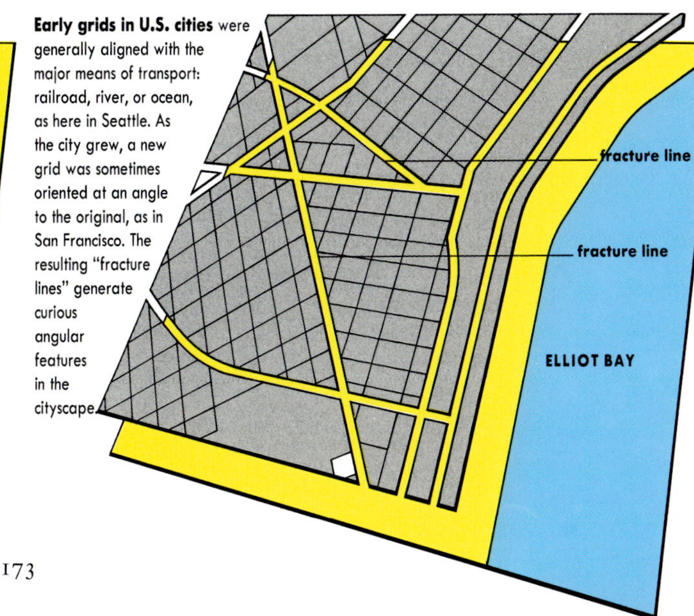

Early grids in U.S. cities were generally aligned with the major means of transport: railroad, river, or ocean, as here in Seattle. As the city grew, a new grid was sometimes oriented at an angle to the original, as in San Francisco. The resulting "fracture lines" generate curious angular features in the cityscape.

THE ISLAMIC CITY
Order within chaos

DESPITE THE CHAOTIC APPEARANCE IT presents at first glance, especially in relation to some of the newer, more spacious and formal parts of the city that surround it, the Islamic city is an entirely rational structure. Its notoriously narrow streets provide vital shade, they keep down winds and dust, and use up little valuable building land.

In fact, there is a clear logic underlying the city's layout, one that is announced in the holy book of Islam, the Koran, and codified by the various schools of Islamic law. Although there are regional differences, most towns and cities that have developed under the influence of Islam at any time in the last 1,300 years show surprisingly similar features. These apply to hundreds of settlements in a broad swath of land from Seville, Granada, and Córdoba in southern Spain in the west, to Lahore in Pakistan in the east. Elements of these ideas can be found in cities as far away as Dar es Salaam in East Africa and Davao in the Philippines.

Although Islamic cities have generally been allowed to grow piecemeal, sometimes over many centuries, every new building or street has been constructed in line with certain basic regulations governing the rights of others and the pursuit of the virtuous life in the densely crowded city environment.

The main guiding principles of Islamic city planning recognize the need to maintain personal privacy, specify responsibilities in maintaining urban systems on which other people rely, such as keeping thoroughfares or wastewater channels clear, and emphasize the inner essence of things rather than their outward appearance. This last principle applies as much to the decoration of houses as to more purely spiritual issues.

The major elements of the Islamic city are easily described. At the city's heart lies the Friday mosque, or *Jami*, typically the largest structure in the city. A number of smaller mosques are often found toward the periphery. It is rare, however, for other mosques to rival the *Jami* in height.

Close to the *Jami* are the main *suqs*, the covered bazaars or street markets that are generally specialized in function. Within the *suqs*, trades are located in relation to the *Jami*. Closest in are those tradespeople who enjoy the highest prestige, such as booksellers and perfumers. Farthest away are those who perform the noxious and noisy trades, such as coppersmiths, blacksmiths, and cobblers. The

Public thoroughfare (*above*): if the road is more than 11 ft 6 in (3.5 m) wide, doors may be placed opposite each other.

Second order street (*above*): doors must not be opposite each other.

Windowsills (*below*) must be at least 5 ft 9 in (1.75 m) above the street.

THE ISLAMIC HOUSE

Privacy is fundamental to the construction of the traditional Islamic city. Above all, women must be protected from the gaze of men. Doors must not face each other across a minor street, and windows are small, slitlike openings above normal eye level. Roofs are surrounded by parapets to prevent neighbors from being overlooked. Waterspouts must not discharge rainwater close to a neighbor's house.

The size of the streets themselves conforms to a relatively simple rule of thumb. Each public thoroughfare must be wide enough to allow two fully laden camels to pass each other, and high enough to allow a laden camel through unhindered. The result is that most through routes have a minimum width and height of about 11 ft 6 in (3.5 m).

The Islamic faith, based on the teachings of the prophet Mohammed (570–632), developed in the Arabian peninsula in the 7th century AD. About one-seventh of the world's population now follow Islam.

Arabs are traditionally a wandering people – even their name means "nomad" – but they also have a long history of city-building. Mohammed seems to have brought some order to principles of city-building that were already established at the time of his birth.

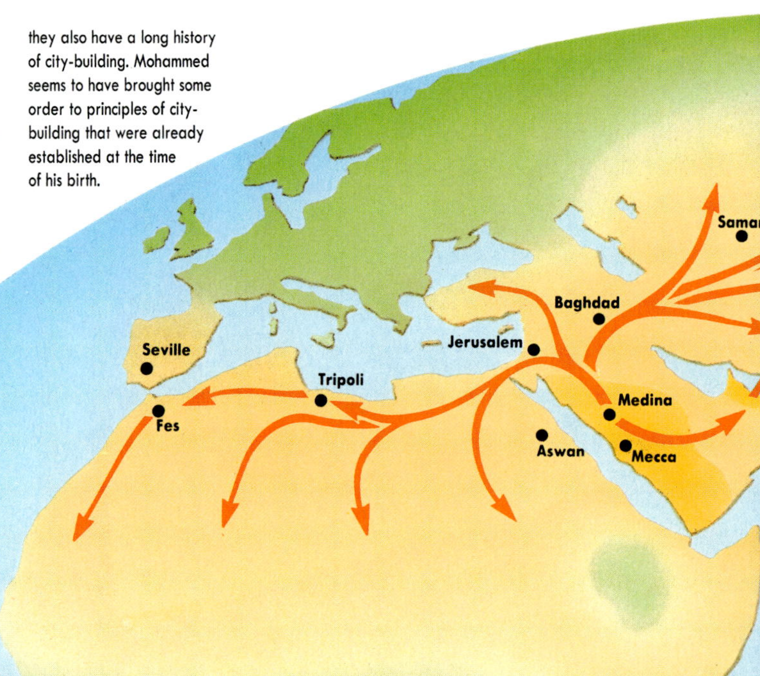

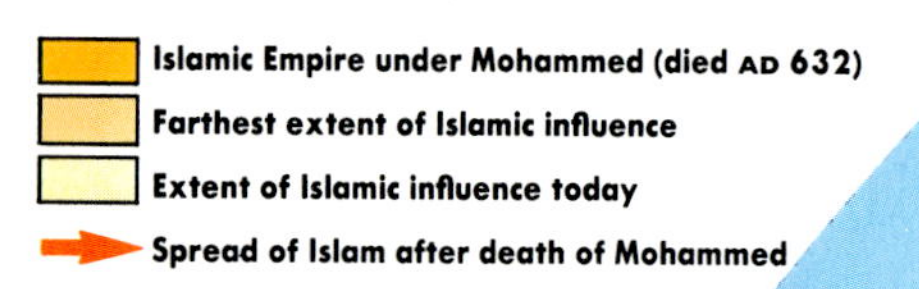

neutral tradespeople, such as dressmakers and jewelers, who occasion no physical offense, act as an intermediary buffer.

Attached to the ramparts, on which are located several towers and gates, is an immense fortified structure, the *Kasbah*. Usually perched on the highest ground, it was a place of refuge to which the sovereign or governor retreated when the main city had fallen to an enemy, or was in the throes of civil war. The *Kasbah* contained not only the palace buildings and the barracks, but also its own small mosques, baths, shops, and even markets.

Everywhere else within the city is filled with cellular courtyard houses of every size and shape, tied together by a tangle of winding lanes, alleys, and cul-de-sacs. Housing is grouped into quarters, or neighborhoods, that are defined according to occupation, religious sect, or ethnic group.

The most important residential unit in the Islamic city, the courtyard house clearly demonstrates the application of the various principles of Islamic city planning. Outside walls lining a street are usually left bare and are rarely pierced by windows. If windows are necessary, they are placed high above street level, making it impossible to peer in. Entrances are L-shaped, and doorways opening onto the street rarely face each other, thus preventing any direct views into the house.

In hot climates, courtyards with trees and water fountains provide shade, but they also provide an interior and private focus for life sheltered from the public gaze. But within the courtyard and the house itself the appearance of plainness often gives way to lavish displays of wealth and decoration. A vividness inside parallels the emphasis in the Koran on the richness of the inner self compared to more modest outward appearance.

At first sight the winding lanes, alleys, and cul-de-sacs of the *medina*, or old city, of Fes (the largest in Morocco) appear haphazardly thrown across the terrain, but a closer look reveals a greater degree of organization.

The widest, or first order, streets of an Islamic city usually radiate outward, linking the core with gates in the city wall. These provide the main axes of movement through the city.

Slightly narrower, second order streets serve the major quarters and define their boundaries, as well as acting as short cuts between the primary routes.

Finally, narrow third order minor alleys and lanes provide access within quarters and are used primarily by those living in the neighborhood.

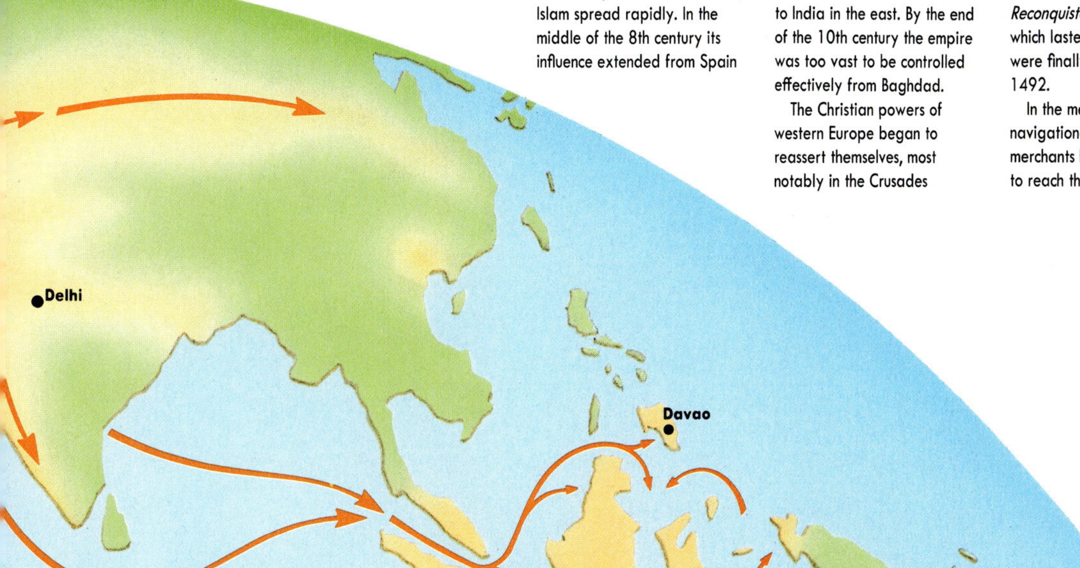

Carried by armies and traders, Islam spread rapidly. In the middle of the 8th century its influence extended from Spain and northern Africa in the west to India in the east. By the end of the 10th century the empire was too vast to be controlled effectively from Baghdad.

The Christian powers of western Europe began to reassert themselves, most notably in the Crusades (1099–1291) and the *Reconquista* in Spain, which lasted until the Moors were finally expelled in 1492.

In the meantime, the navigational skills of Arab merchants had enabled them to reach the East Indies, where they left their mark to such a degree that Indonesia now has the largest Islamic population in the world.

The map shows the extent of Islamic influence at various times, and some of the widespread cities where Arab influence can be seen.

FROM STEEPLE TO SKYSCRAPER

St. Giles, London, 1500–2000

St. Giles-in-the-Fields lies in the heart of London. Where once there were fields, now there are houses, shops, and offices; where those fields were traversed by paths and tracks, now run streets and busy thoroughfares. Once part of one of the worst slums in the capital, synonymous with the grime and squalor of Victorian cities, St. Giles, through continuing cycles of deterioration and regeneration, demonstrates some of the many changes that a city may experience in the course of its history.

In the 18th century, a stretch of open countryside separated St. Giles from the City of London to the east and aristocratic Westminster to the southwest. But the great growth in London's population – it tripled between 1550 and 1650 to reach 375,000 – increased pressures to build outward into the surrounding countryside. By 1720 the area was completely built over with no vestiges of open ground remaining.

The fields to the south of the church, which had remained open until the 1690s, were laid out to form an area that became known as the Seven Dials, a planned development of seven streets radiating from a central point.

This development marked the pinnacle of prosperity for the district. From then on its trajectory was downward. After the Great Fire of 1666, new aristocratic estates were planned farther to the west. Those who could afford it were enticed away, leaving empty and decaying houses that were ripe for less salubrious uses.

As a result, St. Giles developed a reputation for poverty and debauchery. By 1750, one in four houses was registered as a gin shop, in addition to the 82 "twopenny houses," or cheap lodging houses, some of which were also brothels. A transient population of beggars, thieves, and prostitutes flocked to the area; drunkenness, cruelty, gambling, whoring, begging, and crime were rife.

Once begun, the downward spiral was difficult to halt. Although the population continued to rise steadily, almost no new houses were built, and overcrowding reached extraordinary levels. At the same time, the narrow and twisting lanes that had once been field paths and boundaries constituted a serious obstacle to traffic.

There was therefore a strong case for street improvements that would ease congestion and at the same time rid the district of its slum dwellers. The slum area known as the Rookery was completely destroyed in order to make way for New Oxford Street in the 1840s, but the poor merely shifted to neighboring streets. In 1831, the average number of persons per house had been 12.7, but after the clearances some houses with no more than four rooms frequently lodged from 50 to 90 people nightly.

For William Shakespeare, London was not one but three cities, as this map, published in 1572 by his contemporaries Georg Braun and Frans Hogenberg, shows. In the center is the City of London itself, surrounded by its walls and crammed to bursting with houses. To the west the line of the Strand links the City with Westminster, home of Parliament and the royal court. Across the Thames River to the south is Southwark, noted for its bear-baiting, brothels, and theaters. Shakespeare's Globe Theater opened here in 1599.

At the time this map was produced, London was spilling out along the main roads, and housing was beginning to devour the surrounding countryside. The fields of St. Giles, a small hamlet to the northwest, were already being covered with houses. After the Great Fire destroyed most of the City in 1666, the population moved in even larger numbers to new areas outside the walls, and places such as St. Giles were swallowed up in the process.

Interest in town mapping developed in the 1500s, at a time when European cities were growing rapidly and paper and printing were both readily available. This map combines a true survey, or flat plan, with a bird's-eye view of house elevations. Figures in the foreground, depicting typical Londoners in the costume of the time, add further interest.

The area was, however, changing in line with surrounding districts. Toward the end of the 19th century, warehousing, shops, and businesses began to push the remaining population out. Meanwhile, the growth of Oxford Street to the west and Tottenham Court Road to the north as major shopping streets increased pressures to develop the area. In the 1960s the last vestigial streets of the Rookery were bulldozed to make way for a new skyscraper, Center Point.

Change did not stop there. With the removal of Covent Garden fruit and vegetable market in the 1970s, many properties in the surrounding streets began to fall into disuse, and the area entered another downward spiral. Only in the 1980s, with the revitalization of inner-city areas, has there been yet another change. Covent Garden is now a high-quality shopping and tourist center, and the surrounding streets in St. Giles have also undergone a renaissance.

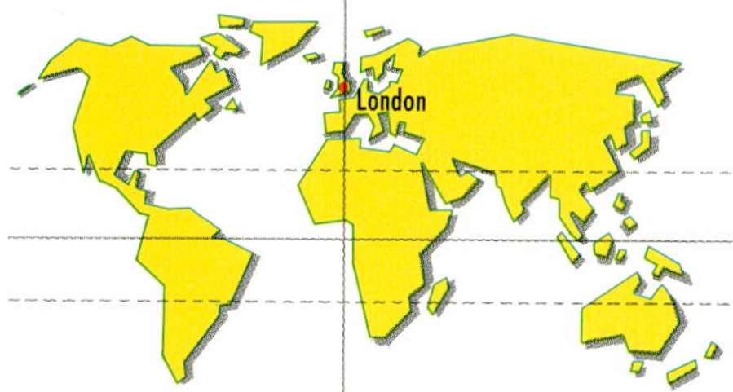

The beautiful 18th-century parish church of St. Giles-in-the-Fields is now dwarfed by Center Point, a 1960s skyscraper. The only indication that the area was once open countryside is the name itself.

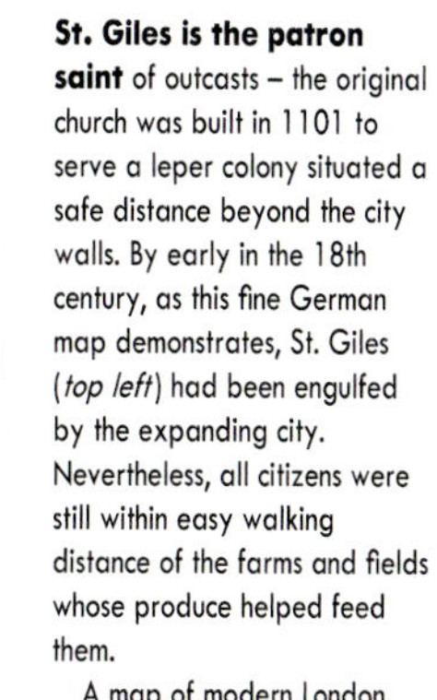

St. Giles is the patron saint of outcasts – the original church was built in 1101 to serve a leper colony situated a safe distance beyond the city walls. By early in the 18th century, as this fine German map demonstrates, St. Giles (*top left*) had been engulfed by the expanding city. Nevertheless, all citizens were still within easy walking distance of the farms and fields whose produce helped feed them.

A map of modern London drawn on the same scale, roughly 2 in to the mile (2 cm to the km), would require a sheet of paper with sides 5 ft (1.5 m) long.

RINGS, FINGERS, AND BELTS

Patterns in city growth

SINCE THE 19TH CENTURY, CHANGES IN transport systems have been the single most important factor in determining the spread of built-up areas.

Before industrialization and the development of mass transport, most cities remained small. Generally limited in size by the time it took to walk across them, few exceeded a radius of 4 miles (7 km). But with the Industrial Revolution, the development of new transport technologies and routes, and population growth, things began to change.

European preindustrial cities usually grew in rings, rather like a tree. As populations increased, walled cities in particular became overcrowded, until pressure on space was so great that they were forced to extend their boundaries. Sometimes new walls were built for protection – in the course of 700 years, Paris had five outer walls, the last of which more or less defines the boundaries of the city of Paris today.

Industrialization and the development of factories in or around city centers meant more advanced systems of transport were needed to move people, resources, and finished goods. Older routes into the city became congested, so roadways were widened, and new means of transport – railroads, canals, tramways, and so on – were developed. The result was an entirely different pattern of urban growth.

Radial routes, above and below ground, spread out from the heart of the city, although their path was often modified by topography or the wishes of important landowners.

As the center became saturated, additional industrial and residential growth generally took place in ribbons or fingers along these radial routes. At first, wedges of open land were left between. Only in this century, with the spread of subway systems, bus routes, and then mass car ownership, have these wedges been filled.

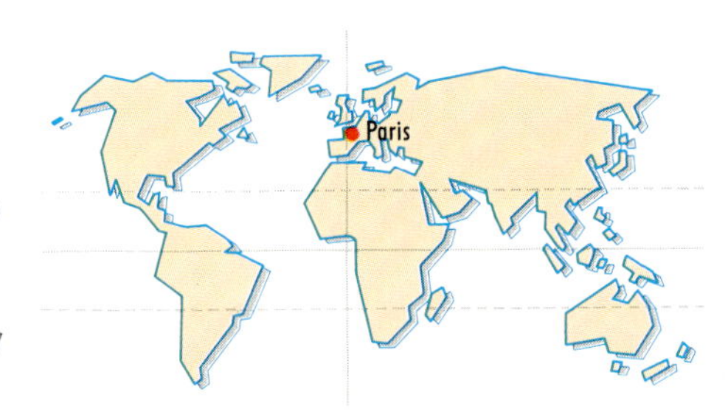

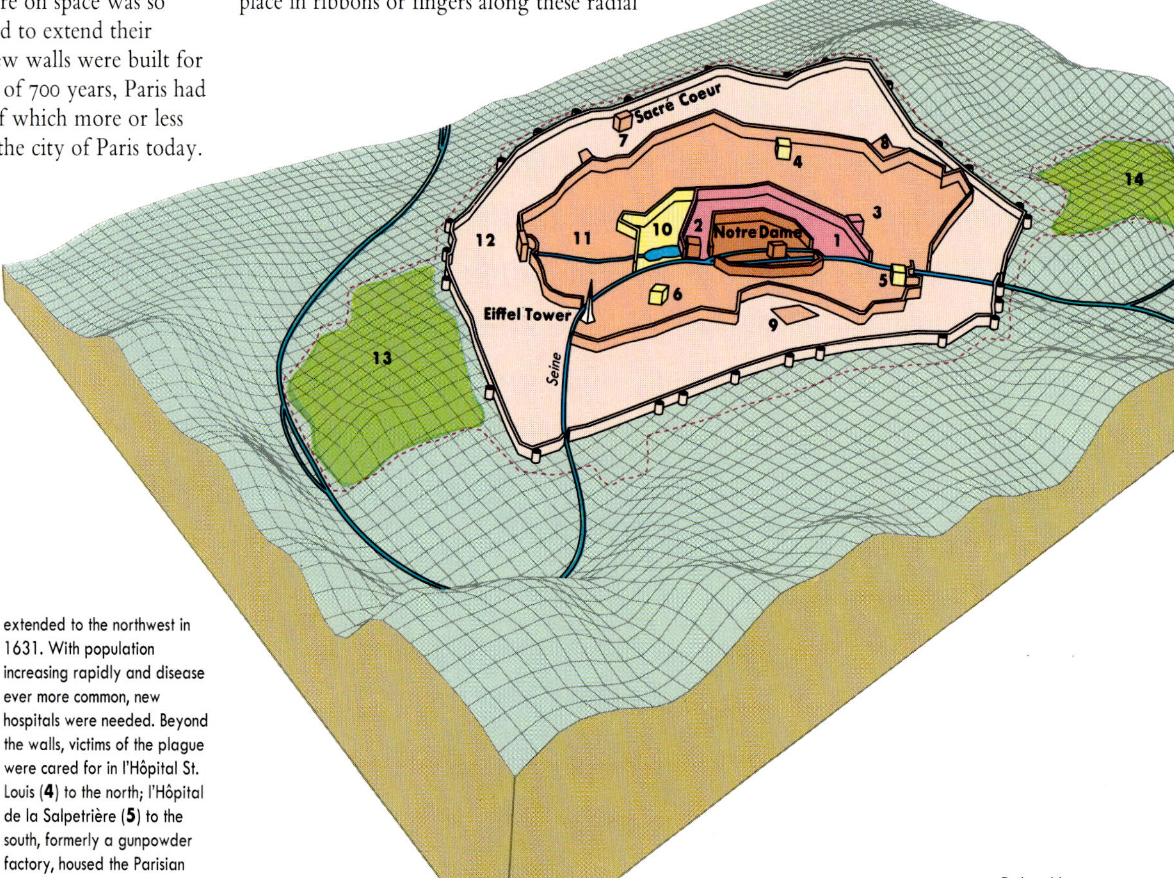

The history of Paris is told by the lines of its city walls. At each phase of its growth, extensive land users – such as hospitals, cemeteries, and gardens – and noxious activities – such as tanning yards and slaughterhouses – were pushed to the edge of the built-up area, outside the existing wall. But as growth continued, new walls were built, and industry and housing ate into the open spaces on what had been the edge of the city. The hospitals, cemeteries, and gardens were surrounded, islands in a sea of buildings.

By 1210, Paris had outgrown its original site on the Île de la Cité (**1**), and walls had been built to defend areas on both sides of the Seine. Just outside the wall, a fortress on the site of what is now the Louvre (**2**) protected the western edge of the city.

As population grew and the city expanded, so a new wall became necessary, and in 1370 the northern defenses were extended. A new fortress, the Bastille (**3**), guarded the eastern approaches. Beyond the walls, skilled craftsmen who refused to abide by the rules of the Paris guilds set up their workshops. The Louvre was built as a royal residence after the fortress, enclosed by the new wall, lost its military function and was demolished.

As late as 1575, the walls still enclosed farmland and open space, but continued growth led to the walls being extended to the northwest in 1631. With population increasing rapidly and disease ever more common, new hospitals were needed. Beyond the walls, victims of the plague were cared for in l'Hôpital St. Louis (**4**) to the north; l'Hôpital de la Salpetrière (**5**) to the south, formerly a gunpowder factory, housed the Parisian poor; and l'Hôpital des Invalides (**6**) on open ground to the west was for army casualties.

By the middle of the 18th century, Paris's population had swelled to nearly 500,000 and new walls were again needed. The Farmers General wall, named after a company of tax farmers or collectors that levied customs duties on goods entering the city, was built in 1791. At about the same time, six million corpses from Parisian cemeteries were moved to catacombs outside the city to create much-needed space within. In the course of the next 30 years, the cemeteries of Montmartre (**7**), Père Lachaise (**8**), and Montparnasse (**9**) were laid out beyond the walls.

To the west, an aristocratic district ran from the Louvre and Tuileries (**10**) palaces along the Champs-Elysées (**11**) to the Arc de Triomphe (**12**). No noxious trades, railroads, or unsightly uses of land were allowed to impinge on this sector.

The final phase of wall-building occurred in 1841, when the enclosed area was doubled. Industry, railroads, and economic development had brought growth to outlying villages such as Montmartre and Passy, and these were enclosed by the new wall. A cannonball's distance away, the wall was reinforced by 16 forts forming a ring that has marked the city's official limits since 1859.

By then, Baron Haussmann, the architect of modern Paris, had started to remodel the center. He converted the line of the old walls into wide boulevards and realigned the roads to link up with new railroad terminals – the principal points of entry to the city. He also revolutionized the flow of water and sewage by installing new systems of pipes.

Under Haussmann's direction, thousands of buildings were demolished and reconstructed. Parks were created both within the walls and beyond them, the most notable being the Bois de Boulogne (**13**) and the Bois de Vincennes (**14**). By 1870 the transformation was complete, and Paris could justifiably claim to be the architectural and cultural capital of Europe.

Enclosed by wall built 1180

Enclosed by wall built c. 1370

Enclosed by wall built 1631

Enclosed by wall built 1784–91

Enclosed by wall built 1841–46

Boundary of present city

Former gate in outer wall, now a Metro terminus

In 1575, when Braun and Hogenberg published this map, Paris was one of the world's greatest cities. Its population – 180,000 – exceeded that of London and Amsterdam, its two main rivals.

The city comprised three separate elements. The original site, the Île de la Cité, inhabited in pre-Roman times by the Parisii, was linked to developments on either side of the Seine River. The flatter right bank served as the port, and was the center of commercial and administrative functions; on the left bank, in the Latin Quarter, was the university. High walls encircled the whole. As problems of overcrowding worsened, Parisians spilled over into new suburbs that were already growing outside the city walls.

1870–1914 Construction of the Eiffel Tower in 1889 symbolized French dynamism and creativity during this period. Paris grew rapidly as both industry and population spilled out from the core. In 1870 ten railroads terminated in the city, but by 1914 the number of stations had doubled and the frequency of trains increased tenfold. Buses replaced streetcars in the early 1900s. Most important, the métro system was built between 1875 and 1914. The outcome was that Paris's population, which had grown to over three million, dispersed along transport corridors, and once separate outlying communities were linked to the city center.

FIXATION LINES AND FRINGE BELTS

Urban growth is by no means a simple or smooth process of addition. Development occurs in a jerky, stop-go fashion.

Construction often stops at what is called a "fixation line." This is usually a natural or artificial barrier, such as a river or a lake, a wall or a railroad track, but it may be a more intangible legal obstacle, such as a protected park or the estate of a major landowner. If the demand for new building is insufficient to force construction to jump over these barriers and resume beyond them, development becomes fixed along these lines.

Extensive users of land – hospitals, cemeteries, market gardens, recreational grounds – are often found on the other side of fixation lines. Although essential to the well-being of city dwellers, this type of land use cannot occur in an already intensively built-up area and so it is pushed to the edge. But as pressure for housing mounts, building leaps over this "fringe belt" and resumes immediately beyond. If extensive users of land cannot resist the pressure to redevelop their sites, then more intensive uses such as houses, offices, or factories may take over what was previously a recreational ground.

1950–present Since 1950 Paris has spread out even farther. Over 350 miles (550 km) of new highways, a regional express underground railroad, and the extension of the existing métro have allowed the Paris conurbation to extend 15 miles (25 km) from the center, over an unbroken built-up area that covers more than 450 square miles (1,200 square km).

1918–1950 In 1919 the last of the city walls were removed and modern Paris began to take shape. By 1930 the population was over four million. Central districts were completely built over and huge suburban housing developments grew up along arterial roads and railroads. Communities that had once been separated by open land were linked to the center by continuous ribbons of built-up areas strung along radial transport routes.

City limits
Main transport routes
Built-up areas

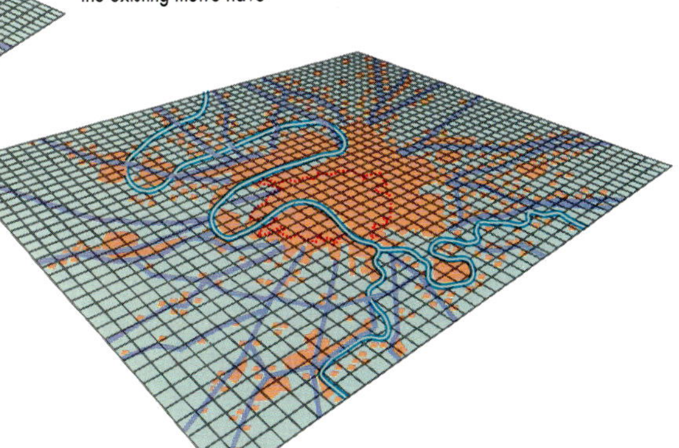

LIVING ON THE FRINGE

The world's suburbs

Few contrasts can be sharper than those between the suburbs on the edges of the world's cities. In the developed capitalist world, the norm is a low-rise, detached, or semidetached house, each on its own neat plot. In socialist cities, high rises generally dominate the skyline. In third world cities shantytowns proliferate. What accounts for these differences?

The cost of housing in the central areas of Western cities is prohibitive for all but the very wealthy. As a result, at night many city centers are practically empty. The majority of the first world's urbanites are in fact suburbanites.

Things are very different on the fringes of socialist cities, where massive and regimented housing blocks are the norm. Often poorly designed, shoddily constructed, and meager in terms of living space and facilities, they are the very antithesis of the affluent Western suburb.

Following the Russian Revolution of 1917, people moved into city centers from the country, settling in houses and apartments confiscated from the wealthy. But the supply soon dried up and overcrowding increased. It became common for several unrelated families to share the same room.

After 1932, the Soviet government imposed a system of internal passports and residence and work permits to stem this one-way tide, and to restrict the size of major cities. The policy had only limited success: Moscow, the most strictly controlled Soviet city, has a population of nine million, and it is growing by 80,000 new residents a year on top of its natural increase.

Under Stalin, grandiose apartment buildings were constructed around squares and along main highways out of major cities such as Leningrad and Moscow. Many featured

A suburb of west London: an archetypal "good neighborhood." Thanks first to public transport systems, and later to the private automobile, millions of people now live far from the city center but commute in to work daily. They can purchase family homes more cheaply, and in more spacious, generally greener surroundings. As a result, suburbs have expanded massively, covering immense tracts of land that even in living memory were given over to farming.

Strong criticism has been leveled at suburban life, mostly by sociologists who claim that the uniformity of suburbs leads to boredom, passivity, and lack of community spirit. Adolescents complain of "nowhere to go," and women tied to the home by domestic duties may find the suburban environment stifling.

On the other hand, as suburbs mature, community life often flourishes. In addition, industry and commerce moving out from expensive or crowded city centers can give greater economic self-sufficiency to an area that was otherwise primarily residential.

Soviet-built apartment blocks in Ulan Bator, Mongolia, overshadow the felt tents which still house many of the people.

Drab high-rises like this are typical of the fringes of many socialist cities. Socialist suburbs are, in principle, built in an integrated way, so that home, place of work, and amenities are close to one another. Few people own automobiles, and what travel is necessary is usually by public transport.

In reality, many planned suburbs remain half-built, and lack adequate infrastructure and amenities.

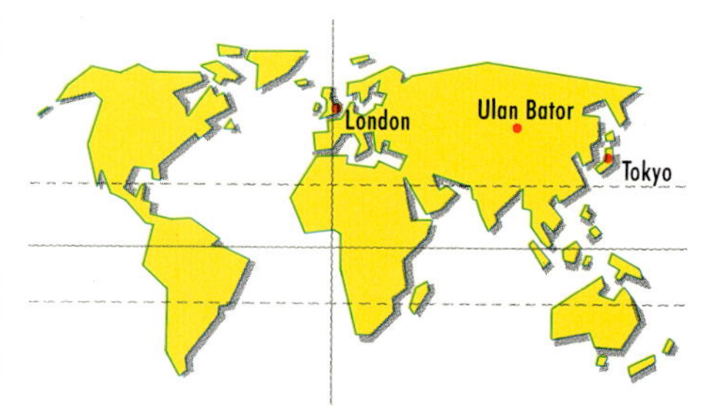

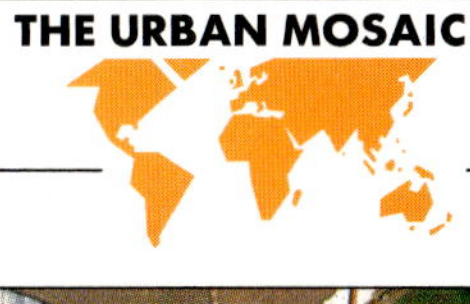

monumental sculpture and fine architectural detailing. Yet inhabitants gained access to their homes not through the showy front entrances but via dark, litter-strewn rear stairways. Entire families were allocated just one room each. The destruction of many cities during the Second World War added to the housing shortage.

By the 1960s, multistory blocks were being constructed from prefabricated concrete units on the edges of towns and cities throughout the Soviet Union. There are no regional differences in style, and facilities and features are highly standardized. Although rents are low – only around 3 percent of wages – each unit has an average floor space of a mere 200 square feet (20 square m). Nevertheless, there are long waiting lists for new apartments – families may wait 20 years or more for a place of their own. Two million such units are now being completed every year in the hope of providing a separate apartment or house for each family by the year 2000.

Many migrants to third world cities begin city life renting expensive, poor-quality accommodation near the center. But most soon escape from these slums by illegally squatting on land on the city's outskirts. Often this land is highly unsuitable; there are few facilities, and only the flimsiest of materials – often the detritus of city life, such as cardboard, tin sheeting, or plastic – to build with. Yet people have little choice. Migration from the countryside, coupled with high rates of natural growth in the city, has invariably outstripped the capacity of governments and private builders to construct low-cost housing.

Are these shantytowns slums of despair, or do they offer hope of a way out of urban homelessness? Are they affluent suburbs in the making, as some have rather surprisingly claimed? The answer varies from place to place, but many shantytowns are indeed a prelude to something better.

Over time, more permanent and weatherproof materials such as concrete blocks and zinc sheeting come to replace cardboard and plastic. Collective pressure on municipal authorities can bring supplies of clean water and electricity. Legal title to the land may be agreed, transforming one-time squatters into owner-occupiers.

After a few years, many households are equipped with TVs, refrigerators, washing machines, and even automobiles. Streets have names, and doors are numbered, so mail can be delivered. Inhabitants open shops and provide services for fellow shanty dwellers. In time, this part of the shanty is incorporated into the city proper. As the city continues to grow, the worst scenes of deprivation and desperation are once again transferred to its outer limits.

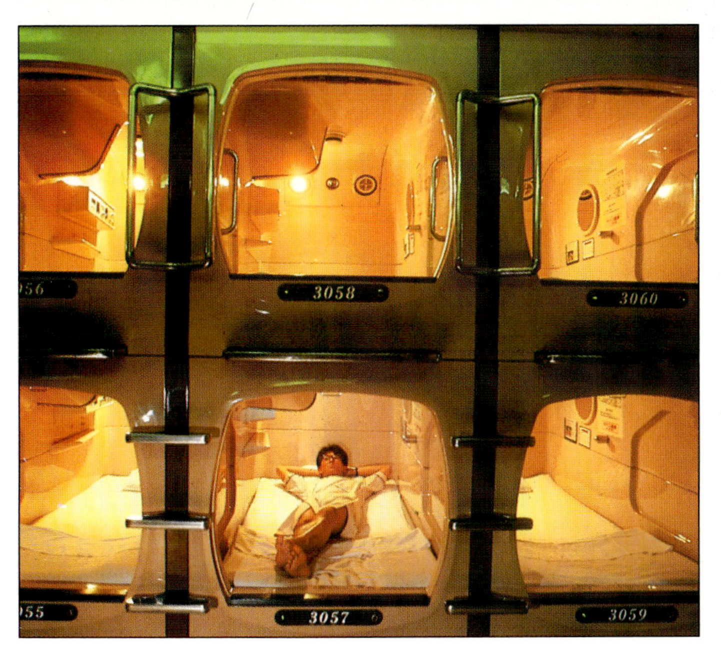

"Capsule hotels" have grown up as a partial solution to the Tokyo commuter's problems, allowing an employee working long hours to buy cheap overnight accommodation.

Property prices in Tokyo make it impossible for any but the most affluent to live in the central districts. The result is a commuter zone that stretches along surface and underground railroad tracks 20–30 miles (30–50 km) from the center. Here, the people of Tokyo can purchase housing that they could not otherwise afford. But they pay for it in terms of time – the majority of Tokyo's commuters travel for at least two hours every day.

THE 4-D CITY
Conquering time and space

SUCH IS THE INTENSITY AND DYNAMISM of urban life that the modern city thrusts out in every direction: not only outward, but upward, downward, and even into the night.

For over 600 years, skylines in European cities were largely dominated by city walls, castles, and the soaring spires of great cathedrals. Except for those who were responsible for building and maintaining these magnificent structures, few people ever ventured to a greater height than four or five stories.

All this changed in the 19th century as European and North American cities became the focal point for the concentration of wealth, power, and people. As land prices soared, the sky became an open frontier free for anybody to colonize. The result is the skyscraper.

Early tall buildings, such as the Great Pyramid of Cheops at Giza in Egypt or Gothic cathedrals like Notre Dame in Paris, had to rely on a massive base, thick walls, or flying buttresses to keep their immense weight from splaying out. Weight at the top was kept to a minimum, which is why they generally tapered to a hollow dome or spire.

If buildings were to create usable space above five or six stories, this major problem had to be solved. Nothing was gained if height could only be achieved at the expense of an enormous base. Chicago's Monadnock Building of 1891 reached 16 stories, probably the maximum possible for a building with load-bearing walls. But the walls at the base were nearly 6 feet (2 m) thick, and windows were tiny. The superstructure had to be made strong but very much lighter. This was solved after the middle of the 19th century by using an iron or steel frame to take the strain.

Instead of bulky masonry or brick walls having to support the full weight of the building, reinforced concrete and glass were increasingly used. From now on walls were only used as cladding – for weather-proofing, or esthetic purposes. As the historian of the city Lewis Mumford wrote in 1938, "The age of crustacean building has given way to the age of vertebrates, and the wall, no longer a protective shell, has become a skin."

The first skyscrapers were by today's standards modest, but the potential for growth was already there. Each increment in height threw up new demands. In 1852 Elisha Otis had already solved one problem when he installed the first passenger elevator – without this device, five floors is the maximum that most people can manage.

In time, novel constructional techniques were devised using prefabricated units, and self-climbing cranes with which to assemble them. Vital advances have also been made in internal communications, heating and ventilation, and coping with such hazards as hurricane-force winds or the prospect of being hit by a plane. The Trans-America Center in San Francisco (1972), built rather like a tree with a strong flexible central spine, successfully withstood the force of the 1989 earthquake.

Tall buildings brought their own problems, among them the theft of light at sidewalk level. In 1920, New York passed an ordinance that skyscrapers must be "stepped back." Among the first buildings designed to meet this requirement was the Chrysler Building (1930), the first to top 1,000 feet (300 m). Such was the passion to earn the accolade of the world's tallest building that its 123-foot (37.5-m) spire was constructed in great secrecy within the upper floors, and pushed up at the last minute.

Since that time, new buildings have gone much higher, and have provided greater floor space. The twin-towered World Trade Center (1973) has floors the size of a football field. The Center is basically a rectangular tube, with all the stiffness in closely packed vertical rods.

Today, the greatest height is achieved using the principle of the bundle of sticks. At ground level, Chicago's 1,454-foot (443-m) Sears Tower consists of nine distinct structures. These taper off at different heights until only two remain at the top. Buildings 500 floors high are now readily conceivable.

However, the main obstacle to increasing height is not technological but economic. So much capital is tied up during construction that interest charges cost more than the building itself. One way of reducing building time is to use prefabricated elements, a practice recently developed in much-publicized high-tech buildings like the Hong Kong and Shanghai Bank in Hong Kong and the Pompidou Center in Paris. Their "exoskeleton" construction makes the interior space more adaptable, and the structure is exposed, in much the same way as the flying buttress was centuries ago.

San Francisco
Chicago
New York
Buffalo
London
Paris
Giza
Tokyo
Hong Kong

The 24-hour dynamism of the city's center brings with it sky-high land prices. These push cities not only upward but also deep underground. Ever since the late 19th century, large cities such as London, Paris, and New York, and more recently Singapore and Frankfurt, have built subway systems to bring people into the city to work and shop. Montreal also has underground shopping malls.

In Tokyo, where urban land is ruinously expensive, engineers have conceived a complete subterranean city of homes, offices, stores, and hotels. In theory, its inhabitants will rarely need to come above ground. They will look out of their "windows" at high-resolution television pictures taken on the surface. The Taisei Corporation's "Alice Cities"– named for the Alice who went underground through a rabbit hole – may well become reality in the early 21st century.

Here in Tokyo, as in almost every other city in the world, the conquest of the fourth dimension – time itself – is complete. During the day, the city center is filled with workers and shoppers. Come early evening, many of these return home to the suburbs, only to be replaced by new cohorts keen to sample the entertainment that is on offer.

Concert halls, movie houses, theaters, restaurants, clubs, and bars swing into action. When these people have had their fill of the fun, they too make for their beds. But the city itself never actually sleeps.

Growing armies of people maintain vital services, or run machinery so valuable it must be kept in perpetual motion. Truckers and trash collectors, journalists and printers, mail sorters and fire fighters, police and health workers toil through the night, not to mention the legions of maintenance staff whose running repairs are needed simply to restore the battered fabric of the city ready for the next morning.

The relative heights and shapes of a number of famous buildings are contrasted in the illustration (*right*). The insignificance of Notre Dame Cathedral against the Sears Tower is remarkable, considering that the cathedral, by human scale, is an extremely imposing building.

Sears Tower, Chicago (1974). 1,454 ft (443 m). The world's tallest building, with 110 stories. Composed of 9 sections built to varying heights, thereby giving greater lateral resistance to wind pressure.

World Trade Center, New York (1973). 1,350 ft (412 m). Densely packed columns of steel in the walls take much of the weight, but leave space only for narrow windows.

Chrysler Building, New York (1930). 1,046 ft (319 m). Tapering shape designed to permit more light through to street level.

Trans-America Center, San Francisco (1972). 845 ft (257 m). Designed to withstand earthquakes, with a strong central support.

Pyramid of Cheops, at Giza, Egypt (*c.* 2650 BC). 480 ft (146 m). The construction of a massive base is one of the simplest ways to support a high building.

Hong Kong and Shanghai Bank, Hong Kong (1985). 520 ft (158 m). The load-bearing structure forms a clearly visible "exoskeleton" on the outside of the building.

Notre Dame, Paris (1344). 225 ft (68 m). Flying buttresses provide the building with external support.

Monadnock, Chicago (1891). 215 ft (65 m). The tallest masonry skyscraper, relying on extremely thick walls at its base.

St. Paul's, London (1710). 365 ft (111 m). Extra height is achieved by surmounting the cathedral with a dome – a structural form that is inherently strong.

Guaranty Building, Buffalo (1895). *c.* 200 ft (60 m). Lightweight but immensely strong, steel frames take the full weight of the structure and allow greater freedom in design.

Skyscrapers have never been without their critics. San Francisco, once called the Baghdad of the Americas for its beautiful domes and minarets, is leading the way by declaring war on the skyscraper. Planners there, as elsewhere, have argued that slablike blocks are ugly, reduce light, create turbulence at ground level, exacerbate street congestion, and simply overwhelm the individual by their sheer size.

For many people, including most volubly the heir to the British throne, Prince Charles, any skyscraper, no matter what its design features, represents an intolerable symbolic victory of Mammon over God. He points, for example, to the way that St. Paul's Cathedral, for 300 years London's tallest structure, is now overlooked and obscured by many corporate headquarters. He hopes that by resisting additional office development in the vicinity, the dominance of its magnificent dome on London's skyline will once more be asserted.

THE SEGREGATED CITY

Behind invisible lines

THE SUPPOSED WEALTH OF CITIES HAS always attracted migrants from near and far, and they have been seen as the cultural melting pots of civilizations. But the sheer size of a city may pose a threat of anonymity which encourages groups to maintain, even to emphasize, their cultural distinctiveness. Far from being a melting pot, the city becomes a social mosaic of separate cultural groups.

In the past, and in some countries still today, the location of particular racial, ethnic, or religious groups has been controlled by law. Jews were often confined in separate walled quarters, or "ghettos," in medieval and renaissance cities across Europe. Similarly, up to around 150 years ago, European merchants in China and Japan were typically kept under virtual house arrest for the duration of their visit. In South Africa, segregation by color is likewise enshrined in law. Countryside and city are divided into exclusive "group areas" defined according to spurious notions of "race."

In many cases, however, residential segregation is freely chosen. People with similar amounts of wealth, or at similar stages in the family life-cycle, tend to live in a belt roughly the same distance from the city center. People generally elect to live where they can afford to, but also where they can be among "their own kind." Such areas are perhaps more usefully described as "enclaves" than ghettos.

The most common form of "voluntary" segregation is still by race and ethnicity. Nowhere is this more true than in the U.S., the recipient over the last few hundred years of people from almost every country in the world. Yet the same processes that have been discerned there apply globally, whether to enclaves of Greeks or Italians in Melbourne, Australia, or Turks and Yugoslavs in Hamburg, Germany.

Immigrants generally arrive with few relevant skills and a weak knowledge of the language

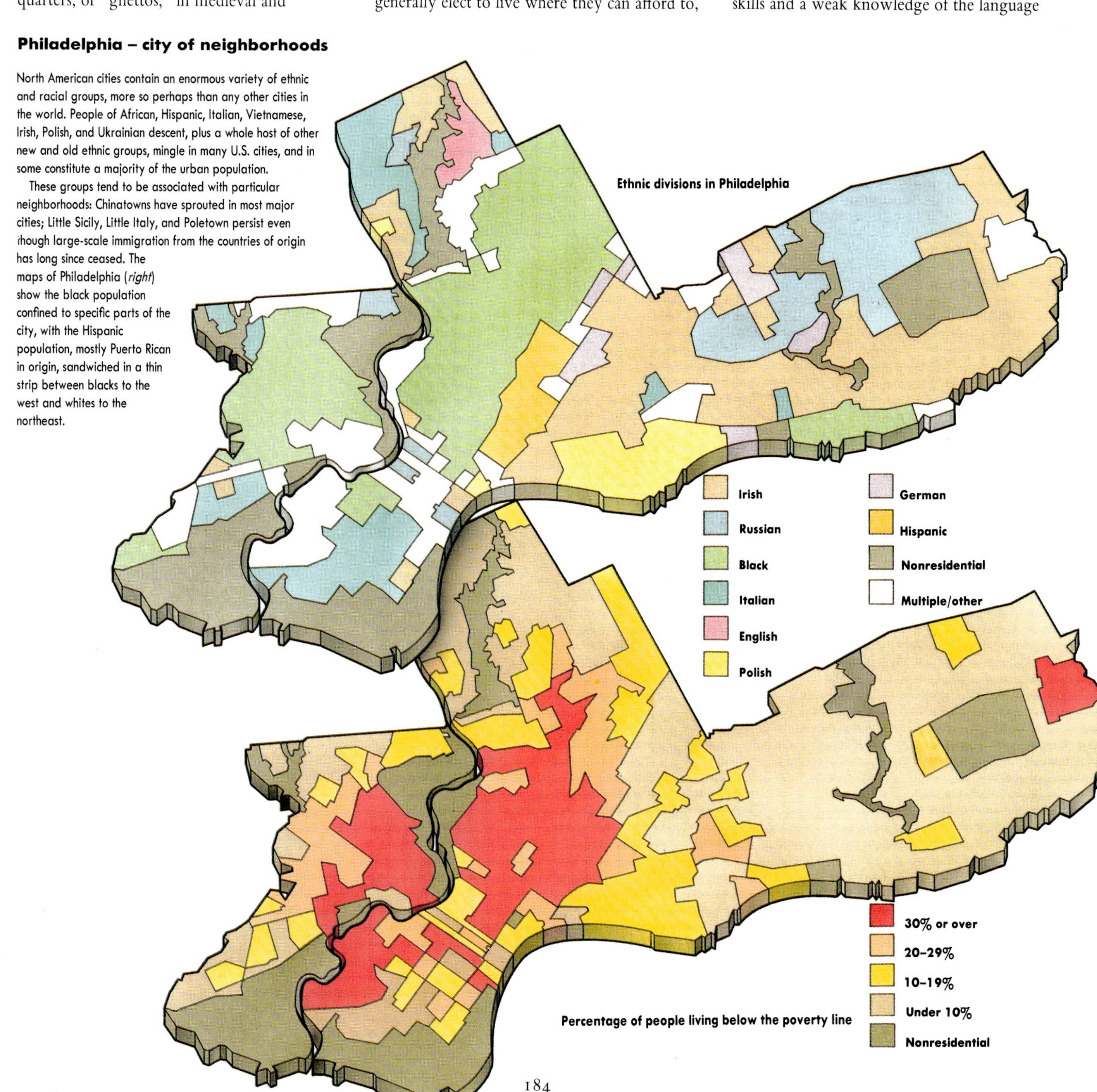

Philadelphia – city of neighborhoods

North American cities contain an enormous variety of ethnic and racial groups, more so perhaps than any other cities in the world. People of African, Hispanic, Italian, Vietnamese, Irish, Polish, and Ukrainian descent, plus a whole host of other new and old ethnic groups, mingle in many U.S. cities, and in some constitute a majority of the urban population.

These groups tend to be associated with particular neighborhoods: Chinatowns have sprouted in most major cities; Little Sicily, Little Italy, and Poletown persist even though large-scale immigration from the countries of origin has long since ceased. The maps of Philadelphia (*right*) show the black population confined to specific parts of the city, with the Hispanic population, mostly Puerto Rican in origin, sandwiched in a thin strip between blacks to the west and whites to the northeast.

Divisions between Roman Catholics and Protestants are deep-seated throughout Northern Ireland, and especially in Belfast. Loyalties may be expressed in a variety of everyday ways, such as choice of newspaper or support of a particular football team, or they may take a more overt form.

On July 12, the anniversary of the Battle of the Boyne in 1690, Protestants adorn their neighborhoods with decorations celebrating their great victory over the Irish. Catholics direct most of their graffiti against the Protestant majority or the British government. The two communities are quite close to each other, but visitors to the city are left in no doubt as to which area they are in.

and customs of their country of adoption. Lacking capital and access to the best-paid work, they move into the poorer areas of the city, where rents are low and where they may be able to take advantage of contacts among fellow immigrants who preceded them from the same country. At first they have to take the worst paying, least pleasant, most insecure jobs.

In time, however, things may improve. Their children, schooled in the host country, acquire the language and other skills that their parents lacked. In the absence of strong legal or informal discrimination against the group, the descendants of immigrants will generally find themselves on a kind of conveyor belt that takes them out of the virtual ghetto they first inhabited and ever farther into the affluent suburbs. As a result, their houses are vacated for the next wave of immigrants.

In Spitalfields, in London's East End, the same neighborhood has been home to generations of newcomers from very different countries of origin. French Huguenots who came to England in the late 17th century to escape oppression at the hands of the Catholics settled there. By the 1890s, the neighborhood was predominantly Jewish, largely populated by immigrants from Russia and eastern Europe. A century later, the area is chiefly home to people from Bangladesh. Curiously, every generation of immigrants has been active in the garment trades. A similar story of migration and succession could be told of New York or Chicago.

Contrary to many people's hopes or fears, this intermingling has rarely produced a fully assimilated, culturally uniform society. Groups typically keep themselves physically and culturally distinct, establishing their own shops, places of worship, social clubs, restaurants, schools, movie houses, and commemorative parades, and celebrating their own holidays. Home "territory" is clearly marked out for insider and visitor alike through shop signs, graffiti, styles of clothing, food snacks on sale, and language spoken.

The community provides services and employment for itself with little recourse to the host society; marriages are generally contracted within the community; and successive generations of children learn to speak their parents' language (although they may never have visited their "home country"). Ethnic clustering may also serve to give a group considerable political power, resulting in a strong electoral base for ethnic representatives. In exceptional circumstances, sectarian military strongholds may develop, as in Lebanon and Northern Ireland.

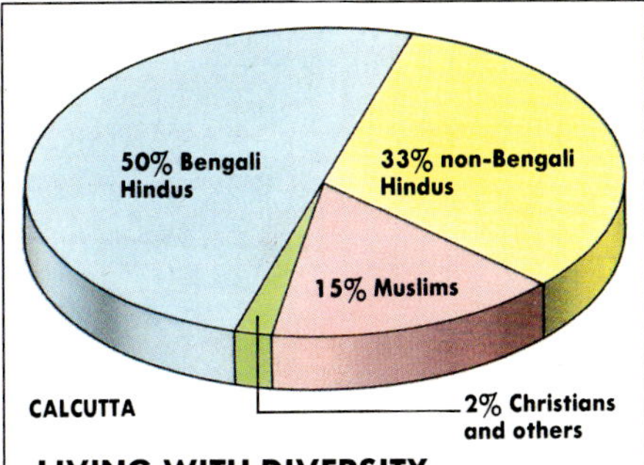

LIVING WITH DIVERSITY

India's largest city has a vast and cosmopolitan population. Dominated by Bengalis, natives of the surrounding region, it includes immigrants from all over India – Sikhs from the Punjab, mainly skilled workers; a prosperous business class from the west and south; and more local immigrants who tend to find work as bearers or laborers. There are also still significant numbers who are of European or Anglo-Indian descent.

Each of these groups tends to be close knit. The huge Hindu population is subdivided by caste, or social status, with each class living in a clearly defined area; the Muslim and Christian sectors are still centered on the same parts of the city as they were in colonial days. Divisions between religious and ethnic groups are not as clear as they once were, but they are unlikely to disappear altogether.

BORN-AGAIN CITIES

Renewal and gentrification

BROWNSTONING IN NEW YORK, whitepainting in Toronto, gentrification in London, chelseafication in Cape Town; elsewhere the same process is being called upgrading, revitalization, trendification. But whatever the name, the regeneration of run-down residential areas is occurring wherever a city succeeds in capturing a share of new "postindustrial" sources of wealth and employment. Areas such as Greenwich Village in New York, Georgetown in Washington, Le Marais in Paris, and Islington in London are all being "recycled" from a faded past to a glittering present.

For most of this century, inner cities in the developed world have been the focus of economic and social decline. As manufacturing industry vacated its traditional locations around city centers, skilled workers and professionals, particularly young people with families, were forced to move to commuter suburbs or follow employment opportunities to new towns and cities. Those left behind – the poor, the unemployed, the old, and recent immigrants – faced a downward spiral of worsening job prospects and a decaying urban fabric.

With the prospect of a bleak future for their property, private landlords often failed to repair buildings. Local authorities neglected public housing and social facilities. Lending institutions were reluctant to provide money for mortgages. Areas they believed to be poor risks were marked out by a red line on their maps – a practice which became notorious as "redlining."

Many inhabitants believed, not always unreasonably, that their neighborhood was being run down deliberately, both to deflate prices and to hasten the start of redevelopment programs. They also resented the speculators who bought properties at rock bottom prices in the expectation of making windfall profits later.

Since 1980, the story has taken a new turn. The tremendous upsurge in service and commercial activity, especially the great rise in centrally located corporate headquarters, has led to an explosion in demand for young professional staff – sometimes disparagingly called "yuppies" (young, urban, professional people). These high-income single people and dual-career childless couples seek centrally located housing: hours spent on congested roads or crowded public transport traveling in from the more family-oriented suburbs are simply not for them. In parallel, old factories, warehouses, and other semiderelict buildings in inner city areas have been recycled into stylish offices, media production facilities, restaurants, and sports clubs.

Gentrification has not always met with the approval of existing inhabitants. Although the influx of wealthier groups can lead to badly needed improvements in the environment, neighborhood rehabilitation is usually accompanied by steep rises in the cost of housing. Poorer members of the community are squeezed out, and those that remain complain of profound changes in the "feel" of the community.

Social tensions between the old and new groups can run high, and in some instances have boiled over into various forms of protest. This ranges from "anti-yuppie" graffiti to vandalism of newcomers' automobiles and property.

The fortunes of a house

The changing fortunes of a neighborhood are mirrored by its houses. A Victorian house in London has passed through several stages of decline and partial rebirth before returning to its former heights.

Built around 1860 for an affluent upper-middle-class family, the house has servants' bedrooms in the attic and a kitchen in the basement with a separate "tradesmen's entrance."

1910 The house's status has changed very little. The property has been handed down to the eldest son. Electricity has replaced gas as a means of street lighting, and the street itself has been paved. The family owns no vehicle of its own – a carriage collects the master to take him to work in the financial district.

1930 The house is now owned by a grandchild of the original occupants. She lives in the suburbs, and has sold the lease. A landlady collects rent from the "respectable families" or "single gentlemen" to whom the rooms are let. The landlady occupies two rooms on the ground floor; the basement is unlived in and unloved. Basic repairs are carried out, but, with only 30 years to go before the lease runs out, it is not worth making expensive improvements.

1950 The former owner of the house has died, and none of her children is interested in this run-down property. They put it in the hands of a rental agency. Sinks and electricity meters are installed in each room; there is a pay phone in the hallway. Interior decoration is basic, and repairs are largely ignored. Tenants range from a medical student on the top floor to a middle-aged Polish couple in the basement.

Mid-1960s The building is falling apart – the rental agency sees no likelihood of a return on investment in repairs (the area will never command higher rents, they feel). Tenants are either transients or recent

1860 **1910** **1930** **1950**

Baltimore, like many other U.S. cities, has undergone an impressive resurgence. In the late 1970s, central areas were run down and decayed. The Inner Harbor, once the hub of Baltimore's commercial and industrial prosperity, was derelict.

Today, high-rise offices, luxury hotels, stores, and apartments jostle for space in the city center. Once blighted neighborhoods have been gentrified by young, wealthy, professional people. Most dramatic of all, the Inner Harbor – the haunt of tramps and drug addicts in the 1960s – has been transformed by a most successful waterfront redevelopment program. Thanks to a whole array of tourist and leisure facilities, including a convention center, arena, festival marketplace, and national aquarium, millions now flock to share in Baltimore's renaissance.

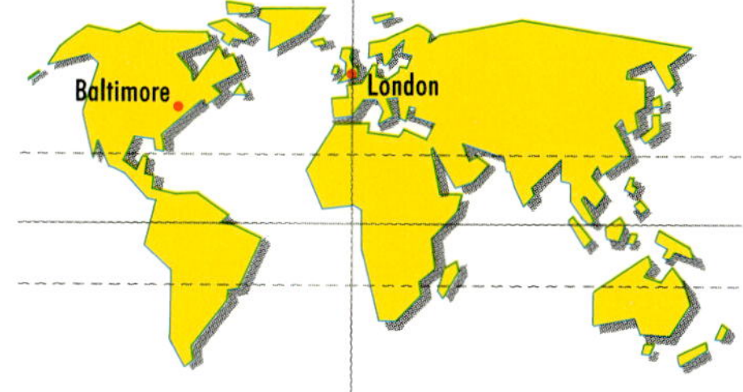

immigrants. Several streets have been demolished to make way for high-rise blocks.

Early 1970s Dereliction is complete. A small fire has made the house uninhabitable, and the owners cannot sell it privately because the area has been "redlined" – banks and mortgage companies see it as a poor risk and refuse loans. The house is sold to the local government, which hopes to redevelop the whole area, but money runs out. Meanwhile, the house is boarded up in an attempt to keep out squatters.

Early 1980s The council sells its empty property to a development company specializing in "period restoration." Whole streets of run-down houses will be converted into "luxury apartments," altering the status of the area. The inside of the house is gutted, the roof repaired. The continuing lack of public money is reflected in the decline of the high-rise apartments behind.

1990 Rather than being turned into apartments, the house is restored as a family home. A dual-income family with two children moves in. The attic – the former servants' quarters – is converted into a self-contained "nanny suite." Metal guards on the lower windows are to deter burglars, while chandeliers, fireplaces, and other "original features" are reinstated to add value. The derelict high-rise has been blown up, and low-rise housing will be built on its site.

And in the future . . . the family may move to the country – they already have a weekend cottage. If they sell both properties, they can buy a Georgian country house. Thanks to computer links and fax machines, they can work from home. They may subdivide their former home into four or five individual apartments to sell; if so, they may keep one.

1960 1970 1980 1990

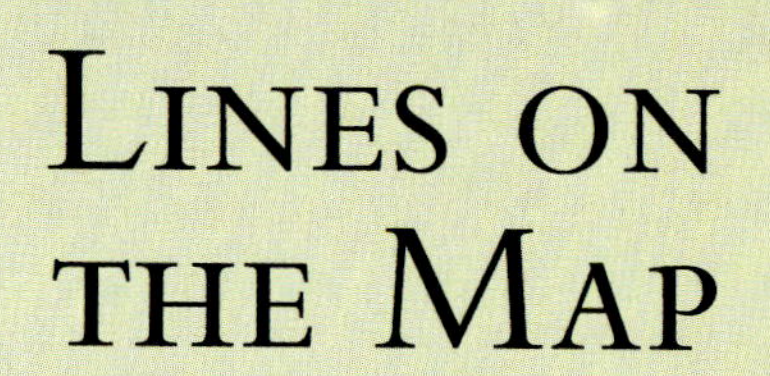

Lines on the Map

The Nature of Boundaries

National borders, which in theory can lie literally anywhere, are defined in fact by the interaction of man and nature. Natural features such as mountains and rivers provide the most logical boundaries. But cartographers' rulers also establish legal borders, and colonialism carves one landmass into separate nation-states. Language and religion regularly define borders, as do local customs and trade. A single tiny island can encompass several nations, and groups of islands may deem themselves one nation. Uninhabited land needs no borders, but wherever people build their homes and put down roots they will eventually create borders as well.

Conflict and Cooperation

The combination of geography and history places the world's nations in a constantly shifting arrangement of allegiances. Smaller countries are inevitably dominated by larger ones, if not absorbed by them completely. Religious differences and territorial claims heighten the tension between nations, and tribal disputes spark internal conflicts. Economic and military might help define superpowers, but topography exerts control over even the most powerful of them. Nations exist without homelands and refugees without homes, while international treaties constantly redefine the meaning of the words friend and foe.

THE NATURE OF BOUNDARIES

IT USED TO BE ASSERTED THAT THE ONLY HUMAN-MADE structure visible from the moon was the Great Wall of China. When Aldrin and Armstrong landed on the moon in 1969, they had other things on their minds and failed to notice if this was true. The superhighways that snake across landscapes (and frontiers), the spread of urban conglomerates like Mexico City, even lakes such as the one made by the Aswan Dam on the Nile, must all rate higher on visibility than the Great Wall.

Nevertheless, the wall keeps one aspect of its reputation intact: it remains in the popular imagination, an ideal image of what a frontier should be. First, it is a solid, highly visible statement that what lies behind it is defended. It existed primarily to keep foreigners out, not to keep people in, like the Berlin Wall. Second, it uses the contours of a mountainous terrain to protect easily defined regions. It has logic behind its alignment; it is not some boundary commission's compromise, drawn regardless of topography.

Frontiers of that kind were basically "Keep Out" notices, whereas the frontiers of the more advanced modern states are increasingly screening devices, with green and red channels for customs and passport controls a formality unless the traveler is an impoverished, unskilled, would-be immigrant or a recognizable drug smuggler. The modern frontier has an elastic quality, too, and is often encountered not on the edge of the national territory but at an international airport, perhaps hundreds of miles from the border.

The Cold War type of frontier, with barbed wire and mines, has been dismantled in Europe, but it still exists elsewhere: in Korea, for example, in Morocco's sand walls in Western Sahara, and at the South African border with Mozambique and Zimbabwe. Israel, too, has its heavily defended frontiers with Lebanon, Syria, and Jordan.

There are 156 sovereign states in the U.N. and a great many are bounded in whole or in part by natural features such as mountains, rivers, and oceans. Some, like Switzerland, Afghanistan, and Nepal, are distinguished by the fact that they are countries in mountainous terrain, while the northern frontiers of India, Pakistan, Italy, and Spain are marked by mountain ranges. Chile, a linear country, has been turned into a cartographical freak by its mountains, the Andes, which separate it from Argentina and force the country into a thin strip along the Pacific coast.

The demarcation of its land frontier by one of the world's most massive mountain ranges has not saved Chile from more or less constant frontier disputes with Argentina and Bolivia over the past 100 years. Unfortunately, quarrels of that kind occur frequently in Latin America, regardless of how frontiers are demarcated.

River frontiers ranging from a mediocre trickle like the Jordan to swelling floods like the Zambezi and the Mekong are frequent, too. The Danube provides frontiers, communications, and irrigated land for no fewer than eight countries before it reaches the Black Sea. In some countries a major river is not the international frontier but is nevertheless the geographical feature that defines it: Egypt and Sudan on the Nile, for instance, and Gambia in West Africa.

And then there are the islands, which account for about a fifth of the U.N.'s membership. Many are micro-states of less than a million people, sometimes more, rather than less, vulnerable because of an isolation that produces poverty and instability. Others, such as Singapore and Iceland, are rich. Indonesia (13,677 islands) and the Philippines (7,100) are important third world nations, and Australia is a power among the producers of commodities.

All islands have one great advantage that landlocked

countries and those with short coastlines do not have: their true economic borders are often far out to sea, well beyond their political boundaries. Iceland fought three cod wars against Britain to gain control of its only significant source of income, the fisheries for 200 miles (320 km) around its coast; Britain is the only oil-sufficient country in the European Community, thanks to the North Sea; New Zealand has excellent gasfields.

In a category all its own is Antarctica, scene at the beginning of this century of the last colonial scramble for territory. Lines drawn from the South Pole to the coast carve up territory the size of the U.S. and Europe combined – roughly 10 percent of the world's land. When they were agreed to, no one was certain whether land or water lay beneath. In some places the land is so depressed by the weight of the ice that it would be far below sea level if the ice melted.

Antarctica has seven claimants to part of its territory, but the claims are not recognized by the U.S. and the USSR, both actively involved in scientific work in Antarctica. At the beginning of the 1990s it seemed unlikely that mining and drilling for oil would ever be permitted in the Antarctic landmass or its surrounding waters. The conservationists' arguments might not have been so persuasive if any minerals worth mining had been found. But even if they had been, the problems created by icebergs, a steep and narrow coastal shelf, and a slowly moving ice sheet make exploitation improbable.

The somewhat surreal nature of Antarctica's frontiers does at least illustrate the fact that without permanent settlement, borders have little meaning. Borders belong to places where there are people, exploitable minerals, rivers, and land that can be farmed and built on. But beyond that the reasons for their existence become hazy. Are they the boundaries of nation-states, composed of people who look alike, speak alike, and acknowledge a common culture? Or are they merely the farthest points at which a ruling group can plant its claims?

A nation has been defined as "an abstraction from a number of individuals who have certain characteristics in common, and it is these characteristics that make them members of the same nation." This, however, does not hold good for the USSR or India, or even for Guatemala or Peru, with their mixtures of Amerindian and European populations. It certainly cannot be applied accurately to African states. In most of them the people look roughly alike and have the same skin color, but they are otherwise a hodgepodge of tribes, customs, and languages.

Kinship between many of these groups is no greater than that between Hungarians and Britons. They are countries that have inherited their frontiers from colonial masters. Imperial

rivalries, the fertility of the land, minerals, and strategic interests all played their part in determining frontiers. Take, for example, the copper belt section of the border between Zaire and Gambia. Belgian and British surveyors drew it to carve up the region's rich mineral reserves. It was incidental that they carved up tribal peoples at the same time.

Language is the most common denominator of nationhood in Europe, helping to keep alive the flame of nationalism in a country like Poland during centuries in which the state as a geographical entity ceased to exist. But a common language does not necessarily mean a common border. German speakers are largely divided between Germany, Austria, and Switzerland; English speakers in Europe between Britain and Ireland, where religion, history, and race divide. Factors other than language can be more important in the making of a state: Switzerland, with four official languages, exists because of common interests between its communities and the fact that the inviolability of its frontiers was guaranteed by more powerful European states in the early 19th century.

If there is a single, generally applicable conclusion to be made about national frontiers, it is that they are visible evidence of the dynamic processes of history. Wars and conquests make them; later wars and insurrections shift and remake them; diplomacy rectifies and confirms them. Tangled, webbed, or precisely straight, they are the skein that joins together the patchwork of many national histories.

DRAWING THE LINE

Adapting the physical fences

THE BOUNDARY FIXED BETWEEN TWO states can never be considered permanent. No matter how seemingly durable, how well established by history or bolstered by fortification, it might always shift or even disappear altogether. Whether it does so depends to some extent on the position and nature of the boundary itself.

Some lines seem to have no clear reason for being in one place rather than another – custom, accident, even mapping error, all play their part. But many do fall into one of a limited number of types, each with its own advantages and problems. The great majority of international boundaries are defined by emphatic geographical features, such as coastlines, mountain divides (and their related watersheds), or rivers and lakes. A substantial minority, including the spokelike divisions that cut up Antarctica, are geometrical.

When rivers or lakes are used to divide states, the boundary is generally taken to be the median line, or the center of the main shipping channel, should this veer toward one of the banks. In a few instances, if boundary treaties have been reached between states of markedly different degrees of power – for example, Namibia and South Africa, which confront each other across the Orange River – the boundary runs along one side. This leaves Namibia in the absurd position of having no formal right to use the river.

But rivers can make uncertain boundaries. The Rio Grande, chosen in 1848 to mark much of the United States–Mexico border, tends to wander across its flood plain. This meant that until offending sections were dammed or canalized in the 1930s, the border had to be redrawn every few years.

Alternatively, watersheds may be used to decide state boundaries, in which case the border is established along a mountain ridge. Mountain ranges may also make boundaries in their own right, although they are often so complicated, so broad, and their terrain so inhospitable that they rarely make good borders. Indeed, some even generate states within themselves, and the boundary may well run along the foothills, a situation that also arises when mountains are used as a barrier against a neighbor's expansion.

Some apparently odd-shaped states owe their boundary lines to the need to secure access to a physical resource – fresh water, say, or a mineral deposit – or, more usually, a transport route such as a sea or river. The massive central African state of Zaire would be landlocked if it did not have a 200-mile (320-km) corridor connecting it to the Atlantic.

Many international boundaries are determined not by obvious physical features but by geometrical lines. The longest such boundary is in North America. Agreed in 1818, the famous 49th parallel separates the greater part of Canada from the U.S. As in this case, almost all geometric boundaries are found in former European colonies – most of them in Africa, where nearly one-third of the boundaries are straight lines. Drawn in the absence of specific knowledge about the territory they crossed, some have subsequently been modified, for example to avoid splitting a settlement. As a result, many lines that look straight in atlases turn out to be much more irregular on larger-scale local maps.

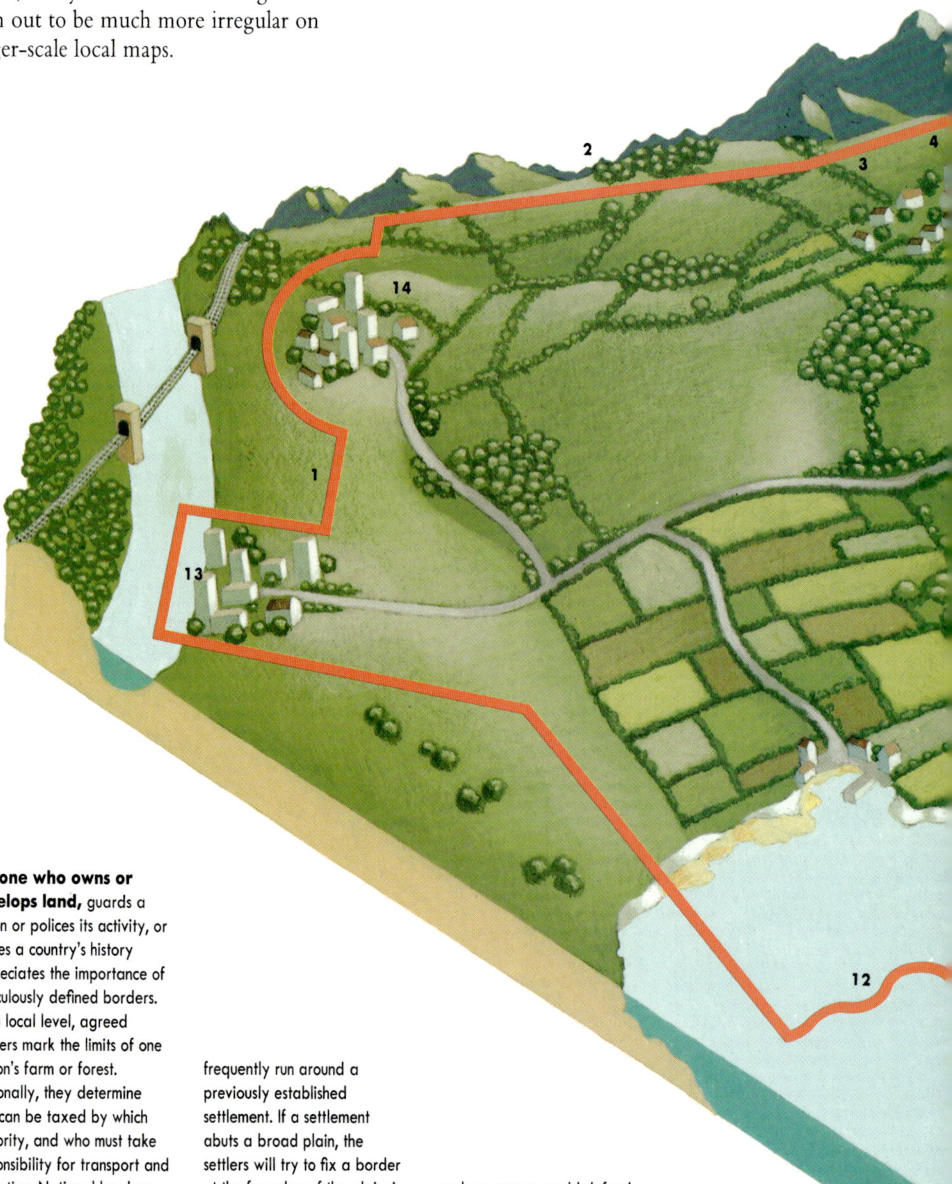

2
3
4
14
1
13
12

Anyone who owns or develops land, guards a nation or polices its activity, or studies a country's history appreciates the importance of meticulously defined borders. On a local level, agreed borders mark the limits of one person's farm or forest. Regionally, they determine who can be taxed by which authority, and who must take responsibility for transport and education. National borders are often established by treaty or annexation, rather than by social or geographical logic. But even negotiated boundaries are likely to take account of natural features, since trade and agriculture so frequently make use of them.

To avoid dividing kinsfolk or separating farmers from their fields, land boundaries frequently run around a previously established settlement. If a settlement abuts a broad plain, the settlers will try to fix a border at the far edge of the plain, in order to preserve its breadth for their own agricultural use or physical protection.

Straight lines, lying at any angle, may be easily drawn across landmasses as borders, but boundaries at sea follow different rules. Until 1950, state sovereignty extended 3 nautical miles (5.5 km) offshore, roughly the distance an onshore cannon could defend.

Today the offshore limit is 12 nautical miles (22 km), in which a nation may forbid navigation, overflying, fishing, and mineral prospecting. When territorial waters overlap, as England's and France's do in the English Channel, a median line generally marks the boundary between the nations.

THE NATURE OF BOUNDARIES

The landscape below shows hypothetical lines that are commonly chosen to divide one state from another. The following are examples of such borders.

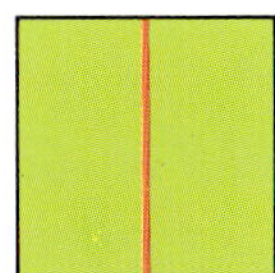

1 Straight borders running north-south
Guatemala/Belize
Libya/Egypt
Angola/Zambia

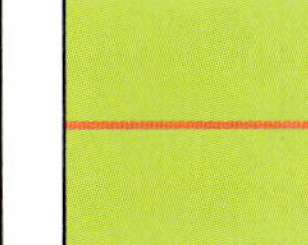

2 Straight borders running east-west
U.S./Canada
Egypt/Sudan
Guatemala/Mexico

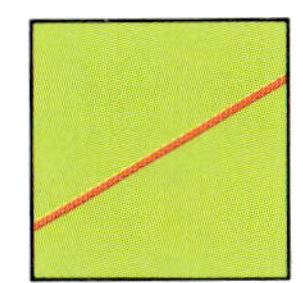

3 Straight borders not running east-west or north-south
Libya/Chad
Algeria/Niger

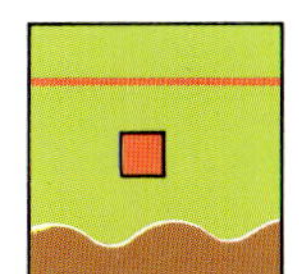

4 Mountain states with borders on foothills
Ethiopia (Sudan)
Nepal (India)
Laos (Vietnam)

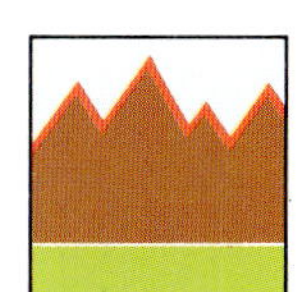

5 Mountain ranges as borders
Argentina/Chile (Andes)
France/Spain (Pyrenees)
Burma/Thailand (Bilauktaung)

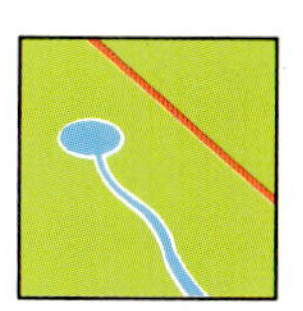

6 Borders to contain river sources
Colombia (Orinoco River)
USSR (Afghanistan River)
Israel (Jordan River)

7 Divided lakes
U.S./Canada (Great Lakes)
Kenya/Uganda (Lake Victoria)
France/Switzerland (Lake Geneva)

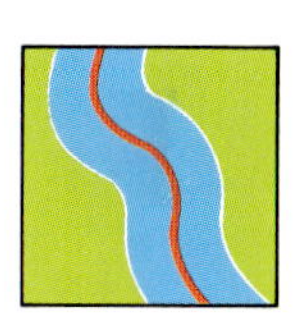

8 Rivers as borders
China/North Korea (Yalu Tumen)
Laos/Thailand (Mekong)
France/Germany (Rhine)

9 Shipping lanes as borders
Sweden/Finland (Aland Strait)
India/Sri Lanka (Palk Strait)
Canada/U.S. (Juan de Fuca Strait)

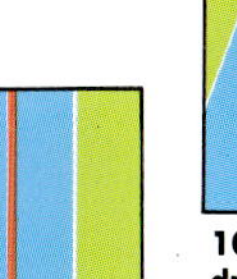

10 Borders drawn to avoid dividing islands
U.S. (Isle Royale)
Kuwait (Bubiyanis)
Japan (Tsushima)

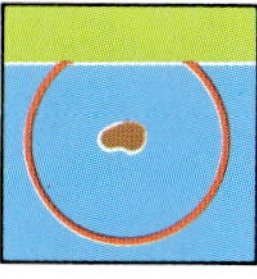

11 Sea limits extended by claiming off-shore islands
Greece (Aegean Islands)
France (Corsica)
Britain (Channel Islands)

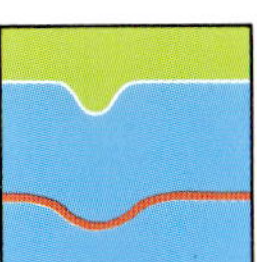

12 The 12-nautical-mile limit following the shoreline
Universal

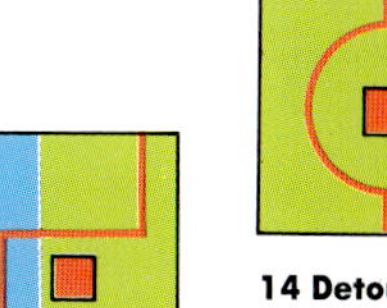

13 Borders marked by corridor access to sea or river
Namibia (Zambezi River)
Zaire (Atlantic Ocean)
Jordan and Israel (Gulf of Aqaba)

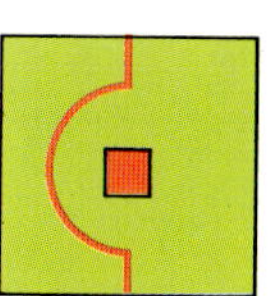

14 Detours to include settlements
Ghana (Burkina Faso)
Sudan (Wadi Halfa)
Pakistan (Nagar Parkar)

WHAT MAKES A BOUNDARY

How nations stake their claims

The entire Korean peninsula has only one nature reserve, a coast-to-coast strip a mere 2½ miles (4 km) wide but 150 miles (250 km) long. It is the most heavily defended plot of land on Earth. For this Demilitarized Zone is one of the few remaining frontiers between two militant economic creeds.

In 1953, after a war in which nearly three million people died, an armistice line was agreed between Communist North Korea and the capitalist south. It was intended to be a temporary boundary pending reunification, with peace talks to be held between the two sides across a table that straddled the line.

Nearly 40 years later, discussions of a sort continue, and may even be making progress. But in the meantime eight million families have been split by the line, which is flanked by heavily mined and intensely monitored fences. One million soldiers are massed north of the border, 650,000 in the south, including 45,000 U.S. troops. A dam has been constructed in the north with a view to flooding the south. There is another in the south to send the floods back.

There are many historical examples of fortified boundaries, from the Great Wall, 1,600 miles (2,575 km) in length, built to separate China and Mongolia, to the Roman Empire's 2,500-mile (4,000-km) *limes*, border defenses that ran the length of the Danube and the Rhine between the Black Sea and the North Sea. More recently, France's 200-mile (320-km) Maginot Line, built between 1929 and 1934, ultimately failed to prevent a German invasion.

But equivalent ramparts are still being built today. Along Western Sahara's curious steplike boundary with Mauritania and Algeria there is a 1,500-mile (2,500-km) wall of sand. It has been built by Morocco's 100,000-strong invading army against the return of Sahrawi guerrillas opposed to their regime.

Likewise, between South Africa and its neighbors Mozambique and Zimbabwe runs the "Snake of Fire" – a lethal, 3,500-volt electrified fence and coils of razor tape constructed to exclude refugees from Mozambique's civil war. Some 200 people were killed here between 1988 and 1990; many more were badly burned or had limbs amputated. These people preferred to try their luck at the fence rather than attempting to cross the Kruger National Park, where lions, once wary of people, have now become accustomed to human prey.

A form of wall, transparent but no less effective, can also be thrown up at sea. French warships and planes police a 72-mile (116-km) exclusion zone around Moruroa Atoll, France's nuclear test site in the South Pacific. It was this blockade that the Greenpeace vessel *Rainbow Warrior* intended to breach when it was sunk by the French in New Zealand in 1985.

Boundaries take many forms, subtle and obvious. Nobody could accidentally stray across the border from North to South Korea, where these guards patrol one of the world's most aggressively maintained frontiers.

The Mai Po marshes in Hong Kong's New Territories act as a sort of no-man's land between Hong Kong and China. As is often the case when boundaries are difficult to patrol, the area has become a nature reserve, and only the birds wander back and forth across the unguarded border.

BERLIN – SYMBOL OF THE COLD WAR

Following defeat in the Second World War, Germany was divided into four sectors that were administered by France, Britain, the U.S., and the USSR. The capital, Berlin, was similarly divided, despite the fact that it was situated 100 miles (160 km) inside the Soviet-controlled eastern region.

In 1948 the USSR began to disrupt contact between West Germany and West Berlin. Then all communications – road and rail connections, water, food, electricity, gas, and coal supplies – were cut. Two million people in the western sectors of Berlin seemed likely to starve unless the whole city was handed over to the Russians. For more than a year all supplies were ferried in by air, by then West Berlin's only supply route, until in May 1949 Stalin agreed to lift the blockade.

In the same year, the three western zones were reunited to form the Federal Republic of Germany, with a new capital in Bonn. The Russian-controlled area of East Germany, with a much smaller population, became the Democratic Republic of Germany.

Geographically part of East Germany, West Berlin remained socially, politically, and

economically emphatically part of West Germany. Massive investment that was poured into it by Western powers and West Germany made it a shopwindow in which to display capitalist prosperity.

During the 1950s thousands of East Germans left their homes for the West, many of them through Berlin. The emigrants – often young, well educated, and highly skilled – were draining the life blood from the East German economy. Drastic action was required to stop it bleeding to death.

On the night of August 12, 1961, a wall was erected between the eastern and western

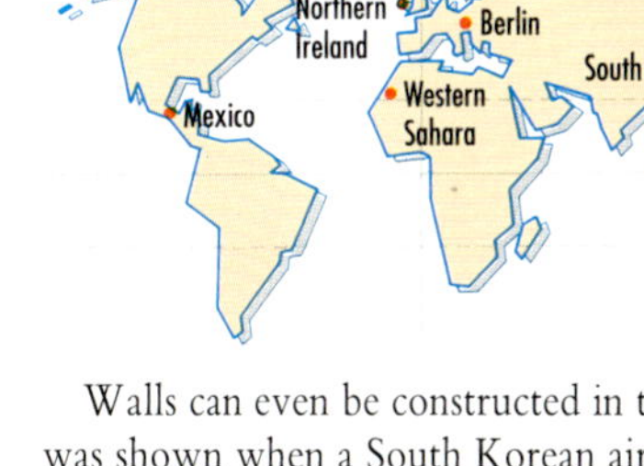

The threat of terrorism makes the border between Protestant Northern Ireland and the Catholic Republic to the south a dangerous area. The British army takes responsibility for the only military frontier in western Europe.

The attraction of the prosperous U.S. leads many impoverished Mexicans to attempt to cross the "tortilla curtain" between the two nations. Guards in helicopters and on horseback restrain the ever-increasing flow of illegal immigrants.

A man-made sand wall 1,000 miles (1,600 km) long acts as a border between Morocco and Western Sahara. In these inhospitable desert conditions, the occupying Moroccan army attempts to control Sahrawi guerrillas fighting for independence.

Walls can even be constructed in the air, as was shown when a South Korean airliner on a flight from New York to Seoul was shot down over Soviet territory in 1983. The Soviet Union claimed the plane was being used to spy on military installations.

International boundaries are, then, sometimes very powerful divides, allocating people to different qualities and styles of life and sometimes stopping the movement of people, goods, and ideas across them. Yet many borders scarcely exist at all, except on a map, like the informal lines between the Benelux countries – Belgium, Netherlands, Luxembourg – or the U.S. and Canada. Fences are deemed unnecessary between such good neighbors.

Elsewhere in the world, where boundaries cross deserts of sand or snow, or mountains or forests, geography dictates they must remain open. Such unpoliced borders allow anti-government Afghan *mujaheddin* guerrillas to cross between Pakistan and their homeland, or Pol Pot's Khmer Rouge to infiltrate Cambodia from military bases in Thailand.

sectors of Berlin to stem the exodus. Initially a barbed wire fence with roadblocks, it was later replaced with concrete and surrounded by a wide no-man's land.

Eventually the wall enclosed all of West Berlin, completely separating it from East Germany as well as dividing the city itself. There were very few crossing points, the most famous being Checkpoint Charlie. These were defended by numerous devices: watchtowers, alarm wires, patrols, prowling dogs, searchlights, mines, and soldiers with orders to shoot to kill.

The demolition of this wall – at which perhaps 200 people died in the course of its 28-year history – was the most evocative symbol of the end of the Cold War and the rebirth of democracy in eastern Europe in the late 1980s. By the time the wall was officially opened in November 1989, thousands of East Germans had emigrated to Austria through newly liberated Hungary, mass demonstrations had occurred throughout the country, and free elections had been promised.

Within a year, almost all traces of the wall had disappeared. "Wallpeckers" had been chipping away at it, collecting souvenirs or pieces to sell. Some large chunks were already installed in major art galleries.

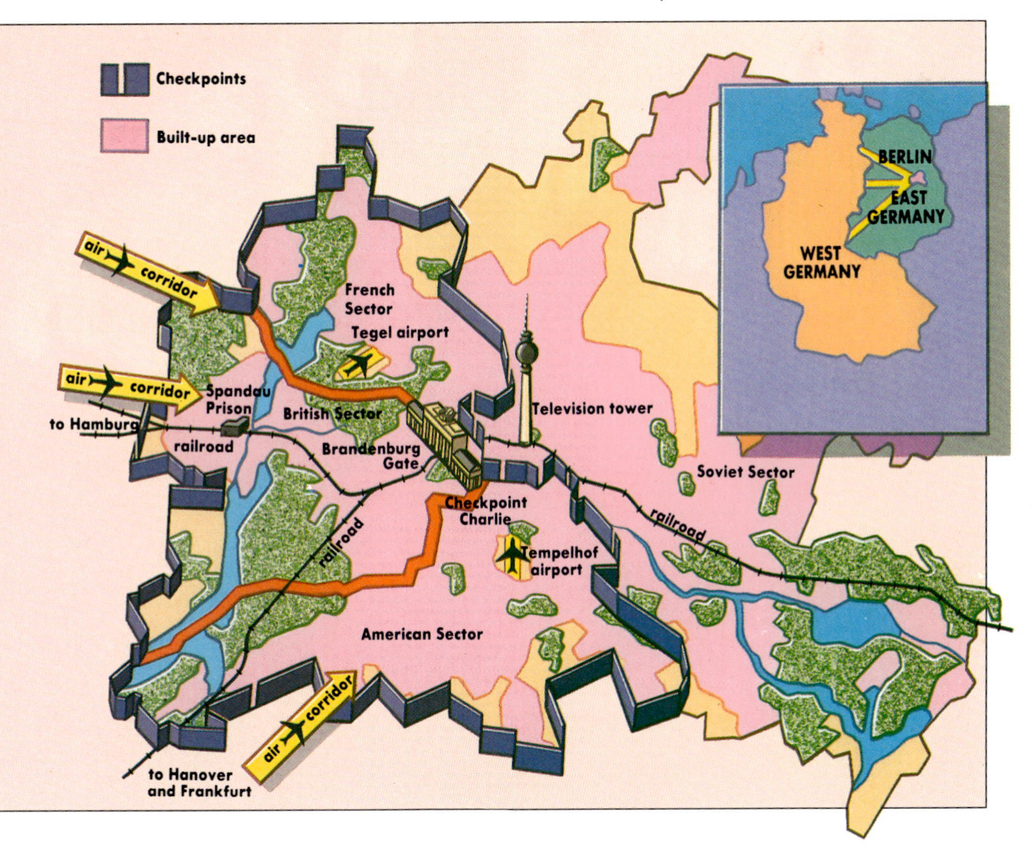

ISLANDS AS NATION-STATES

Sovereignty sealed by the sea

At first glance single islands might be expected to form the most coherent states. After all, they are the most clearly definable geographical forms, both on maps and in the minds of inhabitants and enemies. They share no land boundaries with other states – a common source of friction between neighbors. Their surrounding seas offer a degree of protection from invasion. Yet there are in fact very few island states of any size. Most are incorporated into larger political units, or are subdivided into smaller states or cultural groupings. Islands guarantee neither national unity nor territorial integrity.

Of the world's largest islands – there are 16 of them over 40,000 square miles (100,000 square km) – only four are states in their own right: the great island-continent of Australia, Madagascar, and two others well down the list, Cuba and Iceland. Interestingly, three of the four are budded-off colonies of European powers – Cuba was wrested from Spain in 1898, Australia became politically independent of Britain in 1901, and Iceland only separated itself from Denmark in 1944. Furthermore, there were either very few inhabitants on these islands prior to European colonization, or they were displaced or exterminated early in the encounter, like the Arawak Indians of Cuba.

This budding-off process may be continuing, with immense Greenland following Iceland's example. Having changed hands between Norway and Denmark several times this century, Greenland is now a self-governing part of Denmark, which is only one-fifth its size. Greenland's autonomy is such that in 1982 it was able to withdraw from the European Community, while Denmark itself remained a member.

Several of the world's largest islands are politically divided. The Caribbean island of Hispaniola is split east–west into two separate states: French-speaking Haiti and the Spanish-speaking Dominican Republic. Likewise New Guinea is separated into independent Papua New Guinea and Irian Jaya, currently a reluctant part of Indonesia. Borneo is even more fragmented – part belonging to Indonesia, part to Malaysia, part to Brunei.

Even relatively small islands may be partitioned. Since 1974, the eastern Mediterranean island of Cyprus (3,572 square miles/9,251 square km) has been divided into a northern Turkish sector, about a third of the total area, and a Greek territory occupying the remainder. Sri Lanka (25,332 square miles/65,610 square km), too, seems to be moving toward partition between the Sinhalese majority and the Tamil minority in the north.

Japan, with four major and hundreds of smaller islands, may be culturally homogenous, but island clusters are liable to fragmentation – particularly if a variety of colonial powers have

The Caribbean
The islands of the Caribbean Sea lie in a rough arc from Cuba, the largest, in the northwest, to tiny Grenada in the southeast. Divided Hispaniola, next to Cuba, is the second largest island; Grenada is not as small as St. Kitts and Nevis, but its population, at about 100,000, is less than 1 percent of Cuba's.

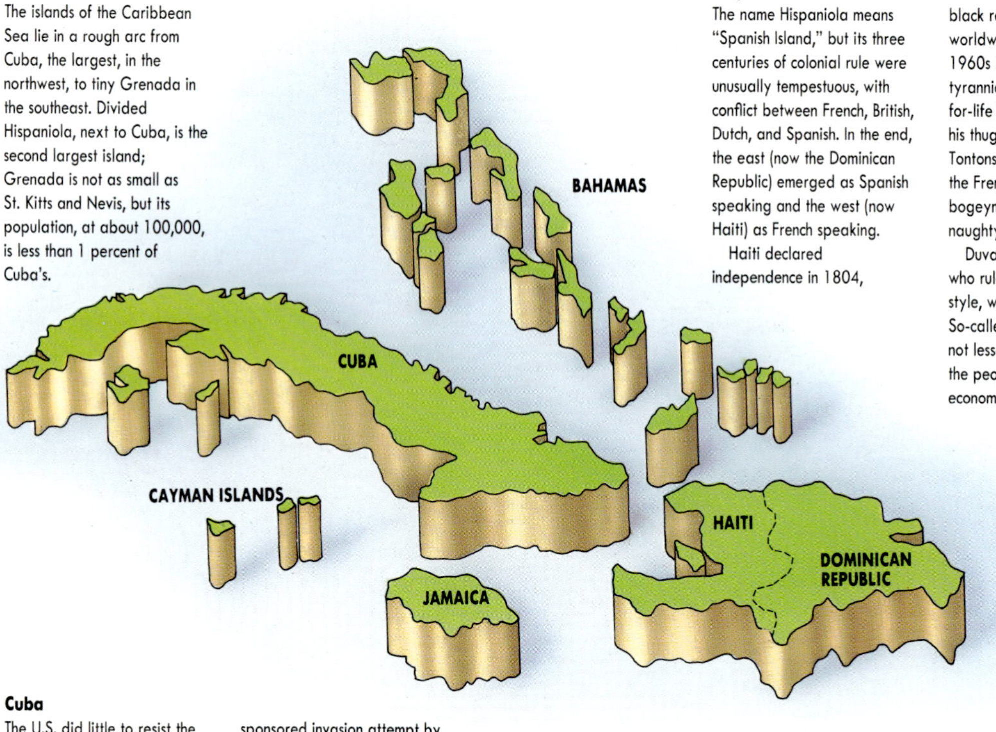

Hispaniola
The name Hispaniola means "Spanish Island," but its three centuries of colonial rule were unusually tempestuous, with conflict between French, British, Dutch, and Spanish. In the end, the east (now the Dominican Republic) emerged as Spanish speaking and the west (now Haiti) as French speaking.

Haiti declared independence in 1804, becoming the world's first black republic. It gained worldwide notoriety during the 1960s because of the tyrannical rule of president-for-life Papa Doc Duvalier and his thuggish secret police, the Tontons Macoutes. The name is the French equivalent of bogeymen, used to frighten naughty children.

Duvalier's son, Baby Doc, who ruled in much the same style, was deposed in 1986. So-called free elections have not lessened the discontent of the people, nor improved the economy. Haitians who can afford it migrate to Florida or the Bahamas.

The Dominican Republic occupies a strategic position between the U.S. and the Panama Canal. In 1965, concern for this sea route and the memory of the recent Cuban revolution led to the U.S. occupying the country and suppressing a socialist attempt at revolution.

Subsequent governments have gone some way along the path to democracy, but a feeble economy and the constant threat of hurricanes mean the Dominican Republic is still one of the poorest nations in the world.

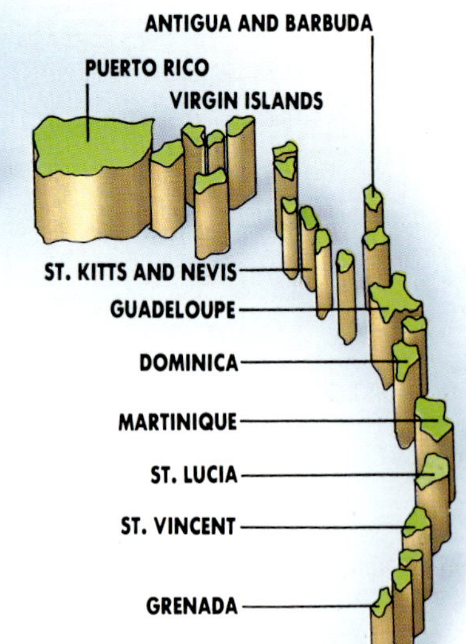

Cuba
The U.S. did little to resist the rise of Fidel Castro in the 1950s because he seemed to be leading a broadly based movement against an unpopular tyrant, President Batista. But once Castro came to power in 1959, he allied himself with the USSR, funded revolutionary movements in Central and South America, and nationalized U.S. and British-owned oil refineries that were refusing to process Soviet oil.

In 1961, U.S. president John Kennedy allowed a C.I.A.-sponsored invasion attempt by anti-Communist Cubans – the incident that became known as the Bay of Pigs. The advance failed, and Cuba promptly proclaimed itself a socialist state. Relations with the U.S. deteriorated still further the following year, when a blockade of U.S. ships prevented Soviet missiles from being sited in Cuba.

Cuba remains closely tied to the USSR, both economically and politically, but has become a power in the Caribbean in its own right.

left a residue of quite distinct cultures on different islands. The Caribbean is the most spectacular example, with its many islands comprising 24 separate states.

The British Isles – with two main islands, Great Britain and Ireland – are fragmented on ethnic, linguistic, religious, and political lines. The root cause lies with the successive invasions of northern European peoples – among them Angles, Saxons, Danes, and Normans – between 1,500 and 900 years ago.

These colonists tended to push earlier peoples into generally less fertile and more mountainous lands in the north and west. The cultural inheritance of the resulting Celtic fringe has never been totally erased. It remains the basis of the political division between the United Kingdom and the Republic of Ireland, and underlies the administrative separateness and distinct national identities of the English, Protestant Northern Irish, Scots, and Welsh.

THE WORLD'S LEADING ISLAND STATES

GREENLAND
ICELAND
U.K.
JAPAN
CUBA
PHILIPPINES
SRI LANKA
BORNEO
NEW GUINEA
INDONESIA
MADAGASCAR
AUSTRALIA
NEW ZEALAND

Only ten nations of any size consist entirely of islands. There are five single islands – Australia, Madagascar, Cuba, Iceland, and Sri Lanka – and five groups of varying complexity – Indonesia (whose largest islands are Sumatra, Sulawesi, and Java), Japan (Honshu), the Philippines (Luzon), New Zealand, and the U.K. Greenland is still a province of Denmark, despite increased autonomy, and New Guinea and Borneo, the third and fourth largest islands in the world, house more than one nation.

Indonesia
The federation of Indonesia has a total landmass that is not much less than that of Greenland, but it is made up of 13,677 islands. Of these, 6,000 are inhabited by a growing population that is the fifth largest in the world. It is a tribute to the efforts of the country's leaders that such a diversity of geography and ethnic groups should be held together as one nation – indeed, the national motto is "Unity in diversity."

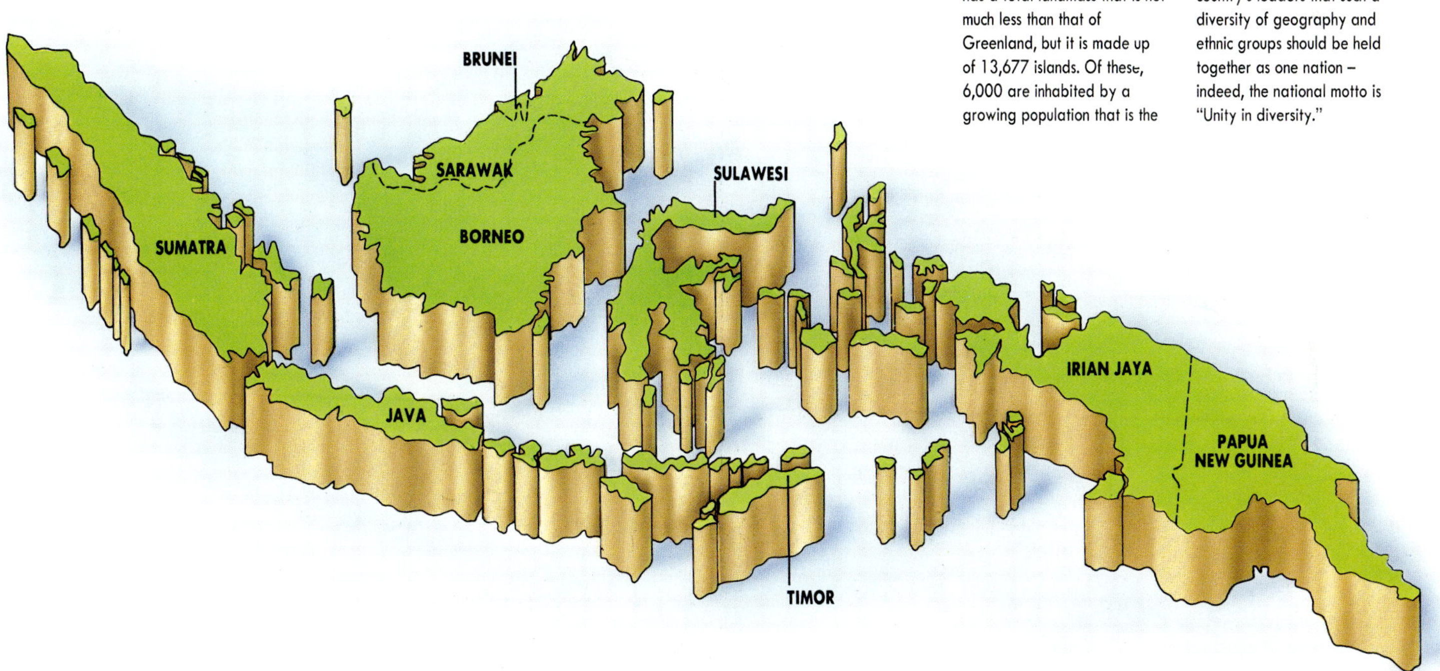

Sumatra
The largest island of the group is Sumatra, nearly as big as Spain and three and a half times the size of crowded Java, where over half the population lives. Another ten million people live on Sulawesi, the strangely shaped island that seems slotted like a piece of a jigsaw puzzle between Borneo and New Guinea.

Two islands in the region are divided into more than one nation. Indonesia occupies three-fourths of the island of Borneo – most of the rest is part of the Federation of Malaysia, and a mere 2,226 square miles (5,765 square km) make up the fabulously wealthy Sultanate of Brunei. To the east, New Guinea is half Indonesian – the province of Irian Jaya – and half the parliamentary monarchy of Papua New Guinea.

Brunei
Brunei is one of the richest countries in the world, and its sultan one of the richest men. At about $21,000 per capita, its gross national product is well ahead of the U.S.'s ($18,500), West Germany's before unification ($18,700), and Japan's ($19,500). The money comes almost exclusively from oil, which is exported to the wealthy countries of east Asia and to the U.S., and which provides an enviable balance of payments surplus.

Timor
Tucked away under the main islands of Indonesia, Timor is deeply divided. The western half, formerly part of the Dutch empire, has belonged to the Indonesian federation since 1950. The eastern end was under Portuguese control for three centuries, until Indonesia annexed it in 1975. This action was prompted by fear that Portugal was about to grant East Timor autonomy, and that this example would inspire the far-flung islands of Indonesia to seek independence for themselves.

Strong resistance to the Indonesian invasion has been fiercely suppressed; many thousands of people have been killed and the guerrilla movement has been forced to retreat to the mountains. Although the United Nations does not recognize Indonesian claims to East Timor, and a public demonstration on the occasion of the Pope's visit in 1989 drew world attention to the problem, the protesters were arrested, and a solution does not appear to be in sight.

Papua New Guinea
The population of Papua New Guinea is estimated at about 3.5 million, but it comprises an extraordinary 700 ethnic groups speaking 700 different languages. Three-fourths of the country is densely covered with rainforest, and most of the people still live in rural villages in the tiny areas of arable land. Tribal warfare still occurs in the remote highlands, although the cannibalism and head-hunting for which the island used to be legendary are largely things of the past.

MOUNTAIN BORDERS
Outlining might with height

GEOGRAPHY'S WALLS – THE EARTH'S mountain ranges – can form strong and enduring international boundaries. In South America, the Andes neatly divide Chile from Argentina for over 2,500 miles (4,000 km), while Brazil's northern boundaries with Venezuela and the Guianas also run along the top of highlands.

In Europe, the Pyrenees for much of their length form the border between Spain and France. The ridge line of the Maritime and Graian Alps to the south of Mont Blanc separates Italy from France.

Similarly, several of Southeast Asia's boundaries are mountain ranges, relics of the farthest limits of ancient, rice-dependent civilizations. Mountains define sections of Burma's borders with India and Thailand, Thailand's with Cambodia, Vietnam's with Laos, and most of the subdivisions that carve up Borneo. Naturally enough, such barriers to communication may be so formidable that mountain geography nurtures profound cultural differences. Distinct languages, diverse ethnic groups, and so on can flourish in relative isolation on either side of the divide.

Yet mountains rarely function as clear-cut boundaries. Even in the Pyrenees the ridge line is not followed very exactly – such a border would wander much too erratically. Mountain ranges, the products of phenomenal forces deep within the planet, are never very tidy places. The high ridges between Norway and Sweden, for example, are so complex, and the terrain is so rugged, that the specific position of the boundary must often be ignored or simplified into a straight line drawn between more clearly defined locations.

Indeed, many ranges – the Rockies of North America, the Urals of the USSR, and the Atlas Mountains of North Africa, for example – do not function as boundaries at all, and none of Africa's or North America's boundaries are determined by ranges.

The reason lies partly in the physical nature of the mountain range itself. With few exceptions, such as the southern Andes and the Pyrenees, the great ranges of the world are not so much single

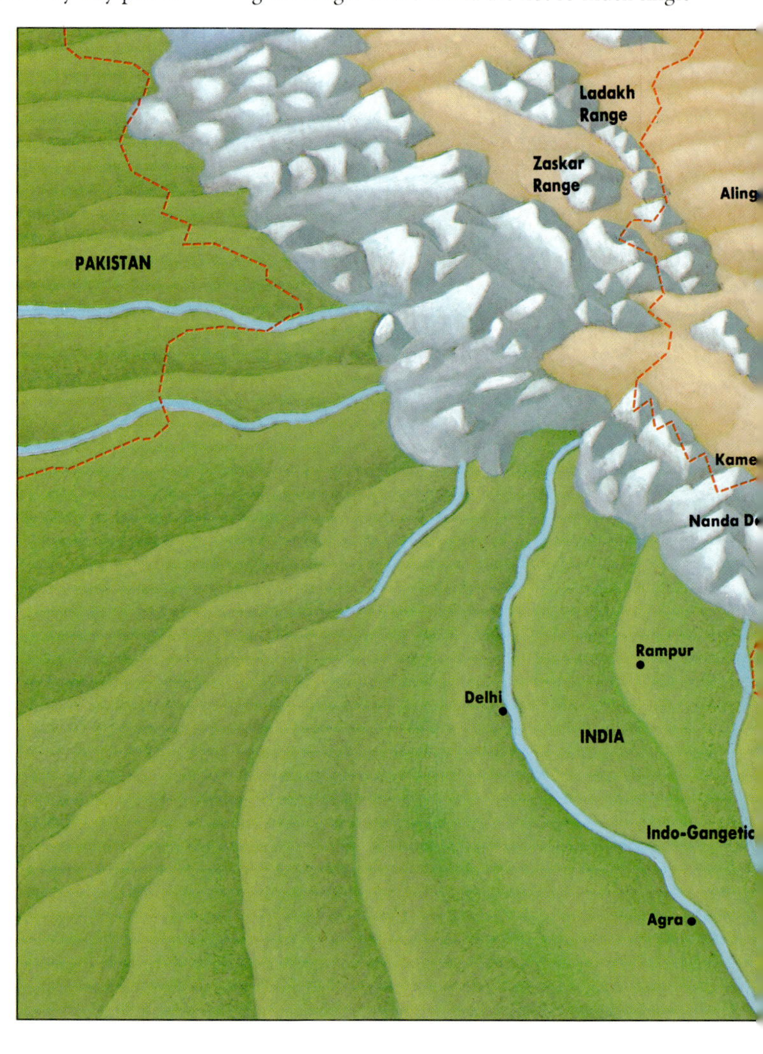

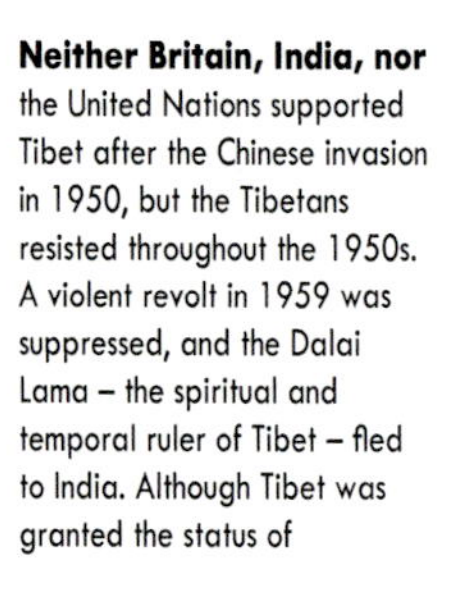

Neither Britain, India, nor the United Nations supported Tibet after the Chinese invasion in 1950, but the Tibetans resisted throughout the 1950s. A violent revolt in 1959 was suppressed, and the Dalai Lama – the spiritual and temporal ruler of Tibet – fled to India. Although Tibet was granted the status of autonomous region in 1965, resentment against the Chinese occupation continued to smolder.

In the 1980s, various religious and economic rights were restored, and overtures made toward the Dalai Lama, but this period of goodwill was short-lived. An anti-Chinese demonstration was violently suppressed in 1987. Martial law was imposed in 1989, and although it was lifted in May 1990, the tension remains. Chinese soldiers patrol the capital, Lhasa, and violence could erupt at any moment.

The photograph shows that even the peaceful Buddhist monks are roused to protest against their Chinese overlords.

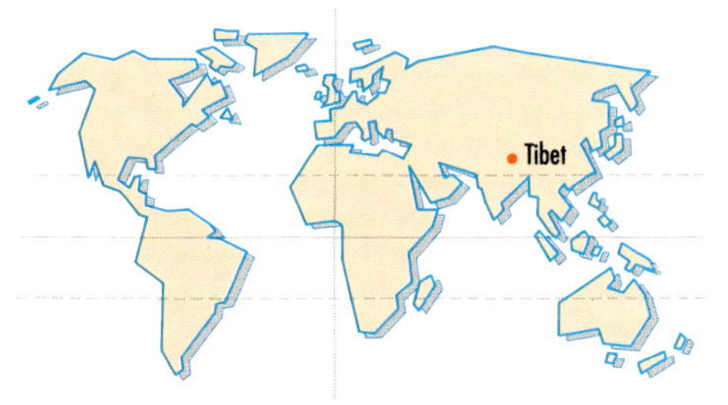

barriers as complex belts of ridges, valleys, and plateaus. Settled enclaves in such upland regions, difficult for a lowland power to invade and rule from afar, often become cultural hearthlands – their self-reliant inhabitants develop unique and durable lifestyles out of which independent nations may be created.

In the highest eastern ridges of the Pyrenees, the tiny nation of Andorra has survived intact for more than a thousand years. Mountains elsewhere in the world have thrown up even more distinctive cultures, like the feudal civilizations of the Aztecs and Incas in Central and South America. Although overwhelmed by the Spanish *conquistadores* early in the 16th century, their respective legacies still contribute to the national identities of Mexico and Peru. Indeed, in both cases the present national territory is largely the result of expansion outward from the original mountain states.

Similarly, Ethiopia's boundaries roughly follow the limits of a series of civilizations that flourished over hundreds, perhaps thousands, of years in the Ethiopian highlands, believed by many to be the center of the great biblical empire ruled by Solomon and Sheba.

Ethiopia's isolation in its mountain fastness allowed it to remain largely independent of 19th-century European colonialism. But the Himalayan nations of Nepal, Sikkim, and Bhutan, together with Afghanistan and Tibet, were more vulnerable, despite their higher elevations. Situated between British India to the south and Russia and China to the north, these became buffer states, absorbing much of the friction between their more powerful neighbors, and hence lessening the likelihood of direct confrontation between the larger nations.

Nevertheless, external pressures from major powers continue to fragment the region, making it one of the political hotspots of the world. The predominantly Muslim state of Kashmir has been the scene of bitter fighting between India and Pakistan since partition in 1949. Afghanistan, invaded by the USSR in 1979, witnessed constant battles between Soviet troops and the guerrilla *mujaheddin* for a decade and is still in a state of turmoil.

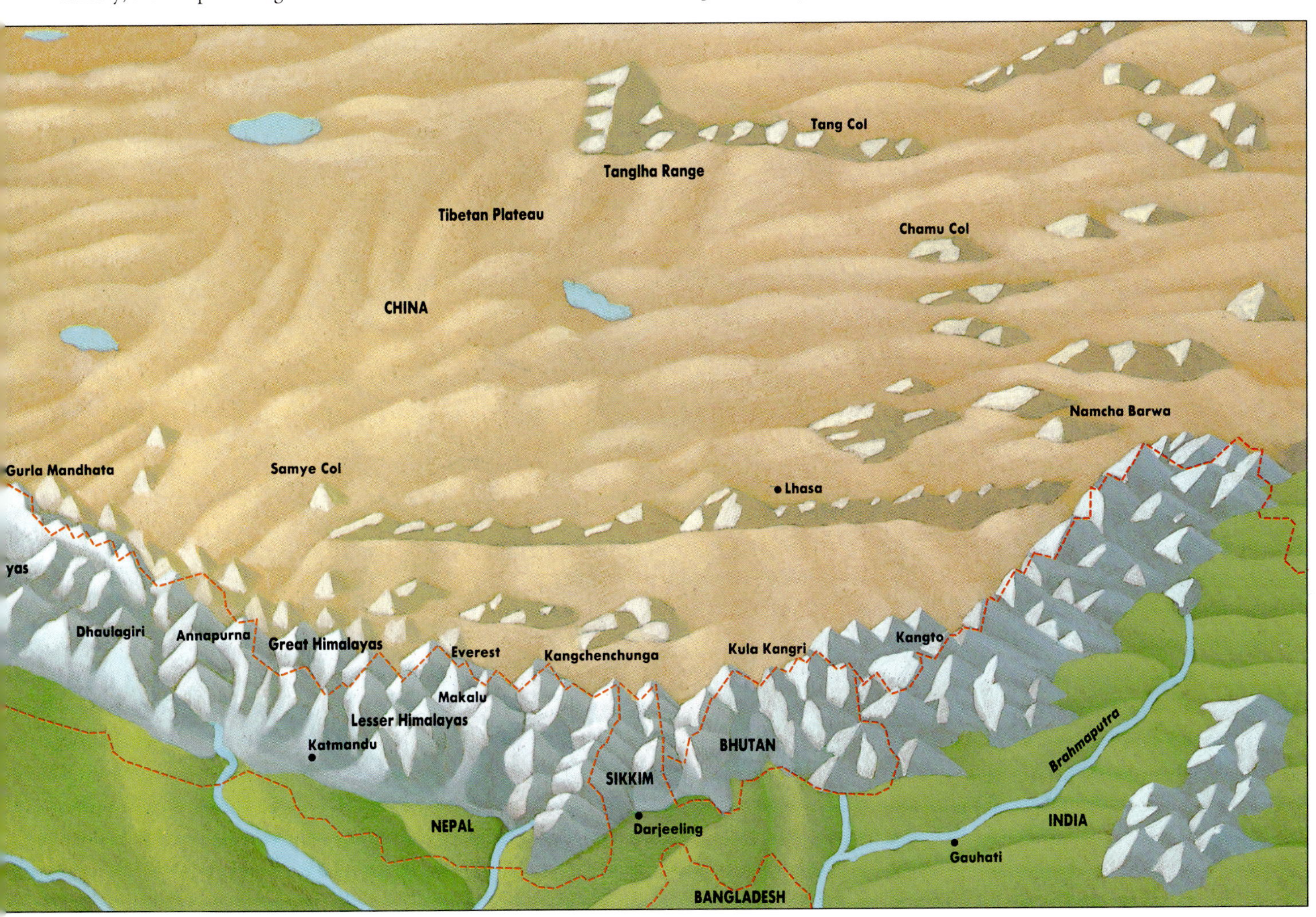

At first glance, it seems that the Himalayas form a simple, unequivocal boundary between India and China. The truth is more complex. For high up in the mountains lies a chain of tiny nations, created by Britain to act as buffers to defend India from Chinese or Russian aggression.

Tibet proclaimed its independence in 1913, although China never recognized the borders and annexed the province in 1950.

Politically, Nepal has been more stable. Independent since the 1950s, it now derives much of its income from visiting mountaineers. But this has brought with it problems of soil erosion, which means that access to the most popular climbs must be strictly controlled.

An uprising against the king in 1990 may lead to a more democratic form of government. But Nepal has no easy access to trade routes, and little mineral wealth. With India imposing trade embargos in the hope of gaining control of its northern neighbor, there is little sign of improvement for Nepal's economy.

Bhutan has been luckier. The tiny Buddhist kingdom cut itself off from the outside world until 1974, and since then has admitted a limited number of tourists. By Western criteria, Bhutan is extremely poor, but most people are self-sufficient farmers, and the king's policy is that "gross national happiness" is more important than gross national product.

THE LONGEST UNDEFENDED BORDER

Canada keeps to the straight and narrow

READING FROM RIGHT TO LEFT, THE border between Canada and the mainland U.S. charts the progress of the relationship between the two neighbors. In the east, the frontier confirms the defeat of the British Crown by the 13 rebellious states in the 18th century. After that, location is allowed to apply its logic and the frontier more or less takes the median line through Lakes Huron and Superior, the stalemate in the war of 1812 having ensured Canada's survival. And, finally, after a nervous squiggle west of the lakes, the frontier makes a bold dash for the Pacific along the 49th parallel, the generals and geographers pushed to one side by diplomats who agreed that the future lay with mutual tolerance.

Geographically, the Canadians may have got the worst of the compromises, most of which were made for them by British negotiators bent on appeasing an expansive U.S., which they saw as a potentially huge market for trade and investment. Economically, though, Canada has done well out of 170 years of peaceful coexistence based in the first place on a common language and similar political traditions.

With a population of 26 million, Canada is a member of the Group of Seven, the seven richest democracies. Between them Canada and the U.S. form the world's greatest bilateral trading partnership, with an annual exchange of goods and services worth around $150 billion.

That trade has shaped Canada's development as much as its climate has. Despite the fact that they inhabit the world's second largest country, eight out of ten Canadians live within 100 miles (160 km) of the U.S., most of them in cities huddled along the border like seekers after the material warmth emanating from the south. Nearly 80 percent of Canada's exports go to the United States, while more than 20 percent of the United States' exports make the journey north.

The success of the European Community (EC) spurred North Americans in the 1980s to remove the last barriers in their commercial relationship, just as Europeans are doing. The Free Trade Agreement signed in January 1988 will eliminate most of the remaining tariffs and restrictions on investment over a ten-year period that began in January 1989.

The Canadian Liberal Party labeled it "the Sale of Canada Act," but lost (to the ruling Conservatives) a general election fought on the issue of its terms. Since then, Canada has encouraged moves to bring Mexico into a free trade arrangement and suggested it should be extended to cover Central and South America.

Open borders may be good for economies, but they are bad news for those seeking to foster a sense of nationhood. Canada, it has been said, has two permanent crises. One is the struggle for political and cultural survival in the shadow of the U.S.; the other is the clash between the traditionally antipathetic cultures of the French-speaking minority (a fifth of the population) in Quebec and the English-speaking majority in the rest of the country. In reality, they are aspects of one crisis – a crisis of identity.

Canada is a confederation more loosely bound than the American federation. Regional loyalties tend to mean more than loyalty to the center, and minorities have been encouraged to retain their distinctiveness. The American melting pot process of assimilation has been rejected in favor of the "salad bowl" theory in which the ingredients remain separate within the context of a uniquely Canadian dressing.

Thus, if the Quebecois are offered special status, fairness dictates that North American Indians and Inuit should be granted ownership of a Texas-sized segment of the Arctic north. Canada remains a monarchy, which pleases the British anglophones but irritates the francophones, so the Queen's role as head of state has been diminished to near invisibility.

The fate of the Meech Lake Accords, which defined Quebec as a distinct society, indicates that the moves to de-Anglicize Canada may have reached their limits. The accords collapsed in mid-1990 when anglophone Manitoba and Newfoundland declined to ratify them.

Formulating the dressing for the Canadian "salad bowl" has required considerable political dexterity. The tasting panel has yet to deliver a verdict on whether it is strong enough to keep Canadians together around the table.

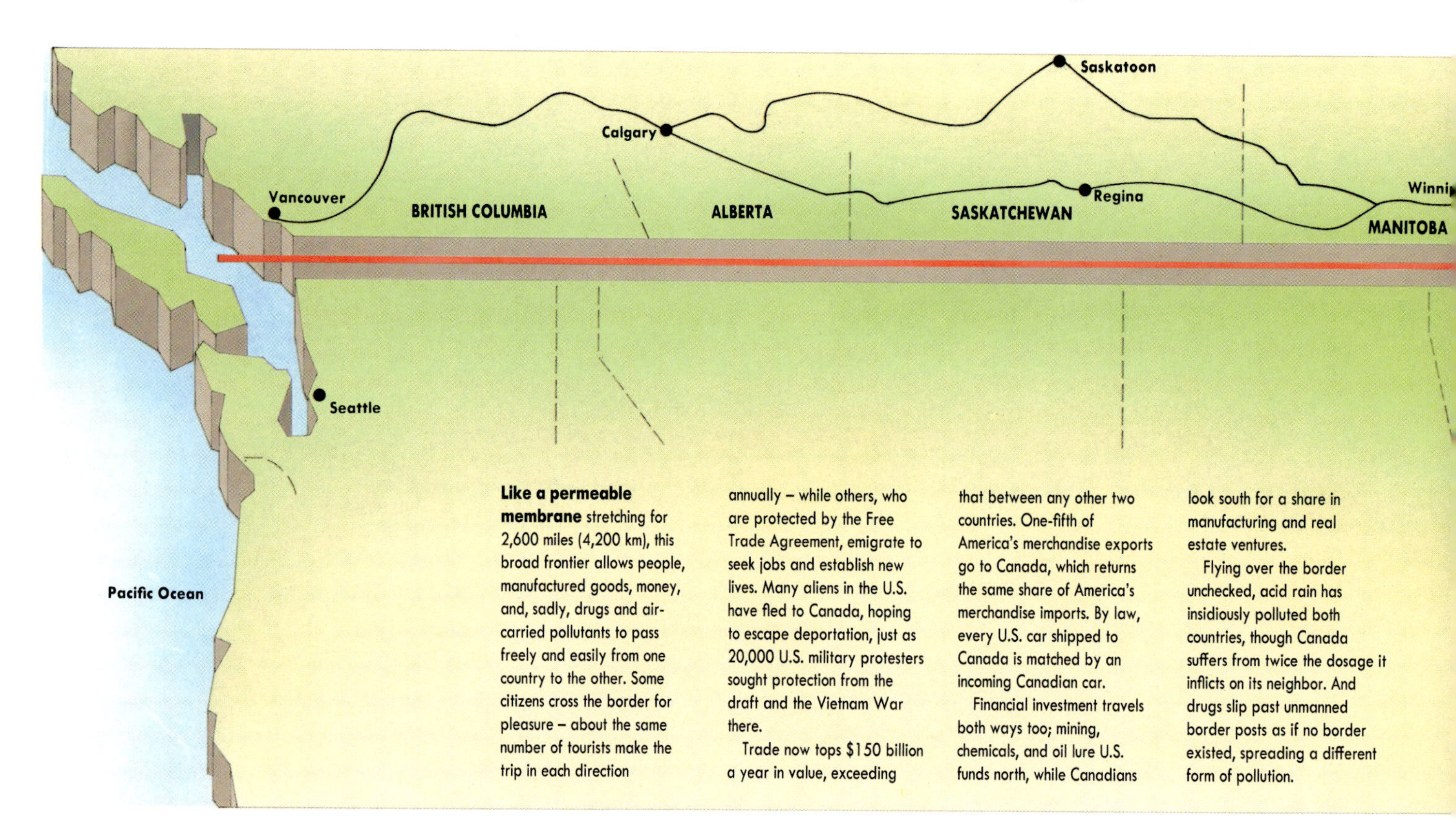

Like a permeable membrane stretching for 2,600 miles (4,200 km), this broad frontier allows people, manufactured goods, money, and, sadly, drugs and air-carried pollutants to pass freely and easily from one country to the other. Some citizens cross the border for pleasure – about the same number of tourists make the trip in each direction annually – while others, who are protected by the Free Trade Agreement, emigrate to seek jobs and establish new lives. Many aliens in the U.S. have fled to Canada, hoping to escape deportation, just as 20,000 U.S. military protesters sought protection from the draft and the Vietnam War there.

Trade now tops $150 billion a year in value, exceeding that between any other two countries. One-fifth of America's merchandise exports go to Canada, which returns the same share of America's merchandise imports. By law, every U.S. car shipped to Canada is matched by an incoming Canadian car.

Financial investment travels both ways too; mining, chemicals, and oil lure U.S. funds north, while Canadians look south for a share in manufacturing and real estate ventures.

Flying over the border unchecked, acid rain has insidiously polluted both countries, though Canada suffers from twice the dosage it inflicts on its neighbor. And drugs slip past unmanned border posts as if no border existed, spreading a different form of pollution.

Unbroken and seemingly endless, this clear path between Maine and the province of Quebec makes the world's longest undefended border a clear, physical reality. Twenty feet (6 m) wide and maintained by the International Boundary Commission, this so-called vista marks the friendly meeting of the nations and some of their peculiar differences as well. In the border region, Canadians enjoy their country's most temperate climate and benefit from a sophisticated combination of transport, flourishing agriculture, and urban development. South of the border, their U.S. neighbors inhabit their country's coldest landscapes and some of its most unyielding soil.

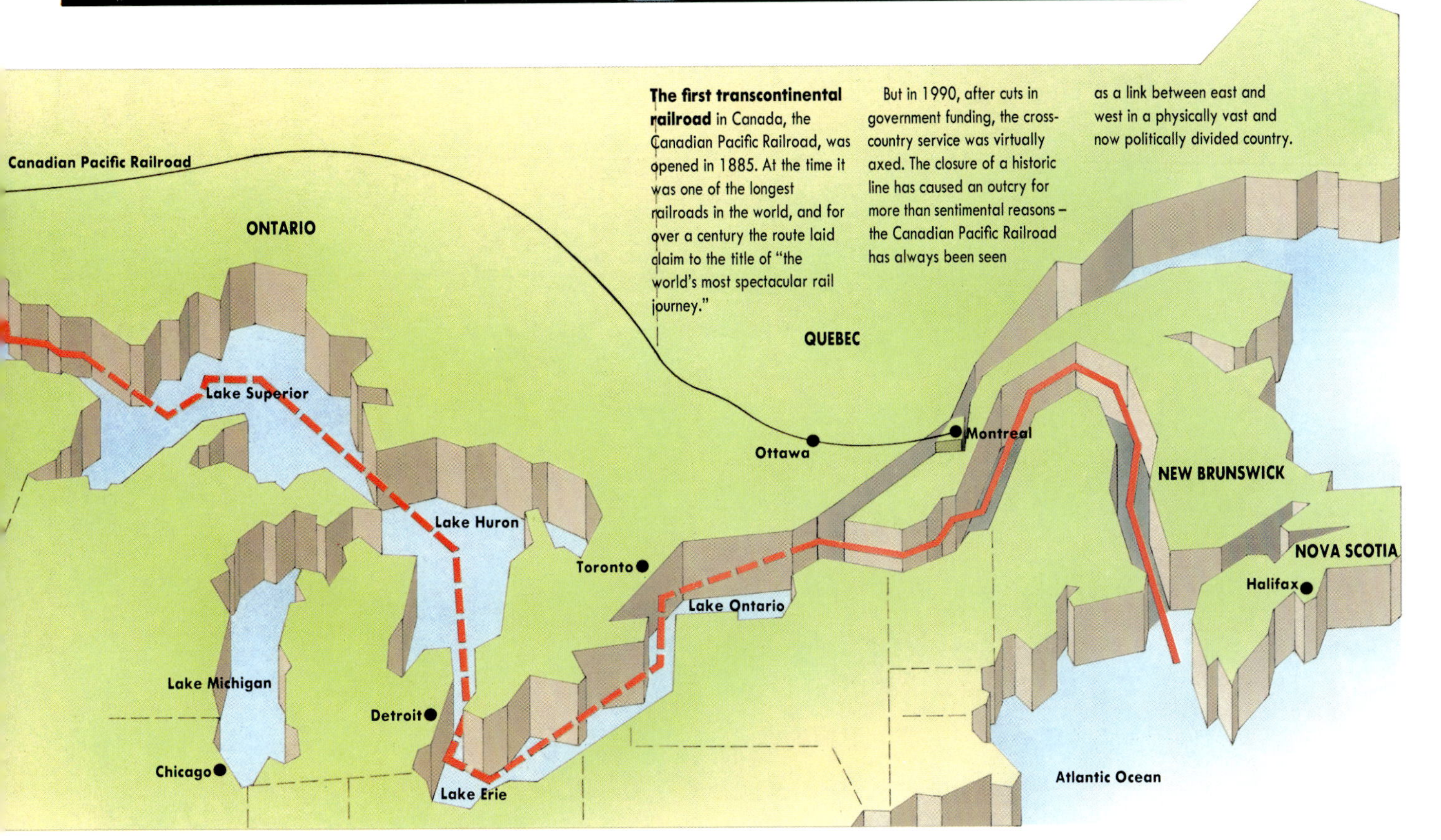

The first transcontinental railroad in Canada, the Canadian Pacific Railroad, was opened in 1885. At the time it was one of the longest railroads in the world, and for over a century the route laid claim to the title of "the world's most spectacular rail journey."

But in 1990, after cuts in government funding, the cross-country service was virtually axed. The closure of a historic line has caused an outcry for more than sentimental reasons – the Canadian Pacific Railroad has always been seen as a link between east and west in a physically vast and now politically divided country.

THE CARVING UP OF AFRICA

Callous lines of colonialism

BETWEEN NOVEMBER 1884 AND February 1885, six nations – Belgium, Britain, France, Germany, Portugal, and Spain – held a conference in Berlin at which they cut up most of Africa and shared the pieces. There were no Africans at the conference. The map was drawn by Europeans, for Europeans. Tiny Belgium gained control of an area of Central Africa 80 times the size of its own country.

Europeans had been trading in Africa for nearly four centuries, yet hardly any were settled there. Only on a few scattered coastal strips was there any significant settlement. The few formal boundaries already in place had chiefly been drawn to separate Europeans – French from Spanish in Morocco, French from British in West Africa, Dutch from British in the south. The arrival of the Belgians and the Germans in Central and East Africa in the 1880s threatened to upset the existing balance of power, in Europe as well as in Africa.

In the absence of detailed knowledge about Africa and African peoples, how were the boundaries to be determined? Physical features were followed, where these were known. But mistakes were easily made in the large areas that remained unmapped. Clearly it was much easier to impose boundaries along lines of latitude or longitude. As a result, more than 30 percent of Africa's boundaries today follow straight lines.

Certainly there was little concern for the inhabitants of the regions. Many Africans were separated from resources they relied on in order to survive, traditional tribal territories were fractured, and very different cultures were arbitrarily grouped together.

For most Africans, initially, the boundaries were of little significance. Lines were not policed, and the colonial culture penetrated very little. People continued to interact at will across the nominal borders – seeking fresh pastures and attending markets. But over time, the imposition of the official, European languages and the development of distinct new social systems hardened the boundaries.

The cultural, economic, and political consequences of the delineation of boundaries may not have been obvious at the time. But although the colonial powers have now gone, swept away by the wind of change that blew across Africa after the Second World War, the boundaries they imposed mostly remain intact.

Many have become sources of conflict, or imposed serious restraints on countries attempting to promote development, forge national unity, or reconstruct more viable states from the territories they inherited. Nevertheless, some states have united, such as the former

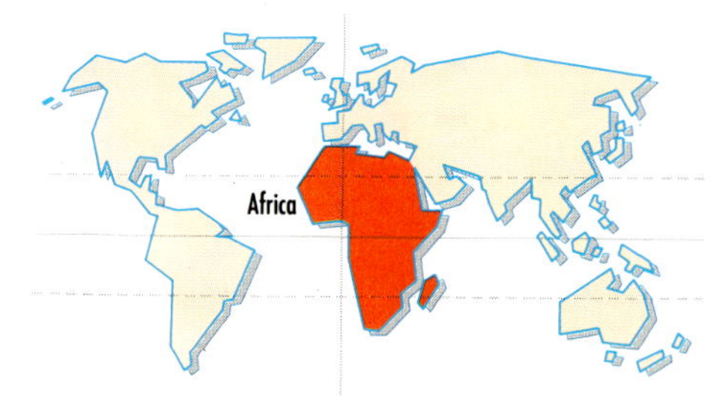

Italian and British Somalilands, which in 1960 became Somalia, and Tanganyika and Zanzibar, which together formed Tanzania in 1964.

Most obviously, the size, location, and resources of states remain obstacles to progress. In some cases states cannot create an effective market: they are simply too small, like Gambia, Guinea, or Sierra Leone, or thinly populated, like Mali, Chad, or Niger. In many others, states are too large, too populous, or too disparate to be governable. Here the move may be toward greater fragmentation, as in attempts by Biafra to secede from Nigeria, or Eritrea and Tigre from Ethiopia.

Such movements are typically described as "tribalist" by state leaders and governments, themselves often recruited from a dominant tribal group. Naturally they are unwilling to cede sovereignty, particularly over territories where mineral resources might be found.

Ashanti gold was one of the main attractions when the European powers first colonized the part of West Africa they called the Gold Coast. Living in the middle of a gold-mining region, the Ashanti used gold or gold plating, worked into elaborate designs and often of ritual significance, for crowns, jewelry, armor, even shoes. The paramount chief or *Asantehene* sat on a golden stool that was the symbol of unity for the Ashanti Empire.

The core of the kingdom was situated in southern Ghana, but the Berlin Conference, with its customary disregard for tribal unity, drew lines that left some Ashanti in what is now the Ivory Coast and some in Togo.

Most Ashanti still live in villages and pay allegiance to local chiefs and the *Asantehene*. The photograph shows Ashanti leaders in traditional dress meeting South African president F.W. de Klerk on a visit to the Ivory Coast in 1989.

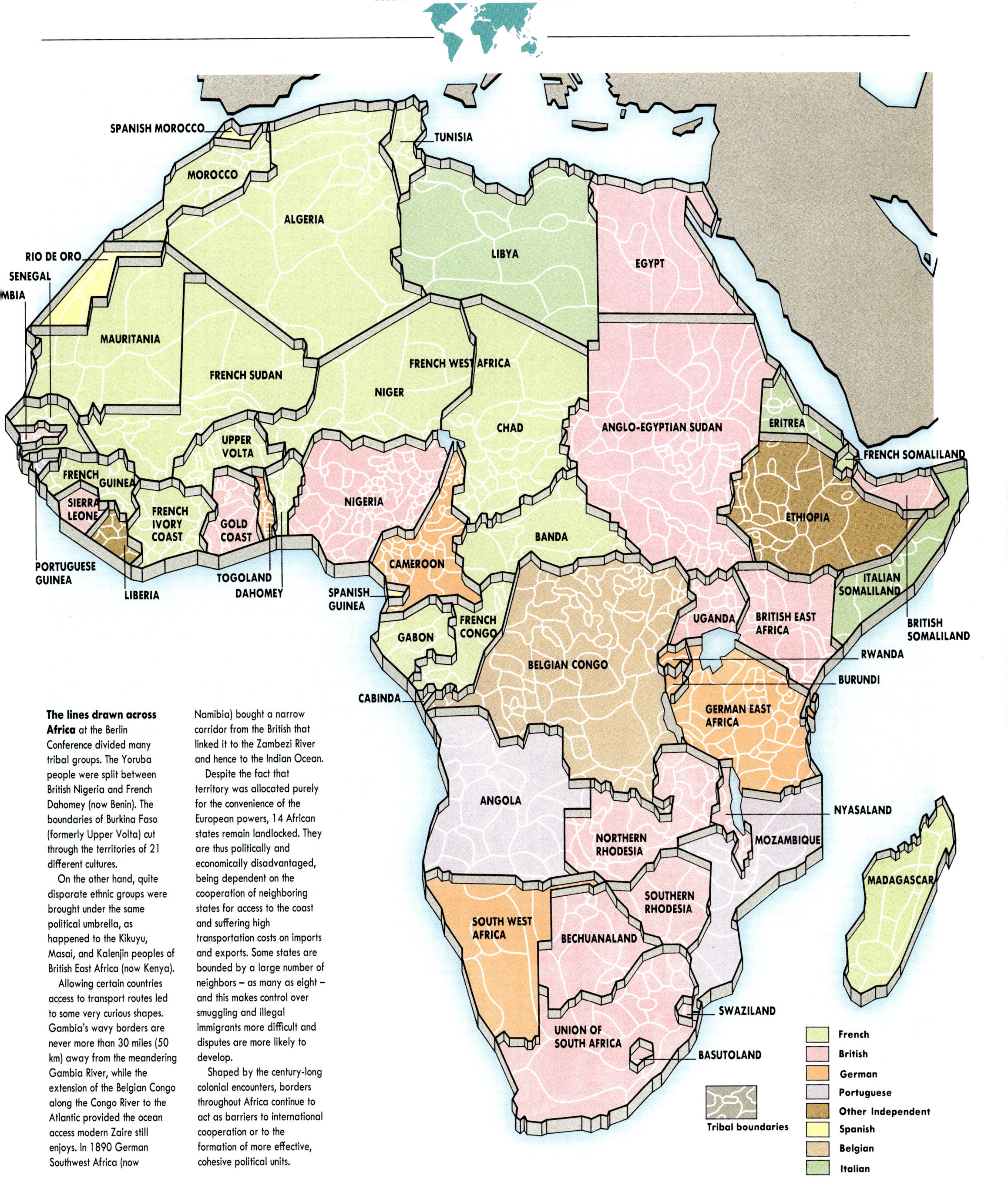

The lines drawn across Africa at the Berlin Conference divided many tribal groups. The Yoruba people were split between British Nigeria and French Dahomey (now Benin). The boundaries of Burkina Faso (formerly Upper Volta) cut through the territories of 21 different cultures.

On the other hand, quite disparate ethnic groups were brought under the same political umbrella, as happened to the Kikuyu, Masai, and Kalenjin peoples of British East Africa (now Kenya).

Allowing certain countries access to transport routes led to some very curious shapes. Gambia's wavy borders are never more than 30 miles (50 km) away from the meandering Gambia River, while the extension of the Belgian Congo along the Congo River to the Atlantic provided the ocean access modern Zaire still enjoys. In 1890 German Southwest Africa (now Namibia) bought a narrow corridor from the British that linked it to the Zambezi River and hence to the Indian Ocean.

Despite the fact that territory was allocated purely for the convenience of the European powers, 14 African states remain landlocked. They are thus politically and economically disadvantaged, being dependent on the cooperation of neighboring states for access to the coast and suffering high transportation costs on imports and exports. Some states are bounded by a large number of neighbors – as many as eight – and this makes control over smuggling and illegal immigrants more difficult and disputes are more likely to develop.

Shaped by the century-long colonial encounters, borders throughout Africa continue to act as barriers to international cooperation or to the formation of more effective, cohesive political units.

THE JEWEL SHATTERED
Dividing the Indian subcontinent

On the stroke of midnight on August 14, 1947, British rule in India ceased and new lines were drawn on the map. The independent states of India and Pakistan were born. Six months later, Ceylon (renamed Sri Lanka in 1972) was also granted independence. The multifaceted, spectacularly beautiful subcontinent, once known as the "Jewel in the Crown" of Queen Victoria's empire, was recut. But the task of reshaping nation-states from the newly created territories has proved formidably difficult.

At the time of independence the subcontinent was populous and poor, home to over 400 million people (the number has since doubled to around 825 million). It had no common language – with the ironic exception of the one that had been imposed, English. There were 14 major languages, seven major religious groups (Hindus, Muslims, Sikhs, Christians, Parsis, Jains, and Buddhists), and about 60 million tribal peoples.

Furthermore, there were 562 independent princely states. Nominally, at least, they were free to make their own decisions about where their future should lie. Kashmir (Hindu-ruled, but mainly peopled by Muslims) and Hyderabad (Muslim-ruled, with a Hindu majority) toyed with complete independence. Both were soon invaded by India.

The major split was between Muslims and Hindus. It became apparent that the only way to prevent Muslim–Hindu war following independence was to partition the country along religious lines. The position of the boundaries was chiefly decided according to the findings of the 1941 census. Territory was allocated according to whether Muslims or Hindus constituted a majority in the census district.

The process seemed in principle to be reasonable enough, but in fact proved impossible to apply consistently. Decisions made at that time continue to reverberate today. Political boundaries were imposed where none had existed before, often in defiance of physical and human geography. Well-established regions like the Punjab, Bengal, and later Kashmir were split on ethnic grounds, but this left vital tributaries and headwaters under Indian control; this fact has contributed to several major wars and many skirmishes over the years.

The new lines also left minorities surrounded by hostile people of a different faith. Violence escalated, causing many to flee to join their coreligionists. About eight million Muslims moved to Pakistan, and a similar number of Hindus and Sikhs to India. In the process, up to two million men, women, and children were killed in intercommunal riots and massacres. Mahatma Gandhi, the most prominent figure in the independence movement, was assassinated by a Hindu extremist opposed to his policy of tolerance toward all religious groups.

Not everyone left, however. Even today India's population is 11 percent Muslim (around 70 million people), while Pakistan's is nearly 2 percent Hindu (1.5 million).

One obvious anomaly of the new order was the divided Muslim state of Pakistan. Its western portion was largely defined by the Indus River and its tributaries, while its eastern portion was carved out of the massive delta at the mouth of the Ganges and Brahmaputra rivers. Yet the two were separated by 1,000 miles (1,600 km) of Indian territory.

As if this were not enough to jeopardize the well-being of the infant state, it soon transpired that Pakistan's two sets of inhabitants had little in common other than their religion. While the west spoke Urdu (among other languages), grew cotton for export, and ate wheat and mutton, the rice- and fish-dependent east spoke mainly Bengali and produced mainly jute. East Pakistan was more populous, poorer, and more prone to famine, typhoons, and flooding.

The Pakistani seat of government, first located in the western port city of Karachi, was moved deep inland to the newly built Islamabad. To the East Pakistanis, it seemed distant and out of touch. Discontent was further fueled by West Pakistan's attempt to impose Urdu as the national language. Eastern moves for independence were opposed through a ruthless campaign that left one million dead and drove an estimated eight million refugees into India to escape the slaughter. Supported by India, the independent republic of Bangladesh was born out of East Pakistan in 1971.

The Indian government, on the other hand, has recognized some of the subcontinent's social divisions and created states based mainly on linguistic regions. Such changes have not satisfied all minorities. Sikhs and Kashmiris in particular still campaign for independence. Nevertheless, given the size and diversity of its population, India's unity is impressive. Hinduism is a strong binding force, and the strength of the country's leaders and their flexibility toward minority groups have persuaded many people that change can occur within, rather than outside, the federal system.

As many as 30 million people left their homes at the time of the partition of the Indian subcontinent. Muslims from India and Sikhs and Hindus from newly created Pakistan trekked hundreds of miles in a search for religious security.

This photograph was taken in September 1947, barely a month after the British had officially quit India. At the time, Old Delhi – a traditionally Muslim city in a region now designated as Hindu – housed 100,000 Muslim refugees. Many sought sanctuary in the old fort of Purana Quilla, shown here, while they awaited the opportunity to make the long journey to Pakistan.

The map gives some indication of the political and religious diversity of the Indian subcontinent. Before partition, tiny areas were controlled by France and Portugal and pocket-size kingdoms were ruled by independent princes, but most of the country was under British rule.

Kashmir

As the Hindu leader of the predominantly Muslim state of Kashmir vacillated about whether to join India or Pakistan, Pathan tribesmen, supported by Pakistan, invaded. Kashmir appealed to the Indian army for help.

The brief war was ended in 1949 by United Nations intervention, and the ceasefire line effectively ceded the northwest to Pakistan. A plebiscite was to follow, by which, given the Muslim majority, Kashmir would almost certainly have passed to Pakistan, but it has never taken place, and the state remains divided as in 1949. Kashmir is still the principal source of discord between India and Pakistan.

Punjab

The Sikh religion was created about 500 years ago to unite warring Hindus and Muslims into a single faith. Most Sikhs are found in the state of Punjab, where their most holy shrine, the Golden Temple at Amritsar, is located.

At independence the Sikh community chose to be part of India, but they were soon dissatisfied with their lack of autonomy, and in 1966 a separate Sikh state was carved out of the Punjab.

These concessions have not satisfied all Sikhs, who continue to campaign for a fully independent nation-state, to be called Khalistan. In 1984, Indian troops violated the Golden Temple, leaving hundreds dead, and in revenge Sikh activists assassinated the Indian premier, Indira Gandhi. The fiercest fighting and greatest bloodshed since independence ensued, with 2,000 Sikhs being killed and 30,000 made homeless in Delhi alone.

PAKISTAN
Islamabad
Karachi
Kashmir
Himachal Pradesh
Lahore
Punjab
Haryana
New Delhi
Uttar Pradesh
Rajasthan
BANGLADESH
West Bengal
Calcutta
Bihar
BURMA
Gujarat
Madhya Pradesh
Maharashtra
Bombay
Orissa
Andhra Pradesh
Karnataka
SRI LANKA
Colombo
Madras
Tamil Nadu
Kerala

Under British rule
Under Indian rule
Under Portuguese rule
Hindu
Muslim
Sikh
Modern state boundaries
Under French rule

Tamils of Sri Lanka

Ever since Sri Lanka became independent from Britain in 1948, its political relations, both internally and with India, have been complicated by the Tamil minority, which composes nearly 3.5 million of Sri Lanka's nearly 20 million people. Subject to a variety of discriminatory measures, including the imposition of the Sinhalese language and Buddhism, the predominantly Hindu Tamils have demanded an independent state within a federation.

Intermittent violence exacted some concessions from the Sri Lankan government, but the heavy-handed treatment of Tamils by security forces caused young Tamils to turn to terrorism, forming groups such as the "Tamil Tigers." At the same time Sinhalese terrorist groups emerged and the situation became increasingly polarized.

In an effort to reduce the violence, in 1987 Sri Lanka met many Tamil demands and even allowed the Indian army to maintain a peacekeeping force on the island. However, this force seems only to have inflamed the rival groups, and many thousands died in attacks and counterattacks.

When the army withdrew in April 1990, nothing had been resolved. A 14-month truce between the Tigers and the government forces came to an end two months later with the deaths of over 600 people in the first few weeks of renewed hostilities.

Bangladesh

Overcrowded, impoverished, and disaster-prone, Bangladesh has boundaries that are practically defined by elevation.

Its tropical rainforest has been cleared to create farming land to feed its population. Bangladesh chiefly occupies the delta where the Ganges, Brahmaputra, and Meghna rivers converge. Its people are vulnerable to flooding, typhoon damage, and waterborne diseases.

For nearly 25 years from 1947, Bangladesh formed East Pakistan. When it became plain that the center of political and economic power had consolidated itself in West Pakistan on the other side of the Indian subcontinent, opposition movements such as the Awami League emerged and thrived. The League won an overwhelming electoral victory in 1970, but were not allowed to take their seats. Civil war erupted and in 1971 an Indian-backed government-in-exile declared independence.

NATIONALISM TRIUMPHS

Has Poland finally taken shape?

During the first half of the 12th century, the German emperor Lothair advanced his frontier to the banks of the Oder-Neisse rivers and forced the Polish monarch Boleslav the Wry-Mouthed into a humiliating submission. Defeat led to a process that was to become familiar over the centuries in the shifting entity known as Poland: disintegration.

The Nazis did not stop at the Oder-Neisse. Conquest and colonization led them along the southern shore of the Baltic and deeper into eastern Europe. They did not withdraw behind the river line until they were defeated in 1945. The USSR annexed territory that had formerly been Polish, and in compensation Poland was given 40,000 square miles (100,000 square km) of Germany, including most of East Prussia.

Millions of German refugees fled westward, forming a political group powerful enough to cause Chancellor Kohl of West Germany to hedge over whether a reunified Germany would promise to honor the Oder-Neisse frontier. But the new Germany that emerged in 1990 showed no desire to redraw the postwar maps, and one of its first acts was to give formal recognition to the border. History had taken a long time to complete its full circle. For the first time in its 1,000-year history, Poland could regard its western flank as permanently secure.

Poland's curse is that it was for many centuries a frontier land whose enemies could come from any of the four points of the compass: Turks from the south, Mongols and Russians from the east, Swedes from the north, and Germans from the west. Apart from the Carpathian Mountains in the south, it had no natural boundaries – and no natural defenses. Its heartland, with Warsaw at the center, is in rich agricultural lands that for a while in the 17th century were Europe's breadbasket. Terrain of that kind is ideal for armies to tramp across, not too hilly and with full granaries on the way.

Despite its vicissitudes, Poland's cultural and political development was precocious and at times brilliant. Catholicism has always been the core in which Poland's nationalism has survived and fought back, but it has been a markedly more tolerant country than most of its neighbors. During its fragile 18th-century age of reason, it had the first codified constitution in Europe and the first department of public education. More than half the world's Jews lived in Poland; peasants were emancipated by landowners and hundreds of thousands of Germans and Russians migrated to what had come to be regarded as a land of the free.

But a constitution that enshrined the people's sovereignty did not find favor with Catherine the Great of Russia. Having installed her lover, the reform-minded Stanislaw Poniatowski, on the throne in 1764, she ordered troops into Poland in 1794 to destroy his reforms. At that

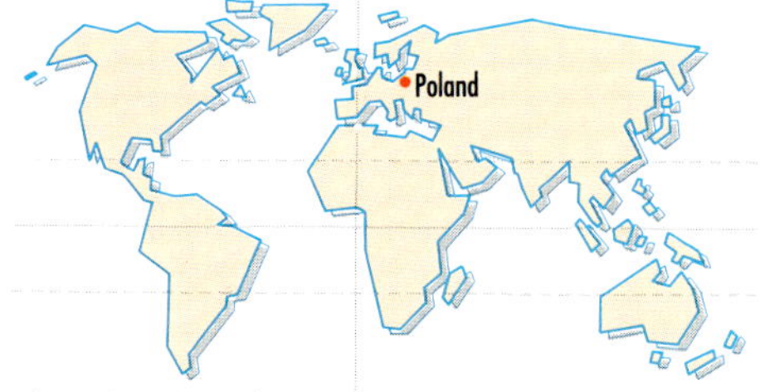

time, Russia and Prussia were the dominant powers in eastern Europe, and, together with the Austrian Empire, they partitioned Poland three times in 30 years at the end of the 18th century.

With its dismemberment in 1795, Poland disappeared as an independent state until 1918. Its 19th-century history is characterized by sporadic revolts, suppression, and mass migration, much of it to the U.S.

So much bottled-up nationalism fed by dreams of ancient glory led to the Poland that emerged after the First World War, which was considerably larger than anticipated by the postwar Treaty of Versailles, thanks to territory wrested from the Soviet Union. In addition, the treaty gave Poland a corridor to the Baltic that divided East Prussia from the rest of Germany. In due course, this gift provided Germany with the pretext for the invasion that started the Second World War – and led to another partitioning of Poland.

The war and 40 years of subjugation to a stifling Communist regime that was finally removed at the end of the 1980s have left marks that are likely to remain well into the next century. Poland's struggles are no longer concerned with frontiers or national survival, but with reviving the economy, cleansing pollution, and stabilizing its new democracy and its institutions. But no country better illustrates the dictum that geography is fate.

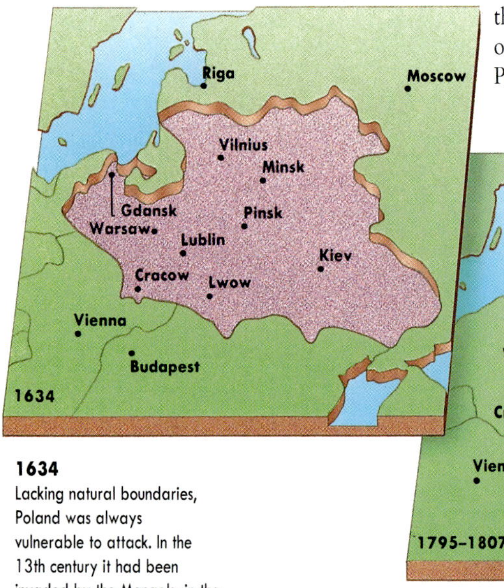

1634
Lacking natural boundaries, Poland was always vulnerable to attack. In the 13th century it had been invaded by the Mongols; in the early 17th century it was simultaneously at war with Sweden, Russia, and Turkey. Victories in all these conflicts meant that in 1634 Poland's territory and status as a European power were at their zenith.

1795–1807
As a result of the intervention of Catherine the Great, Poland effectively ceased to exist throughout the 19th century, despite various attempts at resurrection. Although the Poles retained day-to-day administrative control in the Russian area, repressive rule in the provinces controlled by Austria and Prussia fostered a spirit of nationalism that has characterized Poland ever since.

1812
Napoleon wrested part of Poland from Prussia in the course of his expansion across eastern Europe and briefly established the Duchy of Warsaw. His defeat at Moscow in 1812 (commemorated in Tchaikovsky's famous overture) briefly gave the duchy over to Russian control.

1815–74
The Congress of Vienna, which signaled the end of the Napoleonic Wars, created the Kingdom of Poland. Although the kingdom retained its own system of government, it officially formed part of the Russian Empire. An uprising in 1830 was firmly suppressed, but the secret nationalist movements remained active

Even as a Soviet satellite, Poland had one of the largest percentages of Catholics in the world, and the church continues to exercise political influence.

Poland's joy at the election of Karol Wojtyla as John Paul II, the first ever Polish pope (and the first non-Italian since the 16th century), can be seen by the crowds that greeted him when he visited his homeland in 1983. Two million people crowded into the main square at Katowice, one of Poland's major industrial cities, to hear him pronounce that the right to join a trade union was God-given.

Trade unions in Poland had to be approved by the state and the Communist Party, until the Independent Self-Governing Trades Union, known as Solidarity and led by Lech Walesa, came into being in 1980. Strikes at the Lenin shipyard in Gdansk brought the first major concessions to a trade union movement in the Eastern bloc – the recognition of independent trade unions, the right to strike, the easing of censorship, and the release of some political prisoners.

Solidarity was subsequently suppressed, martial law declared, and Walesa spent nearly two years in prison. World recognition of his achievements came when he was awarded the Nobel Prize for Peace in 1983. In Poland itself he already had a huge following, and at the end of 1990 he became the first president to be elected by the Polish people since the Second World War.

and insurrections continued at intervals throughout the 19th century. The January Insurrection of 1864 was also unsuccessful, and although it led to the liberation of the Polish peasants (following that of the Russian serfs in 1861), it also encouraged the further "Russification" of the Kingdom of Poland.

1921–39
At the end of the First World War, the Treaty of Versailles transferred former Prussian territories from Germany to Poland. It also created the "Polish Corridor," which gave landlocked Poland access to the Baltic Sea via the port of Gdansk. Disputed territories to the east led to conflict between Russia and Poland over the next two years. In 1921, the Peace of Riga established boundaries that remained inviolate until Hitler invaded Poland on September 1, 1939.

1939–45
The advance of Hitler's armies across the Polish frontier precipitated the Second World War. Poland disappeared once again, six million of its citizens were exterminated and much of the land was laid to waste. Poland was caught in the crossfire and suffered enormously.

1945–present
Poland's western frontier was established by an agreement between the victorious powers in August 1945. Only East Prussia, a territory taken from Poland in the 17th century, was restored to Russian control; otherwise the land east of the Oder-Neisse Line was once again part of Poland. Two weeks later, the Polish-Soviet border was established more or less along the Curzon Line, a division suggested in 1919–20 during the earlier conflict between the two nations.

TINY NATIONS
Independence at a price

SCATTERED ACROSS THE WORLD, THERE are about 40 independent states with populations below one million; more than half of them have fewer than 100,000 people. And there are nearly as many tiny countries that are, for the time being at least, under the control of major powers.

Although members of the latter group cannot be counted as states in their own right since they have few, if any, powers of self-determination, some are so geographically distinct that they may well break away in the near future – like France's rebellious "overseas territory" of New Caledonia.

Not all states with small populations are tiny in area. Greenland, for instance, has only 54,000 inhabitants but is larger than Mexico (which has nearly 90 million), or France, Spain, Britain, Italy, and the two Germanies combined (with a total population of 300 million). But at the other end of the scale, Monaco is less than 1 square mile (2.5 square km) in area, and the Vatican City covers one-sixth of a square mile (less than half a square km).

Many of the micro-states are islands or groups of islands located in the Caribbean or the Pacific. Almost all were for some centuries part of European empires. But with the general disruption of these empires during and after the Second World War, and with the active support of the United Nations, many were able to become independent.

How can they survive? In most cases, only with the greatest difficulty. And all have to exploit their unique geographical endowments to provide a service for larger states.

Some have natural advantages from which they can profit. The Seychelles and the Maldives are beautiful and therefore attract high-spending tourists, who flock to their palm-fringed beaches in search of sea and sun. A few possess abundant mineral deposits, or have the requisite climate and soils to grow cherished exotic crops. A handful of states thrive by virtue of the useful – and discreet – financial services they can offer, like the Turks and Caicos Islands, and above all the Cayman Islands. Yet others rely on the simple fact of their uniqueness. Tuvalu and Kiribati make most of their income from issuing postage stamps, which are highly prized by collectors.

Many, particularly those in the Pacific, rely on aid from larger nations, notably the U.S., Britain, and France. But in return they have had to agree to the siting on their territory of military bases, nuclear testing grounds, or hazardous chemical dumps. The local people may not always be happy with the situation, but they have little alternative.

What is the future for these tiny nations? However anomalous in the modern world, some seem durable enough, at least as long as they can continue to provide the service to the rest of the world that sustains them at present. But many are faced with the pressures of rapid population growth, limited economic potential, and environmental threats.

As populations rise, it becomes difficult to provide sufficient food, water, and energy to meet demands. And tourists can add to the problem. Some islands are rapidly running out of natural sources of fresh water, and desalination plants may be more expensive to install and run than any conceivable income to be gleaned from tourism.

Energy needs are also running ahead of supply. Currently most energy is produced by diesel engines, but diesel oil is expensive. Fiji's rivers provide hydropower, but few other islands have suitable falling water. Many are pinning their hopes on renewable sources like solar, tidal, wind, or wave power.

But all seem beset by some difficulty or other – the tidal range is very small in the middle of the Pacific, and storms are an ever-present danger to any generating equipment. Above all, the cost of the equipment may be prohibitive, and the technical sophistication required to run it may be lacking.

Some tiny island states have anticipated huge windfalls from valuable minerals such as cobalt, nickel, manganese, platinum, and oil found beneath their waters. However, none of these minerals is as yet in such short supply as to warrant the expense of mining the ocean floor. The licensing of Japanese and U.S. fishing fleets has also turned out to be less lucrative than at first expected, with local waters being rapidly fished out.

But perhaps the most serious threat of all comes from global warming. Every 1-foot (30-cm) rise in the sea level means that 100 feet (30 m) of beach disappears. Some islands – Kiribati, for instance – are likely to disappear altogether.

POCKET-SIZE PRINCIPALITY

Not all of the world's tiny states are tropical islands. There are six in Europe: Andorra, San Marino, the Vatican, Liechtenstein, Luxembourg, and Monaco.

Even without the ubiquitous tourists, Monaco is three times as congested as Hong Kong. Skyscrapers climb improbably steep slopes, and land is being reclaimed from the sea in an effort to relieve pressure on space.

Monaco's considerable wealth was obvious in the ceremony that attended the funeral of Princess Grace in 1982, shown here. Her husband, Prince Rainier, is a constitutional monarch, ruling over a pocket-size principality whose orchestra is larger than its army. There are a mere 4,500 true Monegasques, and another 24,000 inhabitants, who are predominantly French and Italian. Should Rainier's dynasty, the Grimaldis, die out, the principality will revert to France.

The benign Mediterranean climate and the gambling facilities of Monte Carlo, where roulette was invented, have been attracting the rich to Monaco for more than a century. Its tax laws make it equally attractive for long-term visitors. Monegasques themselves do not pay any taxes, but neither are they allowed to gamble in the casino.

San Marino

Only 24 square miles (61 square km) in area and with a population of 24,000, San Marino nestles in the foothills of the Italian Apennines. Yet it has been an independent nation since the 13th century. Its capital, also called San Marino, is a fortified town 2,000 ft (610 m) above sea level. Its success nowadays is based on tourism.

The *Sammarinesi* form a tightly knit community, with much intermarriage. Their judges and police are, therefore, hired from Italy so that family relationships do not interfere with the course of justice.

Seychelles

These islands are one of many contenders for the title of "the original Garden of Eden." Their greatest source of revenue is tourism.

Strategically situated on the trade route between the Cape of Good Hope and India, the Seychelles were settled by the French and the British and became a republic in 1976. The first president was toppled by a coup, and there have been further attempted coups.

The islands still belong to the British Commonwealth; they are also part of the OAU, and Soviet-made missiles form part of their national defenses.

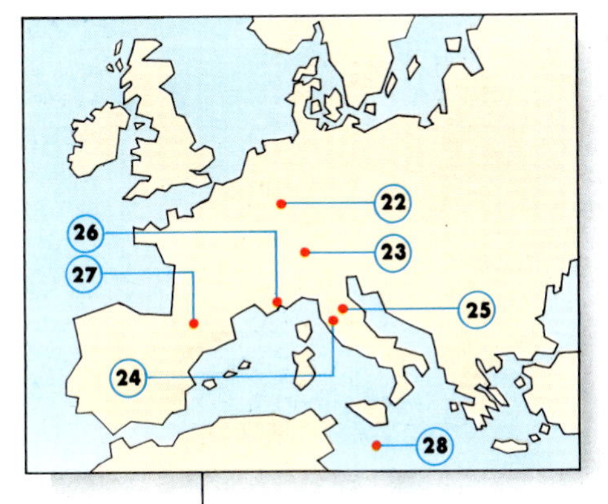

1 Greenland
2 Iceland
3 Bahrain
4 Qatar
5 Bermuda
6 Cyprus
7 Gambia
8 Cape Verde Islands
9 Belize
10 Suriname
11 Guyana
12 Guinea-Bissau
13 São Tomé and Principe
14 Equatorial Guinea
15 Maldives
16 Seychelles
17 Comoros
18 Swaziland
19 Christmas Island
20 Cocos Islands
21 East Timor
22 Luxembourg
23 Liechtenstein
24 Vatican City
25 San Marino
26 Monaco
27 Andorra
28 Malta
29 Bahamas
30 St. Kitts and Nevis
31 Antigua and Barbuda
32 Cayman Islands
33 Dominica
34 St. Lucia
35 St. Vincent
36 Grenada
37 Barbados
38 Northern Marianas
39 Federated States of Micronesia
40 Marshall Islands
41 Kiribati
42 Wallis and Futuna
43 Solomon Islands
44 Nauru
45 Tuvalu
46 Vanuatu
47 Western Samoa
48 Norfolk Island
49 Fiji
50 Tonga

Cayman Islands

A British Crown Colony rather than an independent nation, the Caymans have a population of only 20,000 and an area of 100 square miles (260 square km), but they have become famous worldwide for their attractive offshore banking arrangements. More than 500 major banks and trust companies are registered here, where there is no direct taxation and where confidentiality is guaranteed.

Although revenue from these activities makes a substantial contribution to the budget of the Cayman government, 70 percent of the gross national product derives from tourism. Demand fueled by this industry makes property on Grand Cayman among the most expensive in the world.

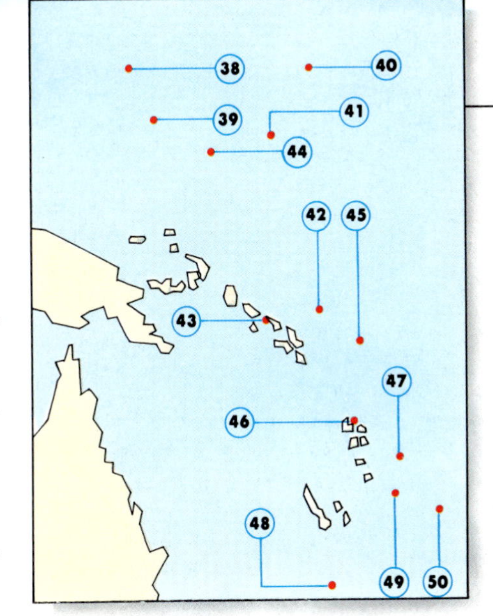

Nauru

At 8 square miles (21 square km), this is the world's smallest republic. There are no taxes, no health charges, no school fees, and food imports are heavily subsidized.

Nauru's wealth comes from its exports of phosphates to be made into fertilizer. But the phosphate supplies are running out and the soil has been stripped away to get at the bird guano deposits beneath.

With only a few years of income remaining, the mines must be landscaped and made fertile again. Otherwise the Nauruans will have to leave the island altogether.

THE NEW COLD WAR

Conflict and collusion in Antarctica

JUST AFTER NOON ON FEBRUARY 1, 1952, a dozen bursts of machine-gun fire rang out over Hope Bay, at the northernmost tip of Antarctica's great peninsula. A small group of British scientists, just arrived with materials to rebuild their fire-damaged station, were surrounded by a platoon of armed Argentinians. One of the Argentinians, an officer brandishing a cocked pistol, ordered the scientists to return to their vessel. He had been instructed, he said, to prevent them from building a base, "using force if necessary."

It was the first and only time that such weapons have been used for offensive purposes on the continent. The issue was one of sovereignty over the peninsula, a kind of handle on the great pan of Antarctica that Britain had long claimed as its own but that projects to within 800 miles (1,300 km) of Argentina (and Chile, too). Britain's stake was based partly on its control of the Falkland Islands, over which Argentina also claimed sovereignty.

Diplomatic endeavors prevented the incident from developing into anything more serious. Nevertheless, it was not forgotten. The conflict over territorial rights in the region continues down to the present.

Antarctica is the coldest, windiest, driest, and most remote of the seven continents. It is centered on the South Pole and lies almost entirely within the Antarctic Circle. Conditions are so harsh over its 5.5 million square miles (14.2 million square km) that there are no permanent inhabitants, only visiting scientists staying in isolated research stations, a small but growing number of tourists, and, on at least one occasion, a pregnant woman brought from Argentina to give birth to a true Antarctican. Even at the height of summer there are only 4,000 or so people on the entire continent.

Yet although few animals or plants are able to survive on the surface, the surrounding seas support huge amounts of shrimplike krill and associated fish, whales, and birds. In the future, these may be a vital source of food for the rest of the world.

No less important, the ice cap and surrounding continental shelf are thought to store great reserves of minerals. Although local conditions make any exploration, let alone exploitation, extremely difficult, guestimates of total mineral wealth are confidently proclaimed. In 1984, for example, the U.S. claimed that Antarctica's total recoverable offshore oil reserves could exceed 45 billion barrels – a quantity which, if true, exceeds Alaska's. Others say that natural gas, coal, and other ore deposits are likely to be immense. In fact nobody can tell yet, and such expectations may turn out to be wildly optimistic.

Antarctica's paradoxical geographical qualities – hostile, yet enticing; isolated, yet vulnerable – go a long way toward explaining the region's current problems. These continue to revolve around the issue of who – if anybody – should own and control the continent.

This is not a simple dispute. Seven nations have long-standing territorial claims to Antarctica, based chiefly on grounds of discovery or geographical proximity. Since these treat the continent and its associated islands as a circular cake to be cut up and portioned out, the subdivisions are strict geometrical sectors. By 1959, five other nations had mining or research stations there, and the threat of international rivalry prompted a far-reaching and effective compromise among the 12.

The Antarctic Treaty, drawn up in 1959 and ratified in 1961, declared that "it is in the interest of all mankind that Antarctica shall continue forever to be used exclusively for peaceful purposes." It froze all claims in the region south of latitude 60 degrees south,

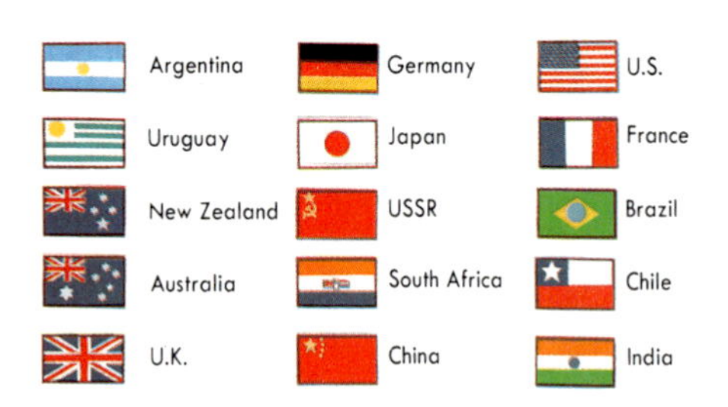

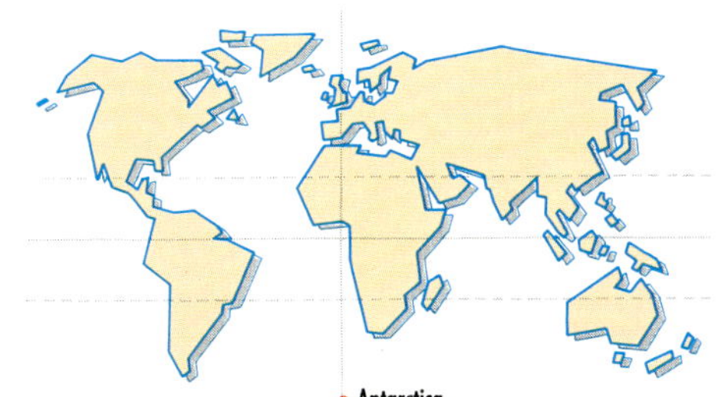

Nevertheless, all previous guarantees for the future of Antarctica are currently open to revision since the 30-year duration of the original treaty elapsed in 1989.

Because of the harshness of Antarctica's environment and the political uncertainties surrounding it, the continent has to a great extent been left in a pristine state. This is changing. Exploitation of the apparently superabundant krill, fish, and whales could provide an important source of protein for human, or more likely animal, consumption. On the other hand, we are so remarkably ignorant about the interactions within the local food chains that it is all too possible that catches will devastate already endangered species like the baleen whale.

On land, oil wells and mines could destroy breeding colonies of seals and birds and irreparably damage the sparse vegetation – a footprint in the Antarctic moss remains for decades. Antarctic storms, icebergs, and pack ice will increase the likelihood and worsen the effects of oil spills, blowouts, and leaks.

Unfortunately, human intervention by both scientists and tourists threatens scientific work, too. Some scientific stations resemble nothing so much as a third world shantytown – corrugated-metal buildings and old storage tanks, surrounded by garbage dumps. At the other extreme, McMurdo Sound is home to Antarctica's largest town, with 1,000 or more summer residents, an immense airfield, four bars, a bowling alley, a gym, and so on. Nevertheless, in 1987 it was found to dump untreated sewage in the sea and burn trash in open-air pits.

The constant arrival and departure of planes means that local snow has become too polluted for scientific study. Elsewhere, tourists are disturbing the breeding patterns of local species, carving graffiti into rocks, and leaving litter. Today, many scientists are demanding that every activity in Antarctica be subject to a full environmental audit of its likely costs and possible benefits.

Little wonder, then, that there are calls, especially from that majority of nonaligned nations who are not party to the 1959 treaty, for Antarctica to be designated a *terra communis*, a common land or world park held by the U.N. on behalf of everybody.

Five nations – Australia, Britain, France, New Zealand, and Norway – have long recognized each other's claims to parts of Antarctica, while two others – Argentina and Chile – both demand roughly the same territory as Britain. Another 15 percent of the continent – the sector facing out toward the eastern Pacific, an area larger than Spain, France, Italy, and Germany combined – is unclaimed.

By 1959, five more states – Belgium, Japan, South Africa, the U.S., and the USSR – had demanded the right to exploit the region's mineral and marine resources, although they advanced no territorial claims. Since then, other nations have also established research stations there.

The map shows the foreign claims to sectors of Antarctica and the scientific stations spread across the continent.

Global politics and economics may underpin all scientific endeavors on Antarctica, but there is no doubt of the interest of the region for science. Because of its distance from any industrial development, Antarctica is a unique control point from which to monitor the planet. It was here, for example, that the hole in the ozone layer was first detected. Ice cores, too, are of great significance, since they contain historical records against which current climate shifts and pollution can be judged.

This satellite photograph was taken by remote sensing equipment; 23 separate images have been put together to produce an overall view of the entire continent.

encouraged scientific cooperation, and granted full freedom of movement. Perhaps most significant of all, it prohibited any kind of military operation. Antarctica became the world's first nuclear-free zone.

In time, many other countries have become signatories to the treaty, including some – among them Brazil, China, India, and Uruguay – that have mounted sufficient research ventures to earn the right to special consultative status, which guarantees full participation in Antarctica's affairs. Since then a number of new treaties have been agreed to, including one in 1980 governing marine fishing rights in a rather larger area within the Antarctic Convergence, the shifting line that separates the near-freezing water of the Antarctic from the warmer waters to the north.

In 1988 at Wellington, New Zealand, 20 treaty nations agreed to forbid any future oil drilling unless sanctioned by every other signatory. Australia and France, however, have called for a perpetual ban on mineral extraction.

Conflict and Cooperation

The revolutionary tide in world affairs that was so evident in the 1980s has swept unabated into the last decade of the century. Western Europe, the epicenter of two world wars, quickened its pace toward political and economic union. Germany was reunited. The Soviet Union conceded defeat in the Cold War and was reduced to begging for food parcels as its empire began to crack at the seams under the stress of economic crisis. China's totalitarian regime remained shaken and humiliated by the democratic upwelling in Tiananmen Square. Japan, with an economy second in size only to that of the U.S., became the most powerful force in world trade and investment, and also the largest donor of aid to the third world. The Kuwaiti crisis saw unprecedented cooperation between the United States and the Soviet Union in the U.N. Security Council.

The trends within these developments are in some ways contradictory. Western Europe's moves toward a greatly strengthened political and economic union were not enthusiastically endorsed by every member of the Economic Community, but on the whole they indicate a willingness to consider a shift of power from sovereign institutions to centralized authorities. The Soviet Union, on the other hand, demonstrated a reverse process as *glasnost* and a drop in already low living standards produced a flare-up of long-repressed nationalism within the 15 republics. Four of them – the three Baltic republics and Georgia – sought independence from the Soviet Union, and the rest loosened the Kremlin's centralized control and debated their futures.

Signs of dissolution could be seen in India, too, the world's second most populous country as well as its largest democracy. In this case, though, the worst conflicts were between castes and religions. Militant Hindus clashed with Muslims, and Sikh extremists in the Punjab waged a merciless fight for their own independent state, Khalistan. Equally dangerous to India's cohesion was the violent reaction of the urbanized Hindu higher castes to plans to end their monopoly of government clerical and administrative jobs.

The ending of the Cold War enabled the Conference on Security and Cooperation in Europe (which brings together all the NATO and Warsaw Pact countries as well as a number of neutral and nonaligned states) to move rapidly toward agreement on reducing military strengths in Europe. A similar effect was seen in the U.N., which for the first time in its history was able to work in the harmonious, cooperative way envisioned by its founders.

As far as the European Community was concerned, its economic success enabled France and Germany, the EC's core states, to marshal a demand for rapid moves toward economic and political union. At the same time thoughts turned toward the likely demise of NATO (created in 1949 under United States leadership to confront the Soviet Union) and the need for a new defense arrangement based on the Community's members alone.

Austria, Sweden, Norway, Finland, and even Switzerland either made overtures toward the Community or gave serious thought to doing so. The Community is essentially western European, but there is nothing in the Treaty of Rome to say it should confine itself to countries west of a certain line. It had reached beyond western Europe's confines to attach Greece. If Greece was eligible, why not Turkey, Cyprus, Malta, and Morocco, or indeed, the former Soviet bloc countries of Hungary, Czechoslovakia, and Poland?

The idea of commonwealths of nations, even of a world commonwealth, is not new. It was the inspiration behind the founding of the League of Nations, which preceded the United Nations. The commonwealth created out of the independent states of the British empire was shaped in the same philosophical mold. The British Commonwealth at its zenith had preferential tariffs, war cabinets, common defense arrangements, and a recognized center of authority, but it became too large (50 states) and too diverse in almost every respect to have much political or economic meaning.

Somewhat similar problems face the European federalists as

they hurry toward building a binding political and economic structure. The traditionally Atlanticist British fear a federal Europe that would weaken its ties with English-speaking North America while spending more and more of its taxpayers' money on reconstructing eastern Europe for the benefit of German exporters. One result of a federated Europe might turn out to be a polarized capitalist world, with rival centers of power in Europe, Japan, and the U.S. The alternative vision is of a loosely knit commonwealth of sovereign states potentially open to all between the Atlantic and the Urals. The criticism in that case is that it would lack the political muscle to match its economic weight.

AT THE BEGINNING OF THE 1990S THE WORLD IN some important respects looked better and safer than it had for a great many decades. A new era had opened in eastern Europe. The threat of a catastrophic nuclear war between the superpowers had all but vanished. Viewed from a different perspective, though, the scene was not so encouraging. In much of Africa and Latin America living standards were falling, not rising as they were in the developed nations.

The World Bank predicted a growth of more than 80 percent in the populations of the poorest countries between 1987 and 2025. India's projected growth, for example, was from 798 million to 1,366 million, Pakistan's from 102 million to 286 million. AIDS, other diseases, war, and famine seemed likely to intervene before figures of that order were reached, but they pointed to the prospect of the north-south divide between poor nations and rich nations deepening.

Combined with the disruption likely to be caused by the global warming predicted in 1990 by the scientists of the Intergovernmental Panel on Climate Change, the demographic and economic trends suggested huge increases in the numbers of refugees and would-be migrants and more regional wars and civil strife. Although there was a new mood of optimism about the future of the U.N. and the ability of nations to cooperate, it was questionable whether the international community would be able to devise and implement an effective strategy for handling a world crisis of such unprecedented severity.

Events in the Gulf demonstrated clearly enough that developed nations would not be able to stand aside when third world disputes threatened their interests. Iraq's unsavory regime was propped up by the Soviet Union and the West in its war with Iran because it was in their interests to do so. The support continued even after Iraq's use of chemical weapons in war and against its own Kurdish minority. It was only when Iraq's leader, Saddam Hussein, annexed Kuwait and appeared ready to occupy Saudi Arabia's oil fields to obtain control of the world's largest reserves that Western nations were forced to end their support and confront him militarily.

The ending of the Cold War has made crisis management easier, but it has not meant the end of crises caused by threats to strategic and other interests. The criminal involvement of General Manuel Noriega in drug smuggling and the threat he posed to the security of the Panama Canal led to the U.S. invasion of Panama at the end of 1989 and the arrest of Noriega. In the new era in which third world instability replaced the Soviet threat, even the western Europeans, long reluctant to cooperate militarily outside NATO's European arena, began to think of creating a joint rapid reaction force to protect their worldwide strategic interests.

CRISIS IN THE USSR/1

The impossible empire

Although widespread fears of Russian expansionism used to abound, the modern Union of Soviet Socialist Republics (or Soviet Union) is no larger than the empire Lenin and the Bolsheviks seized from the tsars in 1917. In fact, it is a little smaller, though its influence has until recently extended into eastern Europe, China and Korea, Southeast Asia, and even farther afield to Cuba, Central America, and Africa. Yet two centuries ago Russia's empire stretched as far as California, and Russia only parted with its last North American colony in 1867, when it sold all of Alaska to the U.S. for a mere $7,200,000.

Nevertheless, the Soviet Union today is still by far the world's largest state – for the present, at least. Its 10,000-mile (16,000-km) land border links it to 12 neighbors, from Norway in the west to North Korea in the east, more than any other state. The size of Canada, the U.S., and Mexico combined, the Soviet Union covers 8.6 million square miles (22.3 million square km) or one-sixth of the Earth's entire landmass. It stretches 10,000 miles (16,000 km) east-west, or about halfway around the world, and about 3,000 miles (4,800 km) north-south. Leningrad, in the west, is no closer to the eastern port of Vladivostok than it is to New York.

The immense distances, together with the hostility of much of the physical environment – icy tundra, marsh, and coniferous forests in the north, deserts and mountains in the south and east – have made massive problems of settlement, resource exploitation, and internal communications. Even today there are few good roads in Asia, and trains take over a week to cross the country.

The Soviet Union's east-west orientation adds to its difficulties. The country stretches over 11 time zones, so when people in the Soviet Far East are arriving at work, Soviet Europeans are going to bed. This is an inordinate problem in a command economy, where all decisions have been made centrally in Moscow.

Given all these apparent drawbacks, why did Russia's empire grow so large? Basically, it had either to grow or face oblivion. From the tiny landlocked core area that surrounded Moscow in 1500, Russians needed to expand in every direction in search of agricultural land, goods to trade, and above all seaports through which to transport them. They also needed defensible borders, since their lands were regularly overrun by, among others, Vikings, Mongols, Poles, Swedes, French, and Germans.

Much of the initial expansion moved north, to unite other Russian principalities and to reach Arctic waters. By 1584, construction had begun on the White Sea port of Arkhangelsk (Archangel), which would open up trade with Britain and aid in the exploration of Siberia.

Access to the Baltic Sea itself was effectively thwarted by Poland, Lithuania, and Sweden until, in 1703, Peter the Great succeeded in building St. Petersburg (modern Leningrad) as his new capital and "window on the world."

Here, as elsewhere in the West, Russia's expansion, in spite of occasional advances into Poland, was generally limited by the numerical strength and technological sophistication of its European neighbors. In spite of every effort to acquire the latest military and industrial skills, Russia was always roughly a generation behind.

The Russian tsars had more luck with expansion when they sent their armies downriver, along the Dnieper, the Don, and the Volga toward the less well-defended south. By 1800, Russia had acquired the Ukraine, with its great grain belt and industrial potential, and gained access to the Black Sea.

During the next century it absorbed the area immediately beyond the Caucasus Mountains, between the Black Sea and the Caspian Sea, and the predominantly Muslim region of central Asia up to the borders of modern Turkey, Iran, and Afghanistan. As a result, Russia became the world's biggest exporter of grain, at least until the U.S. took over that position at the end of the 19th century.

Nevertheless, Russia regained the lead by 1914 because its rich black earth steppes are so

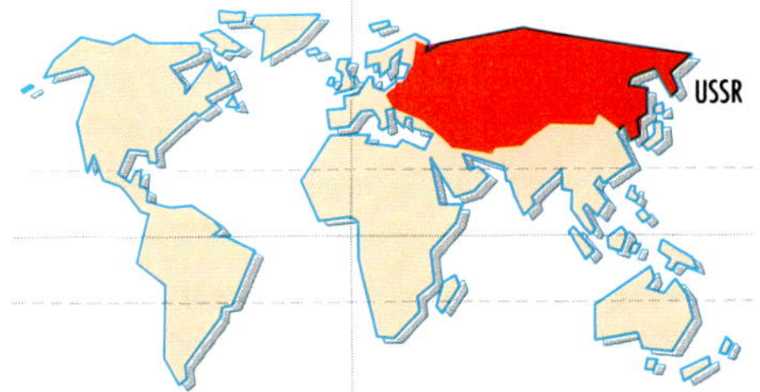

One of the USSR's greatest problems is communication between the various parts of its scattered realm. Fur trappers in the vast reaches east of the Urals suffered terrifying hardships, traveling overland from Moscow or by boat down the River Ob during the brief summer. Mobility was easier in the winter when rivers were iced over: transportation was by ski or sled, the hazards of waterfalls and rapids were reduced, and the nuisance of mosquitoes was lessened.

Transportation is of prime importance in a state the size of the USSR. A great railroad system was developed in the 19th century, with preferential rates for long-distance freight to aid delivery of grain to ports and frontiers. Inland waterways have been developed to supplement the rail and river network. The fact that most of the major rivers run north-south has combined with distance and harsh conditions to make travel from west to east extraordinarily difficult.

Hunger has always been commonplace, starvation not unusual. Yet until the end of the 19th century, Russia was the world's largest grain exporter. Much of the grain was grown on small peasant farms, and the state profited by taxing the peasants in grain. The liberation of the serfs and the introduction of communal farms did something to reduce the chronic food shortage, but productivity has remained low throughout the 20th century. This is due in part to unfavorable weather conditions and in part to limited availability of machinery and fertilizers.

The USSR's food crisis in the early 1990s also has its roots in political difficulties: the individual republics have been accused of !ocal chauvinism – withholding food that should, according to Moscow, be part of a state-controlled stock for widespread distribution.

much larger than their U.S. prairie equivalent – a fact that may have heightened U.S. hostility to the infant Communist republic after 1917. Some believe that, given better organization of production and transportation, the region could again become the breadbasket of the world.

Further expansion to the south was permanently ruled out. Although Russia has for centuries craved outlets for its military and merchant vessels in the Mediterranean or the Indian Ocean, Britain and, recently, the U.S. have always thwarted such hopes, partly by forging alliances with Turkey (which controls the vital straits through the Bosporus and Dardanelles), with Persia (modern Iran), and with Afghanistan.

The only direction in which Russia had more or less free passage was eastward into Siberia, across the Ural Mountains, the traditional boundary between Europe and Asia. Here the obstacle was not so much people – the lands were thinly populated, then as now, by scattered nomadic tribes who lived by hunting, fishing, and reindeer herding – as the environment.

Siberia is the coldest place on earth outside of Antarctica. Pioneers here – mostly trappers and traders, but later peasants, prospectors, and political prisoners – experienced an almost endless catalogue of difficulties, including mountains higher than any in Europe, forests, permafrost, marshes, mosquitoes, and waterways that flow north rather than east.

Yet having crossed the Urals in 1582, Russians had reached the Sea of Okhotsk at the edge of the Pacific by 1639. A little over a century later, in 1741, they laid claim to Alaska before descending the length of the west coast of America into California, then nominally under Spanish control. Their southernmost stockade and trading post was at Fort Ross, built in 1812. A mere 60 miles (100 km) short of San Francisco, it was 9,000 miles (14,500 km) from Moscow.

But even in the Pacific Russians were denied their dream of a year-round warm-water port. Vladivostok, acquired from China in the middle of the 19th century, is icebound for several months of the year. In addition, every attempt on the part of Russia to establish itself farther south, for example beyond Korea at the Chinese port of Port Arthur (modern Lu-shun) in 1898, was devastatingly resisted by Japan.

CRISIS IN THE USSR/2

The empire falls apart

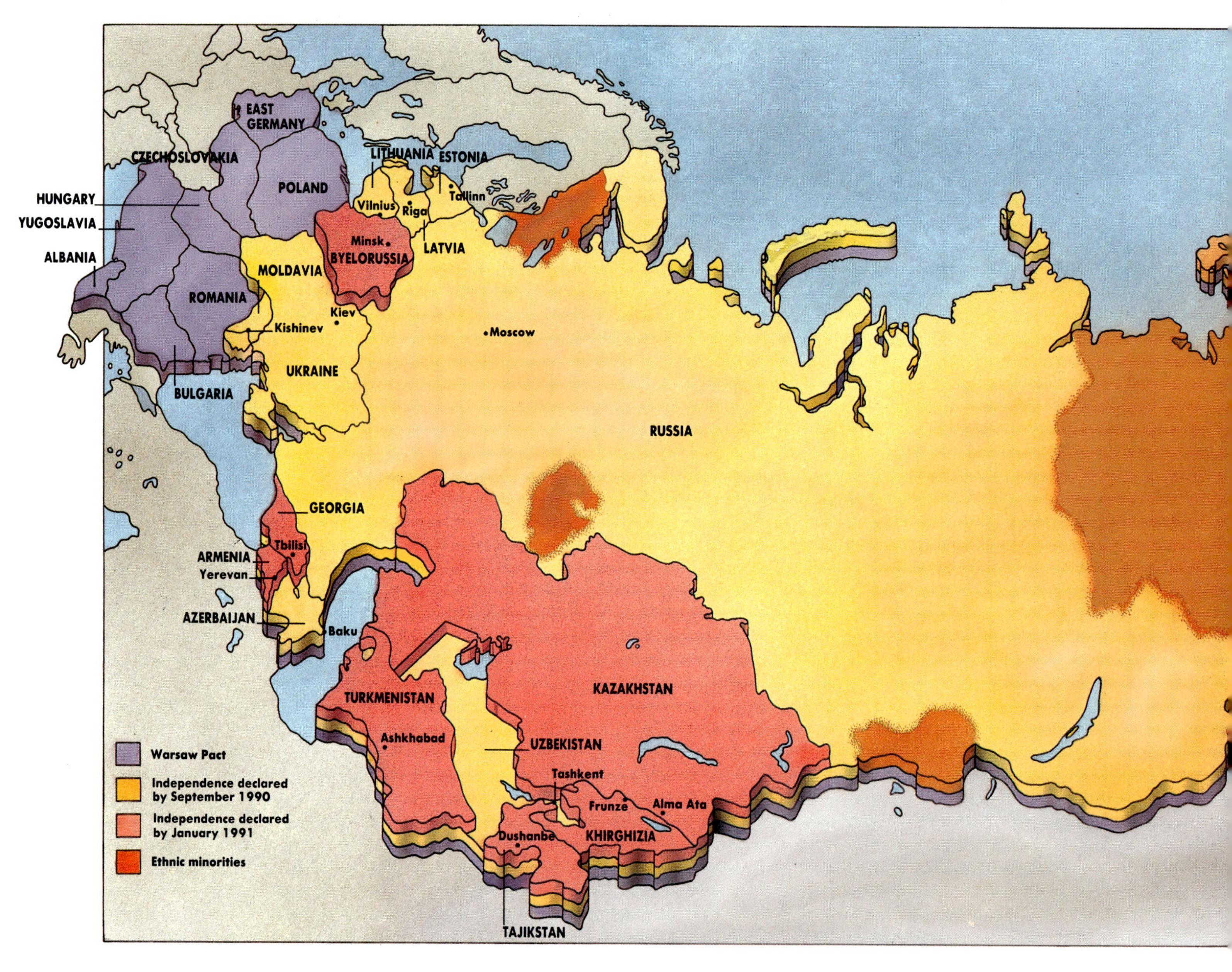

The Russian Soviet Federated Socialist Republic – usually just referred to as Russia – occupies three-fourths of the area of the USSR and is peopled by 60 nationalities, mostly of Slavonic descent. The people of two of the Baltic states – Latvia and Lithuania – are ethnically linked to the Russians, while the Estonians are more closely related to the Finns. All three states became part of the USSR only in 1940, so it is little wonder that they should feel alienated from the rule of Moscow.

Muslims make up 40 percent of the population of the USSR and are a substantial majority in 6 of the 15 republics, notably Azerbaijan. Their ethnic and cultural links with Turkey, Iran, and Afghanistan are stronger than those with Moscow. Many Armenians are Christians, which has brought them into conflict with both their Azerbaijani neighbors and their Communist overlords.

These are only a few examples of the many problems confronting the USSR at the end of the 20th century. It remains to be seen whether Moscow can maintain unity in the face of enormous geographic difficulties, the ever-present threat of famine, and growing moves for independence within the Soviet Socialist Republics.

IN SPITE OF ITS IMMENSE MINERAL, industrial, and agricultural potential, the Russian empire has always been weak. Spanning Europe and Asia, it has a diversity of peoples and environments that has simply been too great to manage. For centuries the empire seemed to be teetering on the edge of economic, administrative, and military collapse. Yet somehow it has survived, more or less intact, both under tsarist and Communist dictatorship. Today, however, things seem to be changing, probably forever.

When Lenin and the Bolshevik Party seized power in 1917, they skillfully used nationalist sentiment to gain support. In return they promised independence. The reality turned out to be very different. Arguing that the "right to divorce did not imply the necessity to divorce," Lenin and Stalin went on to suppress independence movements and even reannex areas that had succeeded in breaking away from the tsar's empire after the First World War. These included the Ukraine, Moldavia, and the Baltic states of Estonia, Latvia, and Lithuania.

Under Stalin, in spite of a rhetoric of self-determination, regional autonomy was often brutally suppressed. Power was concentrated in the hands of the Communist Party in Moscow.

In the period after 1945, expansion of the Russian economic core within the Soviet Union occurred at the expense of its peripheral republics. An obsession with heavy industry, gigantic energy and agricultural projects, along with the testing of nuclear weapons sowed the seeds of economic, environmental, and health disasters. Meanwhile, the movement of ethnic Russians into neighboring republics set up deep cultural tensions, which were aggravated by

The population of the USSR comprises 126 nationalities and as many linguistic groups. They are related to peoples as widely dispersed as Poles and Mongolians. The man and woman on the far right, whose features show their Asian ancestry, are from Tashkent; above is a Ukrainian worker; and right is a Lithuanian girl in national dress.

Moscow's insistence on Russification – the use of the Russian language and official opposition to religion.

In recent years the situation has changed dramatically, and Russia's empire is finally coming apart. The wonder is that it has taken so long.

Demographic changes mean that the Russian population no longer forms a substantial majority of the whole (there are 137 million Russians and 42 million Ukrainians in a total population of 281.7 million). The non-Russian population is growing at a much faster rate, notably Muslims in the Asian south, who now number around 55 million.

Meanwhile, the consistently poor performance of the economy, combined with food shortages, ecological disasters (the most infamous of which was the meltdown of the nuclear reactor at Chernobyl in 1986), and restrictions on personal freedom have fueled discontent. All that was needed was a spark, and that came in 1985 with the election to power of Mikhail Gorbachev.

President Gorbachev immediately set about withdrawing Soviet troops from Afghanistan and implementing the twin processes of *perestroika*, or economic liberalization, and *glasnost*, meaning openness in public affairs. But with the reduction of centralized party authority, ethnic, religious, and nationalist conflicts have been propelled to the fore.

Since 1988, smoldering rivalry between different ethnic groups has erupted in violence in Moldavia, Georgia, Armenia, Azerbaijan, Uzbekistan, and Kirghizia. Adjacent republics are fiercely contesting boundaries. Refugees are flowing across borders and even leaving the Soviet Union altogether.

By December 1990, all 15 of the Soviet Socialist Republics had claimed their sovereignty. Even Russia itself, the most populous of all the republics and by far the largest and richest (since it includes the Ural economic zone and Siberia), has recently declared its intention of placing its own interests above those of the Soviet Union as a whole.

Meanwhile, in eastern Europe, national independence movements that have succeeded in overthrowing repressive governments have transformed the entire balance of power between the Soviet Union and the politically allied nations of the West. As Russia's empire collapses, the future shape of the world, no less than that of the Soviet Union, is being irrevocably transformed.

EUROPE'S EASTERN SHATTERBELT

The fault line opens up

PITY THE SMALL NATION WITH THE geographical misfortune of living next door to a big power or, worse still, trapped between two big powers. For its destiny is either oblivion, swallowed up by its more powerful neighbor, or the role of pawn in a perpetual struggle between two grand masters.

Many such nations – Poland, trapped between Germany and Russia, or Mongolia, now split between China and the Soviet Union, or Uruguay, created to keep Argentina and Brazil from each other's throats – are forced to act as buffers, absorbing friction and setting limits to rival territorial ambitions.

Location places some nations in double jeopardy. They face all the strains associated with acting as buffers – including providing the battlefields on which the big power conflicts are played out – they also find themselves living cheek-by-jowl with equally weak states, prone to coup and countercoup, civil unrest, and economic instability. Such shatter zones – among them southern Africa, Southeast Asia, Central America, and the Middle East – are characterized by geographical, ethnic, and religious fragmentation.

Consider Czechoslovakia, deep within the great shatterbelt that runs through Europe from Finland in to Yugoslavia and Bulgaria. Czechoslovakia was custom built in 1919 to a design dictated by location and approved by the victorious powers in the First World War. As its name implies, it is essentially an association of two distinct Slavonic peoples – the western Czech speakers, mostly Protestants living in the old kingdoms of Bohemia and Moravia, and the eastern Slovaks, chiefly Catholics inhabiting Slovakia.

Czechoslovakia's boundaries were chosen with security in mind. To the south, the massive Austro-Hungarian Empire had been destroyed during the First World War, and the new, smaller states of Austria and Hungary, confined to their ethnic boundaries, provided little threat. Czechoslovakia would also be buffered from any Russian advance in the east by the anti-Communist democracies Poland and Romania.

Much more serious was the risk from Germany, should it adopt expansionist policies in the future. Against this possibility, the new Czech state was ringed to the north and west by a natural defensive wall, the great highland chain that includes the Carpathian Mountains.

But the architects of the new Czechoslovakia had not considered sufficiently the internal stress of ethnic grievances. The new state included three million Germans and many Magyars

Wenceslas Square, Prague, fall 1989: Czechoslovakians gather to celebrate the end of more than 40 years of Soviet rule in their country. Their mood is joyous, optimistic. The chiefly Czech reform movement that went underground in 1968 has reemerged to take power. This time they are confident there will be no tanks.

But the threat of economic chaos in the wake of decades of Soviet domination hangs over them. The picture is much the same in Poland, Hungary, Romania, Bulgaria, Albania, and Yugoslavia, the former buffer states that separated East from West.

In addition to internal unrest and a mounting refugee problem as ethnic minorities strive – often in vain – to have their status recognized, almost every state boundary in the region is bound to come under intense pressure.

(Hungarians), Ukrainians, Poles, Jews, and Romanies (Gypsies).

It coped reasonably well until the 1930s, when world depression set in. Then conflicts escalated. In economically backward Slovakia, many felt that they had exchanged one form of imperial domination for another. The more sophisticated Czechs had taken most of the important positions in government and industry – which therefore remained rooted in the West – and refused to allow much regional autonomy. Meanwhile, dissatisfaction was increasing among German speakers, who had lost power, prestige, and wealth at the hands of the largely Slav government.

External pressure also mounted. Hitler demanded that areas with more than 50 percent German speakers – like the Sudetenland, near the German border – should be annexed. The Czech government protested, but it was out of their control.

Then, as before and since, the life or death of small states was ultimately controlled by the big powers. Chiefly concerned to protect their own interests, Britain and France acceded to Hitler's demands, hoping thereby to appease him. At the same time, other areas of Czechoslovakia were also reapportioned, in the vain hope of satisfying diverse regional interests. Fragments went to Poland and Hungary. In 1939 Hitler took all that was left. Czechoslovakia ceased to exist, and the world was plunged into war.

After 1945, with most of its pre-1938 boundaries restored, Czechoslovakia found itself behind the Iron Curtain with a Soviet-style government. It had become, in effect, a buffer state between the USSR and western Europe, a position it was forced to endure until 1989.

1910
The major empires dominated eastern Europe. Czechs in Bohemia-Moravia and Slovaks in Austria existed reasonably peacefully under the wing of the Austro-Hungarian Empire. But to the south, nationalism was spreading, and the action of a Serb nationalist protesting against Austro-Hungarian rule (by assassinating the Austrian archduke) ignited the First World War.

1930
Bohemia, Moravia, and Slovakia had become one nation. The highly industrialized region around the Bohemian capital, Prague, with its great reserves of coal and iron ore, was united with the rich farmlands of Moravia and the Slovakian port of Bratislava on the River Danube.

Present day
Since the USSR annexed the Ruthenian corridor after the Second World War, it has had a common border with Czechoslovakia – as it has with Poland, Hungary, and Romania. It made use of these borders after the Prague Spring of 1968, when the Czech government introduced social reforms. Within months, Warsaw Pact tanks rolled over the borders to reimpose the Moscow ideology.

Czechoslovakia, like many other states in eastern Europe's shatterbelt, houses many distinct peoples. Even in the celebrations that followed the end of Soviet rule, nationalist aspirations were resurgent.

Particularly vocal demands for autonomy – even an outright break – were heard from the five-million-strong Slovakian minority opposed to what they saw as "Pragocentrism." If their wishes are not met, Czechoslovakia's unity may be extremely short-lived.

NEW WORLD SHATTERBELT
Fragmented Central America

No equivalent area of the world is more shattered into separate – indeed, often warring – states than Central America. With a combined territory roughly the same size as Spain but with less than two-thirds its population, the narrow isthmus between Mexico in the north and Colombia in the south has spawned seven states. Two of them – El Salvador and Belize – do not even stretch from coast to coast, while another two – Guatemala and Honduras – have only a narrow corridor access on one side.

Since 1970 there have been civil wars in Guatemala, El Salvador, and Nicaragua and boundary disputes involving El Salvador and Honduras, Belize and Guatemala, Honduras and Nicaragua, and Nicaragua and Costa Rica. Offshore, Nicaragua, Honduras, Colombia, and the U.S. contest ownership of a number of tiny Caribbean islands. Moreover, with the exception of Costa Rica, every state in the region has been the target of innumerable overt or covert military interventions from powers outside the region, chiefly the U.S., but also Britain, the Soviet Union, and Cuba. Hundreds of thousands of people have been killed or injured and many millions made homeless.

Why should Central America be so fragmented and full of conflict? After all, six of the seven states have a common language and religion from their long association with Spain. (Belize, once the haunt of British pirates and for many years an English-speaking colony, is the exception.) Indeed, when independence from Spain was declared in 1821, the rebels created a new combined state, which they called the United Provinces of Central America.

Yet within only 17 years the association was at an end. Each of the five provinces had become a separate state. (Belize and Panama have somewhat different histories.) A more modest attempt in the 1960s to unite the area in an economic association, the Central American Common Market, also failed within a few years. Why have these countries not become a United States of Central America, along the lines of the U.S., which is, after all, a much more culturally diverse society?

The roots of the conflict lie deep in its geography. Central America is a kind of natural bridge – between North and South America and between the Caribbean and the Pacific. This fact alone has played a large part in attracting the attention of outside powers – originally Spain and more recently the U.S. But Central America has internal difficulties, too. It consists of a highland spine of complex ridges, valleys, and plateaus flanked by narrow, often densely forested or swampy coastal lowlands. Such terrain encourages cultural diversity and makes political control very difficult indeed.

Even the pre-Spanish experience bears this out. Before the 16th century, Central America had for nearly 3,000 years been dominated by a formidable civilization created by Mayan Indians. Yet for all their accomplishments in the fields of architecture, astronomy, and history, the Maya never established a centralized empire: they subdivided the territory into numerous city-states ruled by independent nobles and priests.

Central America is also sharply divided by culture, wealth, and politics. The ethnic composition of the population is extraordinarily complex, with Europeans, predominantly of Spanish descent, Africans, and a great variety of native Indian groups. Intermarriage in most areas has also been common, creating substantial *mestizo* and *mulatto* populations.

Contrasts between the privileged few and the poverty-stricken majority have caused deep divisions. In Guatemala, for instance, 2 percent of the population controls two-thirds of the farmland, while in El Salvador, until recent attempts at reform, just 14 families owned most of the land and ran the country as a feudal fiefdom. Little wonder that leftist and rightist movements flourish.

Add to this the facts that for more than a century the U.S. has considered Central America and the Caribbean to be its own back yard, and that the Soviet Union and Cuba have, at least since the 1960s, been equally keen to see their interests prevail in the region, and there emerges the perfect recipe for chaos. The tragic outcome is that an already fragmented region has become one of the world's most unstable and unsafe shatterbelts.

A massive network of roads, roughly 16,000 miles (25,750 km) long, the Pan-American Highway was first proposed at the Fifth International Conference of American States in 1923 as a continuous link between North and South America. Extending today from Alaska to Chile, it passes through an enormous variety of scenery and climatic conditions, including thick tropical jungles and frigid mountain passes nearly 15,000 ft (4,572 m) high. Designated highways in participating countries are all part of the system, as are branch roads that connect large cities to the primary north-south route.

The photograph shows the highway passing through the inhospitable desert region around Nazca, Peru.

Belize/Guatemala
Formerly called British Honduras, Belize secured independence in 1981. A trip wire force of British troops remains stationed on the border with Guatemala, whose military leaders claim a large part of Belize as their own territory. Guatemala has one soldier for every citizen of Belize, and there are fears that the dispute may escalate.

Honduras/El Salvador
In 1969, a Honduran football team visiting El Salvador for a World Cup game sparked riots in El Salvador and a series of counterriots in Honduras, where Salvadoran immigrants were attacked.

The Salvadoran army crossed into Honduras. In retaliation Honduras mounted an airstrike against El Salvador's major oil refinery. In one week of fighting, 4,000 lives were lost.

Nicaragua
Nicaragua occupies a pivotal position in the heart of Central America. During the 1970s and 1980s, Soviet and Cuban support for the left-wing Sandinista movement, and Communist support for guerrilla movements in El Salvador, challenged U.S. control of the region.

When the Sandinistas came to power in 1979, it seemed as if the USSR would finally gain a foothold on the mainland. In response, the U.S. provided large amounts of aid to the Contra (anti-Sandinista) forces, many of them stationed in Honduras.

Free elections in 1990 ousted the Sandinista government and returned a 14-party coalition. But the Sandinistas remain the largest single party.

Panama
Panama's position has been a crucial one for world trade ever since the canal was opened in 1914, and the U.S. has always been keen to maintain a presence there.

In December 1989, 24,000 U.S. troops invaded Panama to capture the dictator General Manuel Noriega, accused among other things of drug trafficking, corruption, and threatening U.S. access to the canal. He was refusing to accept a continued U.S. military presence in the area after the canal was formally returned to Panama's control at the end of 1999 – in other words, permanent right of unilateral intervention in the country.

Extensive bombing of Panama City, followed by looting, left the capital devastated. Forty U.S. troops and 1,000 Panamanians lost their lives. The action was internationally condemned.

Guatemala
Up to 100,000 people have been killed in Guatemala since the leftist government was deposed by U.S.-backed right-wing exiles in 1954. U.S. support for so-called low-intensity warfare against insurgents kept a repressive military regime in power for 30 years.

Guerrillas, mostly active in the southern highlands, were violently suppressed. In 1977, the U.S. suspended arms sales to Guatemala because of the human rights violations. Five years later, a mainly Indian rebellion was brutally treated by the white minority, the army, and right-wing death squads, leaving 100,000 people homeless.

A civilian government was elected in 1985, but it still faces major economic and political problems. Guatemala has the strongest economy of the region, but has only a small corridor of access to the Caribbean through which it can export goods. If Guatemala could improve its relations with Belize, these transport difficulties would be much reduced.

El Salvador
Until recent attempts at reform, El Salvador was run as a feudal fiefdom by the 14 families who owned most of the land. Social discontent ran high as leftist guerrillas confronted right-wing death squads (often recruited from off-duty soldiers) who killed 10,000 people a year throughout the 1980s. At least 250,000 people fled the country, mainly to Guatemala and Mexico.

U.S. aid supplies 80 percent of the capital San Salvador's income, but reaction in the U.S. against El Salvador's human rights record has brought pressure for this funding to be cut. At present, power lies with the military, but left-wing guerrillas control much of the east and north. Bloodshed continues.

Costa Rica
The "Switzerland of Central America," Costa Rica officially remains neutral in local disputes. Its armed forces were abolished in 1948, and the country has the highest standard of living and the highest literacy rate in the region.

Costa Rica's economy is based largely on coffee, bananas, beef, and sugar. Slumps in world prices for these goods during the 1980s led to its acquiring the area's largest foreign debt. Economic problems have been exacerbated by the arrival of several hundred thousand refugees, of all shades of opinion, from the struggles in Nicaragua and El Salvador; this may yet prove to be a volatile cocktail.

THE NEW SUPERPOWERS

Shifting balances in a changing world

IN 1900, MUCH OF THE WORLD WAS STILL controlled by a dense concentration of powers within Europe. The greatest of these was Britain. As maps of the day, saturated with pink, proclaimed, the sun never set on its vast empire – Canada, New Zealand, Australia, Malaya, India, and many other areas throughout Asia, the Caribbean, and Africa.

Nevertheless, Britain could never entirely eclipse the other powers. True, some were already in decline, like Austro-Hungary and the Ottoman Empire, and Britain could and did profit from their losses. But the economic core states of Europe – France, Italy, Belgium, and Germany – were thriving, as were the U.S. and, later, Japan.

But within 50 years, Europe seemed mortally wounded – largely from self-inflicted injuries. After two world wars, many of the former great powers had collapsed. Japan, Germany, and Italy lost their colonies at a stroke in 1945, but for a while other European powers fought bitter, costly, and futile colonial wars. One by one, Britain, France, the Netherlands, Belgium, Spain, and Portugal had to relinquish their hold over former overseas possessions.

In the meantime, international preeminence inexorably shifted west and east, to the two new superpowers, the U.S. and the Soviet Union. Each gathered around itself a number of other states in military and economic alliances. Although the boundary between the two blocs was explicit in Europe, where an "Iron Curtain" descended from the Baltic to the Adriatic, it was much less distinct elsewhere. Indeed, for the next 40 years states were pulled backward and forward between the two rival powers.

Although the world has frequently been terrified that this covert struggle for supremacy might escalate uncontrollably into all-out nuclear war, the superpowers proved remarkably adept at maintaining the balance of power. As far as the economic core states were concerned, it was a Cold War; only the states of what soon became known as the third world truly felt the heat.

But this global grand design is changing shape, and at an accelerating pace. The simple pattern of an east–west divide has become much more confused, with a variety of new powers emerging in a world where the superpowers are facing problems.

The Communist bloc has been disintegrating for a long time, with states like Yugoslavia, Albania, and above all China going their own way during the 1950s and 1960s. And more recently, poor economic performance, together with the surge of nationalist and democratic movements within the USSR and its satellite states, has caused even more power to slip away from Moscow's hands.

The U.S., too, has its difficulties. With its relative share of world production diminishing fast, U.S. foreign commitments may now be economically, politically, and militarily more than it can bear.

No less momentous changes have been occurring elsewhere. The economies of many of the new powers, often based on access to a

The devastating effects of nuclear warfare, as evidenced by the horrific explosions at Hiroshima and Nagasaki in 1945, have brought every nation's nuclear strength under the closest international scrutiny.

Since each country regards its atomic warheads as crucial elements in its overall defensive strategy, governments tend to treat the size and geographical distribution of their atomic capability as a matter of strict national security. Educated guesses based on the findings of sophisticated technological equipment must, therefore, inform each power of all others' potential might, and no information can be taken at face value.

A militant member of the Organization of Petroleum Exporting Countries (OPEC), Libya has grown rich on the oil discovered there in 1958; its citizens enjoy the highest per capita income of any in Africa. Arab nationalism has spearheaded the country's foreign policy, and belligerence has taken the form of widespread guerrilla activity. Its stacked and silent missiles (*left*) may represent only a fraction of the country's atomic capability.

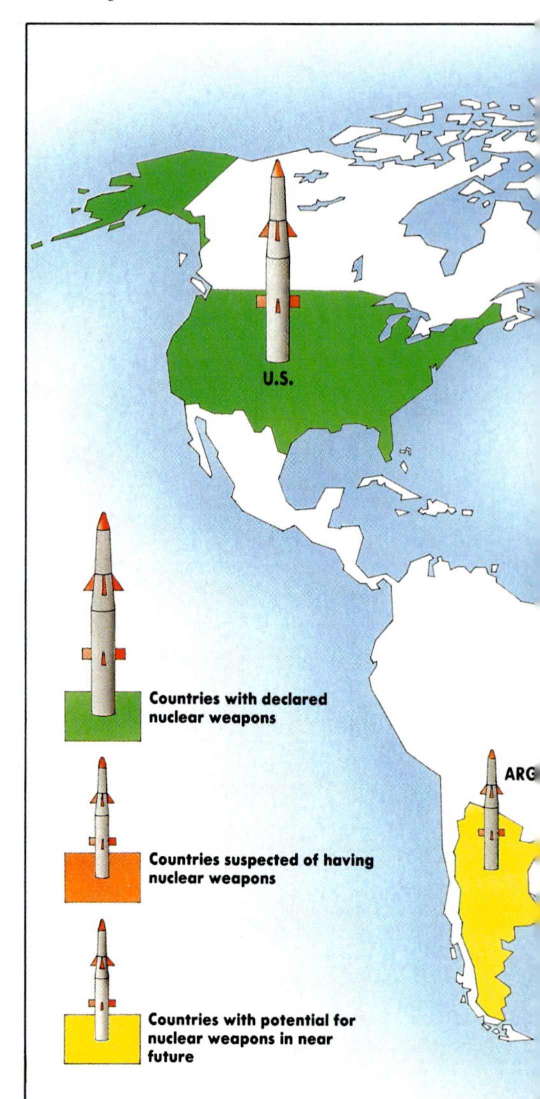

strategic resource such as oil, enable them to act independently of the superpowers. Some have even acquired the nuclear bomb.

The greatest of these new powers are the east Asian giants Japan and China, but regional leaders are emerging in Central and South America (notably Brazil, Argentina, and Mexico), in southern Asia (Pakistan and India), and among the great Islamic states of the Middle East and North Africa (Iran, Iraq, Syria, and Libya). Israel and South Africa also have strong economies, and are formidably well armed.

But perhaps the most surprising development is the phoenixlike reemergence of western Europe from the ashes of the two world wars. It has overcome former national differences to become a powerful and increasingly integrated economic and political entity.

As we enter the 21st century, there is every sign of a multifaceted global power structure emerging. What we cannot know is whether this will provide a more or a less stable world order than the one it has grown out of.

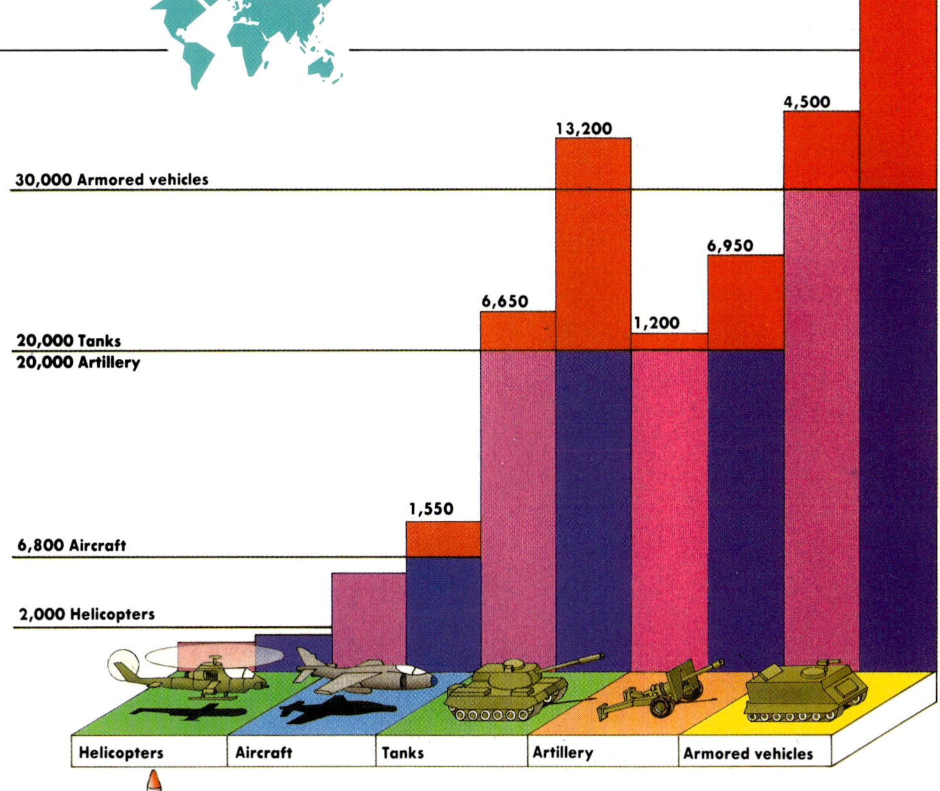

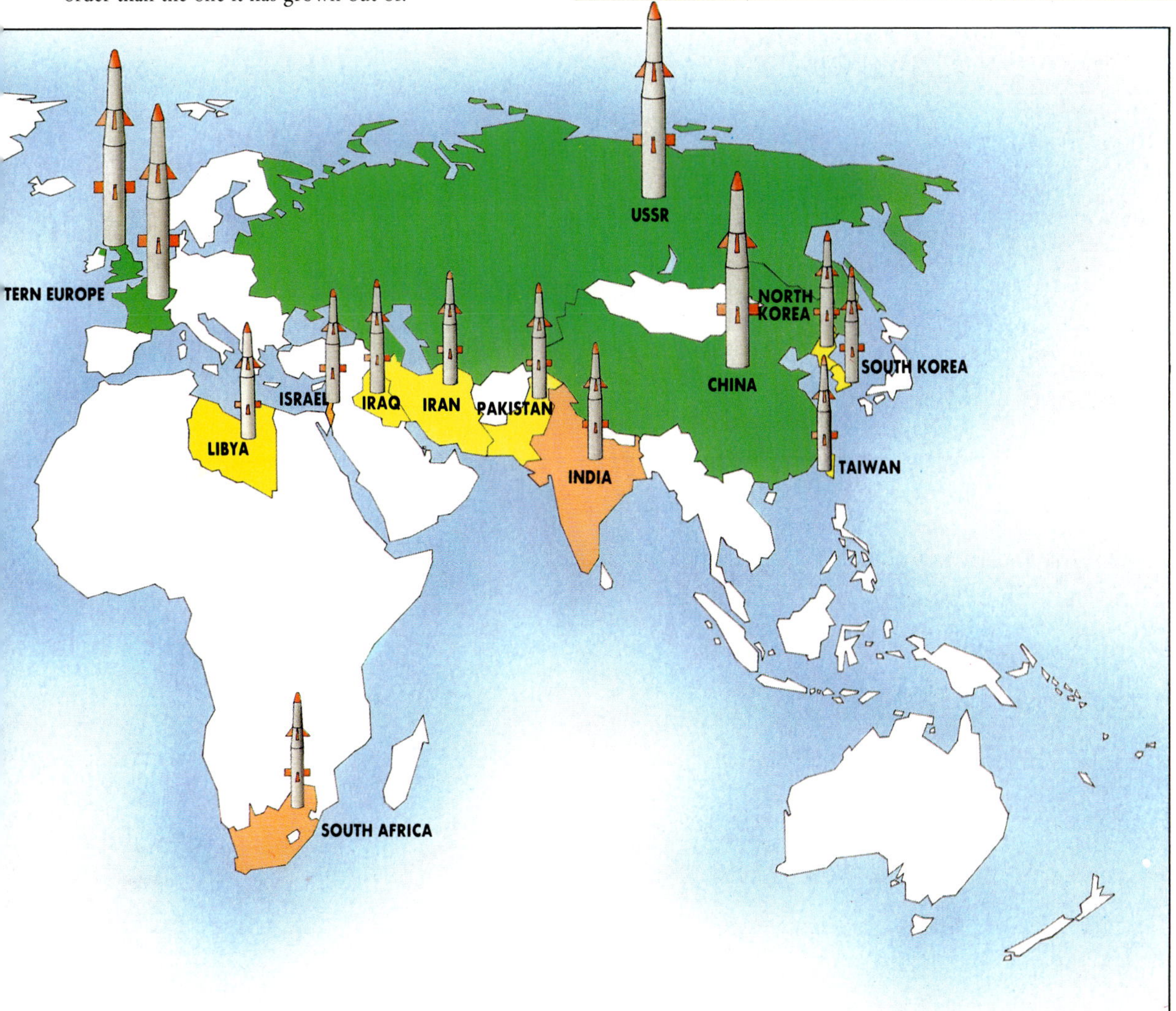

NATO: Canada, U.S., Iceland, Norway, U.K., Luxembourg, France, Portugal, Spain, Germany, Denmark, Belgium, Netherlands, Italy, Greece, Turkey

Warsaw Pact: USSR, Poland, Czechoslovakia, Bulgaria, Hungary, Poland

Agreed reductions

The accord reached in Paris on November 19, 1990, at the Conference on Security and Cooperation in Europe, guarantees an unprecedented reduction in conventional weapons and manpower. This is the largest negotiated reduction since the end of the Second World War.

The 22 member-nations of NATO and the Warsaw Pact signed this Conventional Forces in Europe Treaty. The illustration (*above*) shows their levels of reduction, which should be completed within 40 months of the treaty's ratification by all its signatories.

Members of each alliance are generally permitted to share the distribution of allowed weapons, but no country may adopt more than roughly two-thirds of the alliance's entire allotment of any category of hardware.

HEARTLESS HOMELANDS

South Africa's satellite states

THE FUTURE OF SOUTH AFRICA DEPENDS on the ability of blacks and whites to negotiate a multiracial constitution. If President de Klerk has his way, the constitution will be ready to be tested for approval in the next white elections in 1994. But what sort will it be? And how will it deal with those by-products of apartheid, the black homelands?

There are ten tribal homelands, some of them in one territorial area, others scattered about as if they had been carelessly dropped on the landscape. Bophuthatswana, for example, is in six pieces. Its capital, Mmabatho, is on the Botswana border, separated by more than 200 miles (320 km) from its most far-flung component, which lies close to Lesotho.

Four of the homelands – Transkei, Ciskei, Bophuthatswana, and Venda – are nominally independent, but no state outside South Africa recognizes them. When their citizens travel abroad, they use documents issued by the South African government. All four have their own defense and police forces, but only Bophuthatswana is a success in business.

Tourism, the mining of platinum and rhodium (used in catalytic converters), improved agriculture, and some industrialization have helped to produce its budget surpluses. The other six homelands – KwaZulu, Gazankulu, Lebowa, QwaQwa, KaNgwane, and KwaNdebele – are self-governing.

The 1936 Land Act (now due for repeal) set aside 13.6 percent of the country – roughly the area covered by the homelands – for black occupation. But it was the 1959 Promotion of Bantu Self-Government Act that provided the legislative framework for establishing ethnic "states" within a white-controlled state. Possibly 55 percent of South Africa's 28 million blacks live in them. They are overcrowded and in general the land is badly farmed, partly because it is communally owned and under the control of tribal chiefs (although private ownership is now being introduced in some homelands).

Incomes are on the whole much lower than those earned by blacks on white farms and in the cities. None of the homelands, whether "independent" or self-governing, would survive without massive injections of financial assistance from Pretoria.

The African National Congress will argue in the constitutional negotiations for a unitary state – that is, one with centralized powers that would abolish the homelands. It opposes the idea of power being devolved on a federal or communal basis, which is what de Klerk's National Party wants. The government hopes that the chief ministers of the six self-governing homelands will be on its side at the negotiating table, but they are under extreme pressure to throw in their lot with the ANC. In the four "independent" homelands, too, the movement for reintegration grew during 1990.

However, the case for a federal system is by no means lost on the blacks. The Tswanas of Bophuthatswana are the same people as those of Botswana and, given a choice, might choose to integrate most of their territories into Botswana rather than be swallowed up by a unitary state. Federalism might be a viable compromise.

Even more important in this matter are the Zulus and their leader, Chief Buthelezi, chief minister of KwaZulu. The Zulus are South Africa's largest tribe, and more than 4,000 deaths in the township fighting during the second half of the 1980s has underlined in blood the hostility between the Zulu Inkatha party and the Xhosa-dominated ANC. Federalism would be Buthelezi's preferred path.

Finally, there are the white Afrikaners who have been putting forward their own visionary ideas for independent homelands, into which they would retreat from a black South African government. None of these has so far achieved credibility outside small factions of the white population.

ABORIGINAL RIGHTS

Australia and New Zealand come from the same British imperial background as South Africa, but their "nonwhite" racial groups are relatively small, and there has been no move by the ruling majority to herd them into reserves. Australia's 230,000 Aboriginals (1.46 percent of the population) do, however, hold title to 12 percent of the Australian landmass, and the recent Aboriginal and Torres Strait Island Commission Act recognizes that they were dispossessed of their land by European settlement. The attention paid to their problems has now increased.

The Maoris constitute less than 10 percent of New Zealand's 3.3 million population. Most live in the North Island, many of them in the capital, Wellington, and in the country's largest city, Auckland. The foundation of their rights is the 1840 Treaty of Waitangi. A recent change in the law, which now allows them to raise grievances going back to 1840, has led to Maori claims to the whole coastline, half the fishing rights, and 70 percent of the land. Racial tension increased sharply in the late 1980s.

Bophuthatswana – considered by many to be the most viable of the independent homelands – is the world's largest producer of platinum. It has other mineral and agricultural resources, but a substantial part of its income derives from the tourist resort of Sun City (*left*). Boasting a game park, a casino (illegal in South Africa), a high-class tourist hotel, and "international cabaret," Sun City is an incongruous pocket of glamour in an otherwise drought- and poverty-stricken area.

South Africa

Sun City
Pretoria
Johannesburg
SWAZILAND
Hotazel
Kimberley
Alexander Bay
Bloemfontein
LESOTHO
Durban
SOUTH AFRICA
Cape Town

Citizenship of the homelands is imposed on any African who was born there, lives there, speaks any dialect or language used there, is related to anyone there, or is identified with any part of the population by virtue of his or her cultural or racial background. In other words, almost all black South Africans fall into this category, whether or not they live in the homelands.

- Ciskei (independence 1981)
- Gazankulu
- Bophuthatswana (independence 1977)
- KwaZulu
- Lebowa
- Venda (independence 1979)
- KwaNdebele
- KaNgwane
- QwaQwa
- Transkei (independence 1976)
- ◇ Mineral resources

Living conditions for blacks in the homelands can be exceptionally primitive. Many dwellings are made of rough planks and corrugated iron, while some are built of mud bricks. Plumbing, electricity, and running water are scarce, and paved roads and telephones are virtually nonexistent. Most homeland inhabitants must travel hundreds of miles to obtain the only work for which they are legally qualified.

CHOKEPOINTS

Where trouble waits to happen

LOCATION DICTATES THAT AT SOME POINT between its ports of origin and destination, most of the world's merchant or military shipping must be funneled through narrow waters – the straits, passages, and sounds that surround every continent. Control of these chokepoints is obviously of great strategic importance to any trading nation and has been a critical factor in determining the course of world history.

In some cases international agreement guarantees free passage, but in many others states have gone to war to establish control or secure transit rights. On a global scale, even the threat of closure can cause world powers to step in, as when the Strait of Hormuz became impassable during the Iran–Iraq war of 1980–88. A blockade directed at one country can swiftly cripple its economy and military power.

Typically, these marine chokepoints are natural deep-water channels at the head of a gulf or an enclosed sea, or between two islands, or between an island and the mainland. In a few cases they are artificial, like the Panama and Suez canals, but even here physical geography has largely determined the route taken.

The 16-mile (25-km) wide Strait of Gibraltar, where the Mediterranean meets the Atlantic, has been a vital chokepoint for hundreds of years. Ships have no alternative but to pass between the Pillars of Hercules, the twin rocks that support the tiny military enclaves of Ceuta and Gibraltar.

For nearly 300 years Gibraltar has been under the control not of Spain, which surrounds it, but of Britain. During that time its massive rock has become a honeycomb of gun installations. The African "pillar," Ceuta, is controlled not by Morocco but by Spain. Like the nearby port of Tangier, Ceuta has regularly changed hands.

Gibraltar was only one link in a chain of British-controlled chokepoints on major shipping routes to India, the Far East, Australia, and New Zealand. Although Britain has now lost most of its former naval bases, many of the world's chokepoints remain colonies or dependencies of major powers, or are home to foreign military bases. The U.S. has many such bases on or near chokepoints in the Far East, the Middle East, and the Caribbean – including, of course, the Panama Canal.

Unlike the U.S., which has easy access to both the Atlantic and the Pacific, Russia is hemmed in on every side. All the chokepoints in its approaches to the world's major sea lanes are controlled by rival powers. Its western, Baltic route from Leningrad to the Atlantic is via the narrow Ore Sound, which is governed by Denmark and Sweden. To the southwest, access to the Mediterranean from the Black Sea is restricted by Turkey. To the north, the Arctic Ocean is icebound for much of the year, as is the eastern coastline. Moreover, Vladivostok, the port at the end of the Trans-Siberian Railroad, is in effect closed off from the Pacific by the crescent of islands that makes up Japan.

These chokepoints explain a good deal of Russia's history and international relations. In the game of global power politics, its further expansion has always been checked by alliances between the major Western powers and the states that border these chokepoints.

Most significant were events that occurred in 1905 in the Korea Strait to the south of Japan. This 100-mile (160-km) wide chokepoint became the scene of imperial Russia's most humiliating naval defeat, one that changed the course of world history. The Russian Pacific fleet was besieged by Japan in Port Arthur. In an attempt to relieve it, Russia, with its Black Sea fleet trapped by Turkey and access to the Strait of Gibraltar and the Suez Canal denied by the British, was forced to send the poorly equipped Baltic fleet on a seven-month voyage around the Cape of Good Hope.

Arriving exhausted and much battered by tropical storms, the fleet was surprised by Japanese ships concealed behind the Tsushima Islands that lie midchannel. Within hours the Russian fleet was devastated. News of the defeat so affected morale that the citizens of St. Petersburg (modern Leningrad) rose up. Their revolt was, in Lenin's words, the "dress rehearsal" for the revolution that swept the Bolsheviks to power in 1917.

Consequences of the hostility over the control of chokepoints involve outright confrontation as often as threats. By the end of 1984, Iran and Iraq between them had attacked 31 oil tankers belonging to 14 different nations. Liberia suffered the severest blow, with seven tankers besieged.

Caribbean

The Panama Canal, linking the Pacific and Atlantic oceans, is one of the world's critical chokepoints. About 20 percent of U.S. domestic oil passes through Panama en route from Alaska. Substantial amounts also arrive from Venezuela, Mexico, and Trinidad.

This shipping must pass through one of a number of narrow channels that cut through the chain of Caribbean islands stretching from Cuba in the north to Trinidad in the south.

Today the U.S. has some 40 bases at its disposal in the region, including a huge naval base at Guantánamo in Cuba and installations in Panama itself. The U.S. invasion of Panama in 1989, following previous interventions in Cuba in 1961, in Dominica in 1965, and in Grenada in 1983, illustrates how important control of the Caribbean is.

Suez

When the Suez Canal opened in 1869, Britain moved swiftly to establish control over this new shortcut to the East. Within ten years it had also established bases on strategically situated Cyprus.

Britain was a shareholder in the canal company until 1956, when Egyptian president Nasser nationalized it. In the ensuing conflict with Britain, France, and Israel, Egypt blocked the canal with scuttled ships. Following Arab-Israeli hostilities it was blocked again from 1967 to 1975.

Today, although reopened, the Suez Canal is virtually obsolete. Modern tankers are too big to pass through and now go around Africa via the Cape of Good Hope.

Southeast Asia

The Strait of Malacca, between Sumatra and Malaysia, always important with respect to the Far East trade, is now even more significant as the main route for oil tankers en route from the Middle East to Japan. Access through the strait is controlled by Singapore, which has thrived as a result.

Much of the coastline of northeastern Asia is cut off from the Pacific by a series of partially enclosed seas hemmed in by peninsulas and island chains, stretching from the Aleutian Islands in the north to the Philippines and Indonesia in the south.

Maintaining strong economic and political links with Japan, South Korea, Taiwan, and the Philippines is essential if the U.S. is still to control these vital naval chokepoints that protect its western flank. Should the Philippines make the presence of American military installations difficult to maintain, the U.S. needs a fall-back position. Tiny Palau, which is unusual in the region in having a deep-water bay suitable for submarines, offers just that possibility.

Perhaps the most significant and troublesome chokepoint in recent years has been the Strait of Hormuz, a narrow international waterway lying between Oman and Iran. Its significance lies in the fact that all the oil exported by tanker from the Persian Gulf has to pass through this channel. Although pipelines have been constructed to the Red Sea and the Mediterranean to avoid the chokepoint, 60 percent of Japan's oil and 25 percent of western Europe's still have to pass this way.

When Iran and Iraq began to attack tankers in the strait in 1984, in an attempt to disrupt each other's oil exports, France, Britain, the U.S., and other states cooperated in sending ships to ensure that vital supplies of oil were not disrupted.

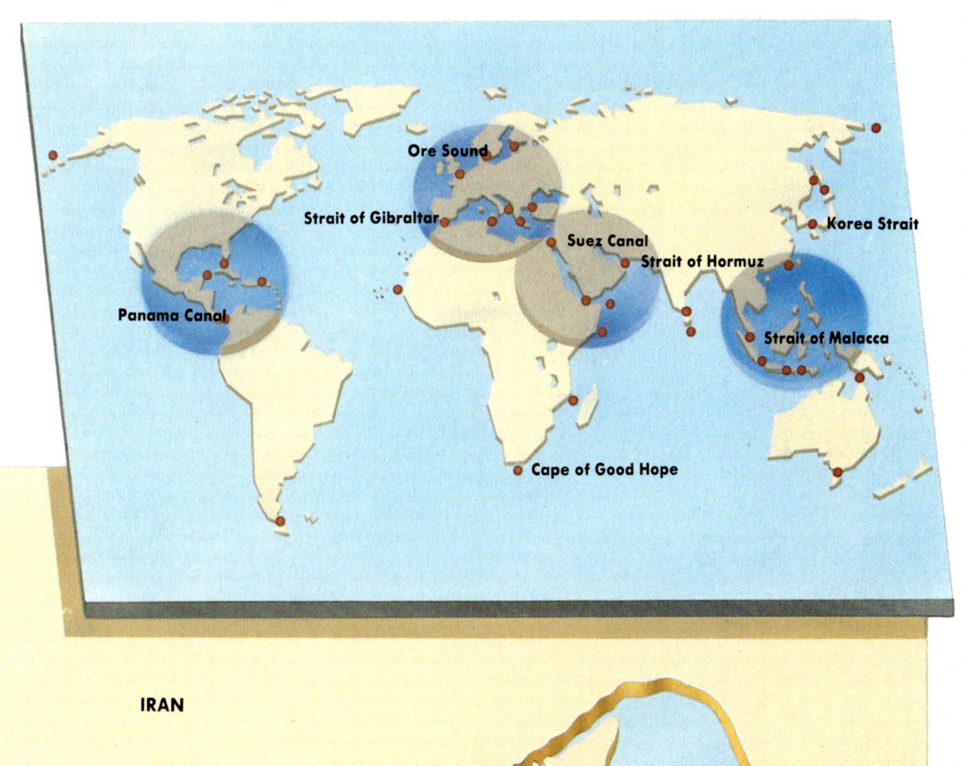

ISLAM, OIL, AND AMBITION

Shifting sands in the Middle East

For most of the last 40 years, world perception of the Middle East has been dominated by three interlinked geographical topics: the bitterly contested borders of Israel and Lebanon, threats to vital transport passages (principally the Suez Canal and the Strait of Hormuz), and the Middle Eastern domination of the world's oil trade. Terrorist atrocities, oil price rises, and superpower conflicts have become attendant problems.

The growth of new regional powers – Saudi Arabia, Syria, Iraq, and Iran – each with a huge standing army and all the sophisticated weaponry that oil money can buy, has fueled fears that confrontations here could easily precipitate war in the entire world.

These disputes have disguised more general problems with the region's geography concerning the arbitrary nature and instability of the state boundaries that were formed after the collapse of the Ottoman Empire. In the years up to and immediately following the First World War, Britain and France took over Turkish rule and encouraged local Arab nationalism. But having fanned the flames, they could not then dampen the resulting fire. Rebellions broke out, and although French and British influence remained strong, a number of new nation-states were created, Egypt (1922) and Iraq (1930) among them.

More such nations appeared after the Second World War; these include Jordan (1946) and a large number of tiny states around the Gulf, like Kuwait (1961), Bahrain, Qatar, and the United Arab Emirates (all in 1971). After more than half a century of Jewish migration into what the immigrants considered the Promised Land of Zion, Israel was officially created in 1949 from land that had once formed part of predominantly Arab Palestine.

Many of these imposed boundaries and manufactured states were poorly conceived and ill-defined, especially those crossing deserts, where today the mineral resources buried beneath the sands provide a source of potential conflict. The Iraq-Kuwait border, it is said, was formalized in 1922 by a British envoy who simply drew a pencil line across a crumpled map. When Iraq challenged the border in 1961 (on the tenuous grounds that Kuwait and Iraq had once been united in the same Ottoman province), British forces were quickly deployed to prevent an invasion. Things were very different in 1990.

Demands have often been heard in the region – the loudest from military dictators like Libya's Colonel Muammar Qaddafi and Iraq's Saddam Hussein – for the formation of a single pan-Arab state. Indeed, Libyan maps display no boundaries and ignore the state of Israel. Yet such demands have been rebuked with vehemence.

The Middle East, excepting Israel and Lebanon, is commonly assumed to be culturally homogeneous, uniformly Arab and Muslim. It is not. Turkey and Iran, for example, where Arabic is the language of religious ritual, are otherwise not racially, culturally, or linguistically Arab at all.

A great number of non-Arab minorities exist within the region, some of considerable size, like the enclave of 18 million Kurds that straddles the borders of Syria, Iraq, Turkey, Iran, and the Soviet Union. Elsewhere the Arabs of North Africa live alongside a predominantly Berber group of nations – Libya, Tunisia, Algeria, Morocco, and Mauritania – and the black peoples of sub-Saharan Africa.

Even Arabs themselves are not culturally united. Many Muslims (indeed, the overwhelming majority) are not Arabs, and many Arabs embrace faiths other than Islam, including Judaism and Christianity. There have also been longstanding conflicts between city-dwellers and nomads, and between Arabs from different tribes. Some of these disputes threaten the very existence of the smallest clan-states, particularly those located around the Gulf. Just as Iraq has claimed and occupied Kuwait, Saudi Arabia once invaded Abu Dhabi, and Iran has demanded Bahrain.

Neither is Islam quite the unifying factor it might seem. There are countless sects in addition

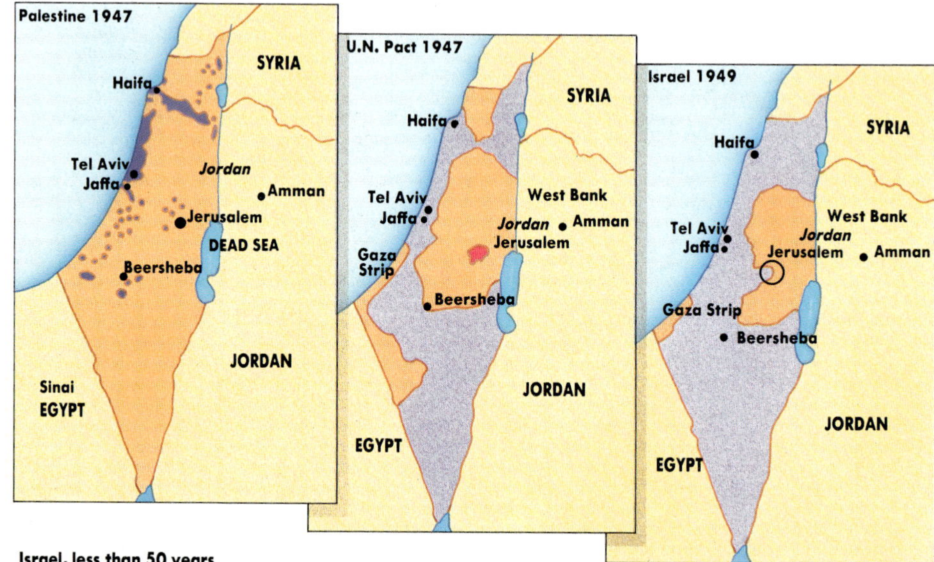

Israel, less than 50 years old, is the first and only state created by the United Nations, yet its boundaries are more disputed than those of any other state in the world.

1947
The influx of Jewish settlers in Palestine and Jordan caused the Jewish population to rise from around 60,000 in 1920 to 650,000 in 1948. By this time social relations with the almost one million Palestinian Arabs were so strained that conflict had become an integral part of life.

In 1947, the newly formed U.N. took control of the area, recommending partition and the creation of a new state, Israel. It was intended that a little over half of the territory go to the Jews.

1949
The new state of Israel was created, owing little to the U.N. proposals. But there were problems. The West Bank, which might geographically have been allocated to Israel, remained part of Jordan.

At its narrowest point, the fertile and highly populated coastal plain of Israel, between the West Bank and the Mediterranean, was less than 16 km (10 miles) wide. Moreover, Jerusalem was divided between Israel and Jordan. Egypt took control of the Gaza Strip, an arrangement Israel accepted on the understanding it was to be temporary.

The passion that triggers violence in the Middle East flares increasingly in Jerusalem and the settlements along the West Bank. In their efforts to establish their own state, the Palestinians are waging an *intifada* to push Israeli occupying forces from the territory they consider rightfully theirs. The Israeli Defense Force has retaliated with what many consider excessive brutality.

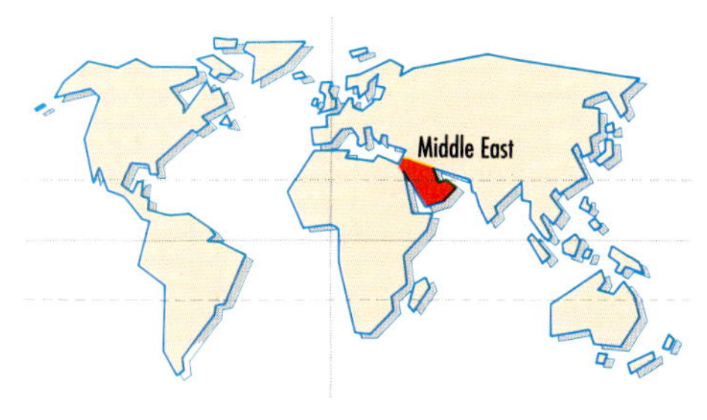

to the fundamental division between Shi'ite and Sunni Muslims. Shi'ites form a majority in the east of the region, while Sunnis form a majority almost everywhere else. Yet clusters of the two groups occur everywhere, and in some states it is the minority religious group that predominates politically. For example, Shi'ite rulers dominate Syria's largely Sunni population, while Iraq's Shi'ite majority is controlled by Sunnis.

Within a few short decades, the Middle East has been transformed from a nomadic and semisubsistent region to an area of immense – and immensely unequal – riches based on oil. Adapting to these changes has posed additional geopolitical problems. There are huge social divisions between the ruling elite and everyone else, whether desert tribespeople, agricultural peasants, or migrant laborers. Islamic militancy, fueled by poverty, is adding an explosive ingredient to an already volatile concoction.

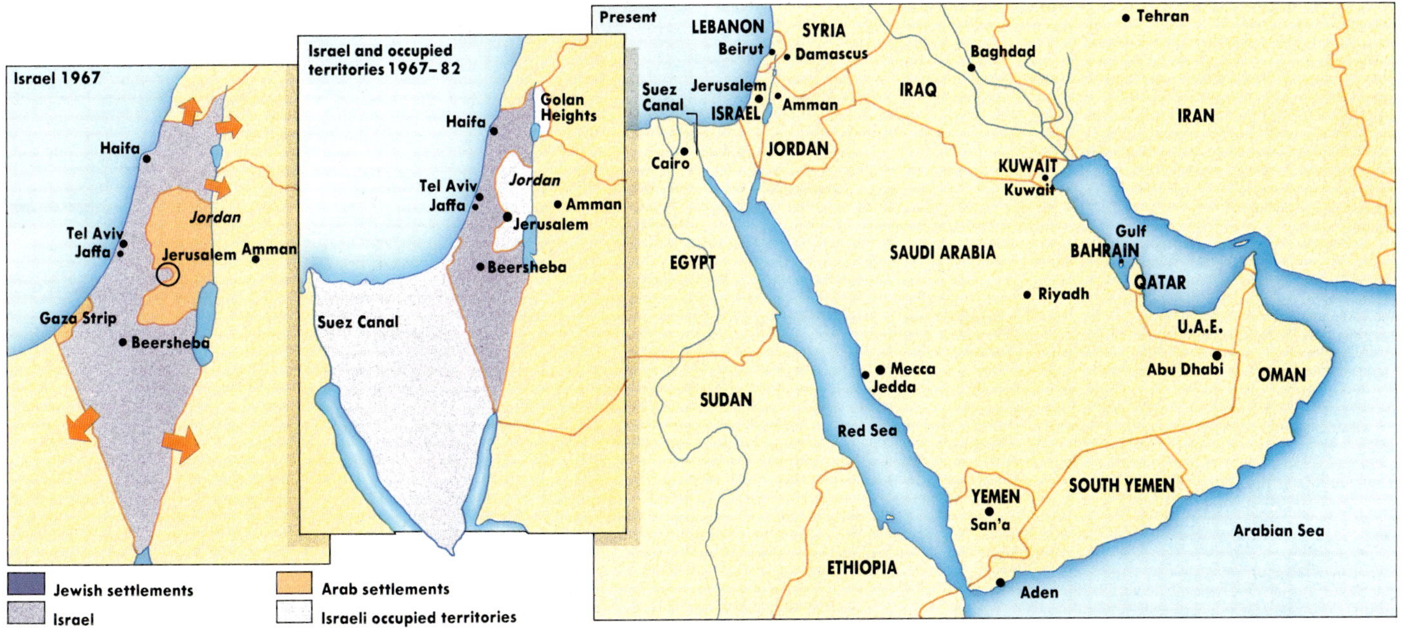

1967
In one of the most astonishing offensives in history, Israel simultaneously attacked Egypt, Syria, and Jordan, taking the Gaza Strip, Sinai, the West Bank (including the whole of Jerusalem), and the Golan Heights of southwestern Syria. In short, after the Six-Day War, the whole of former Palestine was in Israeli hands. The Israelis lost 800 people; Egypt alone lost 10,000. The borders of Israel were now very much more defensible, but they enclosed the large Arab population of the West Bank.

1973
Egypt and Syria attempted to retake land they had lost six years previously. Israel – for whom the conflict became known as the Yom Kippur War – just managed to hold the Golan Heights against the Syrians, but then advanced to within 20 miles (32 km) of Damascus. In the south, Egypt had temporary success, having attacked in five places across the Suez Canal. The Israelis were forced to fall back across Sinai until they reached the pre-1967 boundary, where they dug in.

Present
The situation in Lebanon has been troubled for many years; the arrival of many PLO members ejected from Jordan in 1971 added to the tension. The U.N. now maintains a buffer zone between Lebanon and Israel.

The 1980–88 war between Iran and Iraq began as an Iraqi attempt to establish its dominance. Although Iraq justified its invasion of Kuwait in 1990 by claiming that Kuwaitis and Iraqis were the same people, control of oil was its motivating force.

A NATION WITHOUT A STATE

The plight of the Kurds

IN THE GREAT UPHEAVALS THAT HAVE sundered and at times engulfed the Middle East in the last 100 years, the Kurds have shared with the Armenians the wretched distinction of being the biggest losers. Turks, Arabs, Jews, and Persians created or consolidated their own nation-states as the Ottoman Empire collapsed and other empires waned. Even the Armenians – annihilated as a community by the Turks – have their own autonomous republic in the Soviet Union. But the Kurds have nothing. Only once, in 1946 at Mahabad, in Iran, have they had a state of their own, and then only as a Soviet satellite that Moscow abandoned after the Second World War.

Superficially, the Kurds' political failure seems remarkable. They account for 20 percent of Turkey's population, almost 25 percent of Iraq's, 10 percent of Iran's, 8 percent of Syria's. There are believed to be a million Kurds in the USSR, where they have had more cultural freedom than elsewhere.

A closer look at the map shows why the Kurds have never been able to impose their nationhood on the region, even though they are the world's largest stateless minority, with a population of 21 million or more. They are a mountain people, and the long frontier agreed between the Ottoman and Persian empires in the 17th century (and more or less intact ever since) follows the Zagros and Taurus ranges and divides their homelands.

The mountains have protected them, too, of course, but they have also created a poor and often fratricidal tribal society, divided again by two major (and very different) dialects of Kurdish, a derivative of Sanskrit related to Persian and, more remotely, to the main west European languages. There is still no single Kurdish language. Equally, there is no political party that can unite the nationalist groups in Turkey, Iran, Iraq, and Syria and speak with one voice for all the Kurds.

Unlike the Christian Armenians and the Jews, the mainly Sunni Muslim Kurds have shown little inclination to wander into the wider world outside Islam, and they have never developed a class of rich entrepreneurs capable of financing and promoting their cause. Only in the past two or three decades have significant communities of Kurdish exiles grown up outside the Middle East, the largest and most important of them being the Turkish Kurds who migrated to West Germany in search of work.

In the past, the warlike Kurds were regarded by their neighbors as pests comparable to rats and locusts. Throughout this century, they have been pawns in the struggle between Iran and Iraq, their long-running nationalist rebellions aided when it suited one or the other, betrayed when they were no longer needed.

Most Kurds are Sunni Muslims and as such have suffered religious persecution in Shi'ite-dominated Iran.

The religion of Islam has been closely involved with many of the struggles in the Middle East. The one thing the Muslim nations have in common is a hatred of Israel, but there has also been bitter division within Islam itself, as shown by the conflict between Iran and Sunni-led Iraq.

Many Muslims see their struggle as a holy war, or *jihad*, and the name *mujaheddin*, given to the anti-Soviet resistance movement in Afghanistan, is based on this word. The picture above shows *mujaheddin* guerrillas on guard at the Afghan border.

The regime of President Saddam Hussein of Iraq has been noted for its vast spending on arms and its lack of concern for the rights of its minorities. Both characteristics were horrifyingly demonstrated in 1988, just after the end of the war with Iran. With its foreign debt – largely incurred through arms imports – counted in billions of dollars and its economy in chaos, Baghdad could not afford domestic unrest.

The Iraqis had been known to use chemical weapons before, but now, with more sophisticated technology, whole villages were wiped out and thousands of people killed. The picture (*right*) shows Kurdish refugees arriving at the Turkish border after the gassing in September 1988.

The suppression of any form of national identity in Turkey and the abuse of human and civil rights it involves have been largely ignored by Western governments because of Turkey's importance as a trustworthy NATO ally in a turbulent corner of the world. Even the use of gas by the Iraqi government against its own Kurdish citizens and the disappearance (and presumed murder) of 8,000 civilian members of the Barzani clan did not lead to outright international condemnation and ostracization.

Turkey may contain about half the Kurdish population, but Iraq has always been regarded by the Kurds as the richest and culturally most advanced part of Kurdistan. The valley of the Little Zab, which winds its way toward the Tigris through some of the most beautiful mountains in the Middle East, is claimed by them to be the original Garden of Eden.

It was in Iraqi Kurdistan in the early 1960s that Mullah Mustafa Barzani, having returned to Iraq after years of exile in the USSR after the collapse of the Mahabad Republic, began the longest and bloodiest revolt in Kurdish history, a revolt that was not extinguished until after Iran's defeat in the Gulf War in 1988.

At the beginning of the 1990s, Eden was still recovering from the effects of gassing, widespread carnage and executions, and mass deportations, including the depopulation and destruction of villages in the vicinity of the Iranian border. It is hard to imagine Eden or any other part of Kurdistan mounting a rebellion capable of driving out government troops backed by regimes prepared to take the most ruthless measures. The events of the spring of 1991 reinforce that view.

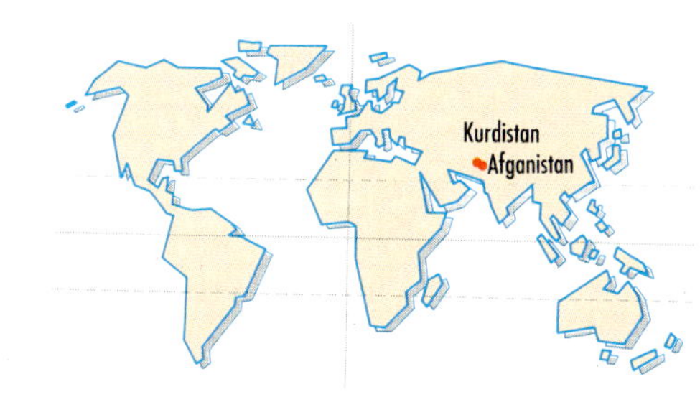

Overlapping large areas of Turkey, Iran, and Iraq, the state of Kurdistan would be a considerable force in Middle Eastern politics if it ever came into being. The fact that the province of Kirkuk, containing about two-thirds of Iraq's oil supplies, would fall within its boundaries is one continuing source of friction between Baghdad and the Kurds.

Ever since Barzani first rebelled against the Iraqi government in 1964, the Kurds have been victims in a power struggle between Iran and Iraq. They were supported by Iran when it felt threatened by Soviet influence in Iraq but suppressed when a Kurdish revolt in Iran was said to have Soviet backing.

The Kurds continued to be victimized throughout the Iran–Iraq war of 1980–88. At the start of the war, 800 Iraqi Kurdish villages were razed to create a security belt along the frontier with Iran.

Iraq's invasion of Kuwait in 1990 has turned world attention away from the persecution of the Kurds, but there are no signs of a happy conclusion to their struggle.

VICTIMS OF FATE

Political and economic refugees

At the end of the 1980s Europe, the birthplace of the political refugee, faced a new twist in a problem it thought it had said goodbye to, more or less, in the 1950s. Although the refugees journeyed, as so often, from east to west, this time their motive for flight was economic rather than political. It could be argued that they were – are – no different from the huddled masses of migrants who had sought an escape from poverty in the U.S. 100 years earlier, but they came unasked and unneeded from countries whose political systems had produced economic collapse.

Behind the floods of East Germans, ethnic Germans from farther east, and Romanians rose the threat of a far greater outpouring from the USSR as travel restrictions were lifted. Soviet officials told their Western counterparts to expect three million refugees a year as economic conditions worsened and the prospect of violent unrest grew.

In 1951, a United Nations convention on refugees defined them as persons with a "well-founded fear of being persecuted for reasons of race, religion, nationality, membership of a particular social group or political opinion" who had been obliged to leave their countries and were unable to return. Refugees were regarded then as a postwar phenomenon, human debris cast up by ideological turmoil.

But the numbers continued to grow as new wars and new persecutions made their contributions. The convention's ruling that refugees were victims of events that had occurred by 1951 had to be changed; the Office of the U.N. High Commissioner for Refugees (UNHCR) widened its umbrella to cover people seeking sanctuary abroad from wars and civil strife.

Counting only those who fall within the UNHCR's categories, the number of refugees has grown from 2.5 million in 1970 to 7.5 million in 1980, to 15 million in 1990. The greatest concentration is in Pakistan, mostly in the Northwest Frontier Province, where more than three million Afghans have taken refuge from the war between government forces and guerrillas. Over 2.8 million more are in Iran. Together, these figures account for a third of Afghanistan's population.

The situation in the Horn of Africa is numerically less dramatic but in human terms far worse. War, persistent drought, and land degradation have combined to produce what the regional governments say is more than two million refugees as a result of migrations between Ethiopia, Sudan, and Somalia (international aid is linked to numbers, so government figures are not always to be trusted).

Ethiopia claims to have given asylum to 700,000 refugees, half of them Somalis and the rest Sudanese. Somalia has what UNHCR officials call a "planning figure" of more than 600,000 refugees from Ethiopia. Sudan estimates that it has 750,000 refugees, most from Eritrea and the neighboring provinces of Ethiopia that have suffered from war and drought. In southern Africa, the wars in Mozambique and Angola have produced perhaps two million refugees, about 800,000 of whom have fled to impoverished Malawi.

To the millions who can be broadly categorized as victims of violence and those in the economic migrant class can be added a new group: the environmental refugees. Many refugees in Africa belong to this category. Some experts have calculated that sea-level rises and other effects of climate change could produce 300 million refugees in the next century.

Population growth and its attendant stresses are also factors in the overall increase. At the beginning of the 1950s, the world's population was under three billion. By the year 2000 it will be six billion. Between 1950 and 1985 the third world accounted for 85 percent of the growth.

Economies have not kept pace. Poverty, knowledge of the huge disparity of living standards between rich and poor nations, war and malign politics, the exhaustion of land resources, the first symptoms of climate change – all have added to the total of people who have looked for security outside their native lands.

Together, the world's refugees constitute a huge and fluid group, blurred around the edges by those who seek betterment as much as security, but growing steadily into a major issue for the end of this century and the next.

Although it has experienced both internal unrest and border disputes, Jordan has been comparatively stable (by Middle Eastern standards) since the end of its civil war in 1971. A poor country with few natural resources, it has taken positive steps toward the improvement of its economy and the reduction of its dependence on overseas aid.

But its economic and political problems are heightened by a constant flow of refugees. This began shortly after the creation of the state of Israel in 1949, when Jordan granted citizenship to almost a million Palestinians. The Arab–Israeli wars of the 1960s brought another 300,000, mostly from the disputed West Bank. With the Iraqi invasion of Kuwait in 1990, yet another wave of refugees, forced to abandon any possessions they could not carry, threw themselves on Jordan's mercy.

The Vietnam War may have ended in 1973, but the consequences are still being felt, not least by the 1.5 million people who have fled to settle in the West. As many as one million of these refugees may have escaped in tiny boats, their only hope of survival to be picked up by larger vessels and taken on to sympathetic host countries.

More stable nations in Southeast Asia have accepted many thousands of refugees from Vietnam and Kampuchea, while others have found asylum in western Europe, the U.S., and Canada.

But the refugees' welcome is not always warm. Most potential host nations place restrictions on numbers. Some keep the would-be immigrants in special camps until their status as political, rather than economic, refugees has been established. In extreme cases, the boat people have found themselves repatriated and left to fend for themselves in their devastated homelands.

TAKING DOWN THE BARRIERS

Europe's new found identity

Ernest Bevin, Britain's plainspoken foreign secretary in the years immediately after the Second World War, dreamed of one day being able to buy a train ticket at Victoria Station in London and going "anywhere I damn well please." In his Europe, the frontiers still rose like walls to greet the traveler. Today, the walls are being flattened under the weight of the growing traffic across western Europe.

Bevin would have approved of the new Europe without frontiers, although he might have been surprised to learn that after mid-1993 he would be able to travel by train from London to Paris through a tunnel under the Channel. Neither Bevin nor anyone else of his generation envisioned this as a likely development.

Despite the advent of high-speed train travel, the average traveler between the European Community's 12 member-states is usually to be found in an airplane, rather than a train. By the end of the 1980s, the number of air passengers was increasing by 10 percent annually, and forecasts predicted that the figure would double by the year 2000 and treble by 2010. Flight delays were increasing, too – by 50 percent in the four years up to the end of 1990. Charles de Gaulle, near Paris, was one of the few major international airports which had room for additional runways.

Europe has undergone previous transportation revolutions. The Romans built a road system whose structure has lasted to this day, and the 19th-century railroad boom created a network that for the first time made it possible to cross Europe from the Atlantic to the Urals other than on horseback or on foot.

The current revolution, though, is remarkable for being political as well as technological. Removing the barriers to the free flow of trade and people within the European Community by the end of 1992 is seen in Brussels, the EC's capital, as meaning more than the commercial goal postulated in the drab phrase "completing the internal market." It is also a step toward a people's Europe, in which the Community hopes to shed its bureaucratic image and emerge as an institution that makes a positive contribution to the quality of life.

Transportation is a major element in the Community's economy, accounting for 7 percent of its gross domestic product. About 60 percent of the Community's trade is between its members, and the largest share of the tonnage is carried by road, with inland waterways (such as the Rhine) next, and the railroads a poor third.

The largest undersea excavation ever attempted, the Channel Tunnel will consist of three tunnels – two for trains and a smaller one for services – 31 miles (50 km) long and 100 ft (30 m) beneath the English Channel. Its completion, which will end Britain's geographical insularity by linking it physically with its fellow members in the European Community, will mark the end of a three-year civil engineering project of staggering proportions.

Advancing 10 ft (3 m) an hour, 11 tunnel-boring machines, each weighing nearly 1,200 tons and costing \$20 million, will have cut through a total of 265 million cubic ft (7.5 million cubic m) of spoil – three times the volume of the Great Pyramid of Cheops.

Trains reaching speeds of 100 mph (160 km/h) will make the underwater journey in 35 minutes and shorten the travel time between London and Paris to a mere three hours.

1952

In order to regulate their coal and steel industries, six European nations banded together initially in 1952. On March 25, 1957, the same nations – Belgium, France, Italy, Luxembourg, West Germany, and Holland – signed the Treaty of Rome, which brought the European Economic Community into existence. As members of the newly created Common Market, the signatories agreed to abolish mutual tariff barriers in order to promote an easy flow of people, goods, and money among member-states and to compete with the massive trade forces of Britain and the U.S.

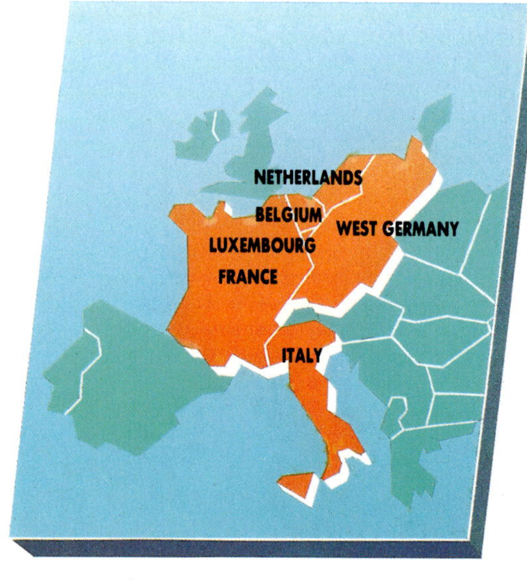

1973

The signing of the Treaty of Brussels on January 22, 1972, should have brought four more nations – Britain, Ireland, Denmark, and Norway – into the European Community. Together with the original six, they would have given the Community a population larger than that of the U.S., a gross national product approaching \$8 billion a year, and 41 percent of world trade. In the end, only the first three became members on January 1, 1973. Norway's membership was barred by the Norwegian people themselves, who voted against it in a national referendum.

The direct costs of frontier formalities account for nearly 2 percent of the value of these goods, so the savings will run into billions of dollars. Add to them the gains from all the other restrictions that have to be removed to complete the internal market, and the sum comes to around $280 billion, or a 5 percent increase in the Community's GNP.

Those are the economic and political incentives. Together with the ever-growing volume of intercommunity trade and passenger traffic, they are sufficient to justify the billions that will be spent over the next two decades on improving the rail and road networks. Brussels and Paris already vie with one another for the position of hub city to the new Europe. If Paris's plans are realized, it will be possible by the year 2005 for air passengers to land at an enlarged Charles de Gaulle airport and start their journey by high-speed train to a wide range of cities from Lisbon to Hamburg.

The most spectacular project is the Channel Tunnel (projected cost $15 billion), whose completion in 1993 should enable the railroads to compete more successfully for the growing (by 7 to 10 percent a year) traffic between Britain and the Continent. Train tunnel projects have become fashionable on the trans-Alpine routes between Italy and Germany, too, as non-EC members Switzerland and Austria have declined to lift restrictions on the ever-growing volume of heavy vehicles that congest their roads and pollute towns and countryside alike.

Denmark is at the center of three plans involving road and rail tunnels and bridges – collectively known as Scanlink – that will join Sweden and Norway, through the island of Zealand, to the rest of Denmark and Germany. At the other ends of the Community, the Autoput route through Yugoslavia to isolated Greece is being improved, and two new railroads and a road across the Pyrenees are under consideration. The Community's political future may have question marks attached to it, but they mean little to those who are building the means to knit it more closely together.

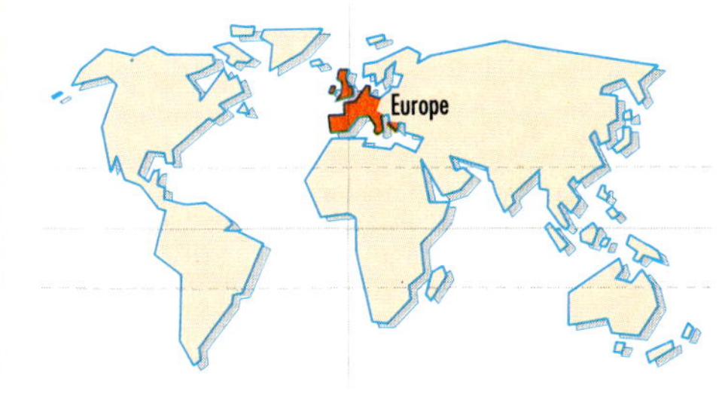

1981
A charter member of the United Nations and a member of NATO since 1951 (except for a six-year period ending in 1980), Greece achieved full membership in the European Community in 1981. Certain factions opposed entry, claiming it would subject Greece to domination by foreign capitalists. But Prime Minister Karamanlis, who signed the membership treaty in 1979, believed that entry could solidify the country's political and social stability.

Membership also broadened the markets for Greek products and opened the country to the EC's financial aid programs.

1986
Neighbors but not always friends, Spain and Portugal had already joined forces to strengthen their economies against European competition by the time they entered the EC in 1986. They had eliminated their trade barriers, and since then have cut inflation, reduced deficits, and shown healthy economic growth. Bilateral investment rose to $500 million by 1989, and both nations have attracted funds from the Community and increased foreign investment.

BEYOND THE NATION-STATE

Pursuing power with pacts and partners

THE INTERNATIONAL ORGANIZATION with the greatest number of member-states is not the United Nations proper, which has 160, but one of its agencies, the Universal Postal Union, with around 200. Begun in 1874 in Switzerland to replace the ludicrous multiplication of separate intercountry agreements, the Union exists to make the whole world into one single postal territory.

It is a prosaic example, perhaps, but one that symbolizes the need for – and the practicability of – international communication and cooperation. How else could a letter stamped and posted in one country be delivered to an address on the other side of the planet?

International systems of transportation or communication of necessity require some sort of international coordination – after all, locomotives need a track of the same gauge on both sides of a border, and telephone lines must connect. Likewise, air or sea traffic controllers not only share the same rules and procedures, they also communicate in a common language – English.

Today, almost every country is joined to others by multilateral ties, including regional or even wider-scale economic, political, and military alliances. These are the organizations – NATO, the European Community, OPEC, and the like – that regularly make the news.

But there also exists a much broader range of cooperative activities and agreements, among them aid and disaster relief, Interpol, extradition treaties, mutual recognition of patents and copyrights, high-tech projects, disease notification, and restrictions governing the abuse of the environment.

International agreements inevitably imply some loss of sovereignty. Even war has its shared rules, many of them encoded in the Geneva Convention of 1949 – history's most widely ratified treaty. Belligerents are forbidden to violate the territory of neutral states or attack civilian populations, or to use any weapons calculated to cause unnecessary suffering – like poison gas or projectiles designed to explode on contact with people – or long-term damage to the environment. Moreover, they must allow representatives of the International Red Cross and the Red Crescent to have access to the sick and wounded.

The global jigsaw of apparently distinct states represented on a political map presents far too fragmented a picture. The nation-state may be here to stay, at least for the foreseeable future. But so, too, are the pressures making for greatly increased regional and even global integration.

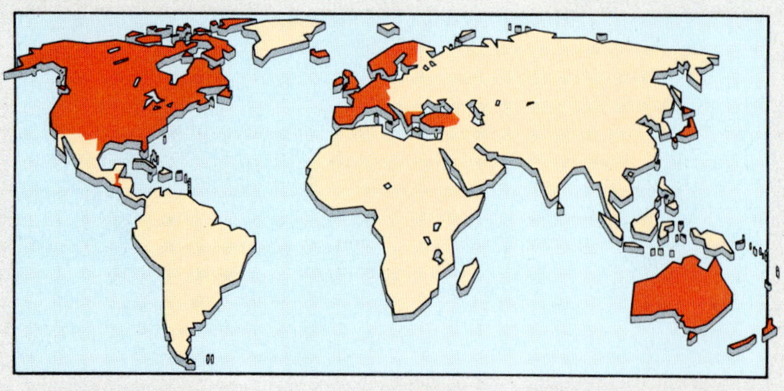

Organization for Economic Cooperation and Development (OECD)
Founded in 1961, the OECD is the body that coordinates economic and social policies between the industrialized nations, its objective being to sustain a high rate of economic growth among its members. It also encourages the flow of aid to developing countries.

The members are Australia, Austria, Belgium, Canada, Denmark, Finland, France, Germany, Greece, Iceland, Ireland, Italy, Japan, Luxembourg, Netherlands, New Zealand, Norway, Portugal, Spain, Sweden, Switzerland, Turkey, the U.K., and the U.S. Yugoslavia has special status but is not a full member.

European Free Trade Association (EFTA)
EFTA was established in 1960 as a trading link between countries excluded from the newly formed European Economic Community (now the European Community). The original members were Austria, Denmark, Finland, Norway, Sweden, Switzerland, and the U.K. Iceland joined in 1970, and Denmark and the U.K. have left to join the EC.

In the 1980s, cooperation between EFTA and the EC grew, tariffs between the member countries were abolished, and customs formalities were simplified.

The current members of EFTA are Austria, Finland, Iceland, Norway, Sweden, and Switzerland.

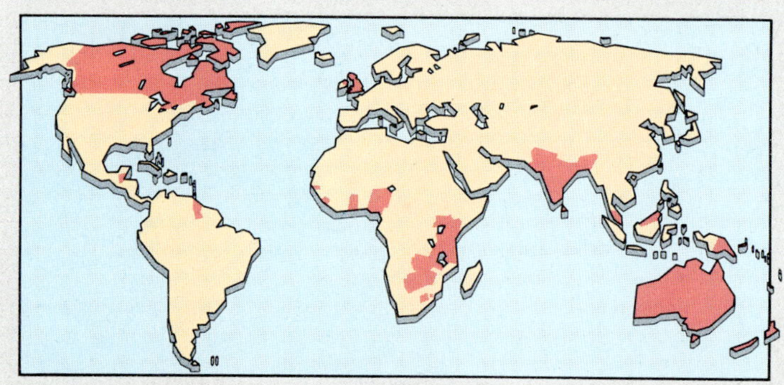

British Commonwealth
Born out of the British Empire, the Commonwealth officially came into existence in 1931. One former British colony, Burma, decided not to join the Commonwealth; the Irish Republic, South Africa, and Pakistan have all resigned.

The members are Antigua and Barbuda, Australia, Bahamas, Bangladesh, Barbados, Belize, Botswana, Brunei, Canada, Cyprus, Dominica, Gambia, Ghana, Grenada, Guyana, India, Jamaica, Kenya, Kiribati, Lesotho, Malaysia, Malawi, Maldives, Malta, Mauritius, Nauru, New Zealand, Nigeria, Papua New Guinea, St. Kitts and Nevis, St. Lucia, St. Vincent and the Grenadines, Seychelles, Sierra Leone, Singapore, Solomon Islands, Sri Lanka, Swaziland, Tanzania, Tonga, Trinidad and Tobago, Tuvalu, Uganda, the U.K., Vanuatu, Western Samoa, Zambia, and Zimbabwe.

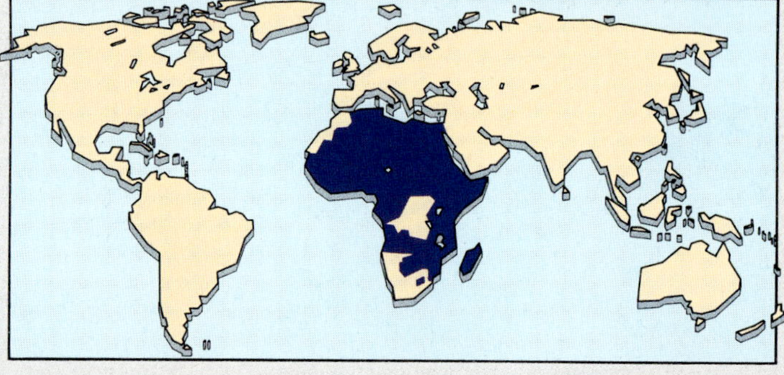

Organization of African Unity (OAU)
The OAU was founded in 1963 to promote African unity and solidarity. Every African state except South Africa and Namibia has been a member, although Morocco and Zaire resigned in the mid-1980s following a dispute over Morocco's annexation of Western Sahara.

The members are Algeria, Angola, Benin, Botswana, Burkina Faso, Burundi, Cameroon, Central African Republic, Chad, Comoros, Congo, Djibouti, Egypt, Equatorial Guinea, Ethiopia, Gabon, Gambia, Ghana, Guinea, Guinea-Bissau, Ivory Coast, Kenya, Lesotho, Liberia, Libya, Malawi, Mali, Mauritania, Mauritius, Mozambique, Niger, Nigeria, Reunion, Rwanda, Senegal, Sierra Leone, Somalia, Swaziland, Tanzania, Togo, Tunisia, Uganda, Zambia, and Zimbabwe.

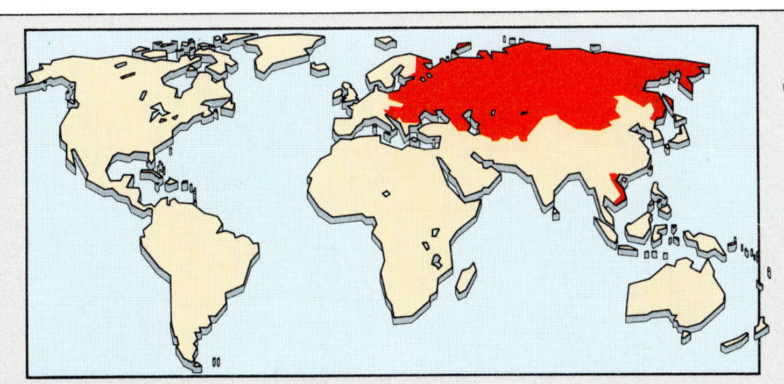

Council for Mutual Economic Assistance (Comecon)
The original members of Comecon were the USSR and five countries of eastern Europe. Its aims are to promote the development of the economies of the member-states within a socialist framework. Albania became a member in 1949 but withdrew in 1961; East Germany joined in 1950, but its membership will presumably lapse in the wake of German unification. In 1962, with the entry of the Mongolian People's Republic, the council was expanded to include widely dispersed states under Soviet influence.

The current members are Bulgaria, Cuba, Czechoslovakia, Hungary, Mongolia, Poland, Romania, the USSR, and Vietnam.

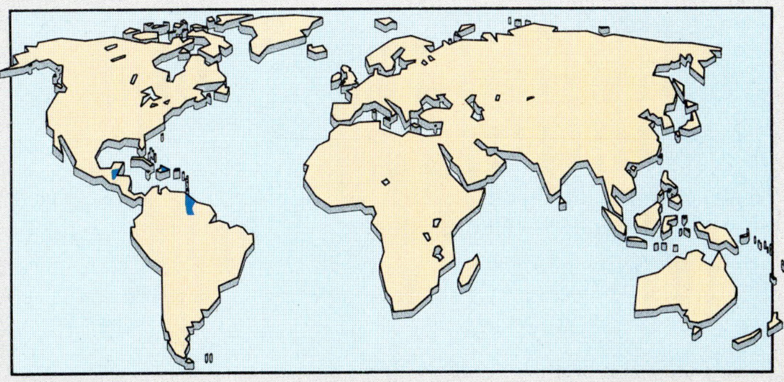

Caribbean Community and Common Market (Caricom)
Caricom is the trade association of the Caribbean nations, with a common economic policy and a joint policy on tariffs to protect members against cheap imports from outside. It also promotes health and other social projects for its less-developed members. It was founded in 1973 and has its headquarters in Georgetown, Guyana.

Full members are Antigua and Barbuda, Bahamas, Barbados, Belize, Dominica, Grenada, Guyana, Jamaica, Montserrat, St. Kitts and Nevis, St. Lucia, St. Vincent and the Grenadines, Trinidad and Tobago. The Dominican Republic, Haiti, and Suriname have observer status.

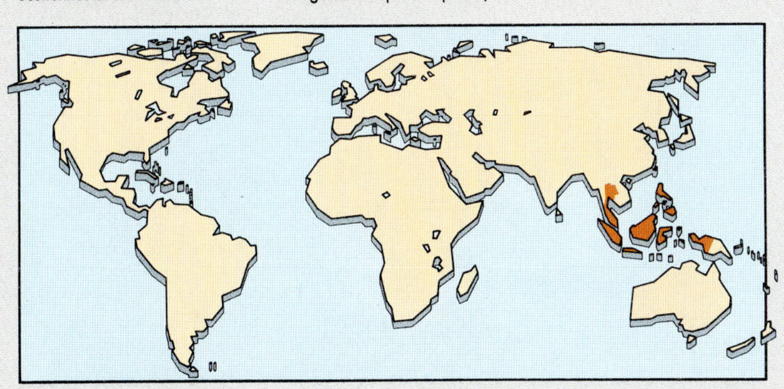

Association of Southeast Asian Nations (ASEAN)
ASEAN was founded in 1967 to promote the economic growth and mutual cooperation of the industrializing nations of Southeast Asia. It is also concerned with maintaining peace in the area. The rapid economic growth of the five founding states in the 1970s (while the developed world was in recession following OPEC's massive increase in oil prices) has made ASEAN an important force in regional affairs. Oil-rich Brunei joined shortly after it became independent in 1984.

The members are Brunei, Indonesia, Malaysia, the Philippines, Singapore, and Thailand.

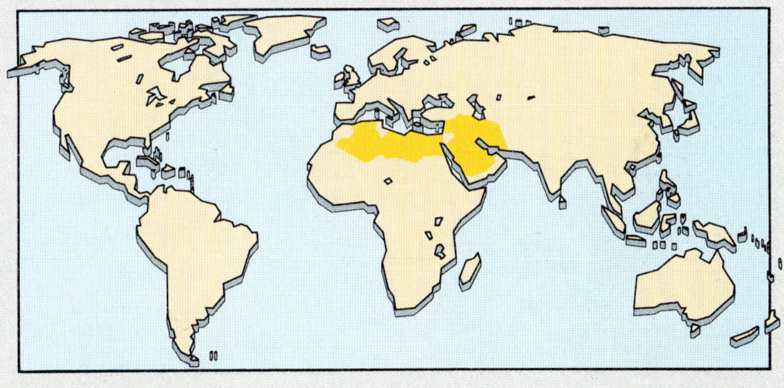

Organization of Arab Petroleum Exporting Countries (OAPEC)
The defined aims of OAPEC are "to promote cooperation in economic activities; to safeguard members' interests; to unite efforts to ensure the flow of oil to consumer markets; to create a favorable climate for the investment of capital and expertise." Founded in 1968, it has its headquarters in Kuwait. As in the Arab League, Egypt was expelled in 1979 and readmitted only in 1989.

Members are Algeria, Bahrain, Egypt, Iran, Iraq, Kuwait, Libya, Qatar, Saudi Arabia, Syria, and the United Arab Emirates.

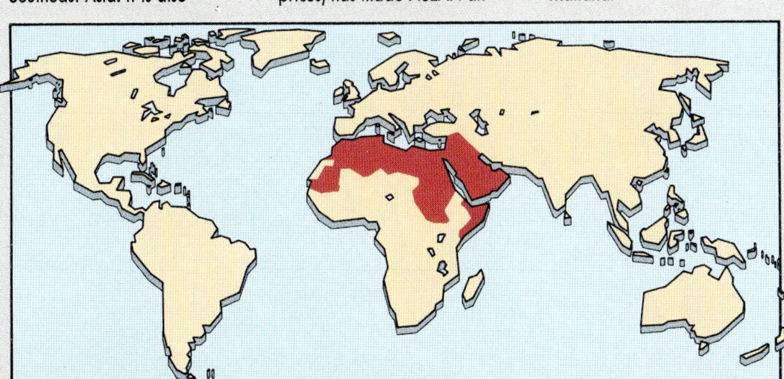

The Arab League
Originally concerned with political, economic, social, and cultural matters, and with mediating disputes between its members, the Arab League now also cooperates over defense. Dissension within the League reflects the problems of the Middle East: in 1964, Jordan protested at the admission of the Palestine Liberation Organization as the representative of all Palestinians; and in 1979, Egypt, a founder member, was suspended after signing a peace treaty with Israel. It was reinstated only in 1989.

Members are Algeria, Bahrain, Djibouti, Egypt, Iraq, Jordan, Kuwait, Lebanon, Libya, Mauritania, Morocco, Oman, Palestine, Qatar, Saudi Arabia, Somalia, South Yemen, Sudan, Syria, Tunisia, the United Arab Emirates, and Yemen.

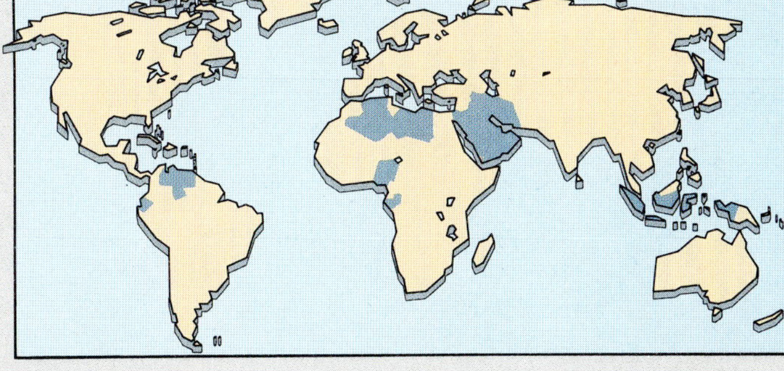

Organization of Petroleum Exporting Countries (OPEC)
Founded in 1961, OPEC caught the world's attention in 1973, when it announced a 70 percent and then a further 130 percent increase in oil prices. OPEC's principal aim is to monitor the supply of oil from its members to world markets. But by 1973 it had become a more political organization, and the price rise was a punishment for the West's role in the Yom Kippur War.

With diminished demand for oil in the late 1980s, OPEC imposed a quota system to prevent the market's being flooded. Iraq's objection to Kuwait's exceeding its quota caused conflict in 1990.

Members are Algeria, Ecuador, Gabon, Indonesia, Iran, Iraq, Kuwait, Libya, Nigeria, Oman, Qatar, Saudi Arabia, the United Arab Emirates, Venezuela, and Yemen.

THE CONTINUING CALL TO ARMS

Not yet a world without war

LEADER FOR A NEW WORLD ORDER

Both the General Assembly and the Security Council of the United Nations monitor and try to maintain world peace.

The General Assembly, shown here, includes every member-state. Each is allowed one vote. The New York-based assembly scrutinizes all matters concerning the U.N.'s administration.

But the General Assembly has limited powers when it comes to action to preserve peace. In the terms of the original charter, the decision for action must be taken by the Security Council, which can meet almost immediately in times of crisis. The council has 15 members.

There are very few independent states that do not belong to the U.N. Some are absent by choice, notably Switzerland; a few others are excluded: South Korea because the Soviet Union objects, North Korea because the U.S. objects. Although it is within the powers of the U.N. to expel members, it has never done so.

AROUND 100 MILLION PEOPLE HAVE been killed in the 250-odd civil and international wars recorded this century. At any one time there are between 10 and 40 wars in progress, involving up to one-fourth of the world's states. Sometimes the dispute is settled after a short and relatively bloodless combat. But in many cases, hostilities continue for years.

Nothing may be gained by either side, but so many lives are lost or ruined, so many people displaced, and such damage done, that participating countries are devastated. In each of six states – Bangladesh, Cambodia, China, Korea, Nigeria, and Vietnam – more than a million people have been killed since 1945. In 1986, International Peace Year, the world spent nearly $1 thousand billion on military activities.

Conflicts are serious enough when they are confined to their own locality. But today, every dispute threatens to spill over onto the regional or worldwide stage.

In 1945, with the memory of the Second World War fresh in every mind, 50 nations met in San Francisco to sign the Charter of the

One of the functions of the United Nations, as defined in its charter, is "to save succeeding generations from the scourge of war." It is the responsibility of the Security Council to determine whether an act of aggression has taken place – or is likely to – and to take appropriate action.

When the U.N. came into being in 1945, protection was particularly needed for territories so small they could not hope to be self-governing, and for the former colonies of defeated powers, such as Germany's African lands, which could easily have been annexed by more powerful neighbors.

At the beginning of the 1990s, the U.N. had peacekeeping forces or observers in Angola, Cyprus, Central America, the Golan Heights, Lebanon, and Kashmir. Other trouble spots such as Cambodia and Western Sahara may require its intervention in the near future.

Its principal role in these areas of conflict is to see that ceasefires or newly negotiated boundaries are respected. The U.N. has no army of its own, but relies on its member-states to provide the military strength it needs. Sweden and Ireland in particular are renowned for their willingness to support U.N. peacekeeping initiatives.

Throughout the life of the U.N., the major powers have largely bypassed it and dealt directly with each other. Only occasionally has the U.N. prevented wars between fractious neighbors, although some progess in the principle of international arbitration can be claimed.

Many people – politicians and general public alike – are skeptical about the U.N.'s role. They claim, reasonably enough, that it has been caught up in big-power rivalry. But with the end of the Cold War, Soviet vetoes of U.S.-backed resolutions will no longer be a matter of course, and greater unity may be seen within the Security Council. Certainly there is no likelihood of an international peacekeeping body becoming superfluous.

The map indicates major conflicts since the Second World War – it is worthy of note that most conflicts in the latter half of the 20th century have taken place south of the Tropic of Cancer.

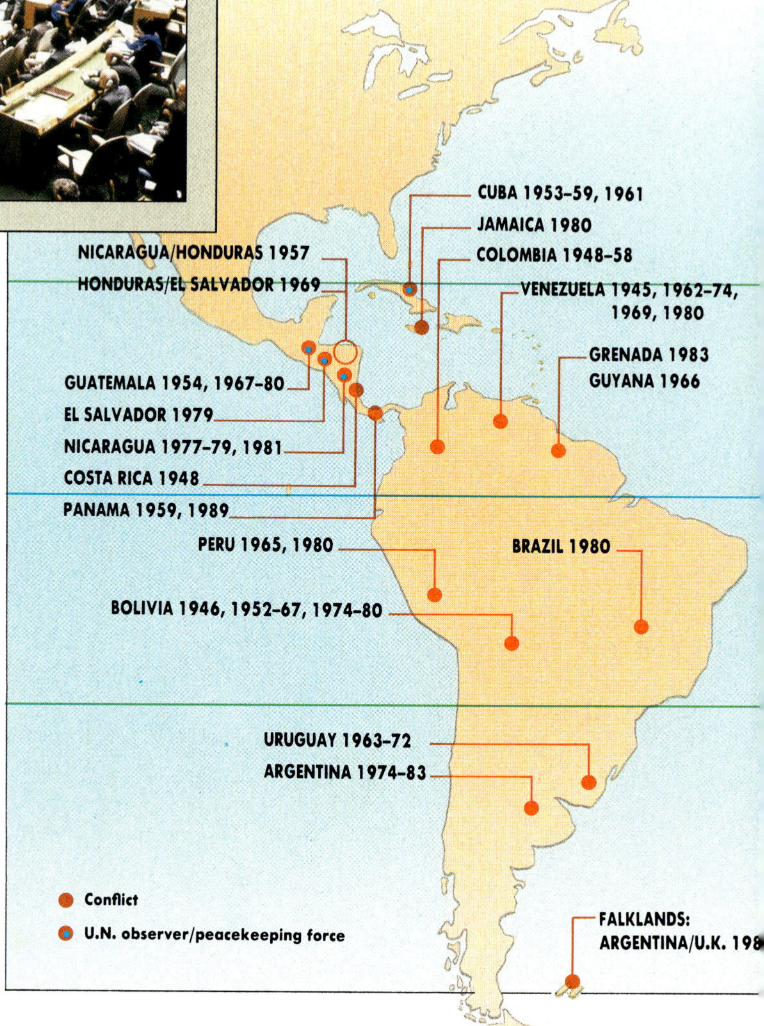

United Nations Organization. Its aim was simple: "to save succeeding generations from the scourge of war, and to reaffirm faith in fundamental human rights, in the equal rights of men and women, and of nations large and small." States were promised the rule of law, freedom from interference in internal affairs, and security from external aggression.

The guarantors of this new global order were the five big powers of the day – Britain, China, France, the Soviet Union, and the U.S. They became the permanent members of the U.N.'s executive body, the Security Council. But every state, regardless of population, area, or economic power, was to be entitled to a single vote in the General Assembly. Membership in the U.N. has grown over the years to 160, from China, with over a billion people, to the latest recruit, Liechtenstein, with fewer than 30,000.

As a last resort, U.N. members are prepared to go to war to preserve peace. The charter stipulates that states must settle disputes by peaceful means, that force, or the threat of force, should not be used except in self-defense or in the common interest. Where agreement cannot be reached, the U.N. should be called upon to mediate – to monitor events, to act as a neutral buffer between disputants, to establish ceasefire lines, to monitor elections, and to negotiate with both parties.

Only in the event of the U.N. failing to bring about some resolution of the dispute can armed conflict legitimately be initiated – a right that Britain claimed when it retaliated against Argentina during the Falklands War of 1982. In cases of extreme delinquency, where there is believed to be a profound threat to world peace, the U.N. is empowered to take collective measures, chiefly economic sanctions or an arms embargo, but also military intervention.

Yet, powerful as it might sound, the U.N. has often found it impossible to intervene effectively to settle disputes. It may be difficult to obtain international agreement for a course of action, sanctions often seem easy to avoid, and the task of low-profile policing of an uneasy truce can prove costly, dangerous, and ultimately futile.

What, then, of concerted military action? The use of force has received Security Council assent on very few occasions, notably during the Korean War (1950–53), in Congo/Zaire when the natives rebelled against their Belgian rulers (1960–64), and after the Iraqi invasion of Kuwait (1990). Yet the U.N. is undoubtedly handicapped by having no standing army of its own, nor even an agreed command structure in case action is authorized. It can only make up a force from the armed services of consenting member-states, and few are willing to subordinate themselves to foreign commanders.

Nevertheless, the U.N. has been involved in a great number of disputes, in both a mediating and a military role. U.N.-coordinated troops have seen action in 18 theaters of war since 1946, when they were first called upon to enforce a ceasefire between Arab countries and Israelis in the Middle East. In 1990, in addition to troops stationed around the Persian Gulf, 11,000 U.N. troops were deployed in eight locations around the world. For all its limitations, the U.N. is the only international peacekeeper the world has at its disposal.

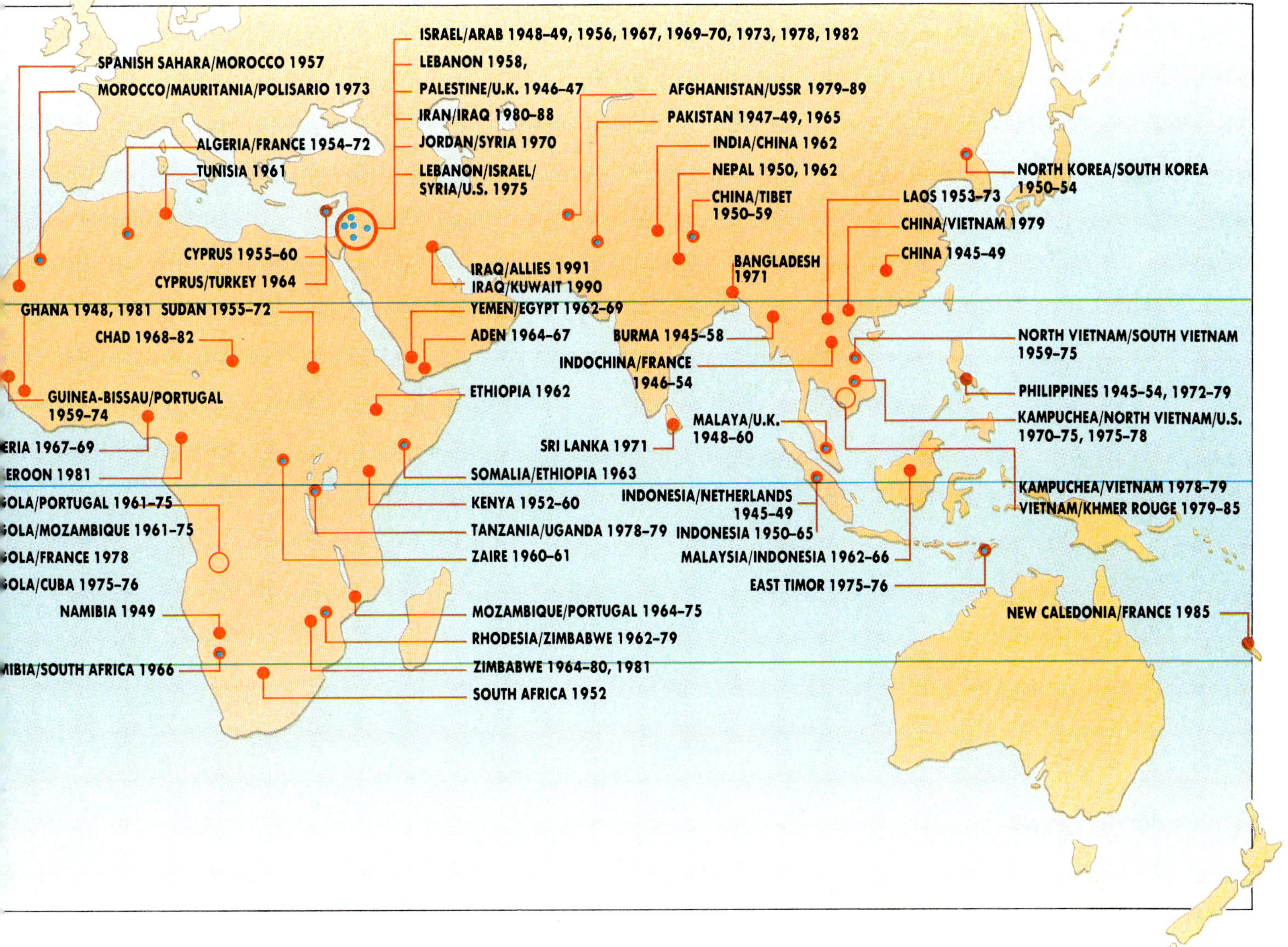

Mapping the World

When the Jesuit priest Matteo Ricci arrived in China in 1601 he found a highly sophisticated people who had been making detailed maps for thousands of years, and yet they believed the world to be square.

In the imperial Chinese view of things, he records in his memoirs, Earth and Heaven are constructed as a series of squares within squares. At the dead center of everything sits the god-emperor in residence within Beijing's square Forbidden City, which is oriented north, south, east, and west.

Around the capital lies a succession of square zones of decreasing civilization: first the directly administered royal domains, then a zone where local rulers send loyal tribute to the emperor, then peoples who have still to be pacified – frontier areas in the process of acquiring the elements of Chinese culture. Around these are ever-larger squares of increasing barbarism and, ultimately, utter savagery.

Ricci had arrived from the outermost square. Moreover, he carried with him a map of the world made by Mercator's friend Ortelius in which China appeared, very small, on the extreme right-hand side. It was too much for China's ruling elite to bear.

But Ricci hit upon a brilliantly simple expedient. He cut the map vertically and reconstructed it with China in the center of things. This somewhat mollified his hosts, who allowed him to stay on in the city, but they could never accept that his flat map was still only a representation of a globe. Indeed, this view was not to be generally accepted in China until the early 20th century.

The idea that the real world is a sphere has a long history stretching back at least as far as the Ancient Greeks 2,500 years ago. Indeed, Greek geometers – the word geometry originally meant Earth measurement – went so far as to estimate its radius and circumference to a quite extraordinary degree of accuracy. This concept, kept alive in the writings of Ptolemy and his successors – Christian, Jewish, and Muslim – eventually influenced the Genoese sailor Christopher Columbus. In 1492 he set sail westward, rather than eastward, to find the Indies, since he would surely end up in the same place whichever way he went.

During the following century Portuguese and later Dutch, English, and other sailors visited just about every coastline as far east and north as Japan. Their findings were duly reported back and represented on printed – no longer hand-drawn – maps. The greatest mapmaker of the 16th century was Gerhard Kremer. He was renamed Mercator ("merchant") in recognition of his efforts to provide more reliable charts for the new ocean-going traders.

By 1569 Mercator had developed a rigorous means of representing the spherical form of the globe on a flat surface. Lines of longitude, which converge at the poles, are splayed out to become parallel with one another, while the already parallel lines of latitude are drawn farther apart with increasing distance from the equator.

Although it had the effect of distorting the sizes and shapes of landmasses, Mercator's projection soon became standard. For the first time it enabled navigators to plot a straight-line course between points on a map and read the direction off as a true bearing.

Maps swiftly came to look like the ones used today. The great sprawling Unknown Southern Land (in Latin *Terra Incognita Australis*), assumed by Ptolemy and many of his successors to join the bottom of Africa with Asia, was finally removed from maps at the end of the 18th century, after Captain Cook spent several fruitless years searching for it. The idea is commemorated in the name Australia for the continent which was visited by Portuguese, Spanish, and Dutch navigators early in the 17th century. Antarctica was not sighted until 1820.

Since Mercator's day great advances have been made in mapping techniques, chiefly in the use of triangulation and, more recently, aerial photography and remote sensing by satellite. Until the 1960s the true distance between London and New York, for instance, was only known to within a mile or so, and the location of the Pacific Islands with an error margin of 2–3 miles (3–5 km). In the last few years the use of satellites has made measurement much more precise. Indeed, it has recently become possible finally to prove the reality of continental drift – the theory that landmasses are moving relative to one another by 1–2 in (2.5–5 cm) a year.

The Earth is very much better known, and certainly very much better mapped, today than ever before. Yet it is hard to escape the feeling that our picture of the planet suffers from some of the same problems that imperial China's square world did. In a sense modern world maps

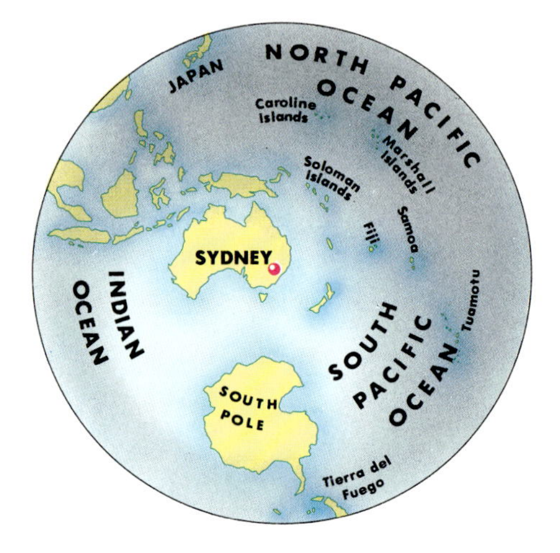

THE IMPORTANCE OF PERSPECTIVE

The world looks different to people depending on where they stand. The vantage points here reveal a variety of national perspectives on neighbors and the rest of the world. The view from **London** *includes most of the world's population. Countries and continents appear to radiate around it, which helps to explain the historical significance of western Europe.*

The view from **Moscow** *cannot be a very comfortable one. Although the Soviet Union has the largest landmass of any country in the world and prolific natural resources, the length of its borders, and the nature of its terrain and climate mean that agriculture, industry, and communications are difficult whatever its political and economic system.*

Viewing the world from **Sydney***, Australia, means looking at vast stretches of water rather than land. Banishment of Britain's criminals to Australia sent them to the other side of the planet, held captive by these oceans. Although Australia has strong historical ties with Britain, it is seeking to establish closer economic ties with countries around the Pacific Rim.*

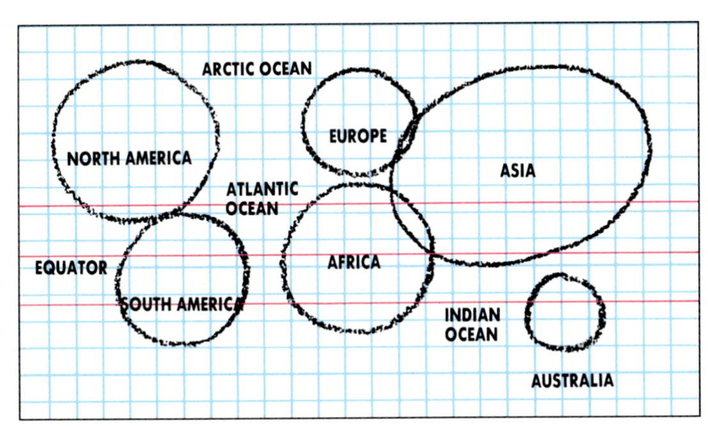

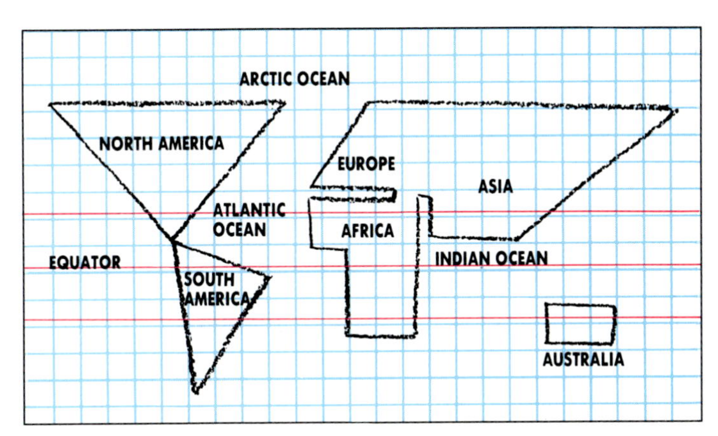

HOW TO DRAW THE WORLD IN 30 SECONDS

Six quickly sketched circles, roughly in the right places and in roughly proportionate sizes, make a working map of the continents. Asia is the biggest, Australia the smallest.

Turn the continents into squares, rectangles and triangles. Remember that the African bulge is over the equator, the Tropic of Cancer underpins Asia, and the Tropic of Capricorn cuts Australia in half.

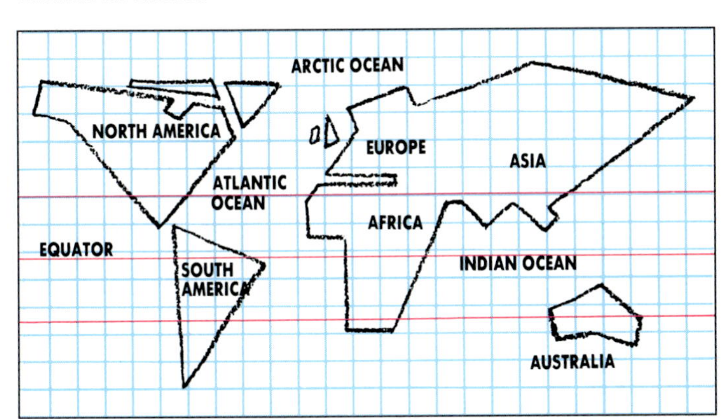

With a few more lines, regional and national identities emerge. India is one more triangle; Scandinavia, the beak of Europe. There is a valid map for making political and economic points.

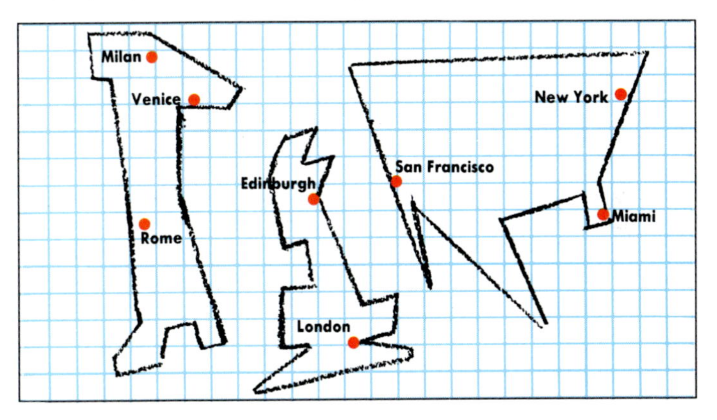

For everyday use, reduce your own country to a simple shape. With important cities as spatial markers, you have the working outline for most nontechnical geographical needs.

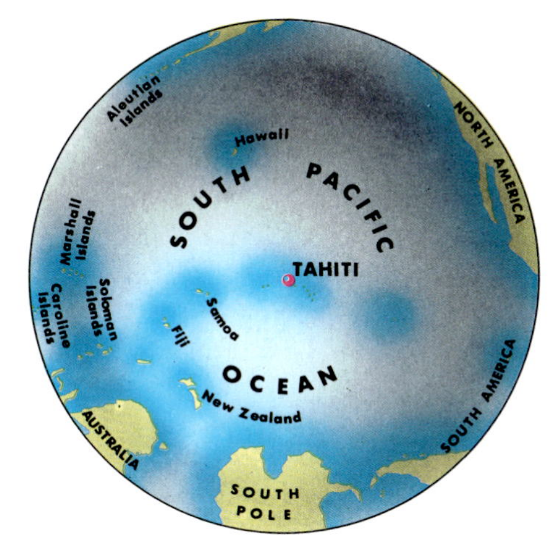

To a ***Tahitian*** *the name "Earth" must sound absurd, since the island lies at the center of a hemisphere of water. From this vantage point only the margins of other continents can be seen. France tests nuclear weapons here; the chorus of protest from Pacific islanders is hardly heard in France's half of the world.*

San Francisco*'s port at Oakland is perfectly positioned for the shipment of goods between the U.S. and Asia. California's economy, like the economies of its trading partners in the Pacific, has boomed. Resources, technology, and mass markets abound in the region, facts which strongly suggest the Pacific Rim will dominate trade in the 21st century.*

Like neighboring China, ***Japan*** *repudiated Western influences until it was forced to recognize them in the last century. Of the two countries, it was the relatively tiny Japan that became the more economically and militarily dynamic – apparently in defiance of its geographical position, terrain, and natural resources.*

Strabo's world view represents the total of the mapmaker's knowledge some 2,000 years ago. It is remarkably accurate, not only for the entire Mediterranean and the Black Sea but also for the Arabian peninsula and western Europe. He correctly places many major rivers. Notice also his assumption that Libya (which we now call Africa) could be circumnavigated.

Strabo did not think of the world as flat but as a globe. The world he shows, which stretches from Iberia (modern Spain and Portugal) on the Atlantic Ocean in the west to India in the southeast, is thought to be a relatively small island-continent in a vast universal ocean.

The great chain of mountains that runs from Turkey in the Mare Internum *due east is not wholly fanciful, as a modern physical map of the world will readily show.*

are Euro-centric, exaggerating Europe's size and importance and subtly understating those of other areas.

Accepted names for continents, seas, rivers, settlements, and so on are generally those given by Europeans, and the very shapes, relations, and orientations of land and sea may reflect European predominance. Regions like the Near, Middle, and Far East, for instance, are only near, middle, and far with respect to Europe.

Take also the position of the baseline from which all measurements of longitude are made, 0 degrees, the meridian of Greenwich just outside London. Greenwich was chosen (in 1884) in large part because Britain was at that time the preeminent maritime power, with much the best naval charts. "Universal time" is measured from Greenwich, too.

As a result, most projections center Europe on the midline of the map, with the rest of the world off to the sides. During the Cold War following the Second World War, this version of the world map had the effect of exaggerating the opposition of North America – to the west – and the Soviet Union – to the east. A different view would have noticed that they are a mere 50 miles (80 km) apart across the Bering Strait.

Today, it is increasingly common for countries to center themselves in their own world maps. But another convention generally remains. Perhaps no less important in generating unconscious attitudes to the world is the habit of orienting maps with north at the top. Many South Americans, sub-Saharan Africans, and Australians, for example, believe this places them in a literally as well as metaphorically "inferior" position, and are drawing "upside-down" maps to compensate.

The world according to Strabo 63BC–21AD

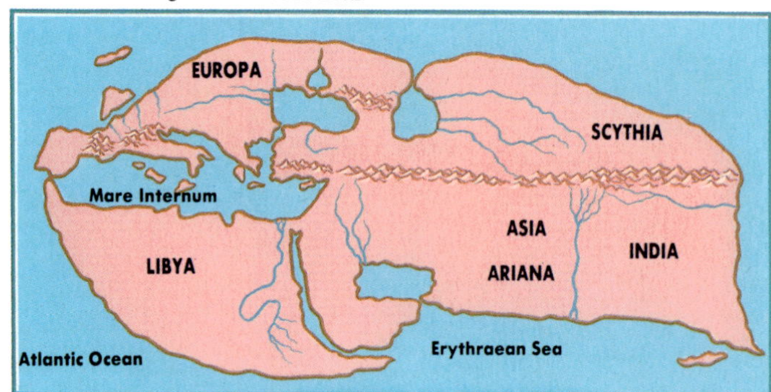

The 12th-century Arabian geographer Edrisi represents the world in much the same way as Strabo and Ptolemy did. Indeed, Ptolemy's writings were kept alive by Islamic scholars in the Middle East, North Africa, and Spain before they were rediscovered during the Italian Renaissance. The picture Edrisi draws of an abundance of islands in the east reflects Arab trading activities: merchants were taking advantage of monsoon winds to travel to India and the Spice Islands.

The world according to Edrisi 1154

The world according to Fra Mauro 1459

Three hundred years later the picture has changed very little, although there is now great confidence that Africa can be circumnavigated. Fra Mauro, who worked in Venice, was kept informed of Portuguese discoveries (although they probably also suppressed some of their most important findings as trade secrets). By this time Portuguese sailors were regularly sailing as far as Sierra Leone in West Africa. Bartolomeu Dias rounded the Cape of Good Hope in 1488.

The First Globe

Greek geographers such as Eratosthenes, Strabo, and Ptolemy, who lived between the 3rd century BC and the 2nd century AD, were no "flat earthers." It was as obvious to them as it is to us that the world is a sphere. Their great predecessor, Pythagoras (c.582–500 BC), had argued for a spherical Earth on the grounds that only the sphere was sufficiently perfect.

But his successor Aristotle (384–322 BC) adduced evidence from the senses, for example the fact that a ship may remain visible from the top of a cliff, although from the shoreline it has apparently disappeared over the horizon. He also noted that the Earth casts a circular shadow on the moon during an eclipse.

The Greek geographers made remarkably accurate estimates of the size of the Earth by measuring the length of the shadow cast by a pole of standard length on midsummer's day. As early as 240 BC Eratosthenes had observed that on June 21 the sun was directly overhead at Syêne (Aswân) in Egypt, yet it was 7.5 degrees off the vertical 500 miles (800 km) due north at Alexandria. Assuming regular curvature, he calculated about 8,000 miles (12,875 km) for the diameter and 25,000 miles (40,230 km) for the circumference (which is just about right).

Eratosthenes also introduced an elementary grid system of longitude and latitude into his maps, with north at the top and east at the right-hand side of the map. It is this system, greatly developed first by Ptolemy and then by Mercator, that we still use today.

Between the time of the Greeks and Columbus, the characteristic view of the world as far as Europeans were concerned was not of a globe but of a flat circle, a mappa mundi *or world plan. For religious reasons, scholars in the Middle Ages interpreted geography in terms of the Crucifixion. In these so-called T-in-O maps, a particularly fine example of which is still to be found in England's Hereford Cathedral* (below), *Jerusalem stands at the center of the circle, with a symbolic cross (in fact a T-shaped interior sea) separating Asia at the top from Europe, bottom left, and Africa, bottom right. Paradise is placed far away to the east across Asia, at the very top of the map, where the sun rises.*

Ptolemy's original maps have disappeared. This 16th-century reconstruction (left) *of one of his spherical projections shows the world that was known at the time. Many parts of Europe and the Middle East are easily identifiable (notice, for example, the British Isles), but south Asia is less recognizable; these errors may have been introduced after Ptolemy. Southern Africa, the Far East, Oceania, and the Americas are missing.*

Notice also Ptolemy's depiction of an immense unknown continent connecting Africa and Asia and enclosing the Indian Ocean. The Tropic of Cancer and the equator, or equinoctial circle, are accurately positioned, although the equator should pass just to the south of Taprobana (modern Sri Lanka).

Over the Horizon, the New World

Willem Janszoon Blaeu, the "Rembrandt of cartography," was the greatest mapmaker of the early 1600s. Working from Amsterdam, already the world's most dynamic port, Blaeu produced many printed maps and atlases, including the *New Geographical and Hydrographical Picture of All the Lands of the World*.

This map reflects the world view of the best-educated people of his day. Although it is more familiar to our eyes than most earlier maps, it is still a curious mix of fantasy and reality. The latest findings are blended with guesswork and a range of mythological allusions, classical references, and decorative devices, including fighting ships, strange sea monsters, and immense flying fish.

Symmetrical images around the map show the Seven Planets as gods (they include the Sun and Moon; this was still an Earth-centered cosmology), the Four Seasons, the Seven Wonders of the Ancient World, and the Four Elements of Fire, Air, Water, and Earth. To the left and right at the bottom of the map are two views of the world from "above" and "below" which carry little information.

Yet the map itself is generally accurate and highly detailed. The projection, the shapes of the continents, and their relative positions hardly look different from those on a modern map. The Old World of Europe, Asia, and Africa is particularly good, but even the Caribbean and South America are well delineated, reflecting the great achievements of the European explorers since 1450.

Coastlines, ports, and islands are most emphasized, but the interiors of continents were still uncertain. Although Asia was well known as a result of trade links to the East along the Silk Road, the apparently detailed interior of Africa, with its complex of rivers and lakes, is largely fanciful.

North America's interior is blank. Its Atlantic and Pacific coastlines are reasonably accurate, particularly in the northeast, where furs and fish were important items of trade with Europe. But absolutely nothing was known about the terrain beyond the coastal ranges to east and west, or the nature of the Arctic coastline.

To the south another great blank, the Unknown Southern Continent *Magallanica*, sprawls across the entire map – neither Australia nor New Zealand were known to Europe until the late 18th century.

A New Geographical and Hydrographical Picture of All the Lands of the World, *engraved by the Amsterdam-based cartographer Willem Janszoon Blaeu in 1606. It was frequently reprinted with small modifications as a result of new discoveries and errors reported by intrepid voyagers. This edition probably dates from around 1620.*

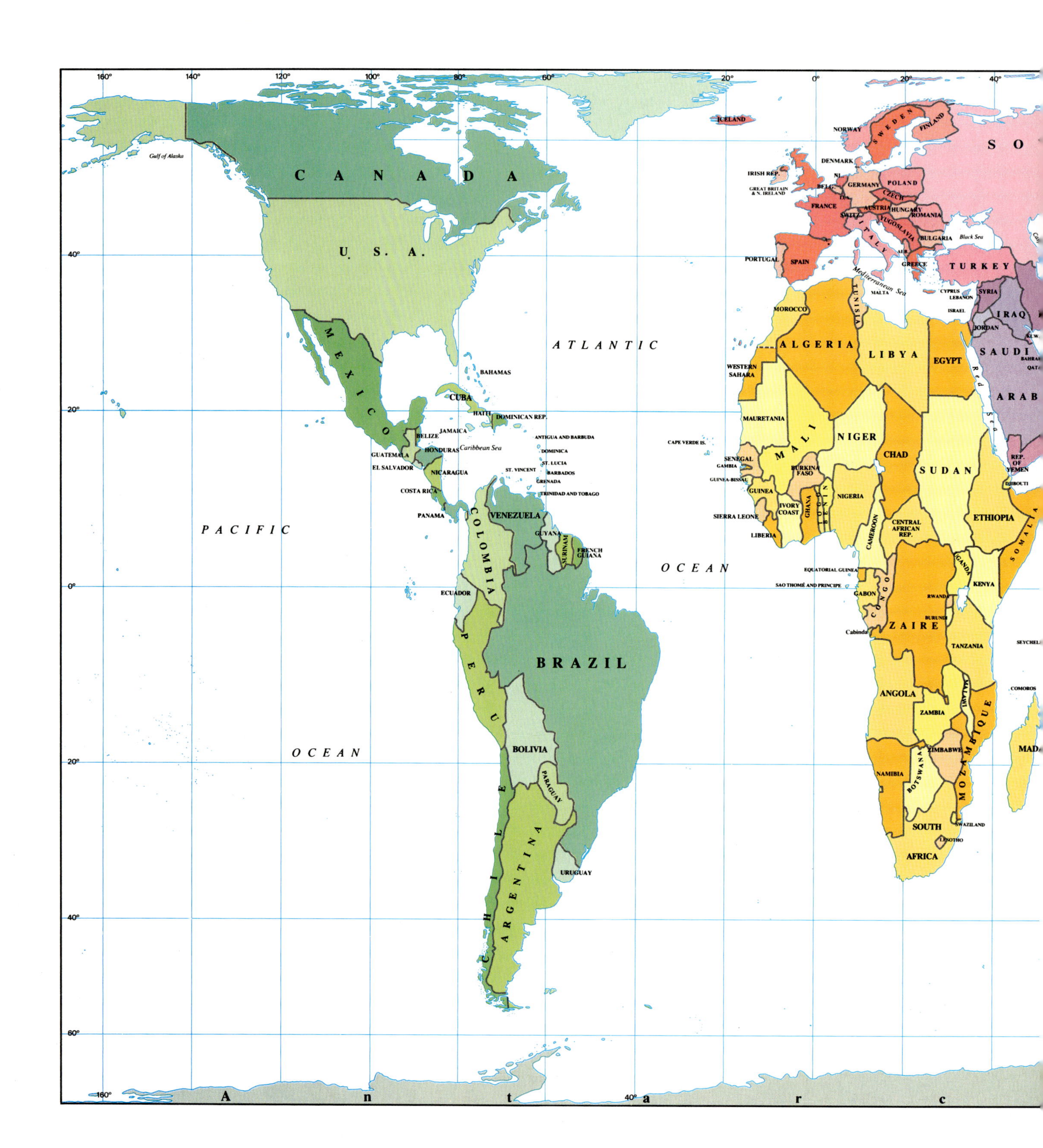
CANADA
U. S. A.
MEXICO
ATLANTIC
OCEAN
PACIFIC
OCEAN
BAHAMAS
CUBA
HAITI
DOMINICAN REP.
JAMAICA
BELIZE
HONDURAS
Caribbean Sea
GUATEMALA
EL SALVADOR
NICARAGUA
COSTA RICA
PANAMA
ANTIGUA AND BARBUDA
DOMINICA
ST. LUCIA
ST. VINCENT
BARBADOS
GRENADA
TRINIDAD AND TOBAGO
VENEZUELA
GUYANA
SURINAM
FRENCH GUIANA
COLOMBIA
ECUADOR
PERU
BRAZIL
BOLIVIA
PARAGUAY
CHILE
ARGENTINA
URUGUAY
Gulf of Alaska
ICELAND
NORWAY
SWEDEN
FINLAND
DENMARK
IRISH REP.
GREAT BRITAIN & N. IRELAND
NL
BELG.
GERMANY
POLAND
CZECH
FRANCE
SWITZ.
AUSTRIA
HUNGARY
ROMANIA
YUGOSLAVIA
ITALY
BULGARIA
Black Sea
PORTUGAL
SPAIN
ALB
GREECE
TURKEY
Mediterranean Sea
MALTA
CYPRUS
LEBANON
ISRAEL
SYRIA
IRAQ
JORDAN
SAUDI ARAB
QAT
TUNISIA
MOROCCO
ALGERIA
LIBYA
EGYPT
Red Sea
WESTERN SAHARA
MAURETANIA
MALI
NIGER
CHAD
SUDAN
CAPE VERDE IS.
SENEGAL
GAMBIA
GUINEA-BISSAU
GUINEA
BURKINA FASO
NIGERIA
REP. OF YEMEN
DJIBOUTI
SIERRA LEONE
IVORY COAST
GHANA
TOGO
BENIN
ETHIOPIA
LIBERIA
CAMEROON
CENTRAL AFRICAN REP.
SOMALIA
EQUATORIAL GUINEA
UGANDA
KENYA
SAO THOMÉ AND PRINCIPE
GABON
CONGO
RWANDA
BURUNDI
ZAIRE
Cabinda
TANZANIA
SEYCHELL
ANGOLA
ZAMBIA
MALAWI
COMOROS
MOZAMBIQUE
MAD
ZIMBABWE
NAMIBIA
BOTSWANA
SWAZILAND
SOUTH AFRICA
LESOTHO
Antarc
SO
160°
140°
120°
100°
80°
60°
20°
0°
40°
20°
0°
20°
40°
60°

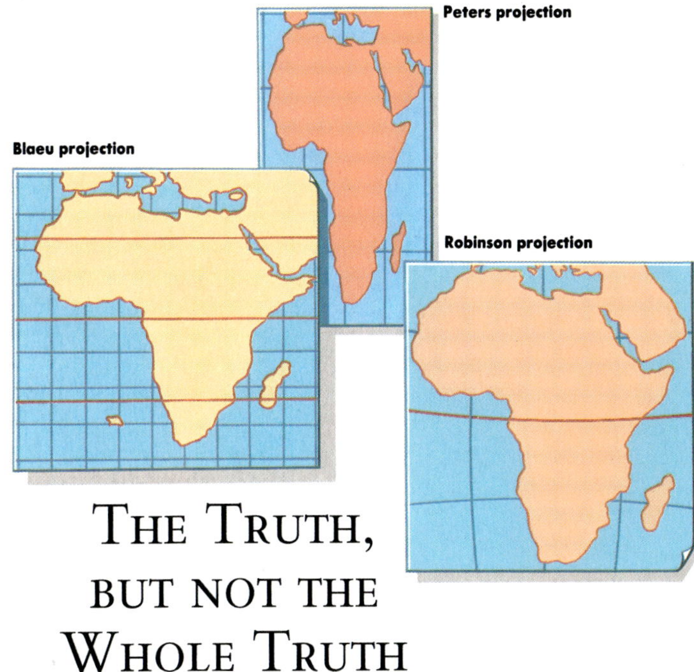

THE TRUTH, BUT NOT THE WHOLE TRUTH

U.S. CARTOGRAPHER ARTHUR ROBINSON describes Arno Peters's map of the world, published in 1973, as "somewhat reminiscent of wet, ragged, winter underwear hung out to dry on the Arctic Circle." But for many, the Peters projection is an almost obligatory antidote to some of the biases inherent in most earlier maps, particularly those following Mercator's principles. Mercator's projection is still the one most widely seen and recognized as the real world.

Compared with a globe, Mercator's representation of the world systematically distorts landmasses. Although his projection is reasonably accurate at the equator, distortion increases with latitude. Take Greenland, for example, which appears much the same size as China, whereas in fact it is only a quarter of the area. Similarly, Europe seems almost the same size as South America, which is twice the area. At the Antarctic most maps omit a major part, which means the equator lies not across the center of the map but toward the bottom.

Mercator's projection underscores a sense of European and North American superiority because it exaggerates their size at the expense of that of many third world countries, most of which are situated near the equator. Peters believes this bias should now be rectified by adopting an area-faithful projection.

In fact there are many other equal-area representations of the world that are less strange to our eyes. Indeed, Arthur Robinson has devised a highly successful one that has recently been adopted by the *National Geographic*. Yet a problem with his map, which the Peters map avoids, is that it has to bend lines of longitude, and hence distorts shapes in a different way.

Which is the true representation of Africa (above)*? In a sense, all are. How a country, a continent, or a sea appears on a map varies with the choice of projection. Blaeu's exaggerates the size of places according to how far they are from the equator, while the Robinson and Peters projections are area-faithful. Yet none looks much like the representation of Africa on a globe.*

The map of the world (opposite) *is a classic Peters projection.*

On a Clear Day, the Infinite Panorama

Until Californian sculptor Tom Van Sant got together with computer specialist Lloyd Van Warren, nobody had ever combined satellite images to produce a single comprehensive image of the real world. The result is a stunning view of our planet in full daylight and crystal clear – it is normally around two-thirds concealed behind clouds. Even at this scale we can readily distinguish the tracery formed by the world's great rivers and their tributaries, such as the Amazon, the Nile, and the Congo; it is even possible to see their outpourings of silt.

Another novel feature is the natural coloring. Satellites do not photograph the world, they take readings of a range of wavelengths, reacting to infrared heat as well as to visible light. This information is sent back to Earth to be analyzed and transformed into images. These can then be colored according to land-use patterns. Conventionally, green and brown areas generally denote not vegetation but asphalt roads, parking lots, industrial sprawl, and so on, while vegetation is usually represented in various shades of red.

In Van Sant's portrait of our planet, colors are once again as we might expect them to be. The seas are blue, white patches are snow and ice, yellow and brown areas are rock and desert, green indicates vegetation. This is also a world seen at its most verdant, with areas shown at the season of maximum plant cover.

The original images, destined for use by the U.S. National Oceanic and Atmospheric Administration in weather forecasting, were taken from a satellite in polar orbit. A single satellite image typically shows only a minute fraction of the whole Earth – in this case an area around 1.5 square miles (4 square km). But as the Earth spins, gradually a complete image of the world can be built up as a kind of mosaic.

Given their multiple demands – low cloud cover, appropriate seasons of the year, and so on – Van Sant and Van Warren had to view literally tens of thousands of images. Having made their selection, the images could be fed into a graphics computer and composited to form a seamless picture.

Van Sant now hopes to improve on his flat map by constructing a GeoSphere, an electronic globe around 21 feet (6.5 m) in diameter mounted in its own special "Earth situation room." The surface of the globe will be sculptured to create relief features and covered in a computer-printed plastic film portraying the world with detailing down to half a mile (1 km) across.

It is not a fanciful idea – a 7 foot (2 m) prototype already exists. Motorized, and using a complex system of projectors and internal lighting linked to live satellite readings, Van Sant's globe will be able to simulate an extraordinary range of phenomena, either in real time or using stop-frame photography, all within 30 minutes of their actual occurrence. These include the movement of weather systems, whale migrations, oil tanker spills, and forest destruction. With an ability to move in from a whole-Earth context to regional high-resolution displays, and thence to high- and low-altitude aerial photography, it will be possible to perform a continuous zoom from the whole Earth to an individual standing on the surface of the Earth.

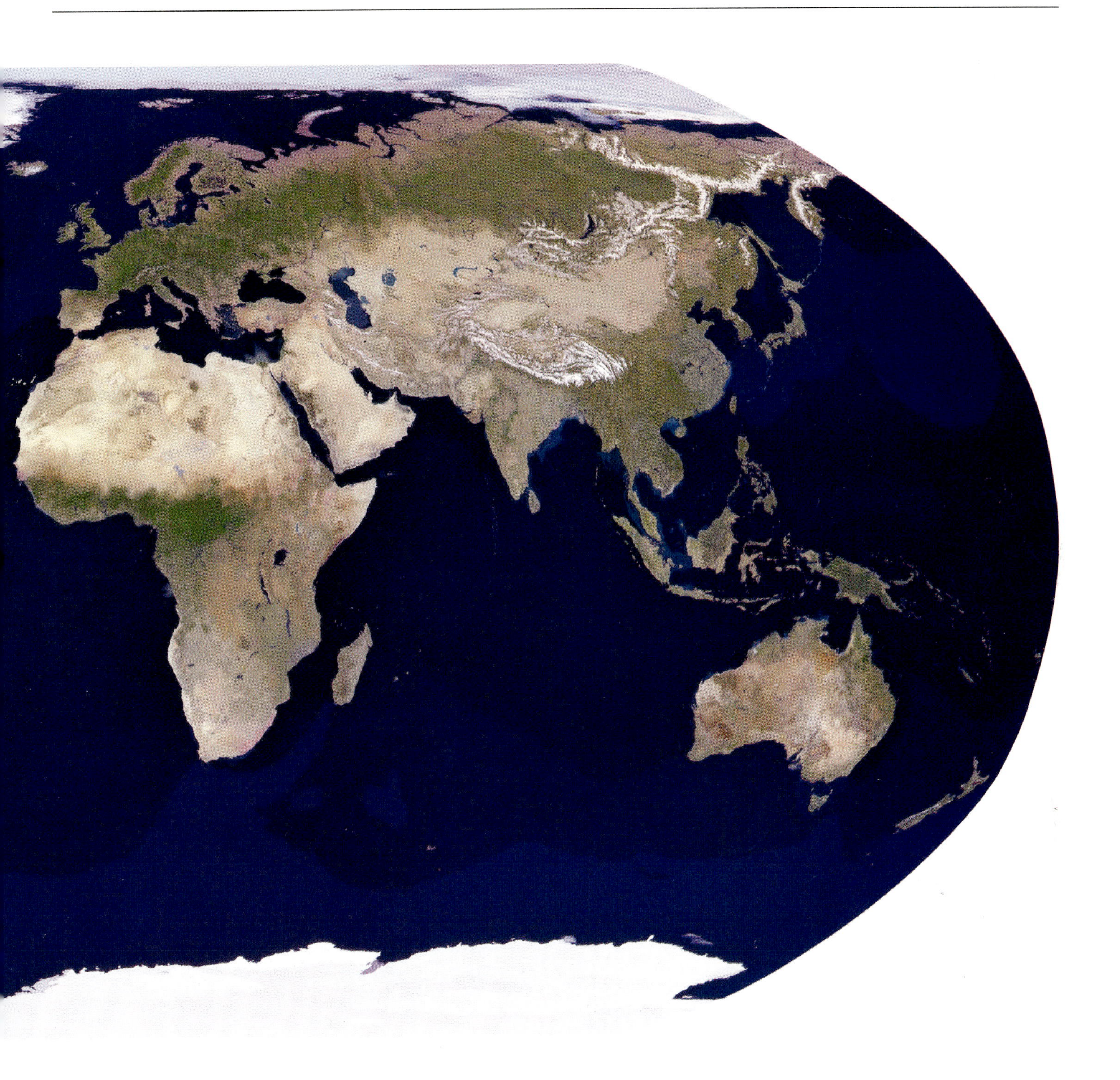

Van Sant confidently expects his sphere will have the same impact on future generations that the very first views sent back from space in the 1960s had on him. "I believe that international peace, as well as good resource management, is dependent on our planetary, or whole-Earth, awareness, as opposed to our everyday, flat-Earth consciousness." Clearly the GeoSphere's promise as a tool for monitoring, analyzing, and communicating global phenomena is unparalleled in human history.

Compiled from thousands of 1.5 square mile (4 square km) satellite photographs by Californian artist Tom Van Sant and computer specialist Lloyd Van Warren, this is the first portrait of a cloudless Planet Earth. Colors have been rendered as natural as possible, and vegetation is seen at its seasonal peak.

CONFIRMATION FROM SPACE

REMOTE SENSING FROM SATELLITES IN EARTH-ORBIT is revolutionizing our view of the world. Images can now be produced that are more accurate, and more speedily updated, than any previous land-based methods, making them a basic tool in any number of fields, including weather forecasting, town planning, ecology, mineral prospecting, and pollution monitoring.

Mapmakers have been using aerial imagery at least since the end of the 18th century, when French scientists went aloft in hot-air balloons and drew what they saw. A century later their successors strapped tiny cameras to pigeons. Aerial photography developed rapidly during both world wars, for obvious reasons. But the most significant advances came in the 1960s and 1970s, after giant strides had been made in rocket and satellite technology.

Whereas photographs had to be returned physically to the ground, new systems were developed to code information as radio waves that can be beamed back to receiver stations on Earth where interpretation can take place. Nor are sensors still restricted to the visible light spectrum, which meant they could only be used during daylight. Most can also "view" the territory beneath using invisible wavelengths. They can take the temperature of the Earth below by measuring infrared or heat waves, and can see through clouds using radar microwaves.

Although the greater height at which satellites operate makes for some losses in detail, they view a far greater area more frequently than an airplane ever could. And of course satellites can overfly any territory with impunity.

Like communications satellites, whole-planet meteorological satellites are placed in geostationary orbits, appearing motionless in the sky over the equator at an altitude of 22,500 miles (36,200 km). But Earth observation satellites must come much closer.

The French-Belgian-Swedish-owned SPOT and the U.S. Landsat satellites are in polar orbit, flying on a north-south path between 375 and 930 miles (600 and 1,500 km) above the surface, and gradually traversing every part of the Earth.

The first two Landsat satellites revisit the same spot every 18 days, but SPOT has tilting sensors than can repeat coverage every 2½ days. Landsat can resolve objects down to about 100 feet (30 m) across, SPOT to around 33 feet (10 m) – certainly good enough to detect houses, fields, and roads.

Although richer countries dominate the technology, many less-developed nations have vigorous remote sensing programs. India, for example, both leases rights to monitor incoming data from Landsat and SPOT and sends up its own satellites. Indian scientists have mapped and now monitor the country's mineral resources, crops, pests, surface water, forests, and soil conditions, and track the monsoon and snowmelt in the Himalayas.

Digital Terrain Modeling (DTM) or computer mapping creates the appearance of a three-dimensional view of a land surface by plotting a regular grid that is offset by a height value (z) at the intersection of the longitudinal (x) and latitudinal (y) lines.

To prepare the database, the mapmaker simply draws a grid on a contour map, establishes elevations then programs them point by point until all the elevation figures have been converted into height values on the grid.

The plotted image can be manipulated by the computer so that it can be viewed from a variety of angles, as though the viewer were traveling in an aircraft over and around the site. The amount of perspective distortion can be adjusted to give the appearance of being close to, or far away from, the site. The height values can be exaggerated to emphasize elevations.

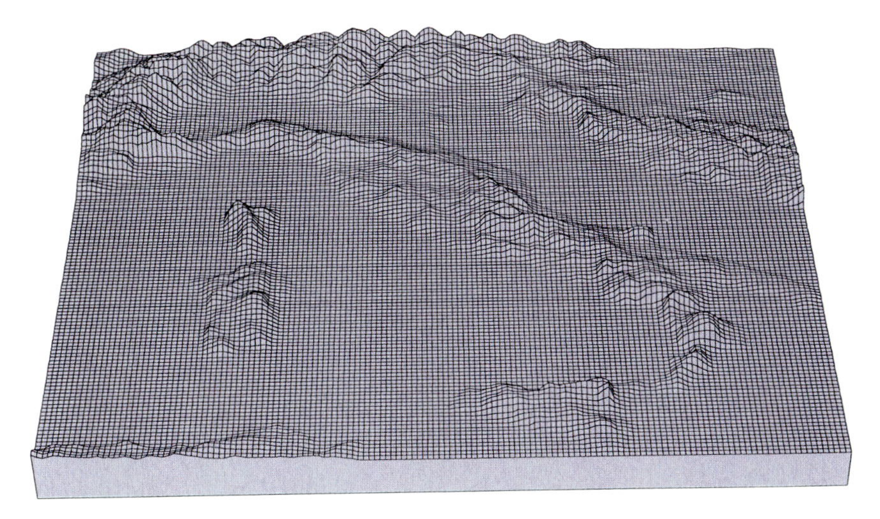

In these examples of Italy and the surrounding areas, plot 1 (above) *is viewed directly from the south, above Libya. The heights are not exaggerated but the mountain ranges are still visible.*

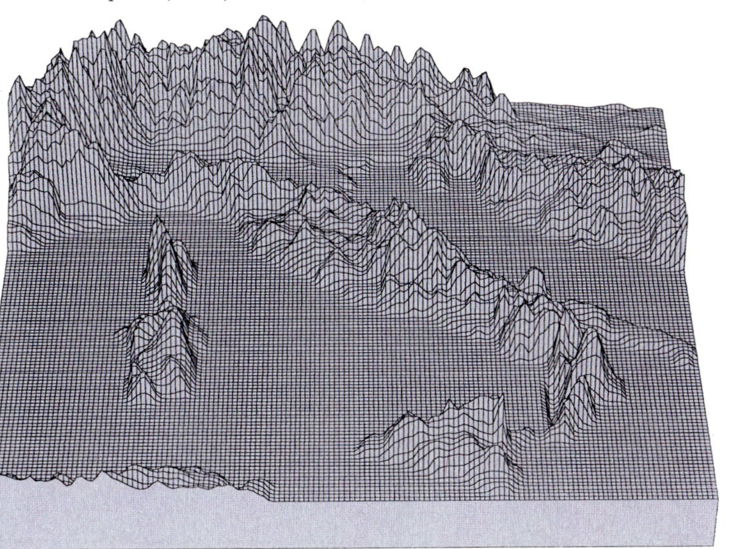

Plot 2 (left) *is taken from the same viewpoint but has exaggerated height values, making the mountains seem much more prominent. The plot area has been reduced to accommodate these higher height values.*

Plot 3 (below) *takes a viewpoint from above Greece, but has also calculated an extra value between each of the existing data points. This gives the appearance of greater detail. The plot has also been stretched from north to south to give a more recognizable image of Italy.*

Plot 4 (below left) *shows Italy viewed from high above Morocco, with the heights exaggerated once again. The Apennines, Alps and Dinaric Alps can all be seen. Contrast these schematic renderings with the satellite picture of the region* (opposite page).

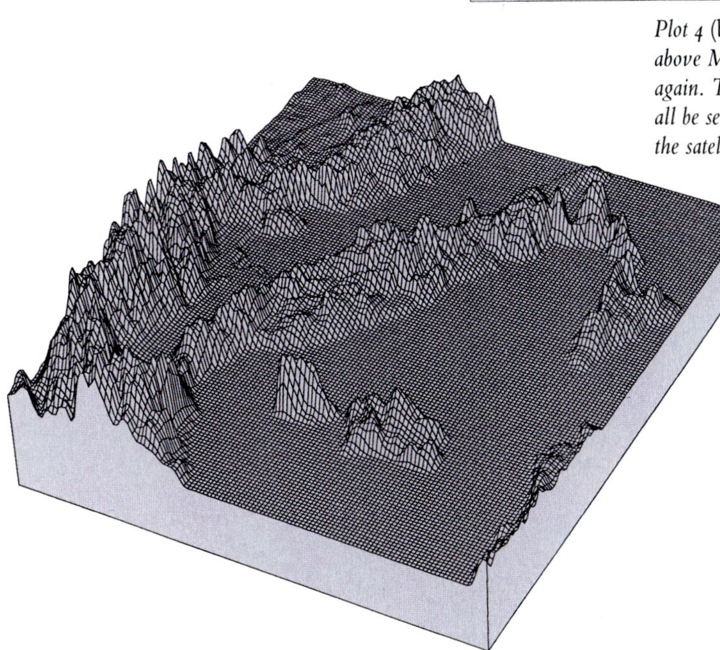

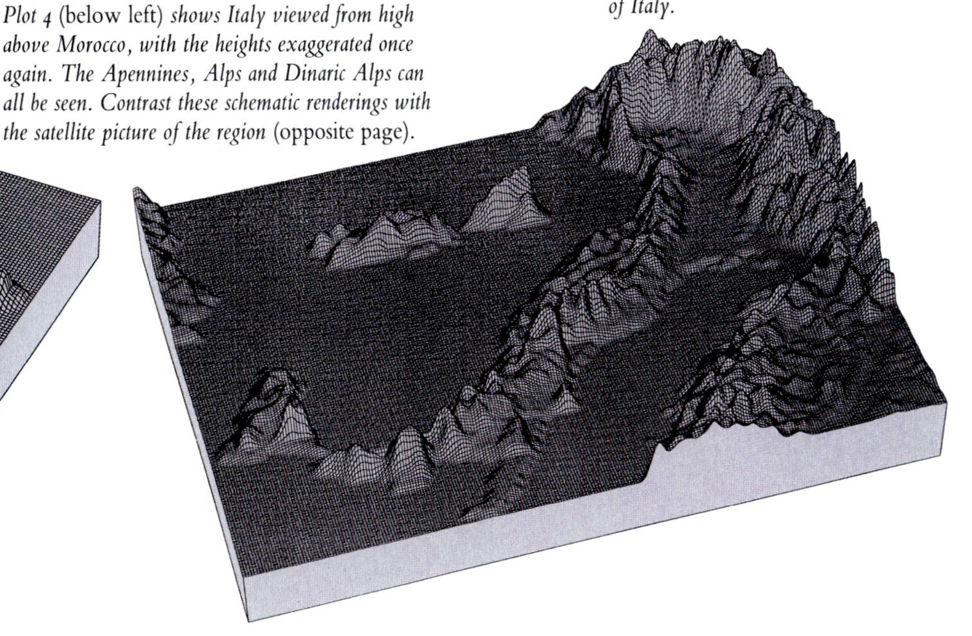

THE WORLD OF NATIONS AND STATES

CEAN
UNION OF SOVIET SOCIALIST REPUBLICS
Caspian Sea
Ulan Bator
MONGOLIA
CHINA
Beijing
NORTH KOREA
Pyongyang
Seoul
SOUTH KOREA
JAPAN
Tokyo
TURKEY
SYRIA
Damascus
Baghdad
Amman
IRAQ
JORDAN
Tehran
IRAN
Kabul
AFGHANISTAN
Islamabad
PAKISTAN
New Delhi
NEPAL
Kathmandu
Thimphu
BHUTAN
KUWAIT
Kuwait
BAHRAIN
QATAR
Doha
Riyadh
U.A.E.
Muscat
SAUDI ARABIA
OMAN
YEMEN
San'a
INDIA
BANG.
Dhaka
BURMA
VIETNAM
Hanoi
LAOS
Vientiane
HONG KONG (U.K.)
Taiwan (China)
PACIFIC
OCEAN
Rangoon
THAILAND
Bangkok
CAMBODIA
Phnom Penh
Manila
PHILIPPINES
Andaman Islands (India)
SRI LANKA
Colombo
DJIBOUTI
Djibouti
Ababa
ETHIOPIA
SOMALIA
Mogadishu
MALDIVES
BRUNEI
MALAYSIA
Kuala Lumpur
SINGAPORE
KENYA
Nairobi
INDONESIA
PAPUA NEW GUINEA
Port Moresby
SOLOMON ISLANDS
Honiara
Jakarta
SEYCHELLES
Dodoma
ZANIA
INDIAN
OCEAN
COMOROS
ilongwe
WI
BIQUE
MADAGASCAR
Antananarivo
MAURITIUS
FIJI
Suva
AUSTRALIA
Canberra
NEW ZEALAND
Wellington
ALB. – ALBANIA
AUST. – AUSTRIA
BANG. – BANGLADESH
BEL. – BELGIUM
BULG. – BULGARIA
CZECH. – CZECHOSLOVAKIA
DEN. – DENMARK
HUN. – HUNGARY
ISR. – ISRAEL
LEB. – LEBANON
L. – LIECHTENSTEIN
LUX. – LUXEMBOURG
NETH. – NETHERLANDS
S. – SWITZERLAND
U.A.E. – UNITED ARAB EMIRATES

THE WORLD THAT NATURE INTENDED

OCEAN
Sea
Ural Mountains
Lena
Bering Sea
Volga
L. Baikal
Amur
Danube
Aral Sea
Gobi
Black Sea
Caspian Sea
Tien Shan
Sea of Japan
Huang He
Tigris
Euphrates
Indus
Himalaya
Brahmaputra
Chang Jiang
Sea
The Gulf
Ganges
PACIFIC
Tropic of Cancer
Nile
Red Sea
Rub al-Khali
Mekong
OCEAN
Bay of Bengal
Blue Nile
Ethiopian Highlands
South China Sea
White Nile
Zaire (Congo)
Equator
L. Victoria
L. Tanganyika
INDIAN
L. Malawi
Great Barrier Reef
Zambezi
OCEAN
Great Dividing Range
Kalahari Desert
Tropic of Capricorn
Great Australian Desert
Drakensberg
Darling
Murray
Cape of Good Hope
HERN OCEAN

BIBLIOGRAPHY

Adam, A. & Dunlop, S. *Village, Town and City* Heinemann Educational, London, 1976

Alexander, John W. & Gibson, Lay James *Economic Geography* Prentice Hall, New Jersey, 2nd ed., 1979

Allen, J. *et al.* (compilers) *Geographical Digest 1990–91* Heinemann-Philip Atlases, Oxford, 1990

Andreae, Bernd *Farming, Development and Space* Walter de Gruyter, Berlin and New York, 1981

Baedeker's Rhine Automobile Association, Basingstoke and Jarrold, Norwich, 1985

Baedeker's Switzerland Automobile Association, Basingstoke and Prentice-Hall, Englewood Cliffs (n.d.)

Baedeker's Venice Automobile Association, Basingstoke and Prentice-Hall, Englewood Cliffs, 1987

Baines, John & Málek, Jaromír *Atlas of Ancient Egypt* Phaidon, Oxford, 1984

Barnaby, Frank (ed.) *Future War* Michael Joseph, London, 1984

Barr, Pat *Foreign Devils* Penguin, Harmondsworth and Baltimore, 1970

Barry, R.G. & Chorley R.J., *Atmosphere, Weather and Climate* Methuen, London, 3rd ed., 1976

Baynham, Simon *Africa from 1945: Conflict in the 20th century* Franklin Watts, New York, 1987

Beddis, Rex *The Third World: Development and Interdependence* Oxford University Press, Oxford and New York, 1989

Bergman, E.F. *Modern Political Geography* W.C. Brown, Dubuque, 1975

Berry, Brian J.L. *Geography of Market Centers and Retail Distribution* Prentice-Hall, Englewood Cliffs, 1967

Binford, Lewis R. *In Pursuit of the Past* Thames & Hudson, London and New York, 1983

Bonavia, David *The Great Cities: Peking* Time-Life Books, Amsterdam, 1978

Borg, Astrid *The Cultural History of Russia* Aurum Press, London, 1984

Bowman, K. *Agriculture* Macdonald Educational, London, 1985

Broek, J.O.M. & Webb, J.W. *A Geography of Mankind* McGraw-Hill, New York, 1968

Bromley, Ray & Gerry, Chris (eds.) *Casual Work and Poverty in Third World Cities* John Wiley, Chichester and New York, 1979

Brunn, S.D. & Williams, J.F. (eds.) *Cities of the World* Harper & Row, New York, 1983

Brunsden, Denys & Doornkamp, John (eds.) *The Unquiet Landscape* Douglas David & Charles, Vancouver, 1972–3

Burtenshaw, D. *et al.* *The City in West Europe* John Wiley, Chichester and New York, 1981

Burton, Jan *et al.* *The Environment as Hazard*, Oxford University Press, Oxford and New York, 1978

Campbell, Christy *Nuclear Facts* Hamlyn, London, 1984

Canby, Courtlandt *Archaeology of the World* Chancellor Press, London, 1980

Carr, M. *Patterns: Process and Change in Human Geography* Macmillan, London, 1987

Carter, Harold *The Study of Urban Geography* Edward Arnold, Melbourne and Baltimore, 3rd ed., 1981

——*An Introduction to Urban Historical Geography* Edward Arnold, London and Baltimore, 1983

Catchpole, B. *A Map History of Russia* Heinemann Educational, London, 1974

Ceram, C.W. *A Picture History of Archaeology* Thames & Hudson, London and New York, 1958

Chaliand, Gérard & Rageau, Jean-Pierre *Strategic Atlas: World Geopolitics* Harper & Row, New York, 1985

Chandler, David P. *A History of Cambodia* Westview Press, Boulder, 1983

Chomsky, Noam *Turning the Tide* Pluto Press, London, 1986

Cipolla, Carlo M. *The Economic History of World Population* Penguin, Harmondsworth and New York, 4th ed., 1967

Clark, David *Post-Industrial America* Methuen, London and New York, 1984

Clarke, William R. *Explorers of the World* Aldus Books, London, 1984

Cole, J.P. *Development and Underdevelopment* Methuen, London, 1987

——*The Development Gap* John Wiley, Chichester and New York, 1981

The Collins Atlas of World History Collins, Glasgow, 1987

Cooke, Alistair *Alistair Cooke's America* BBC Books, London, 1973

Cottrell, John *Mexico City* Time-Life Books, Amsterdam, 1979

Cumming, W.P. *et al.* *The Discovery of North America* Elek Books, London, 1971

The Daily Telegraph World Atlas Telegraph Publications, London, 1988

Davidson, Basil *The Story of Africa* Mitchell Beazley, London, 1984

Dawson, John A. *Teach Yourself Geography* Hodder & Stoughton, London, 1983

de Blij, Harm J. *Systemical Political Geography* John Wiley, Chichester and New York, 2nd ed., 1973

de Blij, Harm J. & Muller, Peter O. *Geography: Regions and Concepts* John Wiley, Chichester and New York, 5th ed., 1988.

Dethier, Jean *Down to Earth* Thames & Hudson, London and New York, 1982

Dickenson, J.P. *A Geography of the Third World* Methuen, London, 1983

Dodwell Marketing Consultants *Industrial Groupings in Japan 1988–89* Dodwell Marketing Consultants, Tokyo, 1988

Douglas, Jan *The Urban Environment* Edward Arnold, London and Baltimore, 1983

Douglas, Dr. Mary *et al.* *Man in Society* Macdonald, London, 1964

Duckham, A.N. & Masefield, G.B. *Farming Systems of the World* Chatto & Windus, London, 1970

The Economist Atlas Hutchinson Business Books, London, 1989

Egli, E. *Switzerland* Paul Haupt, Berne, 1978

Elliot, James *The City in Maps* British Library, London, 1987

Fairbank, John K. *et al.* *East Asia* Allen & Unwin, London, 1973

Farnie, D.A. *The English Cotton Industry and the World Market, 1815–1896* Clarendon Press, Oxford, 1979

Forde, C.D. *Habitat, Economy and Society* Dutton, New York, 5th ed., 1963

The Gaia Atlas of Planet Management Pan, London, 1985

Gallion, Arthur B. & Eisner, Simon *The Urban Pattern* Van Nostrand Company, New York and London, 4th ed., 1980

Gaskell, T.F. *The Gulf Stream* Cassell, London, 1972

Gerhardt, P. *et al.* *The People Trade* International Broadcasting Trust, London, 1985

George, Susan *A Fate Worse Than Debt* Penguin, Harmondsworth, 1988

Gerster, Georg *Grand Design* Paddington Press, London and New York, 1976

Gilbert, A. & Gugler, J. *Cities, Poverty and Development* Oxford University Press, Oxford and New York, 1982

Gilbert, Martin *American History Atlas* Weidenfeld & Nicolson, London and New York, 1968

Girouard, Mark *Cities and People* Yale University Press, New Haven and London, 1985

Gould, Peter & White, Rodney *Mental Maps* Penguin, Harmondsworth, 1974

Grant, M. *Ancient History Atlas* Weidenfeld & Nicolson, London and New York, 3rd ed., 1986

Green, Francis & Sutcliffe, Bob *The Profit System* Penguin, Harmondsworth, and Viking Penguin, New York, 1987

Guest, Arthur *Man and Landscape* Heinemann Educational, London, 1974

Gugler, Josef & Flanagan, William G. *Urbanization and Social Change in West Africa* Cambridge University Press, Cambridge, 1978

The Guinness Book of Records, 1990 Guinness Publishing, Enfield, 1989

Haggett, Peter *Geography: A Modern Synthesis* Harper & Row, New York, 3rd ed., 1983

Hakim, Besim Selim *Arabic-Islamic Cities* KPI, London, 1986

Hall, Peter *The World Cities* Weidenfeld & Nicolson, London and New York, 3rd ed., 1984

Hall, P. & Markusen, A. (eds.) *Silicon Landscapes* Allen & Unwin, Boston, 1985

Harpham, Trudy *et al.* (eds.) *In the Shadow of the City* Oxford University Press, Oxford and New York, 1988

Harpur, James & Westwood, Jennifer *The Atlas of Legendary Places* Bloomsbury, London, and Weidenfeld & Nicolson, New York, 1989

Harris, Walter D., Jr. *The Growth of Latin American Cities* Ohio University Press, Athens, 1971

Hellen, J.A. *North Rhine – Westphalia* Oxford University Press, Oxford and New York, 2nd ed., 1983

Herbert, David T. & Thomas, Colin J. *Urban Geography: A First Approach* John Wiley, Chichester and New York, 1982

Hills, C.A.R. *World Trade* Batsford, London, 1981

Hogben, Lancelot *Columbus, the Cannon Ball and the Common Pump* Heinemann, London, 1974

Holford, Ingrid *The Guinness Book of Weather* Guinness Superlatives, Enfield, 2nd ed., 1982

Honeygold, D. *International Financial Markets* Woodhead-Faulkner, Cambridge, 1982

Hoskins, W.G. *The Making of the English Landscape* Hodder & Stoughton, London, 1955

Hudson, F.S. *Geography of Settlements* Macdonald & Evans, London, 1970

Hughes, Robert *The Shock of the New* BBC Publications, London, 1980

Huntington, Ellsworth *Civilization and Climate* Yale University Press, New Haven, 3rd ed., 1924

Images of the World Collins-Longman, Glasgow and Harlow, and Rand McNally, New York, 1983

Information Please Almanac Houghton Mifflin, Boston, 1985

Insight City Guides *Beijing* APA Publications, Hong Kong, 1989

International Institute for Environment and Development, World Resources Institute *World Resources 1987* Basic Books, New York, 1987

Jackson, Karl D. *Cambodia 1975–78: Rendezvous with Death* Princeton University Press, Princeton, 1989

James, Preston E. & Webb, Kempton *One World Divided* John Wiley, Chichester and New York, 3rd ed., 1980

Jellicoe, Geoffrey & Jellicoe, Susan *The Landscape of Man* Thames & Hudson, London and New York, 1975

Johnston, N.J. *Cities in the Round* University of Washington Press, Seattle and London, 1983

Jordan, Terry G. & Rowntree, L. *The Human Mosaic* Harper & Row, New York, 2nd ed., 1979

Kidron, M. & Segal, R. *The State of the World Atlas* Pan, London, 1981

——*The New State of the World*

Atlas Pan, London, 1984
Kidron, M. & Smith, D. *The War Atlas* Pan, London, 1983
Kiernan, Ben *How Pol Pot came to Power* Verso, London, 1985
Klare, Michael T. & Kornbluh, Peter (eds.) *Low-Intensity Warfare* Methuen, London, 1989
Knapp, Brian *Britain in Today's World* Unwin Hyman, London, 1988
Knapp, Brian *et al. Challenge of the Human Environment* Longman, Harlow, 1989
Knox, Paul *Urban Social Geography: An Introduction* Longman Scientific & Technical, Harlow, and John Wiley, New York, 2nd ed., 1987
Knox, Paul & Agnew, John *The Geography of the World Economy* Edward Arnold, London and New York, 1989
Kreye, O. *et al. Export Processing Zones in Developing Countries* International Labor Organization, Geneva, 1987
Lanegran, David A. & Palm, Risa *An Invitation to Geography* McGraw-Hill, New York, 1978
Langer, William L. (compiler & ed.) *An Encyclopedia of World History* Harrap, London, 5th ed., 1972
Lapo, G. *et al. Moscow: capital of the Soviet Union* Progress Publishers, Moscow, 1976
Legum, Colin *Africa Handbook* Penguin, Harmondsworth, rev. ed., 1969
Lewis, D.L. (ed.) *The Growth of Cities* Elek, London, 1971
Lowry, J.H. *World City Growth* Edward Arnold, London, 1975
Mabogunje, H.J. *The Development Process* Hutchinson, London, 1980
McEvedy, C. *The Penguin Atlas of Recent History (Europe since 1815)* Penguin, Harmondsworth and New York, 1982
Macfarquhar, Roderick, and the editors of the Newsweek Book Division *The Forbidden City* Reader's Digest, London, in association with Newsweek, New York, 1972
McGee, T.G. *The Urbanization Process in the Third World* Bell, London, 1971
Matthews, Rupert O. *The Atlas of Natural Wonders* Facts on File, Oxford and New York, 1988
May, John *The Greenpeace Book of Antarctica* Dorling Kindersley, London, 1986
Mazrui, Ali A. *The Africans: a Triple Heritage* BBC Publications, London, 1986
Meinig, D.W. *The Shaping of America: Volume I: Atlantic America, 1492–1800* Yale University Press, New Haven and London, 1986
Mellor, R.E.H. *Eastern Europe* Macmillan, London, 1975
Menen, A. *Venice* Time-Life Books, Amsterdam, 1976
Middleton, Dr. Nick *Atlas of World Issues* Oxford University Press, Oxford and New York, 1988
——*Atlas of Environmental Issues* Oxford University Press, Oxford and New York, 1988
The Mitchell Beazley Atlas of Earth Resources Mitchell Beazley, London, 1979
The Mitchell Beazley Concise Atlas of the Earth Mitchell Beazley with George Philip, London, in association with Rand McNally, New York, 1973
Moore, Robert *Living in Venice* Macdonald, 1985
Morgan, W.B. & Munton, R.J.C. *Agricultural Geography* Methuen, London, 1971
Morris, J. *Among the Cities* Viking, New York and London, 1985
Muir Wood, Dr. Robert *Earthquakes and Volcanoes* Mitchell Beazley, London, 1986
Napier, William *Lands of Spice and Treasure* Aldus Books, London, 1971
The National Geographic Picture Atlas of our World National Geographic, Washington, 1979
The National Geographic Society *Vanishing Peoples of the Earth* National Geographic, Washington, 1968
Nalivkin, D.V. *Hurricanes, Storms and Tornadoes* A.A. Balkema, Rotterdam, 1983
Newby, Eric *The Mitchell Beazley World Atlas of Exploration* Mitchell Beazley, London, 1975
Nixon, Brian *World Contrasts* Bell & Hyman, London, 1986
Normanton, Simon *Tibet: the lost civilization* Hamish Hamilton, London, 1988
Omond, Roger *The Apartheid Handbook* Penguin, Harmondsworth, 2nd ed., 1986
Overman, Michael *Water: solutions to a problem of supply and demand* Open University Press, Milton Keynes, rev. ed., 1976
Oxford Illustrated Encyclopedia: Volume I: The Physical World Oxford University Press, Oxford and New York, 1985
Palen, J. John *The Urban World* McGraw-Hill, New York, 2nd ed., 1981
Palmer, R.R. *A History of the Modern World* Knopf, New York, 1950
Paris Atlas 1900 Librairie Larousse, Paris, 1989
Paullin, Charles O. *Atlas of the Historical Geography of the United States* Carnegie Institution of Washington and the American Geographical Society of New York, 1932
Pears Cyclopedia, Pelham Books, London, 98th edition, 1989
Peters, Arno *Atlas of the World* Longman, 1989
Philbrick, A.K. *This Human World* John Wiley, Chichester and New York, 1963
Philip's Modern School Atlas George Philip, London, 83rd ed., 1987
Pounds, N. *The Ruhr* Greenwood Press, New York, 1968
The Rand McNally Family World Atlas Rand McNally, Chicago, 1988
The Rand McNally New International Atlas Rand McNally, Chicago, 1988
Rapoport, Amos *Human Aspects of Urban Form* Pergamon Press, Oxford and New York, 1977
Raw, M. *Understanding Human Geography* Bell & Hyman, London, 1986
Reader, John *Man on Earth* Collins, London, and University of Texas Press, Austin, 1988
The Reader's Digest Great World Atlas Reader's Digest, London, 1962
The Reader's Digest Guide to Places of the World Reader's Digest, London and New York, 1987
Readings from the Scientific American (various authors) *Cities: Their Origin, Growth and Human Impact* W.H. Freeman, San Francisco, 1973
Rockett, Mel *Themes in Human Geography* E.J. Arnold, Leeds, 1987
Royston, R. *Modern Cities* Macdonald Educational, London, 1987
Scott, Peter *Geography of World Agriculture: Volume 9: Australian Agriculture* Akadémiai Kiadó, Budapest, 1981
Segal, Gerald *The Simon & Schuster Guide to the World Today* Simon & Schuster, New York and London, 1987
Semple, Ellen Churchill *Influences of Geographical Environment* Constable, London, 1933
Short, John R. *An Introduction to Urban Geography* Routledge & Kegan Paul, London, 1984
Simpson, E.S. *The Developing World: An Introduction* Longman Scientific & Technical, Harlow, and John Wiley, New York, 1987
Sinclair, Keith *A History of New Zealand* Penguin, Harmondsworth, rev. ed., 1988
Sivin, Professor Nathan *et al.* (editorial board) *The Contemporary Atlas of China* Weidenfeld & Nicolson, London, and Houghton Mifflin, Boston, 1988
Smith, David M. *Geography, Inequality and Society* Press Syndicate of the Cambridge University Press, Cambridge and New York, 1987
Smith, Peter J. *Encyclopedia of the Earth* Century Hutchinson, London, 1986
Storry, Richard *A History of Modern Japan* Penguin, Harmondsworth, rev. ed., 1965
Summerson, John *Georgian London* Barrie&Jenkins,London,rev.ed., 1988
Symons, Leslie *The Soviet Union: A Systematic Geography* Hodder & Stoughton, London, 1983
Tames, Richard *The Muslim World* Macdonald, London, 1982
Temple, Robert K.G. *China, Land of Discovery and Invention* Patrick Stephens, Wellingborough, 1986
Thomas, William J., Jr. (ed.) *Man's Role in Changing the Face of the Earth* University of Chicago Press, Chicago and London, 1956
Time-Life Books (eds.) *Scandinavia* Time-Life Books, Amsterdam, 1985
The Times Atlas of the World Times Books, London, in collaboration with Bartholomew, Edinburgh, comprehensive edition, 1967
The Times Atlas of World History Times Books, London, rev. ed., 1986
Trager, James (ed.) *The People's Chronology* Heinemann, London, 1979
Treharne, R.F. & Fullard, H. (eds.) *Muir's Historical Atlas: Ancient, Medieval and Modern* George Philip, London, 9th ed., 1962
Trevor-Roper, Hugh (ed.) *The Golden Age of Europe* Thames & Hudson, London and New York, 1968
Tuan, Yi-Fu *Topophilia* Prentice-Hall, Englewood Cliffs 1974
Tudge, Colin *The Encyclopedia of the Environment* Christopher Helm, London, 1988
Turnbull, Colin *The Forest People* Pan, London, 1961
——*The Mountain People* Jonathan Cape, London, 1973
United Nations Centre for Human Settlements (Habitat) *Global Report on Human Settlements* Oxford University Press, Oxford and New York, 1987
The Usborne Book of World Geography Usborne, London, 1984
Vance, J.E., Jr. *This Scene of Man* Harper's College Press, New York, 1977
Vidal de la Blache, P. *Principles of Human Geography* Constable, London, 1926
Wallraff, Gunter *Lowest of the Low* Methuen, London, 1988
Watson, Jack B. *Success in 20th Century World Affairs since 1919* John Murray, London, 1977
Waugh, David *The World* Thomas Nelson, Walton-on-Thames, 1987
Westwood, Jennifer (ed.) *The Atlas of Mysterious Places* Weidenfeld & Nicolson, London and New York, 1987
White, P. *The West European City: a social geography* Longman, Harlow and New York, 1984
Whitfield, Dr. P. *et al. The Atlas of the Living World* Weidenfeld & Nicolson, London, and Houghton Mifflin, Boston, 1989
Whynne-Hammond, Charles *Elements of Human Geography* Allen & Unwin, London, 2nd ed., 1985
Wombwell, Paul *The Globe: representing the world* Impressions Gallery of Photography, York, 1989
The World Book Atlas World Book Encyclopedia, Chicago and London, 1977
The World Development Report World Bank, Washington, 1989
Zahran, M. *Challenges of the Urban Environment* Beirut Arab University, Beirut, 1973
Zewen, L. *et al. The Great Wall* Michael Joseph, London, 1981

INDEX

Numerals in italics refer to illustrations and photographs.

V

W

Y

Z

ACKNOWLEDGMENTS

PICTURE CREDITS

l=left; *r*=right; *c*=center; *t*=top; *b*=bottom

1 Ascani/Hoa-Qui; 2*l* P. de Wilde/Hoa-Qui; 2*tr* The Venice Picture Library; 2*br* Robert Harding Picture Library; 3 Mark Segal/Tony Stone Associates; 4*t* E. Valentin/Hoa-Qui; 4*b* Bruno Barbey/Magnum Photos; 5 The Telegraph Colour Library; 6 Georg Gerster/The John Hillelson Agency; 8–9 Edmond van Hoorick; 10 Swiss National Tourist Office; 10–11 Edmond van Hoorick; 12 Nestlé S.A.; 13 Alain Morvan/Frank Spooner Pictures; 14–15 Y. Arthus Bertrand/Ardea; 16–17 S. Jonasson/Frank Lane/Bruce Coleman Inc.; 20–21 National Remote Sensing Centre; 23 Spectrum Colour Library; 24–25 Tony Stone Associates; 28–29 Robert Harding Picture Library; 30–31 Paolo Gori/The Image Bank; 32 E. Valentin/Hoa-Qui; 34–35 David Muench; 36–37 Liz & Tony Bomford/Ardea; 38–39 Don Klumpp/The Image Bank; 40 Sarah Errington/The Hutchison Library; 42–43 Tibor Hirsch/Susan Griggs Agency; 45 Chantilly Musée Condé/Giraudon; 46–47 Robert Harding Picture Library; 49 Michael Friedel/Rex Features; 50–51 Gerhard Gscheidle/The Image Bank; 52–53 Ma Po Shum/Aspect Picture Library; 55 Adam Woolfitt/Susan Griggs Agency; 56–57 Michael Holford; 57 Royal Geographical Society, London/The Bridgeman Art Library; 59 Sebastiao Salgado/Magnum Photos; 61 British Library; 62–63 M. Huet/Hoa-Qui; 64–65 D.C. Lowe/Tony Stone Associates; 68–69 Manchester City Council; 70 Archiv für kunst und Geschichte; 72 The Kobal Collection; 74–75 Sepp Seitz/Susan Griggs Agency; 76*l* The Mansell Collection; 76*r* Mary Evans Picture Library; 77 Topham Picture Source; 78 Frank Spooner Pictures; 79 Julian Calder/Tony Stone Associates; 80 Anthony Suau/Black Star/Colorific!; 85*t* John Massey Stewart; 85*b* Thierry Cazabon/Tony Stone Associates; 86 Steve McCurry/Magnum Photos; 88 Jay Maisel/The Image Bank; 91 Curtis Willocks/Stockphotos; 93 Anthony Suau/Black Star/Colorific!; 94 Richard Kalvar/Magnum Photos; 98 Earth Satellite Corporation/Science Photo Library; 99 Adam Woolfitt/Susan Griggs Agency; 101 Robert Harding Picture Library; 103 David Beatty/Susan Griggs Agency; 104 Robert Harding Picture Library; 107*t* Abbas/Magnum Photos; 107*b* Airphoto/Susan Griggs Agency; 112 Gunter Heil/Zefa Picture Library; 112–13 Geoff Tompkinson/Aspect Picture Library; 114 John Moss/Colorific!; 117 Grandadam/Explorer; 118 Scianna/Magnum Photos; 121 J. Kyle Keener/Katz Pictures; 122 David Hughes/Bruce Coleman; 125 Novosti/Katz Pictures; 127 Carp/*Stern*; 128–29 Georg Gerster/The John Hillelson Agency; 134–35 Orion Press/Zefa Picture Library; 135*t* Annet Held/Arcaid; 135*b* François Perri/Colorific!; 136*t* Burt Glinn/Magnum Photos; 136*b* Saharoff/Hoa-Qui; 138–39 Roland & Sabrina Michaud/The John Hillelson Agency; 140–41 Leo Meier/Weldon Trannies; 142–43 Earth Satellite Corporation/Science Photo Library; 144–45 Wendy Watriss/Susan Griggs Agency; 152 Paulo Fridman/Colorific!; 155 Paul Chesley/Photographers Aspen; 158–59 Roy Garner/Rex Features; 160 Earth Satellite Corporation/Science Photo Library; 161 Bodleian Library, Oxford MS. Bodley 264 fol 218R; 163 John Lawlor/Tony Stone Associates; 165 Steve Benbow/Impact Photos; 168–69 The Telegraph Colour Library; 169 Peter M. Miller/The Image Bank; 170–71 Bill Weems/Woodfin Camp; 171 Jason Shenai/Susan Griggs Agency; 172–73 JAS Photographic; 175 Georg Gerster/The John Hillelson Agency; 176–77 The Bridgeman Art Library; 177 David Paterson; 179 Royal Geographical Society; 180 Peter Charlesworth/JB Pictures/Katz Pictures; 180–81 Julian Calder/Tony Stone Associates; 181 Lewis/Network; 182–83 Pictor International; 185 S. Franklin/Magnum Photos; 187 Richard Bryant/Arcaid; 188–89 Sarah Leen/Matrix/Katz Pictures; 194*t* M. Setboun/Rapho/Network; 194*c* Spectrum Colour Library; 194*b* Chip Hires/Frank Spooner Pictures; 195*tl* Frank Spooner Pictures; 195*tr* Pacemaker Press International; 195*b* Eddie Adams/Frank Spooner Pictures; 198 Asupi/Impact Photos; 201 Sarah Leen/Matrix/Katz Pictures; 202 Popperfoto/AFP; 204 Popperfoto; 207*l* Frank Spooner Pictures; 207*r* Mark Cator/Impact Photos; 208 Rex Features; 210–11 The Telegraph Colour Library; 216*tl* Eve Arnold/Magnum Photos; 216*tr* Jay Dickman/Matrix/Katz Pictures; 216*bl* Eve Arnold/Magnum Photos; 216*br* Rex Features; 218–19 Eric Bouvet/Frank Spooner Pictures; 220 Georg Gerster/The John Hillelson Agency; 222 P. Habans/Sygma; 224 Liba Taylor/The Hutchison Library; 225 Topham Picture Library; 226 Photo News/Frank Spooner Pictures; 228–29 John Reardon/Rex Features; 230 Steve McCurry/Magnum Photos; 231 Mark Cator/Impact Photos; 232 D. Erwitt/Magnum Photos; 233 Rex Features; 234–35 QA Photos; 238 Markel/Liaison/Frank Spooner Pictures; 242–43 Royal Geographical Society, London/The Bridgeman Art Library; 243 By permission of the Dean and Chapter of Hereford; 244–45 Royal Geographical Society; 246–47 Peters Projection/Oxford Cartographers; 248–49 Tom Van Sant/GeoSphere Project; 250–51 the Telegraph Colour Library.

MAPS AND ILLUSTRATIONS CREDITS

l=left; *r*=right; *c*=center; *t*=top; *b*=bottom

8 Chapman Bounford; 9 Paul Selvey; 10 Matthew Bell; 11 Paul Selvey; 12 Trevor Hill; 17 Euromap; 18–19 Matthew Bell; 20 Euromap; 21 Matthew Bell; 22 Matthew Bell; 24 Matthew Bell; 25 Euromap; 26–27 Matthew Bell; 29 Matthew Bell; 30 Matthew Bell; 31 Euromap; 32 Euromap; 33 Matthew Bell; 35 Euromap; 36 Matthew Bell; 37 Euromap; 39 Euromap; 41 Martin Woodford; 43*t* Euromap; 43*b* Matthew Bell; 44 Matthew Bell; 47 Euromap; 48 Euromap; 50 Matthew Bell; 51 Euromap; 52 Matthew Bell; 53 Euromap; 54*t* Euromap; 54*cb* Matthew Bell; 57 Euromap; 58 Euromap; 60 Matthew Bell; 61 Euromap; 63*t* Euromap; 63*b* Matthew Bell; 66–67 Susan Beresford; 68 Euromap; 70 Euromap; 71 Trevor Hill; 72–73 Trevor Hill; 74*t* Trevor Hill; 74*cb* Euromap; 75 Euromap; 77 Trevor Hill; 78 Euromap; 79 Paul Selvey; 81 Trevor Hill; 82–83 Susan Beresford; 84–85 Euromap; 86–87 Trevor Hill; 88–89 Euromap; 90 Euromap; 91 Paul Selvey; 92 Euromap; 93 Simon Roulstone; 94 Trevor Hill; 95 Simon Roulstone; 96–97*b* Anthony Cowland; 97*t* Paul Selvey; 98–99 Trevor Hill; 100–101 Euromap; 102–3 Euromap (maps), Trevor Hill (illustration); 105 Richard Manning; 106 Trevor Hill; 108–9 Susan Beresford; 110–11 Simon Roulstone; 114–15 Trevor Hill; 116–17 Euromap; 118–19 Trevor Hill; 120–21 Euromap; 122 Paul Selvey; 123 Richard Manning; 124 Euromap; 125 Trevor Hill; 126–27 Trevor Hill; 130–31 Susan Beresford; 132–33 Euromap; 135 Euromap; 137*tb* Euromap; 137*c* Anthony Cowland; 138 Anthony Cowland; 139*l* Anthony Cowland; 139*r* Euromap; 140–41*b* Martin Woodford; 141*t* Euromap; 142 Martin Woodford; 145*l* Euromap; 145*r* Aziz Khan; 146–47*t* Pavel Kostal; 146–47*b* Martin Woodford; 148–49 Susan Beresford; 150–51*t* Euromap; 150–51*br* Martin Woodford; 153*t* Euromap; 153*b* King & King; 154–55 Trevor Hill; 155*t* Euromap; 156–57 Euromap; 158 Euromap; 161 Anthony Cowland; 162 Chapman Bounford; 163*l* Trevor Hill; 163*r* Euromap; 164 Martin Woodford; 165 Euromap; 166–67 Susan Beresford; 168 King & King; 169 Euromap; 170*t* Euromap; 170*b* Trevor Hill; 171 Trevor Hill; 172 Martin Woodford; 173*t* Euromap; 173*b*Martin Woodford; 174*t* Anthony Cowland; 174–75 Martin Woodford; 177 Euromap; 178–79 Chapman Bounford; 180 Euromap; 183*t* Euromap; 183*b* Anthony Cowland; 184–85 Trevor Hill; 186–87 Anthony Cowland; 187*t* Euromap; 190–91 Susan Beresford; 192–93*c* Neil Breedon; 192–93*tb* Paul Selvey; 195*t* Paul Selvey; 195*b* Martin Woodford; 196 Mick Saunders; 197*t* Paul Selvey; 197*b* Mick Saunders; 198–99*c* Neil Breedon; 198–99*b* Paul Selvey; 200–201*b* Ian Howatson; 201*t* Paul Selvey; 202 Paul Selvey; 203 Martin Woodford; 205*t* Richard Manning; 205*b* Paul Selvey; 206*t* Paul Selvey; 206–7*b* Martin Woodford; 209 Mick Saunders; 210 Matthew Bell; 211 Paul Selvey; 212–13 Susan Beresford; 214–15 Trevor Hill; 214*r* Paul Selvey; 219*t* Paul Selvey; 219*cb* Matthew Bell; 221*c* Trevor Hill; 221*t* Paul Selvey; 222–23 Ian Howatson; 225 Paul Selvey; 227 Trevor Hill; 228–29 Martin Woodford; 229*t* Paul Selvey; 231*t* Paul Selvey; 231*c* Matthew Bell; 234–35 Paul Selvey; 236–37 Paul Selvey; 238–39 Martin Woodford; 240–41*b* Lovett Johns Cartographers; 241*t* Dick Bateman; 242 Matthew Bell; 247 Matthew Bell; 251 Chapman Bounford; 252–53 Euromap; 254–55 Euromap. Special thanks to the Venice in Peril Fund for references. Also thanks to Dr. Tony Warnes and Dr. Rita Gardner for advice.

A MARSHALL EDITION

EDITORIAL DIRECTOR:	RUTH BINNEY	ART DIRECTOR:	JOHN BIGG
PROJECT EDITOR:	MARILYN INGLIS	ART EDITORS:	MIKE ROSE
SENIOR EDITOR:	CAROLINE TAGGART		PETER LAWS
CONTRIBUTING WRITER:	DAVID ADAMSON		DAVID ALLEN
RESEARCH:	HEATHER MAGRILL	ART ASSISTANT:	PAUL TILBY
	JAZZ WILSON	PICTURE EDITOR:	ZILDA TANDY
	GABRIELLE FERNEE	PICTURE RESEARCH:	RICHARD PHILPOTT
DESK EDITORS:	LINDSAY MCTEAGUE	PRODUCTION:	BARRY BAKER
	ROGER FEW		JANICE STORR
	BARBARA NEWMAN		NIKKI INGRAM
	CAROL HUPPING	INDEX:	KATHIE GILL